Current Research in Geochemistry

Current Research in Geochemistry

Editor: Joseph Clarke

MURPHY & MOORE
www.murphy-moorepublishing.com

Murphy & Moore Publishing,
1 Rockefeller Plaza,
New York City, NY 10020, USA

Visit us on the World Wide Web at:
www.murphy-moorepublishing.com

ISBN: 978-1-63987-146-9 (Hardback)

Cataloging-in-Publication Data

Current research in geochemistry / edited by Joseph Clarke.
p. cm.
Includes bibliographical references and index.
ISBN 978-1-63987-146-9
1. Geochemistry. 2. Earth sciences. 3. Chemistry. I. Clarke, Joseph.
QE515 .C87 2022
551.9--dc23

Table of Contents

Preface

Geochemistry is an integrated field of chemistry and geology. It deals with the tools and principles of chemistry in order to explain the mechanisms behind major geological systems such as the Earth's crust and its oceans. The domain of geochemistry extends beyond the Earth, covering the entire Solar System, and has made crucial contributions to the understanding of a number of processes. A few of such processes are mantle convection, planet formation, and the origins of granite and basalt. Some of the subfields of geochemistry are aqueous geochemistry, biogeochemistry, cosmochemistry, isotope geochemistry, organic geochemistry, photogeochemistry and regional geochemistry. This book outlines the processes and applications of geochemistry in detail. It aims to shed light on some of the unexplored aspects of geochemistry and the recent researches in this field. As this field is emerging at a rapid pace, the contents of this book will help the readers understand the modern concepts and applications of the subject.

This book has been the outcome of endless efforts put in by authors and researchers on various issues and topics within the field. The book is a comprehensive collection of significant researches that are addressed in a variety of chapters. It will surely enhance the knowledge of the field among readers across the globe.

It gives us an immense pleasure to thank our researchers and authors for their efforts to submit their piece of writing before the deadlines. Finally in the end, I would like to thank my family and colleagues who have been a great source of inspiration and support.

Editor

1

Structure and properties of forsterite-$MgSiO_3$ liquid interface: molecular dynamics study

Fumiya Noritake* and Katsuyuki Kawamura

Abstract

The mechanical properties of partially molten rock, such as their permeabilities and viscosities, are important properties in geological processes. We performed molecular dynamics simulations in terms of structures and diffusivities in forsterite-$MgSiO_3$ liquid interfaces to obtain the nanoscale dynamic properties and structure of the interface. The characteristic structure of the forsterite-$MgSiO_3$ liquid interfaces was observed in the simulations. In the layered structure of the altered surfaces, Si-rich and Mg-rich layers exist alternately in the vicinity of the crystal-liquid interfaces. The layered structure might be formed by the strength difference between Si-O covalent bonds and Mg-O ionic bonds. The difference in the layered structure, indicated by the thickness of the $MgSiO_3$ liquid film, might be caused by the difference in degrees of freedom of the configuration in the liquid film. The two-dimensional diffusivity of oxygen atoms parallel to the interface is controlled by two factors. One factor is the thickness of the liquid film, which decreases oxygen diffusivity with decreasing film thickness. The other is the composition of the sliced layer, where oxygen diffusivity increases with increasing Mg/Si ratio. The effect of the crystal-liquid interface found in this study is negligible in texturally equilibrated rocks. However, the interface can affect the melt flow in deformed samples because a grain boundary melt film with a thickness of several nanometers exists stably in deformed partially molten rock.

Keywords: Solid-liquid interface; Molecular dynamics simulation; Forsterite; Silicate liquid

Background

Knowledge on viscosity and permeability of partially molten rocks is important for understanding volcanism and the thermal history of the earth. To understand the results obtained by experiments and to estimate physical properties in extreme conditions that are difficult to reproduce in laboratory experiments, it is necessary to know the local structure and the properties of silicate crystal-liquid interfaces. In particular, knowledge on the nanoscale structure and properties of silicate crystal-liquid interfaces might be useful in estimating the properties of rocks containing a small degree of melting. Hiraga et al. (2002) reported the presence of nanoscale melt films using high-resolution electron microscopy and energy dispersive X-ray profiling in scanning transmission electron microscopy. The properties of melt in such thin regions are considered to be different from properties in a bulk melt because of the effect of crystal surfaces. For instance, the lateral self-diffusivity of water to a crystal surface is different from a bulk surface in the case of a water-brucite surface (Sakuma et al. 2003), a water-muscovite mica surface (Sakuma and Kawamura 2009), and others. The dynamic property anomalies on solid-liquid interfaces affect the properties of bulk rock, e.g., permeability (Ichikawa et al. 2001).

Molecular dynamics simulations are widely used to investigate the physical properties and structures of crystals, liquids, gasses, and interfaces. In these simulations, we set the initial positions and velocities of all atoms; then, the atoms are forced to move according to given force fields under a proper ensemble. Molecular dynamics simulations are useful methods for investigating the nanoscale structure and properties because they give us the trajectory of each atom. In this study, the structure and properties of the forsterite-$MgSiO_3$ liquid interface are investigated by application of molecular dynamics simulations. It is essential to know the structure and physical properties of forsterite-$MgSiO_3$ liquid interfaces because forsterite is the liquidus mineral of primordial magmas.

* Correspondence: f.noritake@s.okayama-u.ac.jp
Graduate School of Environmental and Life Science, Okayama University, 3-1-1, Tsushimanaka, Kita-Ku, Okayama 700-8530, Japan

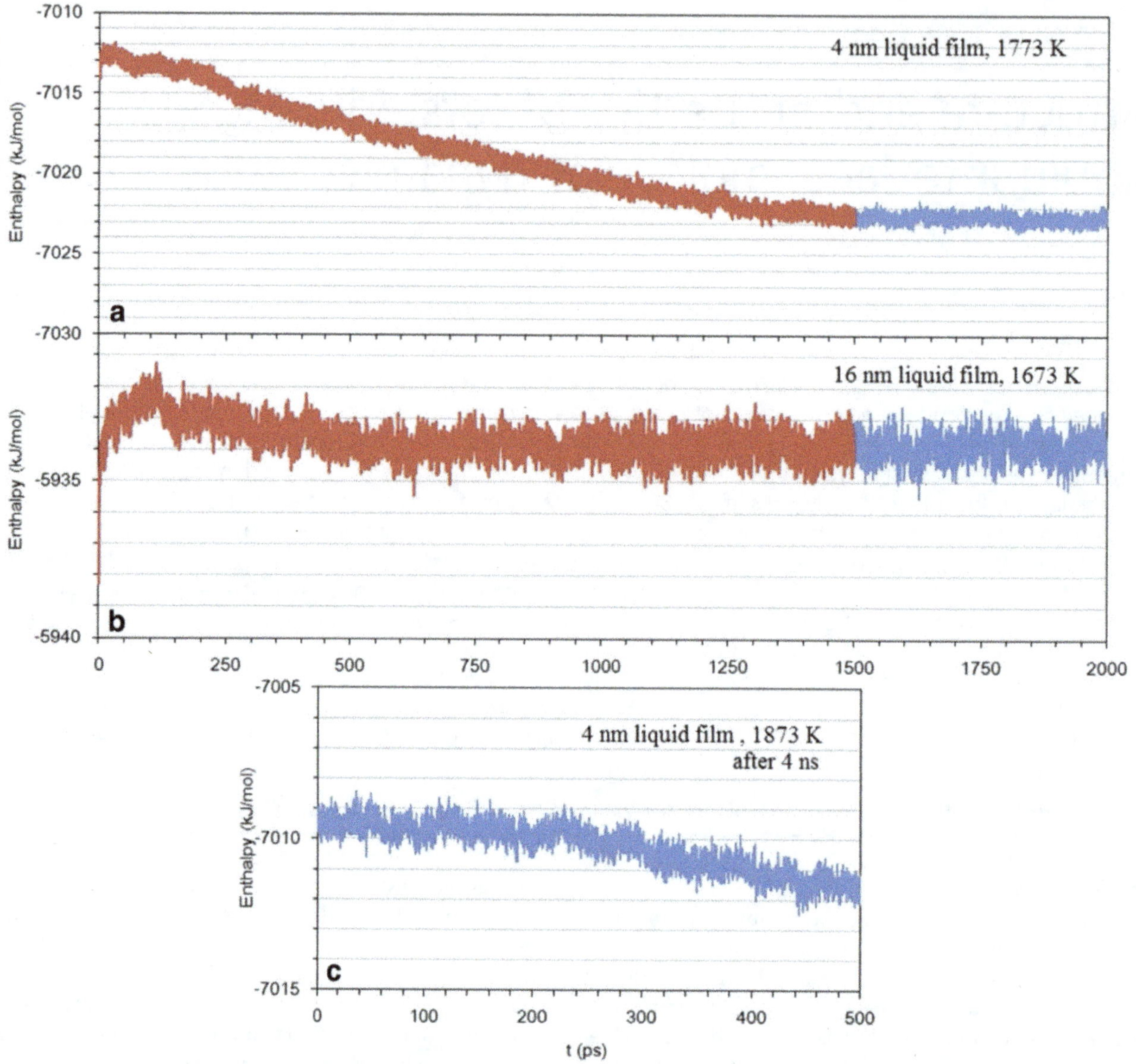

Figure 1 Plots of enthalpy of the simulated systems as a function of time. The enthalpy is plotted as molar enthalpy of **(a)** $Mg_{1.5459}Si_{1.0000}O_{3.5480}$, **(b)** $Mg_{1.2425}Si_{1.0000}O_{3.24345}$, and **(c)** $Mg_{1.5459}Si_{1.0000}O_{3.5480}$.

Methods

Molecular dynamics simulations were performed with the NPT ensemble at ambient pressure and various temperatures by application of the MXDORTO code (Sakuma and Kawamura 2009). The initial structure contains 21,440, 26,440, 43,440, and 63,440 atoms in which sheets of $MgSiO_3$ liquid consisting of 8,000, 15,000, 30,000, and 50,000 atoms (approximately 4-, 7-, 16-, and 28-nm thickness, respectively) are sandwiched between the (010) surfaces of forsterite. First, we simulated the equilibrated $MgSiO_3$ liquid film in a vacuum, starting with a randomly generated structure and randomly generated velocities of atoms through 0.5 ns (1,000,000 steps) at 1,973 K and

Table 1 Inter-atomic potential parameters

	Potential parameters and their values			
Atomic parameters	Q	a (Å)	b (Å)	c (kJ mol^{-1} Å^3)
Si	2.23607	0.99759	0.0830	0.000
O	−1.11803	1.81819	0.1539	27.400
Mg	1.11803	1.09881	0.0470	40.000
Pair parameters	D_1 (kJ mol^{-1})	β_1 (1/Å)	D_2 (kJ mol^{-1})	β_2 (1/Å)
Si-O	668,428	59.636	−105,335	45.514
Mg-O	28,960.0	5.0	−1,556.8	2.24
3-body parameters	f (kJ mol^{-1})	θ_0 (deg)	r_m (Å)	g_r (1/Å)
Si-O-Si	0.0006	147.0	0.170	168.0

Table 2 Comparison of lattice constants between MD simulations and experiments

	Forsterite		Enstatite	
	MD	Exp[a]	MD	Exp[c]
a (Å)	4.8709	4.7534	18.921	18.230
b (Å)	10.2820	10.1902	8.840	8.817
c (Å)	6.1249	5.9783	5.388	5.181
Density (g cm^{-3})	3.0467	3.2275	2.960	3.203
K_0 (GPa)	128	129[b]	95.0	95.8[d]
K'	5.69	4.2[b]	11.0	14.9[d]

MD, molecular dynamics; Exp, experiments. [a]Fujino et al. 1981. [b]Duffy et al. 1995. [c]Morimoto and Koto 1969. [d]Angel and Hugh-Jones 1994.

quenched to 300 K. Second, we simulated bulk forsterite with 13,440 atoms (11 × 5 × 8 unit cells of forsterite), starting with the given experimental crystal structure, which was obtained by an experiment (Fujino et al. 1981) and with randomly generated atom velocities; then, we cut along the (010) surface. Finally, we combined forsterite sheets along the (010) surface and an $MgSiO_3$ liquid film. We defined the z-axis perpendicularly to the (010) axis. Maintaining isobaric and isothermal conditions, we performed the relaxation up to 1.5 ns with a time step of 0.5 fs. Then, the statistical averages of the structure and physical properties were obtained from the velocities and coordinates of each atom in the simulations, through 500 ps. We confirmed that relaxation has been achieved by a plot of enthalpy versus time (Figure 1). The enthalpy of the system fluctuated around a constant value after 1.5 ns of relaxation (Figure 1a,b). The enthalpy of the system in the process of melting decreased, although 4 ns elapsed in the simulation (Figure 1c).

We imposed periodic boundary conditions in all directions. The Ewald summations were applied for evaluating the long-ranged Coulomb interactions. The pressure and temperature were kept constant at a given value through a scaling procedure. All the atoms were moved by application of the Verlet algorithm under a time interval of 0.5 fs. The function of the inter-atomic potential model was the same as that used in our previous work (Noritake et al. 2012):

$$U_{ij(rij)} = \frac{1}{4\pi\varepsilon_0}\frac{Z_iZ_je^2}{r_{ij}} + f_0(b_i + b_j)\exp\left(\frac{a_i + a_j - r_{ij}}{b_i + b_j}\right) + \frac{c_ic_j}{r_{ij}{}^6} + D_{1ij}\exp\left(-\beta_{1ij}r_{ij}\right) + D_{2ij}\exp\left(-\beta_{2ij}r_{ij}\right) \quad (1)$$

where $U_{ij}(r_{ij})$ is the potential energy, ε_0 is the permittivity of a vacuum, Z_i is the effective charge of the i-th

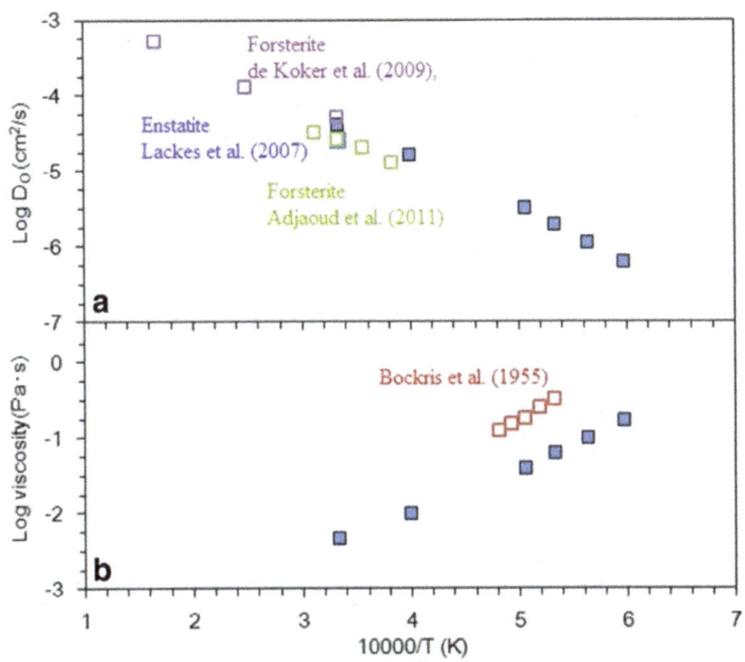

Figure 2 Comparisons of self-diffusion coefficients of oxygen and viscosities. (a) Comparison of diffusion coefficients of oxygen $MgSiO_3$ bulk liquid with other molecular dynamics studies. Closed squares indicate our study, and purple, green, and blue open squares indicate the data of forsterite liquid from de Koker et al. (2009), Adjaoud et al. (2011), and enstatite liquid by Lacks et al. (2007), respectively. **(b)** Comparison of calculated viscosities with experimental results. The experimental data are taken from Bockris et al. (1955). Viscosities are calculated using the Einstein-Stokes relation and the effective radius reported by Lacks et al. (2007).

atom, e is the elementary charge, r_{ij} is the distance between the i-th and j-th atoms, and f_O is the constant 4.1865 kJ Å^{-1} mol^{-1}. The values a_i, b_i, and c_i are the repulsion diameter, softness coefficient, and van der Waals coefficient of the i-th atom, respectively. The values D_{1ij}, β_{1ij}, D_{2ij}, and β_{2ij} are the parameters for the radial covalent bonds. The first term of this model gives the Coulomb potential, the second term gives the short-range repulsion, the third term is the van der Waals force, and the fourth and fifth terms are the radial part of the covalent bond. The three-body interactions between i, j, and k are represented as follows:

$$U_{kij} = -f\left[\cos\{2(\theta_{kij}-\theta_0\})-1\right]\sqrt{k_{ij}k_{ik}} \quad (2)$$

$$k_{ij} = \frac{1}{\exp\left[g_r(r_{ij}-r_m)\right]-1} \quad (3)$$

where f is the force constant, θ_{kij} is the angle among the atoms k, i, and j, and θ_0, g_r, and r_m are parameters for adjusting the angular part of the covalent bonds. Atomic and pair parameters are given in Table 1. The potential parameters used in this study are the same as those used in Noritake et al. (2012) for silicon and oxygen atoms. We added new parameters for magnesium atoms empirically. We verified the potential parameters by calculating the lattice parameters and compressibility of forsterite and enstatite crystals (Table 2). The bulk modulus, K_0, and the

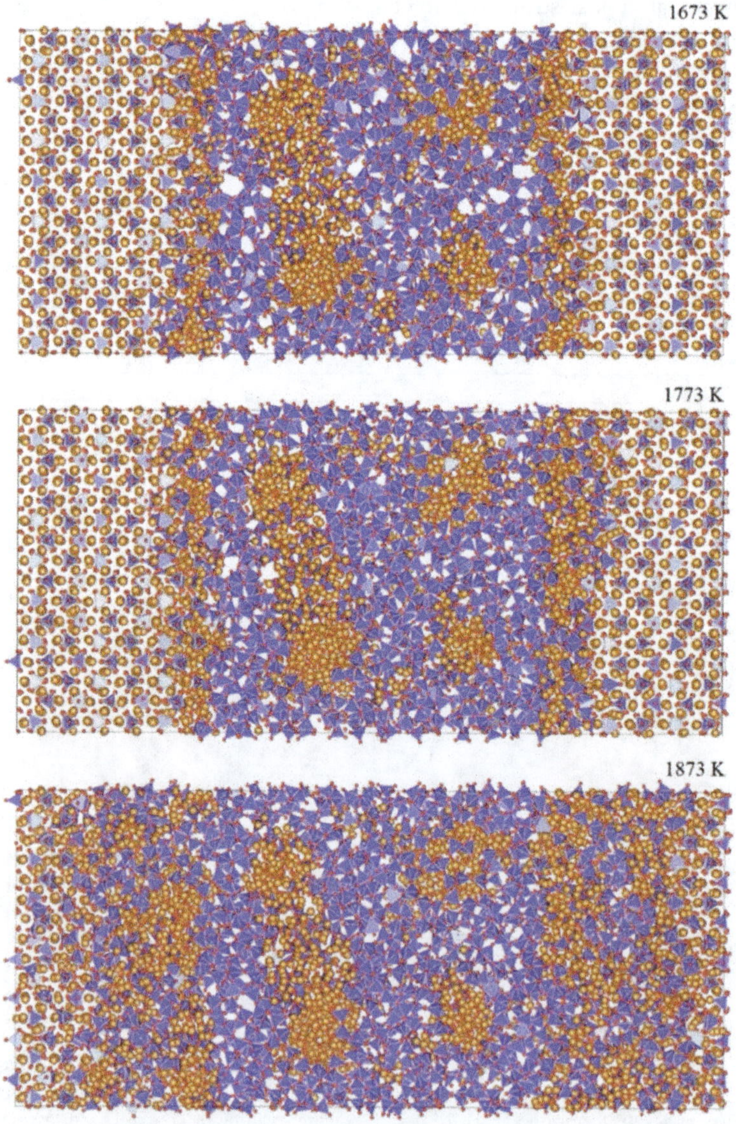

Figure 3 Snapshots of molecular dynamics simulations. The simulations are for the system forsterite-8,000 atom $MgSiO_3$ liquid perpendicular to the forsterite a-axis at various temperatures. The red, blue, and orange spheres represent O, Si, and Mg atoms, respectively.

pressure derivative of the bulk modulus, K', are obtained by fitting the simulated compression results along the third-order Birch-Murnaghan equation of state. The calculated lattice parameters and compressibility show good agreement with the experimental data (Fujino et al. 1981; Duffy et al. 1995; Morimoto and Koto 1969; Angel and Hugh-Jones 1994). We also simulated the bulk $MgSiO_3$ liquid ($N = 8,000$) at ambient pressure to verify our inter-atomic potential model by comparing the self-diffusion coefficients of oxygen with other force fields and Car-Parrinello molecular dynamics (de Koker et al. 2009; Adjaoud et al. 2011; Lacks et al. 2007). The calculated diffusion coefficients show good agreement with other molecular dynamics studies (Figure 2a). Figure 2b shows the calculated viscosity using the Einstein-Stokes relation with the effective

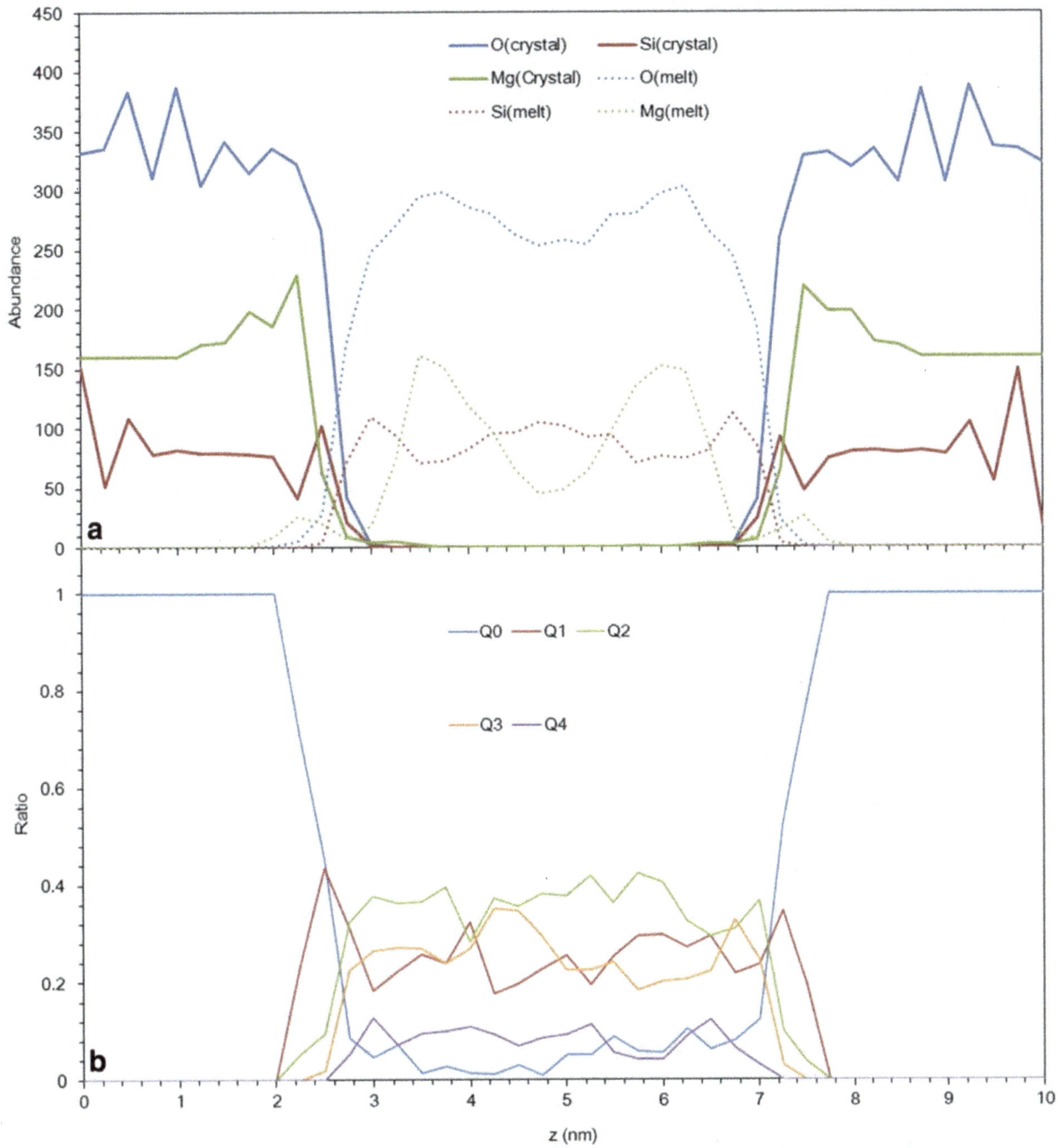

Figure 4 Distribution of elements and *Q*-species at 1,773 K. (a) Distribution of elements perpendicular to the (010) surface of forsterite. **(b)** Distribution of *Q*-species perpendicular to the forsterite surface.

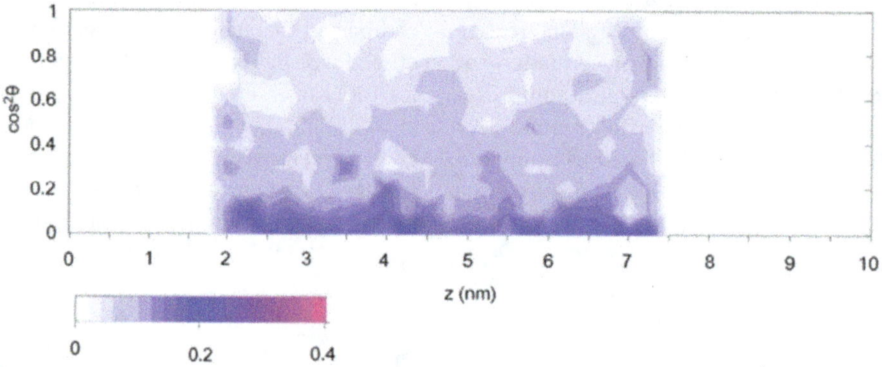

Figure 5 Distribution of the direction of Si-O-Si bridging perpendicular to the forsterite surface. The directions are shown as the $\cos^2\theta$ of the Si-Si vector and *z*-axis.

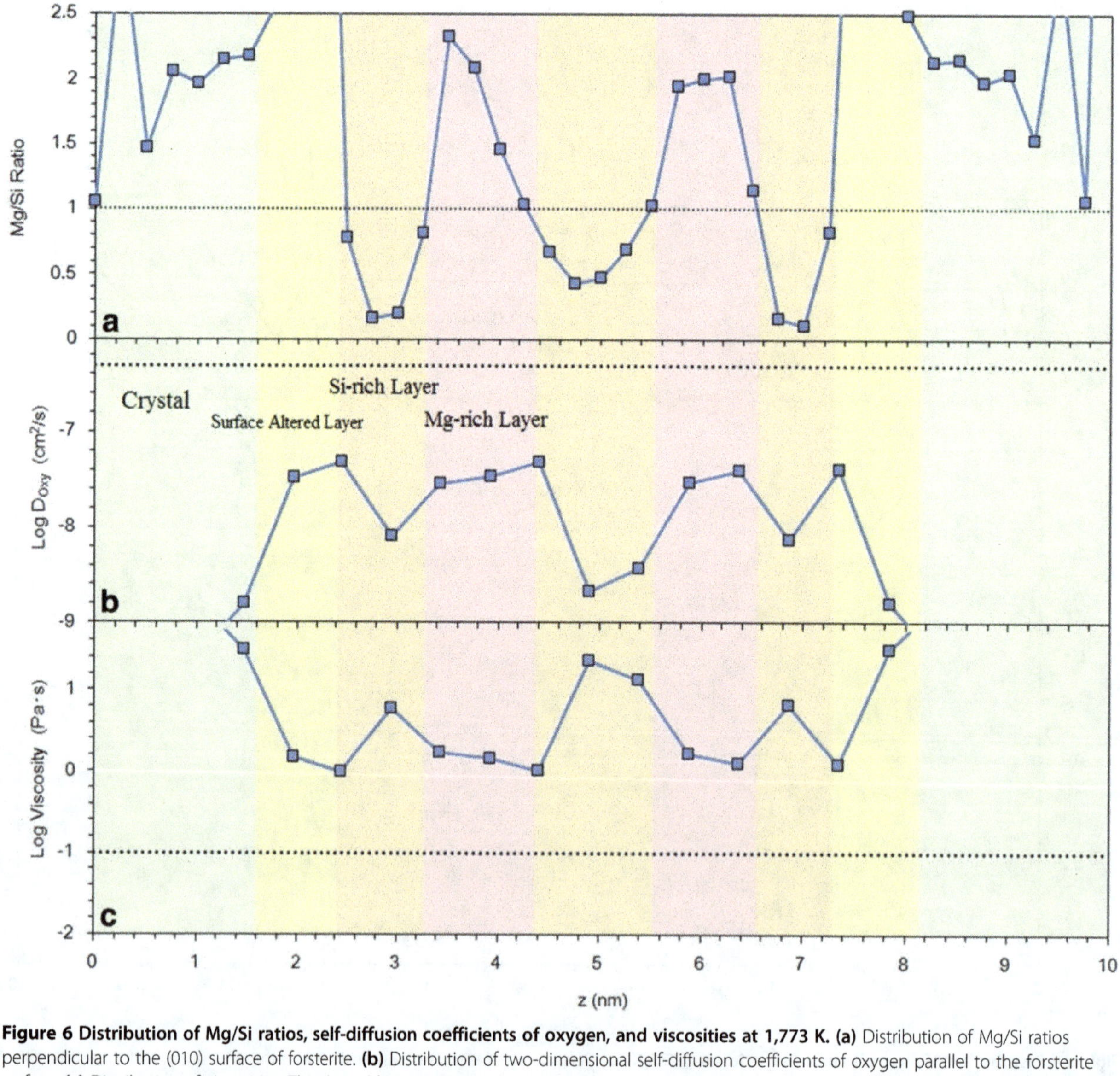

Figure 6 Distribution of Mg/Si ratios, self-diffusion coefficients of oxygen, and viscosities at 1,773 K. (a) Distribution of Mg/Si ratios perpendicular to the (010) surface of forsterite. **(b)** Distribution of two-dimensional self-diffusion coefficients of oxygen parallel to the forsterite surface. **(c)** Distribution of viscosities. The dotted line represents the value of bulk $MgSiO_3$ liquid.

radius reported by Lacks et al. (2007). The viscosity is lower than the experimental data (Bockris et al. 1955) by approximately half an order of magnitude. However, the activation enthalpy shows good agreement with the experimental data.

Results and discussion

4-nm liquid film: the structuralized liquid film

Figure 3 shows snapshots of molecular dynamics simulations perpendicular to the *a*-axis of forsterite using the visualization software VESTA (Momma and Izumi 2011). The liquidus temperature appears to be 1,873 K in Figure 3 because the crystal structure of the forsterite is altered. However, equilibrium is not achieved in a 4.5-ns run at 1,873 K (Figure 1c). The liquidus temperature of the simulated system is approximately 150 K lower than expected from the phase diagram (Jung et al. 2004). Figure 4a shows a distribution of elements perpendicular to the (010) surface of forsterite at 1,773 K. Figure 4b gives a distribution of *Q*-species of Si atoms perpendicular to the forsterite surface at 1,773 K. The threshold length of Si-O bonding for calculating the *Q*-speciation is 0.22 nm. The altered crystal surfaces exist at $z = 2$ to 2.5 nm and 7 to 7.5 nm. The abundance of Mg atoms is higher than that of Si atoms in this region. The nesosilicate structure is disturbed, and 20% to 40% of the SiO_4 tetrahedra form Si-O-Si bridging in this region. In the near crystal surface, the Si-rich layer exists in the liquid domain. The Mg-rich layer exists on the inside of Si-rich layer. Figure 5 shows a distribution of the direction of the Si-O-Si bridging perpendicular to the forsterite surface as a square of the cosine between the *z*-axis and the vector Si-Si calculated as an inner product. The orientation of the Si-O-Si bridging in the sandwiched liquid film is clearly parallel to the crystal surface. Figure 6b gives a distribution of the two-dimensional self-diffusion coefficients of oxygen parallel to the forsterite surface, and Figure 6c shows the calculated viscosities determined using the Einstein-Stokes relation with the effective radius obtained by Lacks et al. (2007). The two-dimensional self-diffusion coefficients of oxygen are consistently lower than that of the bulk $MgSiO_3$ liquid (dotted line) by an order of magnitude. The two-dimensional self-diffusion coefficients of oxygen are strongly related to the Mg/Si ratio of the layer to which the oxygen atoms belong (Figure 6a). In the high Mg/Si ratio layer, the self-diffusion coefficients of oxygen show higher values than those in the low Mg/Si ratio layer.

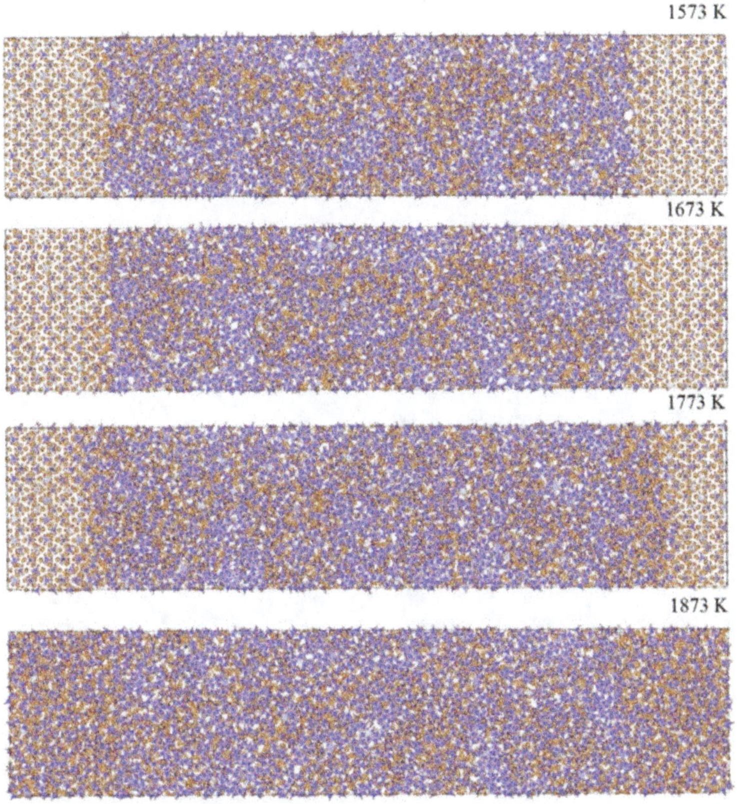

Figure 7 Snapshots of molecular dynamics simulations. Simulations are for the system forsterite-30,000 atom $MgSiO_3$ liquid perpendicular to the *a*-axis of forsterite at various temperatures. The red, blue, and orange spheres represent O, Si, and Mg atoms, respectively.

16-nm liquid film

Figure 7 shows snapshots of molecular dynamics simulations perpendicular to the *a*-axis of forsterite. The volume of the melting part increased at 1,773 K, and the system is completely melted at 1,873 K. Figure 8a shows the distribution of elements perpendicular to the (010) surface of forsterite at 1,673 K. Figure 8b gives the distribution of *Q*-species perpendicular to the forsterite surface at 1,673 K. The altered crystal surfaces exist at $z =$ approximately 3 and 20 nm. The nesosilicate structure is altered, and approximately 20% of the SiO_4 tetrahedra formed Si-O-Si bridging in this region. In the near crystal surface, the Si-rich layer exists in the liquid domain. In the domain between the disturbed crystal surface and the Si-rich layer, Si-O-Si bridging is formed, and Q^1 becomes the dominant species. The Mg-rich layer exists on the inside of the Si-rich layer. The composition of the sliced layer of the $MgSiO_3$ liquid is heterogeneous and oscillates, whereas the sliced layer is distant from the crystal surface (Figure 8a). Figure 9 shows the distribution of the direction of the Si-O-Si bridging perpendicular to the forsterite surface. The orientation of the Si-O-Si bridging in the sandwiched liquid film is clearly parallel to the crystal surface. However, the orientation in the disturbed crystal surface is strongly perpendicular to the crystal surface. Figure 10b gives the distribution of two-dimensional self-diffusion coefficients of oxygen parallel to the forsterite surface, and Figure 10c shows the calculated viscosities. The two-dimensional self-diffusion coefficients of oxygen are all slightly lower than those of bulk $MgSiO_3$ liquids (dotted line). The two-dimensional self-diffusion coefficients of oxygen are strongly related to the Mg/Si ratio of each layer in which the oxygen atoms belong in the liquid region. In the high Mg/Si ratio layers, the self-diffusion coefficients of oxygen show larger values than in the low Mg/Si ratio layers (Figure 10a). However, coefficients in altered crystal surface regions are lower than those in liquid regions, in contrast to the case of the 4-nm liquid film.

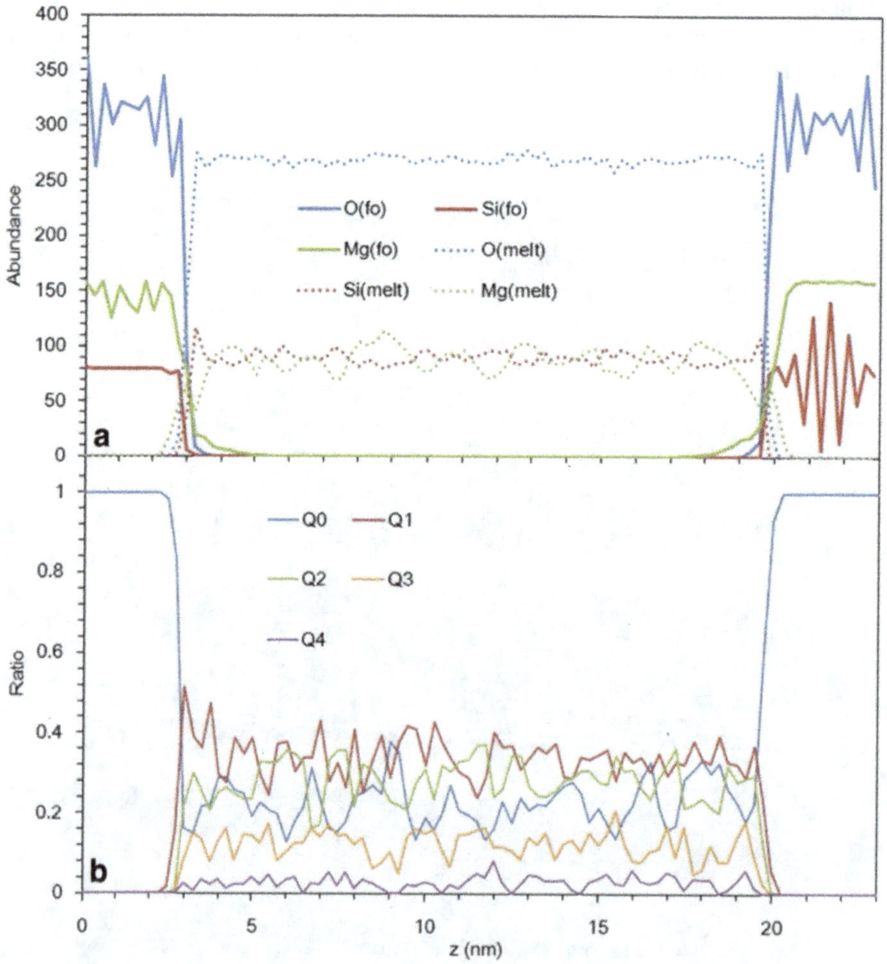

Figure 8 Distribution of elements and *Q*-species at 1,673 K. (a) Distribution of elements perpendicular to the (010) surface of forsterite. **(b)** Distribution of *Q*-species perpendicular to the forsterite surface.

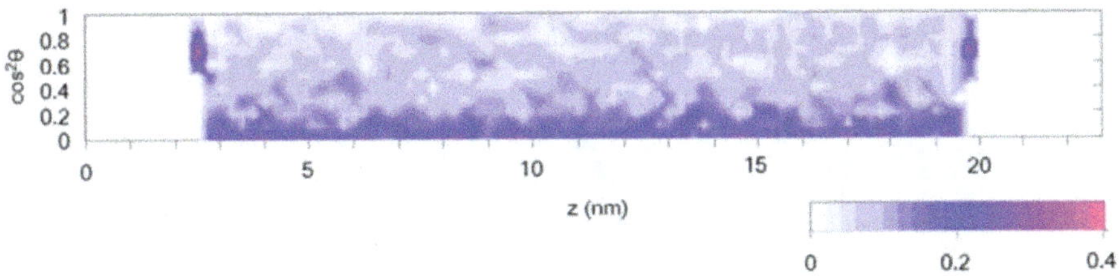

Figure 9 Distribution of the direction of the Si-O-Si bridging perpendicular to the forsterite (010) surface. The directions are shown as the $\cos^2\theta$ of the Si-Si vector and *z*-axis.

Effect of film thickness

We show the result of simulation of the system containing films of 7- and 28-nm thickness. The layered structure is also found in this system. However, the global decrease of self-diffusion coefficients is much smaller than the interface system with the 4-nm liquid film (Figure 11). Figure 12 shows the distribution of the Mg/Si ratio, elements, and the two-dimensional self-diffusion coefficients of oxygen of the system with a 28-nm liquid film. The layered structure is similar to the case of the 16-nm liquid film. The two-dimensional self-diffusion coefficients of oxygen in the middle part ($z = 12$ to 23 nm) of the liquid films are the same as those of bulk liquids. This means that the $MgSiO_3$ liquid film sandwiched by the forsterite (010) surface acts like a bulk liquid in the region 7 nm apart from the crystal surface. Figure 13 shows the calculated

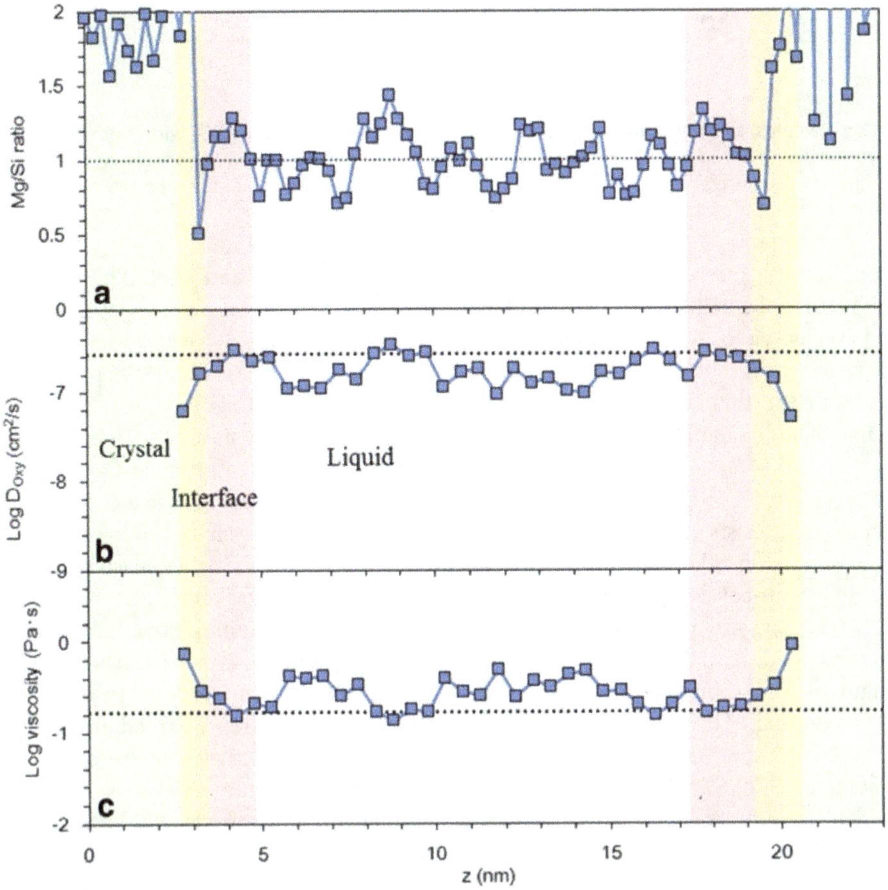

Figure 10 Distribution of Mg/Si ratios, self-diffusion coefficients of oxygen, and viscosities at 1,673 K. (a) Distribution of Mg/Si ratio perpendicular to the (010) surface of forsterite. **(b)** Distribution of two-dimensional self-diffusion coefficients of oxygen parallel to the forsterite surface. **(c)** Distribution of viscosities. The dotted line represents the value of bulk $MgSiO_3$ liquid.

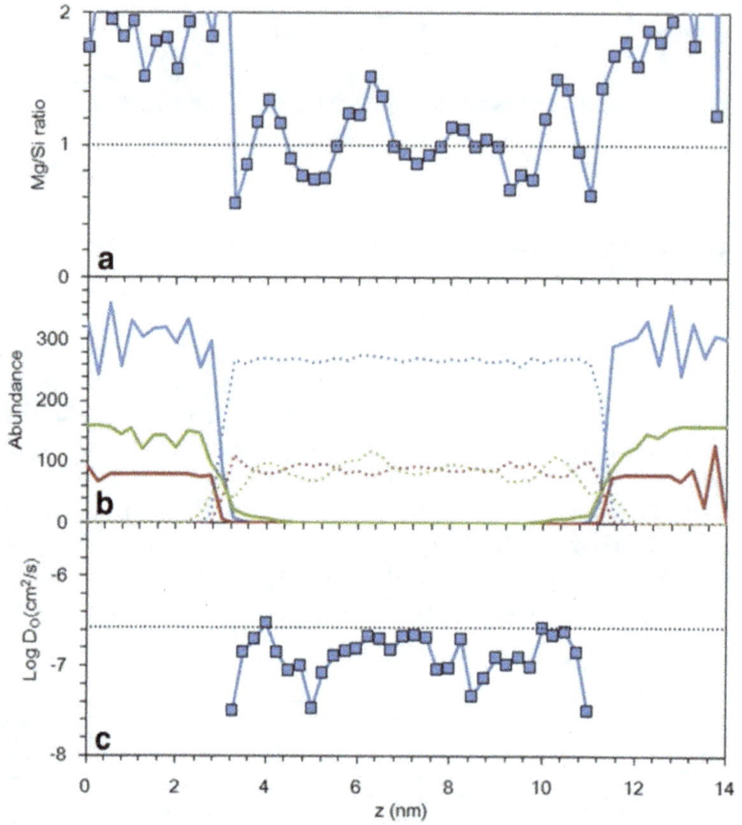

Figure 11 Mg/Si ratios, elements, and self-diffusion coefficients of oxygen for forsterite-$MgSiO_3$ with 7-nm liquid film. (a) Distribution of Mg/Si ratio perpendicular to the (010) surface of forsterite at 1,673 K. **(b)** Distribution of elements perpendicular to the (010) surface of forsterite at 1,673 K. **(c)** Distribution of two-dimensional self-diffusion coefficients of oxygen parallel to the forsterite surface at 1,673 K.

viscosities for various film thicknesses. The average viscosity approaches the viscosity of a bulk liquid with increasing film thickness. The average viscosity in the liquid films of 28-nm thickness is nearly the same as that of a bulk liquid. The average viscosity in the thin (<5 nm) liquid film is much higher than that of a bulk liquid.

General discussion

The layered structure that exists in the liquid films in these simulations is caused by the difference in bond strength between Si-O semi-covalent bonds and Mg-O ionic bonds. Si-O-Si bridgings are excessively formed between the altered crystal surface and the Si-rich layer because the strength of Si-O bonding is much higher than that of Mg-O bonding. The excess formation of Si-O-Si bridging in these compositions produce free oxygen atoms. Formation of Si-O-Si bridging depletes the non-bridging oxygen as an acceptor of magnesium atoms and concentrates silicon atoms. The excess free oxygen atoms concentrate magnesium atoms. Consequently, an Si-rich layer is formed in the nearest surface region of the liquid domain, and an Mg-rich layer is formed on the inside of the Si-rich layer on the liquid film (Figures 6a and 10a). The Q-speciation reflects the Mg/Si ratio on the Si-rich layer because Q^4 species increase in the case of the 4-nm liquid film. However, the Q-speciation does not reflect the Mg-rich layer. The abundance of Q^0 is too small in the case of the 4-nm film (Figure 4), meaning free oxygen atoms are produced. Clustering of the -Si-O- network also occurs. The Q-speciation does not reflect the Mg/Si ratio because the -Si-O- corner shared network makes the cluster. The clustering and layering weaken with increasing film thickness. The weakening can be explained by the degree of freedom in the film structure. The excess free oxygen and magnesium atoms from altered surfaces could not be accepted in the liquid film in the case of a thin film because the degree of freedom in the film structure is small. However, they can be accepted in a liquid film with increasing film thickness. Consequently, the degree of clustering decreases with increasing film thickness (Figures 6, 10, 11 and 12).

The layered structure is also reported by Gurmani et al. (2011). They investigated forsterite-$MgSiO_3$ liquid with a film thickness of up to 8 nm by the classical molecular dynamics method. However, the contrast of

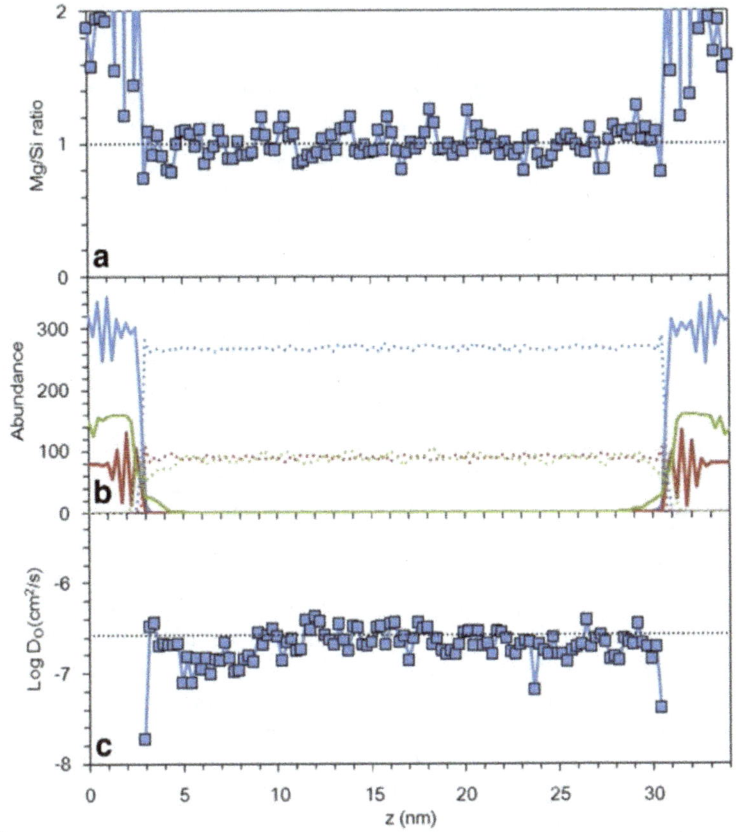

Figure 12 Mg/Si ratios, elements, and self-diffusion coefficients of oxygen for forsterite-$MgSiO_3$ with 28-nm liquid film. (a) Distribution of Mg/Si ratio perpendicular to the (010) surface of forsterite at 1,673 K. **(b)** Distribution of elements perpendicular to the (010) surface of forsterite at 1,673 K. **(c)** Distribution of two-dimensional self-diffusion coefficients of oxygen parallel to the forsterite surface at 1,673 K.

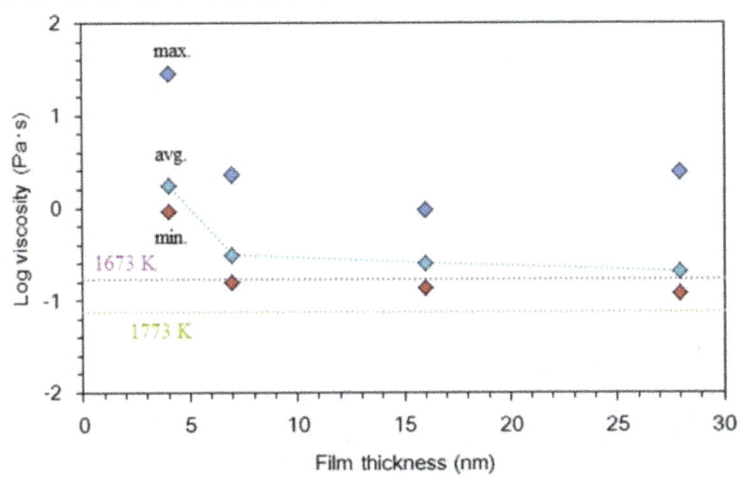

Figure 13 Plots of viscosity as a function of film thickness. Max. and min. denote the maximum and minimum viscosity in the liquid region, respectively. Avg. means the average of the entire liquid film (4 and 7 nm) and the average of the region 5 nm apart from the crystal surface (16 and 28 nm).

composition and decrease of self-diffusion coefficients by the interface was much smaller than in our study. The difference between Gurmani et al. (2011) and our studies is caused by the area of the surface cross section. Horbach et al. (1996) reported the finite-size effects in simulations of silicate glass. According to their study, the system should contain more than 8,000 atoms (4 to 5 nm for each side of the periodic cell) to avoid the size effects. Gurmani et al. (2011) simulated the same system with approximately 2-nm length sides for the cross section of the interface. The length of the cross section of our simulation is approximately 5 nm to avoid the size effects reported by Horbach et al. (1996).

The structure of the altered crystal surface seems to be different because of the thickness of the liquid films. The concentration of magnesium in the altered crystal surface with the 4-nm liquid film is much higher than that with the 16-nm liquid film (Figures 4a and 8a). In addition, the orientation of Si-O-Si bridging in the altered crystal surface with the 16-nm liquid film is strongly perpendicular to the crystal surface compared with the surface with the 4-nm liquid film (Figures 5 and 9). These differences might be explained by the degree of freedom of the configuration of the liquid films. There might be no structural flexibility in the liquid domain to accept the excess magnesium atoms in altered crystal surfaces because there are only five layers in the 4-nm liquid film. In contrast, the concentration of magnesium in the altered crystal surface with the 16-nm liquid film is lower than that with the 4-nm liquid film because of the structural flexibility of the liquid film. The surface becomes well-ordered by acceptation of magnesium in liquid and disordered by existence of excess magnesium atoms.

The two-dimensional self-diffusion coefficients parallel to the crystal surface are dependent on their distance from the altered crystal surface. Those coefficients show larger values in Mg-rich layers and smaller values in Si-rich layers. The regional dependences are simply explained by the composition of the sliced layer. Self-diffusion coefficients of network-forming elements in binary silicate liquids depend on their SiO_2 contents (Keller et al. 1982; Keller and Schwerdtfeger 1979). Self-diffusion coefficients of Si and O atoms decrease with increasing SiO_2 contents in bulk silicate liquids because of the difference in bond strength between Si-O covalent bonds and ionic bonds. Consequently, the self-diffusion coefficients of oxygen atoms decrease in Si-rich layers and increase in Mg-rich layers. The two-dimensional self-diffusion coefficients of oxygen in the 4-nm liquid film are lower than those in the 16-nm liquid film by an order of magnitude (Figures 4b and 8b). The difference in the self-diffusion coefficients might be explained by the structural flexibility of liquid films. The variety of Mg/Si ratios in the 4-nm liquid film is larger than that in the 16-nm liquid film. The degree of freedom of the configuration in the 4-nm liquid film is lower than that in the 16-nm liquid film. The liquid film is strongly structured by the crystal surface in the case of thin liquid films. Consequently, the flexibility of motion of each atom in the 4-nm liquid film is lower than that in the 16-nm liquid film, and mobility is decreased by the lower flexibility.

The oxygen diffusivity is related to shear viscosity by the Einstein-Stokes relation. The Einstein-Stokes relation is stated as

$$D = kT/6\pi r\eta, \tag{4}$$

where D is the self-diffusion coefficient, k is the Boltzmann constant, T is the temperature, r is the radius of diffusing particles, and η is the shear viscosity. This equation has been used with remarkable success in a variety of silicate liquids to relate diffusivity of network forming atoms and shear viscosity (e.g., Oishi et al. 1975; Dunn 1982; Shimizu and Kushiro 1984; Rubie et al. 1993; Lacks et al. 2007). Applying the Einstein-Stokes relation to the results of this study, the viscosity of thin film silicate liquid might be considered to have higher viscosity than bulk liquid by 1 order of magnitude (Figure 6c). In the 16-nm film liquid, the viscosity of the liquid is slightly lower than that of bulk liquid in the near surface region, the same as with bulk liquid in the Mg-rich layer, and oscillates with distance from the crystal surface (Figure 10c).

Implications

The melt segregation velocity in partially molten rock is inversely proportional to the melt viscosity (McKenzie 1989). In texturally equilibrated partially molten rocks, melts reside at the grain corners and edges (see Figure one of Zhu and Hirth 2003). A melt channel exists on the grain edge, and its cross-sectional area is determined by the dihedral angle, the melt fraction, and grain size. Von Bargen and Waff (1986) showed that the minimum channel cross-sectional area of partially molten rock consists of tetrakaidecahedron grains of varying melt fraction and dihedral angle θ. We calculated the viscosity at melt channels using the results of our simulations (Figure 13) and geometrical calculation by von Bargen and Waff (1986). Figure 14 shows the logarithm of normalized viscosity of melts in melt channels versus melt fraction with two grain sizes and two dihedral angles. Waff and Bulau (1979) showed that the distribution of dihedral angles in partially molten ultramafic rocks is between 30° and 47°. The viscosity becomes higher than bulk liquids by 1 order of magnitude at lower melt fractions and 1-μm grain. However, viscosity seems to be constant with varying melt fractions in the case of 10-μm grain size. The

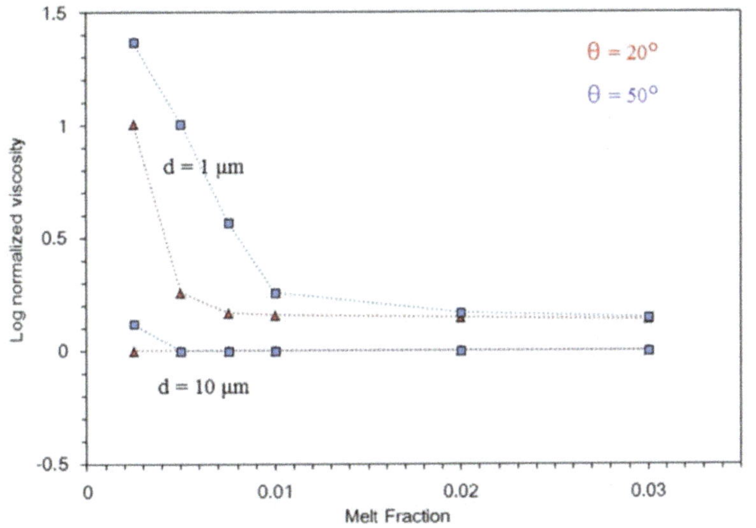

Figure 14 Plots of normalized viscosity at melt channels as a function of melt fraction. *d* and *θ* in the figure mean grain size and dihedral angle, respectively.

effect of the silicate crystal-liquid interface is negligible in melt flow in the mantle conditions, postulating the textural equilibrium, because the range in grain size of the mantle is 1 to 50 mm (Faul and Jackson 2005). However, a melt film of 0.6- to 3.0-nm thickness was found in deformed partially molten olivine-orthopyroxene rocks (de Kloe et al. 2000). Consequently, the crystal-liquid interface can affect the melt flow in deformed samples.

Conclusions

We showed the results of molecular dynamics simulations of forsterite-$MgSiO_3$ liquid, which contribute to our understanding of the nanoscale structure of the interface and diffusivity of atoms in the interface. From these simulations, the characteristic structure of liquid films in forsterite-$MgSiO_3$ liquid interface is observed. We also observe the layered structure of an altered crystal surface, an Si-rich layer, and an Mg-rich layer in the crystal-liquid interface. The layered structure is formed by the strength difference between Si-O covalent bonds and Mg-O ionic bonds. Si-O-Si bridging and free oxygen atoms are excessively formed in the near surface because Si-O bonding is much stronger than Mg-O bonding. The difference in the layered structure indicated by the thickness of $MgSiO_3$ liquid films might be caused by the difference in degree of freedom of the configuration in liquid films. The two-dimensional diffusivity of oxygen is controlled by two factors. One factor is the thickness of the liquid films that decreases oxygen diffusivity with decreasing film thickness because of the decrease of degree of freedom of the configuration in the liquid films. The other factor is the composition of the sliced layer, where oxygen diffusivity increases with increasing Mg/Si ratio because Si-O bonding is much stronger than Mg-O bonding. The degree of Mg/Si layering strongly depends on film thickness and decreases with increasing film thickness. The effect of the crystal-liquid interface is negligible in texturally equilibrated rocks. However, the interface can affect the melt flow in deformed samples because the grain boundary melt film of several nanometer thickness exists stably in deformed partially molten rocks.

Competing interests

The authors declare that they have no competing interests.

Authors' contributions

FN performed the simulations and analyzed the data. Both FN and KK proposed the topic, designed the study, mutually discussed throughout the study, and wrote the paper. Both authors read and approved the final manuscript.

Acknowledgements

We thank the editor and three anonymous reviewers for the valuable comments. This work was supported by the Grant-in-Aid for JSPS fellows, 251856.

References

Adjaoud O, Steinle-Neumann G, Jahn S (2011) Transport properties of Mg_2SiO_4 liquid at high pressure: physical state of magma ocean. Earth Planet Sci Lett 312:463–470, doi:10.1016/j.epsl.2011.10.025

Angel RJ, Hugh-Jones DA (1994) Equations of state and thermodynamic properties of enstatite pyroxenes. J Geophys Res 99:19777–19783, doi:10.1029/94JB01750

Bockris JO'M, Mackenzie JD, Kitchener JA (1955) Viscous flow in silica and binary liquid silicates. Trans Faraday Soc 51:1734–1748, doi:10.1039/TF9555101734

De Kloe R, Drury MR, van Roermund HLM (2000) Evidence for stable grain boundary melt films in experimentally deformed olivine-orthopyroxene rocks. Phys Chem Min 27:480–494, doi:10.1007/s002690000090

De Koker NP, Stixrude L, Karki B (2009) Thermodynamics, structure, dynamics, and freezing of Mg_2SiO_4 liquid at high pressure. Geochim Cosmochim Acta 72:1427–1441, doi:10.1016/j.gca.2007.12.019

Duffy TS, Zha C, Downs RT, Mao H, Hemley RJ (1995) Elasticity of forsterite to 16 GPa and the composition of the upper mantle. Nature 378:170–173

Dunn T (1982) Oxygen diffusion in three silicate melts along the join diopside-anorthite. Geochim Cosmochim Acta 46:2293–2299, doi:10.1016/0016-7037(82)90202-2

Faul UH, Jackson I (2005) The seismological signature of temperature and grain size variations in the upper mantle. Earth Planet Sci Lett 234:119–134, doi:10.1016/j.epsl.2005.02.008

Fujino K, Sasaki S, Takeuchi Y, Sadanaga R (1981) X-ray determination of electron distributions in forsterite, fayalite and tephroite. Acta Crystallogr B 37:513–518, doi:10.1107/S0567740881003506

Gurmani SF, Jahn S, Brasse H, Schilling FR (2011) Atomic scale view on partially molten rocks: molecular dynamics simulations of melt-wetted olivine grain boundaries. J Geophys Res 116:V12209, doi:10.1029/2011JB008519

Hiraga T, Anderson LM, Zimmerman ME, Mei S, Kohlstedt DL (2002) Structure and chemistry of grain boundaries in deformed, olivine + basalt and partially molten lherzolite aggregates: evidence of melt-free grain boundaries. Contrib Mineral Petrol 144:163–175, doi:10.1007/s00410-002-0394-1

Horbach J, Kob W, Binder K, Angell CA (1996) Finite size effects in simulations of glass dynamics. Phys Rev E 54:R5897–R5900, doi;10.1103/PhysRevE.54.R5897

Ichikawa Y, Kawamura K, Nakano M, Kitayama K, Seiki T, Theramast N (2001) Seepage and consolidation of bentonite saturated with pure- or salt-water by the method of unified molecular dynamics and homogenization analysis. Eng Geol 60:127–138, doi:10.1016/S0013-7952(00)00095-8

Jung I, Decterov SA, Pelton AD (2004) Critical thermodynamic evaluation and optimization of the $FeO-Fe_2O_3-MgO-SiO_2$ system. Metall Mater Trans B 35:877–889, doi:10.1007/s11663-004-0082-9

Keller H, Schwerdtfeger K (1979) Tracer diffusivity of Si^{31} in $CaO-SiO_2$ melts at 1600 °C. Metall Trans B 10:551–554

Keller H, Schwerdtfeger K, Petri H, Hölzle R, Hennesen K (1982) Tracer diffusivity of O^{18} in $CaO-SiO_2$ melts at 1600 °C. Metall Trans B 13:237–240

Lacks DJ, Rear DB, van Orman JA (2007) Molecular dynamics investigation of viscosity, chemical diffusivities and partial molar volumes of liquids along the $MgO-SiO_2$ join as functions of pressure. Geochim Cosmochim Acta 71:1312–1323, doi:10.1016/j.gca.2006.11.030

McKenzie D (1989) Some remarks on the movement of small melt fractions in the mantle. Earth Planet Sci Lett 95:53–72, doi:10.1016/0012-821X(89)90167-2

Momma K, Izumi F (2011) VESTA 3 for three-dimensional visualization of crystal, volumetric and morphology data. J Appl Crystallogr 44:1272–1276, doi:10.1107/S0021889811038970

Morimoto N, Koto K (1969) The crystal structure of orthoenstatite. Z Kristallogr – Cryst Mater 129:65–83

Noritake F, Kawamura K, Yoshino T, Takahashi E (2012) Molecular dynamics simulation and electrical conductivity measurement of $Na_2O\cdot 3SiO_2$ melt under high pressure; relationship between its structure and properties. J Non-Cryst Solids 358:3109–3118, doi:10.1016/j.noncrysol.2012.08.027

Oishi Y, Terai R, Ueda H (1975) Oxygen diffusion in liquid silicates and relation to their viscosity. In: Cooper A (ed) Mass transport phenomena in ceramics. Material science research, vol 9. Springer, New York, pp 297–310

Rubie DC, Ross CR, Carroll MR, Elphick SC (1993) Oxygen self-diffusion in $Na_2Si_4O_9$ liquid up to 10 GPa and estimation of high-pressure melt viscosities. Am Mineral 78:574–582

Sakuma H, Kawamura K (2009) Structure and dynamics of water on muscovite mica surfaces. Geochim Cosmochim Acta 73:4100–4110, doi:10.1016/j.gca.2009.05.029

Sakuma H, Tsuchiya T, Kawamura K, Otsuki K (2003) Large self-diffusion of water on brucite surface by ab initio potential energy surface and molecular dynamics simulations. Surf Sci 536:L396–L402, doi:10.1016/S0039-6028(03)00577-6

Shimizu N, Kushiro I (1984) Diffusivity of oxygen in jadeite and diopside melts at high pressures. Geochim Cosmochim Acta 48:1295–1303, doi:10.1016/0016-7037(84)90063-2

Von Bargen N, Waff HS (1986) Permiabilities, interfacial areas and curvatures of partially molten systems: results of numerical computations of equilibrium microstructures. J Geophys Res 91:9261–9276, doi:10.1029/JB091iB09p09261

Waff HS, Bulau JR (1979) Equilibrium fluid distribution in an ultramafic partial melt under hydrostatic stress conditions. J Geophys Res 84:6109–6114, doi:10.1029/JB084iB11p06109

Zhu W, Hirth G (2003) A network model for permeability in partially molten rocks. Earth Planet Sci Lett 212:407–416, doi:10.1016/S0012-821X(03)00264-4

2

Cool, alkaline serpentinite formation fluid regime with scarce microbial habitability and possible abiotic synthesis beneath the South Chamorro Seamount

Shinsuke Kawagucci[1,5*], Junichi Miyazaki[1], Yuki Morono[2], Jeff S. Seewald[3], C. Geoff Wheat[4] and Ken Takai[1]

Abstract

South Chamorro Seamount (SCS) is a blueschist-bearing serpentinite mud volcano in the Mariana forearc. Previous scientific drilling conducted at SCS revealed highly alkaline, sulfate-rich formation fluids resulting from slab-derived fluid upwelling combined with serpentinization both beneath and within the seamount. In the present study, a time-series of ROV dives spanning 1000 days was conducted to collect discharging alkaline fluids from the cased Ocean Drilling Program (ODP) Hole 1200C (hereafter the CORK fluid). The CORK fluids were analyzed for chemical compositions (including dissolved gas) and microbial community composition/function. Compared to the ODP porewater, the CORK fluids were generally identical in concentration of major ions, with the exception of significant sulfate depletion and enrichment in sulfide, alkalinity, and methane. Microbiological analyses of the CORK fluids revealed little biomass and functional activity, despite habitable temperature conditions. The post-drilling sulfate depletion is likely attributable to sulfate reduction coupled with oxidation of methane (and hydrogen), probably triggered by the drilling and casing operations. Multiple lines of evidence suggest that abiotic organic synthesis associated with serpentinization is the most plausible source of the abundant methane in the CORK fluid. The SCS formation fluid regime presented here may represent the first example on Earth where abiotic syntheses are conspicuous with little biotic processes, despite a condition with sufficient bioavailable energy potentials and temperatures within the habitable range.

Keywords: Forearc serpentinite mud volcano, South Chamorro Seamount, Limit of biosphere, Present-days' chemical evolution, Radio-isotope-tracer carbon assimilation estimation

Introduction

Serpentinization of ultramafic rocks that are the major components of the oceanic lithosphere produces highly reductive fluids, in which transformation of inorganic carbon species to organic matter occur (e.g., McCollom 2013). Serpentinization-associated geofluid systems have been recognized as modern analogs of plausible settings for the origin of primordial life (e.g., Russell et al. 2014) as well as ancient microbial communities (e.g., Takai et al. 2006; Ueda et al. 2016). Serpentinization-associated geofluid systems, typically characterized by an abundance of H_2 and/or highly alkaline fluids, have been discovered at a range of tectonic/geological settings on the modern Earth where ultramafic rocks encounter water circulation. The most notable examples include slow-spreading mid-oceanic ridges (Charlou et al. 2002; Konn et al. 2015), ocean core complexs (Kelley et al. 2001; Früh-Green et al. 2003), ophiolites (Barnes et al. 1978; Schrenk et al. 2013), coastal springs (Barnes et al. 1967), onshore volcanic hot springs (Homma and Tsukahara, 2008; Suda et al. 2014), inner trench slope along the southern Mariana forearc (Ohara et al. 2012; Okumura et al. 2016a; Onishi et al. 2018), and forearc serpentinite seamounts (Mottl et al. 2004; Fryer, 2011).

* Correspondence: kawagucci@jamstec.go.jp
[1]Department of Subsurface Geobiological Analysis and Research (D-SUGAR), Japan Agency for Marine-Earth Science and Technology (JAMSTEC), 2-15 Natsushima-cho, Yokosuka 237-0061, Japan
[5]Institute of Geochemistry and Petrology, ETH Zürich, Zürich, Switzerland
Full list of author information is available at the end of the article

Forearc serpentinite mud volcanoes are formed by extensive serpentinization of the mantle, triggered by the injection of slab-derived fluids into the overlying mantle wedge in a subduction setting. These serpentinization products ascend to form the largest mud volcanoes on Earth (Fryer, 2011). Deep-sea serpentinite mud volcanoes are outstanding settings, compared to other serpentinization-associated geofluid systems (Table 1), for studying abiotic organic syntheses (i.e., present-days' chemical evolution) and for finding the boundary between microbiologically habitable and uninhabitable zones (i.e., limit of life and the biosphere) in natural environments of the Earth. For example, the impermeable mud volcano allows for decreased entrainment of photosynthesis-derived organic matter and seawater-derived microbial populations into the internal part of the volcano where pristine mud and fluid ascend from the deep. As such, if we can obtain these pristine fluids contributing to the formation of serpentinite mud volcanoes, then they can be used to address the unique geochemical and microbiological processes associated with these volcanoes.

The South Chamorro Seamount (SCS) is located at the southern end of a serpentinite mud volcano chain in the Mariana forearc, where the Pacific plate subducts beneath the Philippine Sea plate (Fig. 1a). To date, the SCS is the only known location of active blueschist-bearing serpentinite mud volcanism that hosts lush deep-sea chemosynthetic communities sustained by the serpentinite fluid seepage (Fryer and Mottl, 1997). The Ocean Drilling Program (ODP) Leg 195 drilled the summit of the SCS (Site 1200) in 2001 and obtained core samples in order to examine geological and biogeochemical processes associated with subduction (Shipboard Sci. Party 2002). The serpentinite mud porewater, with the exception of taken near the seafloor, exhibited a highly basic pH of 12.5 and had a composition rich in dissolved sulfate and carbonate. The alkaline porewater chemistry is indicative of slab-derived fluids that had reacted with mantle rocks when ascending through the overlying plate (Fig. 1b), but few studies have investigated the origin of the abundant sulfate in the deep (Mottl et al. 2003, 2004; Hulme et al. 2010). In addition, porewater samples from shallow zones ~ 20 mbsf exhibited sulfate depletion and high alkalinity. The chemistry of these shallow fluids has been interpreted to reflect the oxidation of CH_4 (and H_2) derived from the deep by a shallow, sub-seafloor microbial community including sulfate-reducers (Mottl et al. 2003; Aoyama et al. 2018). However, the gas composition of the porewater could not be accurately determined, since the standard ODP sampling retrieval, handling, and on-board procedures resulted in a significant loss of dissolved gases. Microbiological analyses including phospholipid analysis (Mottl et al. 2003), cell count, and cultivation attempts (Takai et al. 2005) were carried out, but the community structure and potential microbial functions could not be determined.

A cased borehole observatory, the Circulation Obviation Retrofit Kit (CORK), was deployed during the ODP Leg 195 at Hole 1200C on the SCS (Shipboard Sci. Party 2002; Wheat et al. 2008). When the observatory seal was opened on 20 March 2003, 2 years after its installation, over-pressurized fluids discharged from the CORK outlet

Table 1 Deep-sea serpentinization-associated geofluid fields

Name of field	South Chamorro Seamount (SCS)	Shinkai Seep Field	Lost City	Rainbow
Location	13°47 'N–146°00'E	11°40'N–140°03'E	30°07'N–42°07'W	36°14'N–33°54'W
Tectonics	Forearc seamount chain	Non-accretionary trench slope	Oceanic core complex, MAR	Non-transform offset, MAR
Driving force	Serpentine mud volcanism	(Hydrothermalism?)	Hydrothermalism	Hydrothermalism
Fluid source	Slab-derived fluid	Seawater penetration	Seawater penetration	Seawater penetration
Approach	ODP, CORK, Dive	Dive	Dive, IODP	Dive
Depth (m)	2960	5800	700–800	2300
T max	2	–	> 91	365
pH	12.5	(Brucite chimney)	10.7	2.8
H_2 (mM)	0.01 (possibly < 40)	–	15	16
CH_4 (mM)	37 (possibly < 65)	–	2	2.5
Note	The strongest alkaline deep-sea geofluid system so far identified. See text for details.	Active alkaline fluid seeping from brucite chimney surrounded by macrofauna colony.	Hot alkaline fluid discharging from tall carbonate chimney, only identified on the modern seafloor at this moment.	Black smoker acidic fluid discharging from sulfide chimney with abundant H2. Similar vent sites have already been found on the modrern seafloor.
Ref	See text	Ohara et al. 2012	Kelley et al. 2001	Charlou et al. 2002
		Stern et al. 2014	Früh-Green et al. 2003	
		Okumura et al., 2016a Onishi et al. 2018	Proskurowski et al. 2008	

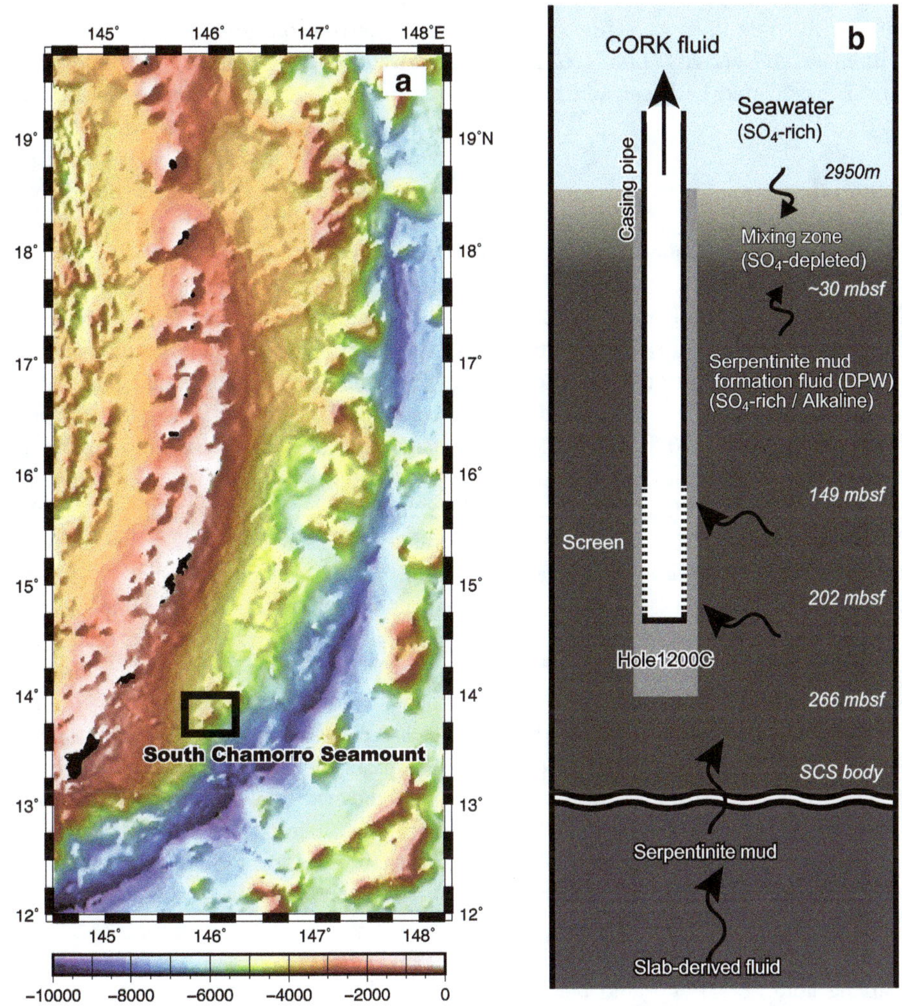

Fig. 1 Topography and schematic illustration of the Mariana forearc and the SCS. **a** Seafloor topography of the Mariana forearc region. **b** Schematic illustration of the ODP Hole 1200C focused in the present study

pipe (this fluid is referred to as the 'CORK fluid' hereafter) at a flow rate of ~ 0.2 L/s. The endmember chemical composition of the CORK fluid, estimated from two fluid samples diluted with entrained ambient seawater, was generally similar to those from the porewater samples from deeper zones of the ODP core (Wheat et al. 2008). Although the deep porewater samples obtained during the ODP Leg 195 in 2001 contained abundant sulfate and negligible sulfide, sulfide was present in the highly alkaline (pH 12.4) CORK fluid. This difference implies that partial reduction of sulfate to sulfide could occur within the SCS mud volcano before the formation fluid discharged through the cased borehole (Wheat et al. 2008). Discharge from the open borehole continued for 37 days, until the observatory was resealed on 28 April 2003. During the 2003 campaign, however, no gas-tight fluid sampling was conducted despite gas species such as H_2, CO, CH_4, and other hydrocarbons are key chemical clues for investigating both abiotic organic synthesis and microbiological habitability/function in an active serpentinite mud volcano.

The present study presents results from four subsequent sample collection campaigns utilizing remotely operated vehicle (ROV) dives with gas-tight fluid samplers (Seewald et al. 2001; Saegusa et al. 2006) and other sampling tools conducted between 2009 and 2012 to collect fresh fluids discharging from the cased borehole of Hole 1200C. The chemistry of the CORK fluid, particularly dissolved gas species, as well as the microbial community composition and function were determined. Carbon and sulfur compositions in the CORK fluids were altered from those from the ODP porewater samples, probably due to changes in the subseafloor hydrological field before and after the scientific drilling event. Analyses of the CORK fluids and the ODP porewater suggested that the sulfate-methane couple likely co-exist in the native SCS formation fluid. However, low biomass and activity were observed environments where physical and chemical conditions, particularly the temperature (~ 3 °C) and bioavailable energy potentials, were suitable for supporting a prosperous microbial community. Multiple lines of evidence, such as low microbial biomass and activity, point to abiotic hydrocarbon synthesis associated with serpentinization occurring beneath the SCS.

Methods/experimental

Site description

In 2001, a series of boreholes were drilled near the seepage sites at the summit of the SCS (ODP Site 1200) (Shipboard Sci. Party 2002). The ODP Hole 1200C was drilled to 266 mbsf prior to casing and the installation of a CORK device. The final configuration was placed at the base of the casing at 202.8 mbsf, providing a screened section from 148.8 mbsf to 202.3 mbsf (Fig. 1b). On 20 March 2003, the CORK monitoring instruments were recovered using the ROV Jason II, and on 28 April 2003 the top of the CORK was resealed by a steel housing holding a downhole pressure sensor (Wheat et al. 2008).

Sampling and analysis

Fluid sampling

CORK fluids were collected for both chemical characterization and analyses of microbial biomass/function during four cruises of the R/V Natsushima with ROV Hyper-Dolphin in January 2009 (NT09-01, HPD#941-947), May–June 2009 (NT09-07, HPD#1007-1010), February 2012 (NT12-04, HPD#1349-1350), and September 2012 (NT12-23, HPD#1433-1434). On 20 January 2009 (HPD#941), the valve on the original-type CORK was opened. The discharging fluid (#941) appeared blackish in coloration, probably due to suspension of sulfide minerals that had accumulated during the past 6 years. The likely source for this blackish fluid is the discharging fluid that occupied the pipe casing when the valve was closed in 2003. After initial samples were collected, the steel plug was replaced with a polyvinyl chloride (PVC) tube and a ball valve. This PVC tube was inserted past the latch holes in the CORK body, sealing the borehole from the bottom seawater. Thus, fluids collected from the top of the CORK can be considered pristine CORK fluids with minimal or no entrainment of the bottom seawater (Wheat et al. 2008). Fluid collected during subsequent dives appeared to be colorless and clear at the mouth of the CORK but the coloration changed to whitish after venting, most likely due to the precipitation of brucite from seawater-derived Mg^{2+} and CORK fluid-derived OH^-.

CORK fluids were collected using several types of fluid samplers, including WHATS, Bag, and Isobaric Gas-Tight (IGT) samplers (Seewald et al. 2001; Saegusa et al. 2006). The whitish fluids were also collected into pre-evacuated stainless-steel bottles with radio-isotope-labeled tracers for metabolic activity measurements (see below). The inlet nozzles for WHATS and Bag samplers were inserted as deep as possible into the borehole in order to minimize seawater contamination. For fluids collected using IGT samplers, a polycarbonate reservoir was attached to the CORK ball valve. The reservoir was flushed with several volumes of CORK fluid before the IGT sampler snorkel was inserted into the reservoir through a rubber diaphragm to collect pristine CORK fluid without seawater entrainment.

A special sampling device, named the "Deep Sampler," was also used during cruises NT12-04 and NT12-23 to collect fluid within the borehole at subseafloor depths where the screened casing is positioned (Fig. 2). The Deep Sampler consisted of a 100 mL gas-tight and pressure-preserved titanium bottle and a valve connected to a winch on the seafloor with a rope. The open titanium bottle-chamber-valve units were lowered to a depth of 179 mbsf within the depth range of the screens, before a messenger weight was dropped from the seafloor to slide the bottle unit to the chamber-valve unit for fluid sampling (a few seconds to 10 s time) and in the end sealing the fluid in two of the chambers (Fig. 2). The temperature was monitored during fluid sampling, which increased from ~ 1.7 °C at the seafloor to 3.3 °C at 170 m below the seafloor (thermal gradient of 9.4 °C/km). Samples collected by the Deep Sampler and the other samplers did not show significant differences in their chemical composition, implying that fluid alteration and seawater contamination during the 12 h ascent through the cased hole (Wheat et al. 2008) was negligible.

Geochemistry

Immediately upon recovery of the ROV, fluid samples were subsampled for liquid and gas analyses. For liquid phase analysis, the fluid sample was withdrawn into disposable plastic syringes for filtration, and then introduced into acid-washed HDPE bottles and glass vials. Shipboard measurements included pH, alkalinity, chlorinity, and hydrogen sulfide. Shore-based analyses with devices such as ion chromatography (IC), inductively coupled plasma atomic emission spectrometry (ICP-AES), and ICP mass spectrometry (ICP-MS) were conducted using techniques similar to those carried out for the ODP 1200 porewater samples (e.g., Mottl et al. 2003, 2004), thus minimizing analytical biases among these studies. Dissolved gas extraction from fluid samples collected with the WHATS sampler was conducted using previously reported techniques (e.g., Kawagucci et al. 2016; Miyazaki et al. 2017). To extract the total inorganic carbon (mainly CO_3^{2-}) as CO_2 for abundance and isotope analyses, the fluid sample was acidified with ~ 3 g of solid sulfamic acid (H_3NSO_3) in the extraction line. While the total gas content was determined barometrically during the gas extraction, the relative abundance of H_2, CH_4, CO, CO_2, C_2H_6, and C_3H_8 in the whole gas phase were determined by gas chromatography using a helium ionization detector. The

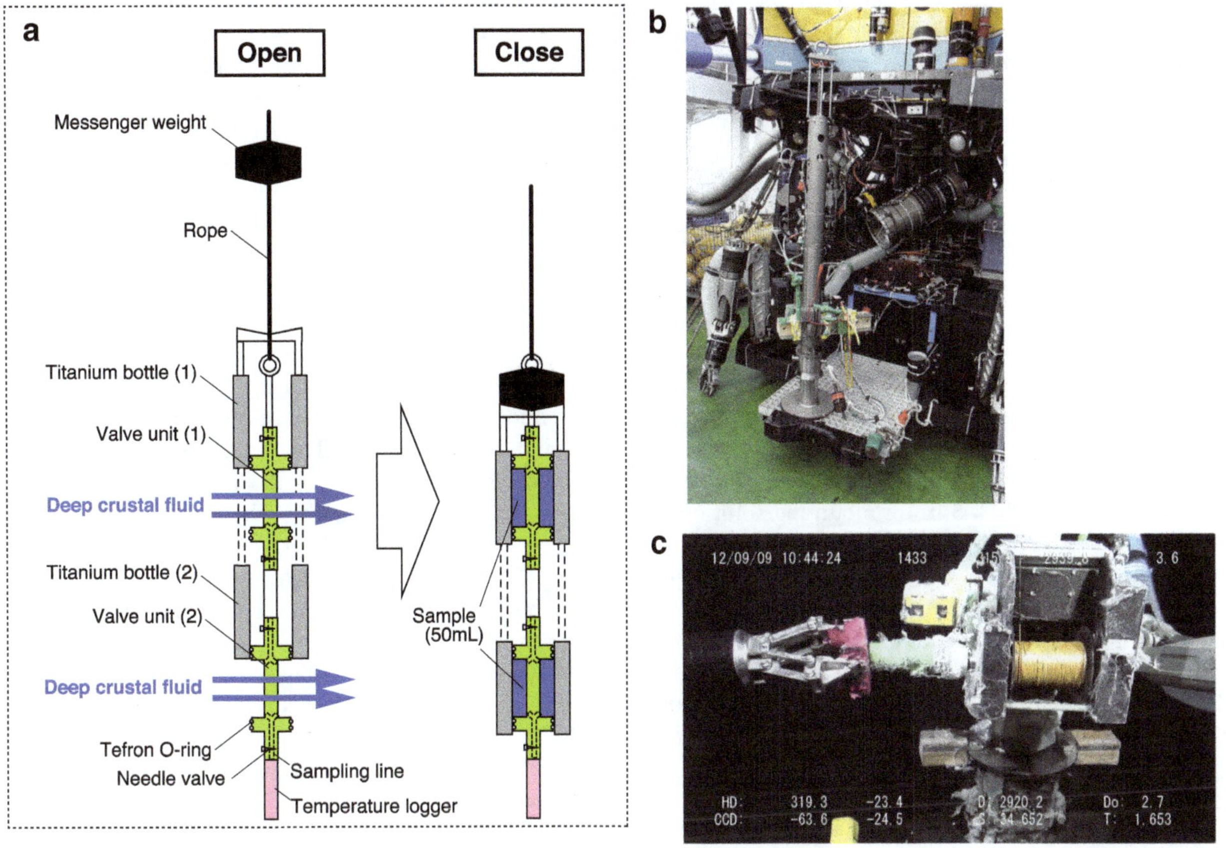

Fig. 2 A schematic illustration and photographs of the "Deep Sampler" used for CORK fluid sampling. **a** The open titanium bottle, chamber, and valve units were lowered to the screened depths (170 mbsf) through the cased hole. Then, a messenger weight was dropped from the seafloor along the rope, resulting in closing the bottle unit to the chamber-valve unit for fluid sampling. **b** A photograph showing the Deep Sampler being prepared in the payload of the ROV Hyper-Dolphin. **c** A photograph showing the on-site operation of the Deep Sampler from 170 mbsf by manipulation of reeling in the support line

stable isotopic compositions of H_2, CH_4, CO_2, and C_2H_6 were determined using the continuous flow method with DELTA XP Advantage and/or MAT253 Isotope Ratio Mass Spectrometer (Thermo Fisher Scientific) in JAMSTEC in accordance with previously published techniques (e.g., Okumura et al. 2016b; Shibuya et al. 2017). In the present study, isotope ratio and isotope fractionation between two molecules is represented by conventional δ and α notations as $\delta^{13}C_A = [(^{13}C_A/^{12}C_A)/(^{13}C_{STD}/^{12}C_{STD}) - 1]$ (presented in permil scale) and $\alpha C_{A-B} = (\delta^{13}C_A + 1)/(\delta^{13}C_B + 1)$. Samples collected using IGT samplers were analyzed at sea for dissolved H_2 and CH_4 using gas chromatography with thermal conductivity detection, following headspace extraction.

Microbiology

During cruises NT09-01 and NT09-07, microbial cells in subsampled fluids (110–370 mL) were collected on-board the ship onto 0.2 μm pore 47 mm polycarbonate membranes (K020A047A, ADVANTEC, Tokyo Japan) placed in line holder (KS-47, ADVANTEC, Tokyo Japan). The cells were fixed by placing a fixation solution (2% paraformaldehyde-PBS solution) onto the membranes and kept overnight at 4 °C. Then, the membrane was washed by PBS, partially dried by placing on filter paper, and kept at − 20 °C for later cell counting. The fluid sample taken on the cruise NT12-23 was immediately fixed by adding 1/5 volume of 10% Formalin Neutral Buffer Solution (Wako, Japan) and stored at 4 °C until further processing. Since white precipitates formed in fixed fluid samplers during storage at 4 °C, separation of cells was conducted according to Morono et al. (2013), with some modifications. In brief, for 3–15 mL of fixed solution, 1/10 volume each of detergent mix (100 mM EDTA, 100 mM sodium pyrophosphate, 1% [*v*/*v*] Tween 80) and methanol were added and treated in an ultrasonic bath (Bioruptor UCD-250, Cosmo Bio, Tokyo Japan) at 160 W with cycles of 1 min ON and 1 min OFF for 20 min. The treated samples were then carefully layered onto step gradient heavy density solution composed of 67% (*w*/*w*) sodium polytungstate (SPT1, SOMETU, Berlin, Germany), 80%, 50%, and 30% (*w*/*v*) Nycodenz and centrifuged for 120 min at 10,000×*g* at 25 °C. Light density upper layers were recovered, and 10% of the suspension was passed through a 0.2 μm pore 25 mm polycarbonate membrane to trap microbial cells. The filters were stored at − 20 °C until counting. From Bag sampler (NT09-01) and Deep Sampler (NT12-23), about 10 L and 40 mL of fluids, respectively, were filtered with 0.22 μm-pore-sized cellulose acetate filters to collect microbial cells. Filters were stored at − 80 °C for microbial 16S rRNA gene clone sequencing analysis.

Microbial cell densities in CORK fluid samples were determined using a semi-automated cell counting system (Morono et al. 2009). In brief, trapped and fixed cells collected on a polycarbonate membrane were stained with SYBR green I staining solution (1/40 [*v*/*v*] SYBR Green I in TE buffer). The stained filter was washed with 1 ml of TE buffer and then mounted on a glass microscope slide with 3–5 μL of mounting solution (2:1 mixture of VECTASHIELD mounting medium H-1000 and TE buffer). Microscopic fluorescence image acquisition (at 525/36 nm [center wavelength/bandwidth] and 605/52 nm by 490 nm excitation) was performed automatically using a fluorescence microscope equipped with an automatic slide handler (Morono and Inagaki, 2010). The resulting images were analyzed using the macro of the Metamorph software (Molecular Devices, CA, USA) to identify and count microbial cells on the membrane.

Phylotype composition of 16S rRNA genes were determined using clone sequencing in the laboratory based on a previously published methodology (Nunoura et al. 2012) with some modifications. DNA assemblages were extracted from microbial cells on filters using the UltraClean Soil DNA Isolation Kit (MO BIO Laboratories, Carlsbad, CA, USA). Amplification of 16S rRNA genes from the extracted DNA assemblage was accomplished by a Polymerase Chain Reaction (PCR) method with LA Taq polymerase (Takara Bio, Otsu, Japan) using a primer set of 530F/907R. The amplified 16S rRNA gene fragments were applied to electrophoresis and cleaned up with a QIA quick Gel extraction Kit (QIAGEN, Valencia, CA, USA). The 16S rRNA gene clone libraries were constructed using the TA cloning kit (Invitrogen, Carlsbad, CA, USA). The nucleotide sequences of randomly selected clones were determined with a BigDye Terminator version 3.1 sequencing kit (Applied Biosystems, Foster City, CA, USA) using M13M4 (5′-GTTTTCCCAGTCACGAC-3′) and M13RV (5′-TGTGGAATTGTGAGCGG-3′) primers on an ABI PRISM 3730xl Genetic analyzer (Applied Biosystems, Foster City, CA, USA). The sequences were assembled with Sequencher Ver. 4.8 software (Gene Code Corporation, Ann Arbor, MI, USA). The most similar reference sequences were retrieved from RDP and the GenBank database. The nucleotide sequences obtained were deposited to DDBJ (DNA data bank of Japan) with accession numbers LC279286-LC279362.

Radio isotope tracer experiment to estimate potential metabolic rates

Potential activities of microbial assimilations of bicarbonate, methane, carbon monoxide, formate, acetate, and leucine (autotrophic, methanotrophic, carboxydotrophic, and heterotrophic activities, respectively) were estimated by radio isotope tracer experiment using CORK fluid samples obtained during the expeditions NT09-01 and NT09-07. The CORK fluid was collected in multiple evacuated 150 mL stainless-steel bottles, each connected to a three-way valve and a sampling nozzle. All surfaces of the bottle, valve, and nozzle were treated with SilicoNert coating (Silcotek Corporation, Bellefonte, PA, USA) to avoid direct contact between the fluid and stainless steel. The sampling was conducted in duplicates for each of the radio isotope tracers. Prior to the seafloor sampling by the ROV, each of the radio-isotope-tracers ($NaH^{14}CO_3$, $^{14}CH_4$, ^{14}CO, $H^{14}COONa$, $^{14}CH_3{}^{14}COONa$, ^{14}C-leucine) was placed in two bottles (with and without formaldehyde at a final concentration of 4%) as solution ($NaH^{14}CO_3$, $H^{14}COONa$, $^{14}CH_3{}^{14}COONa$, ^{14}C-leusin and formaldehyde) or gas ($^{14}CH_4$ and ^{14}CO) in sealed glass ampoules. The final radioactivity level was 0.25 MBq per bottle. In order to assess the methanotrophic activity, three treatments were used: (i) only $^{14}CH_4$ added, (ii) $^{14}CH_4$ and O_2 (a final concentration of 40 μM O_2) added, and (iii) $^{14}CH_4$ and Na_2SO_4 and Na_2S (final concentrations of 10 mM and 0.5 mM, respectively) added. The additional O_2 or Na_2SO_4 and Na_2S were introduced either as a gas or as a solution sealed in a glass ampoule. The final concentrations of ^{14}C-labeled tracers in the bottles were estimated to be about 0.1 mM, 0.4 mM, 0.4 mM, 0.1 mM, 0.1 mM, and 0.1 mM for bicarbonate, methane, carbon monoxide, formate, acetate, and leucine, respectively. After placing the radio isotope tracer into the bottle with three stainless-steel balls, the bottle was evacuated with a vacuum pump. The bottle was then sealed and shaken, which resulted in the stainless-steel balls rapturing the glass ampoules containing the radio-isotope-tracers and formaldehyde. At the seafloor, the sampling nozzle of the radio-isotope-tracer sampler was placed in the discharging CORK fluid and sufficiently flushed. Fluid sample was collected by opening and closing the valve with the ROV manipulator. Immediately after recovery of the ROV, sample bottles were incubated at 4 °C for 49–123 days until measurement in a shore-based laboratory. The bottles containing 4% formaldehyde were used as negative controls (i.e., with no microbial activity).

As a positive control for methane assimilation, stationary phase cells of a gammaproteobacterial methanotroph, *Methylomarinum vadi*, (Hirayama et al. 2013) were incubated in its medium (Hirayama et al. 2007) at 37 °C for 1 day under the same substrate condition (0.4 mM) as the aerobic methanotrophic activity measurement, as described above. The cell density of *M. vadi* was adjusted to be 2.0×10^6 cells in the bottle (1.3×10^4 cells/mL). As a positive control for bicarbonate assimilation, natural deep-sea diffusing hydrothermal fluid (average fluid temperature of 57 °C) from the Iheya North field (Takai et al. 2004), sampled in July–August 2010 using the same method described in previous sections of the present study, was incubated at 60 °C for 5 days under the same substrate condition (10 mM). The cell density of microbial community in the CORK fluid was 1.0×10^2 cells/mL, thus each incubation bottle contained 1.5×10^4 cells.

CORK fluid samples (approx. 150 mL) were incubated with radio-isotope-tracers for 49–123 days at 4 °C and then depressurized and filtered with a 0.22 μm-pore-sized cellulose acetate filters. Each filter was fixed with 10 mL of 4% (*w*/*v*) paraformaldehyde/PBS solution (pH 7.2) for 30 min, then thoroughly rinsed with filter-sterilized 0.1 M HCl in 3% (*w*/*v*) NaCl solution to remove carbonate precipitates. The filter was slurried in 20 mL of Hionic Flour cocktail (PerkinElmer, Waltham, MA, USA), and the radioactivity measured using a Tri Carb 2910TR liquid scintillation counter (PerkinElmer, Waltham, MA, USA). The assimilation rates of the tracers were calculated from concentrations and radio-isotope ratios given as the substrate in the initial state and the radio-isotope abundance of the filter. Blanks were evaluated by non-labeled batch incubation and used for calibration.

Results

pH, chlorinity, and major and minor ions

Concentrations of major and minor ions in the fluids were nearly identical in all samples, regardless of the sampling device used for collection (Additional file 1: Table S1). The chemistry of conservative components in CORK fluids were quite similar to those of deep (> 20 mbsf) porewater samples (DPW) obtained from the ODP Holes 1200 (Additional file 1: Table S1). For example, the fluid chlorinity (503–510 mmol/kg) of CORK fluids was distinguishable from seawater level (540 mmol/kg) and identical to those of the DPW (510 ± 5 mmol/kg) (Fig. 3a). CORK fluids ranged from 12.1 to 12.3 in pH values, which were slightly lower than the highest pH of 12.5 ± 0.1 in the DPW (Fig. 3b). Slight differences between CORK fluids and the DPW in alkali metals (K, Li, Rb) (Additional file 1: Table S1) have been interpreted to be due to sampling artifacts during the core recovery and porewater extraction of the ODP serpentinite samples (de Lange et al. 1992; Wheat et al. 2008). Alkali earth metals (Mg, Ca, Sr) were depleted in the CORK fluids, as well as the DPWs (Fig. 3c). Seawater entrainment during the seafloor sampling cannot be evaluated precisely, because any entrainment of Mg from seawater would be precipitated immediately as brucite. Nevertheless, the nearly identical and high concentrations of B, for example (Fig. 3d), are consistent with a scenario where mixing with ambient seawater during sample collection was minimal. Combined, these data are consistent with the CORK fluids being derived from a deep-sourced formation fluid that flowed into the borehole through screens located at a depth range of 149 and 202 mbsf.

Sulfate, alkalinity, sulfide, and ammonium

Sulfate concentrations in the discharging CORK fluids were 0.4–1.7 mM, significantly lower than the concentration of 28 mmol/kg in the DPW and the seawater (Fig. 3e). The alkalinities of these fluids were much higher (119–126 meq/kg) than in the DPW (62 meq/kg) (Fig. 3f) and comparable to the value estimated from the previous CORK fluid observation (104 meq/kg) (Wheat et al. 2008). Both sulfate concentration and alkalinity in the CORK fluid from Dive #941 were in-between values of the DPW and other CORK fluids. These results point to a stoichiometry of 2 meq increase of alkalinity with 1 mmol decrease of sulfate in the CORK fluids when compared to the DPW (Fig. 4a). Dissolved reduced sulfur (sulfide) in CORK fluids was significant despite negligible sulfide concentrations detected in the DPW and showed a temporal decreasing trend: ~ 24 mM in 10 days, 21 mM in 4 months, and ~ 10 mM in more than 36 months (Fig. 3g). The sum of sulfate and sulfide concentrations in the CORK fluids were not higher than 28 mM, suggesting quantitative conversion in addition to the removal of dissolved sulfur to the solid phase (Fig. 4b). Ammonium concentrations were identical at 0.2–0.3 mM among all CORK fluids and the DPW (Additional file 1: Table S1). The highly variable alkalinity with the constantly low ammonium concentrations implies that the alkalinity variation do not result from the decomposition of sedimentary Redfield organic matter (Redfield, 1958) but from that of N-depleted organic carbon such as CH_4.

Gas composition and stable isotopes

Methane concentrations in the CORK fluids were variable between 5.5–6.1 mM in 10 days, 9.2–37.8 mM in 4 months, and 16.1–37.4 mM in more than 36 months after the valve was opened, respectively (Fig. 3h). CH_4 concentrations in CORK fluids were significantly higher than those observed in the DPW (< 2 mM). The differences strongly suggest that CH_4 was degassed from the DPW during core recovery and sample processing during the ODP Leg 195 (Mottl et al. 2003). This claim is consistent with a solubility-controlled CH_4 concentration of ~ 2 mM at room pressure and temperature (Wiesenburg and Guinasso jr, 1979). The CH_4 concentrations in CORK fluids are the highest among any deep-sea serpentinization-associated geofluid systems

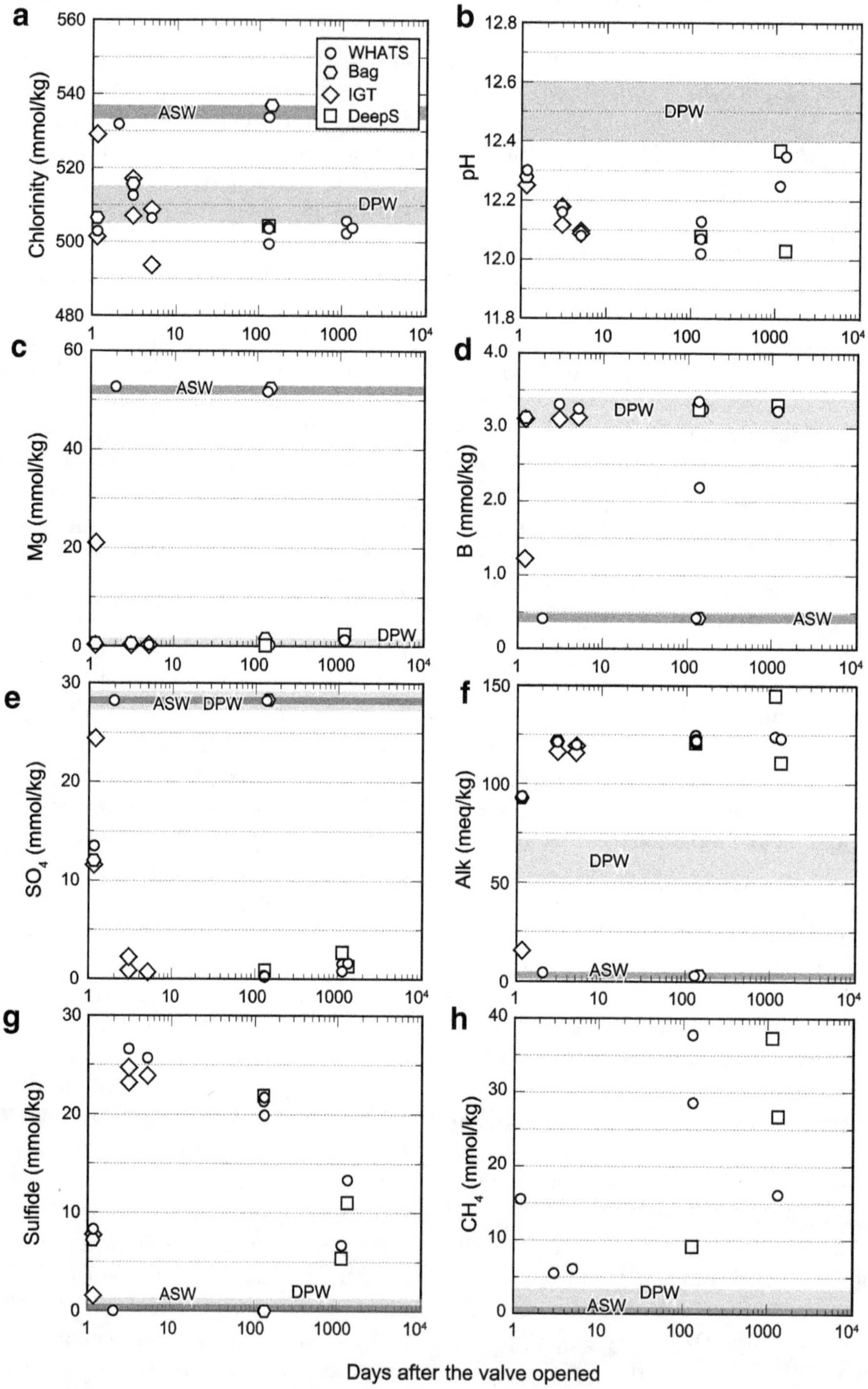

Fig. 3 Time-series of fluid chemistry since the valve opening. Light and dark gray bars represent the deep part of the ODP porewater at Site 1200 (DPW) (Mottl et al. 2003) and ambient seawater (ASW) composition, respectively. Each symbol represents a type of fluid sampler used: WHATS (circle), Bag (hexagon), IGT (diamond), and Deep Sampler (square). Time (*x* axis) is in logarithmic scale

reported to date (< 2 mM) (Charlou et al. 2002; Kumagai et al. 2008; Schrenk et al. 2013; Konn et al. 2015; McDermott et al. 2015; Seyfried et al. 2011). Ethane concentrations were also variable between 0.01 and 0.23 mM. Relative abundances of CH_4 to C_2H_6 (C_1/C_2) were ~ 27 in 1 day, 216–3178 in 10 days to 4 months, and 231–285 in ~ 36 months after the valve opening. Molecular hydrogen concentrations in the CORK fluids ranged between 0.003 and 0.107 mM and were the lowest among values repoted from deep-sea serpentinization-associated geofluid systems

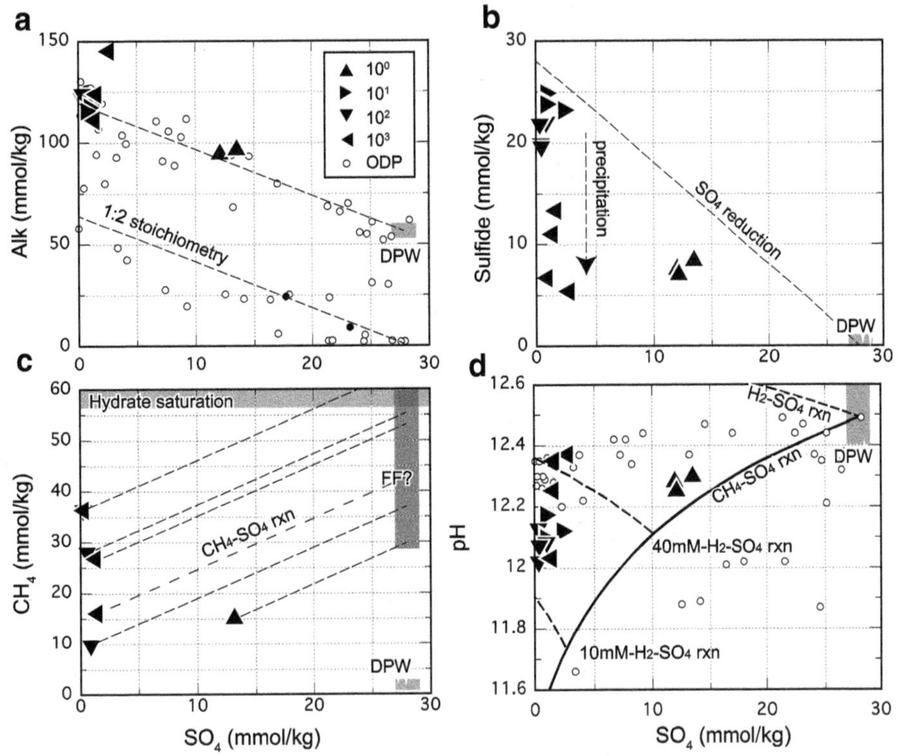

Fig. 4 Relationships between sulfate and alkalinity, sulfide, CH_4, and pH. Filled triangles represent the CORK fluids from the present study, classified by the number of days after the valve was opened. Filled and open circles represent the CORK fluid observed in 2003 and DPW, respectively, shown for comparison. Light gray zones represent the estimated endmember formation fluid composition of the ODP porewater (DPW) or a solubility of CH_4 hydrate clathrate under the SCS condition (only shown in panel **c**). Broken lines in panels (**a**), (**c**), and (**d**) represent ideal changes in the chemical composition based on selected reactions. The dark gray zone in panel (**c**) represents possible CH_4 concentrations of the SCS formation fluid (FF) without post-drilling alteration. The solid curve in panel (**b**) represents sulfate consumption through CH_4-SO_4 reaction from the porewater-based endmember. Broken curves in panel (**b**) represent sulfate consumption with 10, 40, and > 112 mM of H_2, in addition to the CH_4-SO_4 reaction

to date (> 1 mM) (Charlou et al. 2002; Kumagai et al. 2008; Schrenk et al. 2013; Konn et al. 2015; McDermott et al. 2015; Seyfried et al. 2011). Relative abundances of CH_4 to H_2 (CH_4/H_2) ranged from 333 to 1350 for the CORK fluids. This CH_4/H_2 ratio is higher than those of seafloor geofluids associated with serpentinization in other geologic settings (< 1). Total dissolved inorganic carbon (DIC), extracted and quantified as CO_2, ranged from 17 to 51 mM, which is consistent with the estimated carbonate alkalinity of the DPW (45 meq/kg: Mottl et al. 2003). The potentially insufficient acidification of DIC samples prior to the gas analysis may result in the underestimation of DIC concentrations in the CORK fluids.

Carbon and hydrogen isotopic ratios of methane ($\delta^{13}C_{C1}$ and δD_{C1}) and ethane ($\delta^{13}C_{C2}$ and δD_{C2}) were uniform among the CORK fluids, with values of − 37‰, − 117‰, − 11‰, and − 86‰, respectively, regardless of concentration (Additional file 1: Table S1). The $\delta^{13}C_{C1}$ value of CORK fluids is significantly lower than other serpentinization-associated hydrothermal fluids reported to date (> − 18‰) (Fig. 5a) (Charlou et al. 2002; Kumagai et al. 2008; Schrenk et al. 2013; Konn et al. 2015; McDermott et al. 2015). The higher $\delta^{13}C_{C1}$ value in the shallow ODP porewater observed previously (− 11 ± 5‰ at 15 mbsf: Mottl et al. 2003) is significantly different from the CORK fluid and is likely attributed to ^{13}C enrichment in the remnant CH_4 due to microbial methanotrophic activity in the fluid-seawater interface. The $\delta^{13}C_{C2}$ value of the CORK fluids falls into the range of $\delta^{13}C_{C2}$ values observed in serpentinization-associated hydrothermal fluids (− 14‰ to − 2‰) (Konn et al. 2015). The δD_{C2} value of the CORK fluids (− 86‰) is higher than that of the Lost City field (~ − 150‰) (Fig. 5b). Total DIC of the CORK fluids showed $\delta^{13}C_{DIC}$ values of − 2.9‰ to + 1.4‰. The $\delta^{13}C_{DIC}$ values close to + 0‰ suggest that the DIC originates from sedimentary carbonate (~ + 0‰) rather than photosynthesis-based organic matter (~ − 25‰) (Sano and Marty 1995). Hydrogen isotope ratios of H_2 (δD_{H2}) were − 787‰ to − 761‰, which can be attributed to H_2-H_2O hydrogen isotope equilibrium at an in situ temperature of 2 °C (Horibe and Craig, 1995) when using the δD_{H2O} value of + 12‰ observed in the DPW (Mottl et al. 2003).

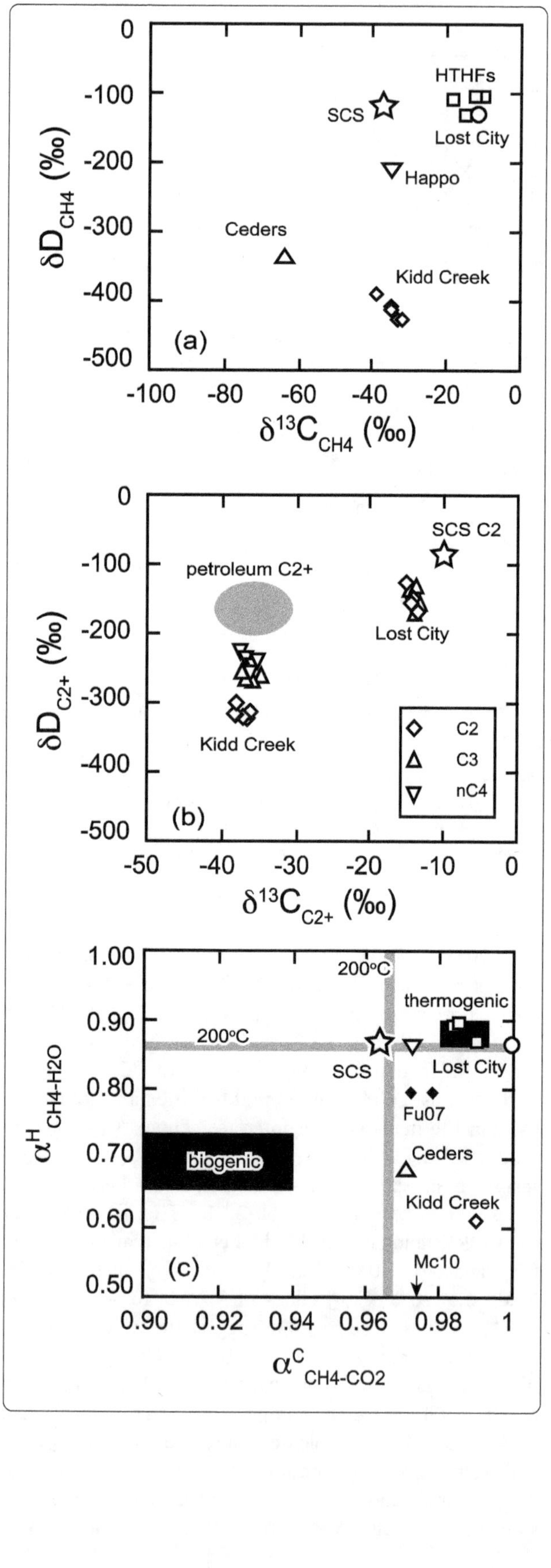

Fig. 5 Stable isotope composition of CH_4, C_2H_6, and the relevant molecules. Open and filled symbols, respectively, represent data from observations of hydrocarbon-enriched geofluid systems and experiments simulating methanogenesis. Data sources are following: high-temperature hydrothermal fluids (Konn et al. 2015; Kawagucci et al. 2016 and references therein), Lost City (Proskurowski et al. 2006; 2008), Happo (Suda et al. 2014), The Ceders (Wang et al. 2015), Kidd Creek (Sherwood-Lollar et al. 2002), petroleum field (Burruss and Laughrey 2010), magnetite experiments (Fu et al. 2007), Fe experiment (McCollom et al. 2010), thermogenic experiment (Seewald et al. 1994) (assuming the CH_4-H_2O isotope equilibrium), and microbial incubation (Okumura et al. 2016b)

Microbial cell density

Microbial cell densities of CORK fluids were estimated by epifluorescence microscopy (Table 2). In the blackish fluid discharging at the time when the valve was first opened (Dive #941), we measured a microbial cell density of 2.0×10^3 cells/mL. This is much lower than the cell density in the ambient seawater (2.2×10^4 cells/mL) sampled during the same cruise (Table 2). The microbial cell densities in the clear CORK fluid sampled 4 days and 4 months after the valve opening decreased to $4.4–6.6 \times 10^2$ and 1.0×10^2 cells/mL, respectively (Table 2). These results suggest that, before the valve opening, certain microbial communities maintained their populations and functions in some of the microhabitats in the stagnant CORK fluid of the sealed borehole and wellhead structures. Such local microbial populations in the stagnant fluid system were flushed out by the newly created discharging fluid flow after the valve was opened. After more than 36 months, however, the microbial cell densities in the CORK fluid were $4.1–7.6 \times 10^3$ cells/mL (Table 2).

Microbial phylotype compositions

The 16S rRNA gene phylotype compositions of microbial communities in CORK fluid samples at the time that the valve was opened, 3 days later, and 45 months later were examined. In total 89, 88, and 44 clones were sequenced for the microbial communities in the CORK fluid samples at the time of valve opening, 3 days later, and 45 months later, respectively (Fig. 6). The phylotype compositions were significantly different among the fluid samples.

In the blackish discharging fluid (2.0×10^3 cells/mL of microbial cell density), the gamma-proteobacterial phylotypes dominated the clone library and consisted almost entirely of a single phylotype closely related with *Methylobactor*, a methanotrophic endosymbiont of *Bathymodiolus* mussels living in the SCS seepage sites (Fig. 6). The alpha-proteobacterial and actinobacterial phylotypes were the second most abundant groups in the library. Other components were phylotypes of *Betaproteobacteria* and *Firmicutes*. In the

Table 2 Microbial cell density in the CORK fluid and ambient seawater samples

Cruise ID	HPD dive#	Time after valve opening	Sample type	Cell count (cells ml^{-1})
NT09-01	941	0 day	Bag	2000
NT09-01	943	2 days	Bag	440
	943	(ambient seawater)	Bag	22,000
	945	4 days	Bag	660
NT09-07	1007	5 months	Bag	101
	1009	5 months	Bag	139
NT12-04	1349	45 months	WHATS	7600
	1350	45 months	Deep sampler	4100

CORK fluid obtained 3 days after opening ($4.4–6.0 \times 10^2$ cells/mL of microbial cell density), the gamma-proteobacterial phylotypes, especially the *Methylobactor* phylotype, significantly decreased in the relative abundance and alpha-proteobacterial phylotypes also decreased in the relative clonal abundances (Fig. 6). In contrast, phylotypes of *Deltaproteobacteria, Firmicutes* (*Alkaliphilius*), *Actinobacteria, Deinococcus-Thermus*, and *Archaea* increased in relative abundances. Most archaeal phylotypes were related to extremophilic members such as hyperthermophiles and extreme halophiles that do not inhabit the cold and highly alkaline serpentine formation fluid regime of the SCS (Fig. 6). Although this simplified interpretation may lead to misunderstanding since the microbial cell density was decreased by fluid replacement during the 3-day discharge of the CORK fluid, phylotypes that increased in clonal abundances 3 days after the valve opening more likely represent truly indigenous microbial populations entrained by the serpentinite formation fluid derived from the deep. In the CORK fluid obtained by the Deep Sampler 45 months after the valve opening ($4.1–7.6 \times 10^3$ cells/mL of microbial cell density), the microbial phylotype composition changed drastically (Fig. 6). Most of the predominant phylotypes found in the fluids were different from those sampled 3 days after the valve was opened. At 45 months after, phylotypes of *Deltaproteobacteria, Epsilonbacteria, Planktomycetes, Thermus--Deinococcus*, and unclassified bacterial group dominated the library. It is still unclear why the CORK fluid obtained by the

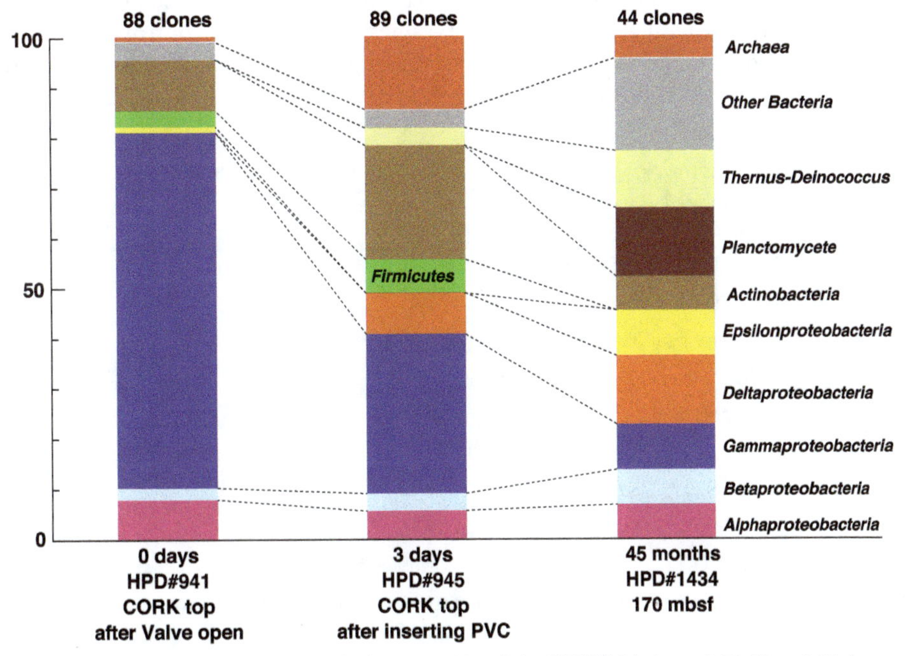

Fig. 6 Phylotype compositions of 16S rRNA genes in microbial communities of the CORK fluids. In total, 88, 89, and 44 clones were sequenced for the microbial communities in the CORK fluid samples at the time of valve opening (0 day), 3 days, and 45 months after the valve was opened, respectively

Deep Sampler 45 months after had a higher cell density and a different microbial phylotype composition compared to the initial phase of CORK fluids. Nevertheless, it should be noted that the sampling scheme by the Deep Sampler was different from those by other fluid samplers, with the Deep Sampler (5 cm in diameter with the bottle being 1 m in length) was delivered to 170 mbsf through the cased borehole. During the descent, the bottle could have hit the casing wall and detached accreted materials which likely contain microbial populations. Thus, the microbial community abundance and composition in the fluid obtained by the Deep Sampler may include, in addition to the microbial communities entrained from the formation fluid ("pelagic") flow, microbial populations specifically colonizing the steel casing pipe wall (Nakagawa et al. 2006; Nigro et al. 2012).

Microbial carbon assimilation activity

Microbial methane-, carbon monoxide-, formate-, acetate-, and leucine-assimilations (methanotrophic, carboxydotrophic, and heterotrophic activities) were examined with the CORK fluid samples obtained 1 to 6 days after the valve opening, and the bicarbonate-, methane-, formate-, and acetate-assimilation activities were examined with the CORK fluid samples obtained 5 months after the valve opening (Table 3). *Methylomarinum vadi* is a marine aerobic methanotroph optimally growing at 37 °C with a generation time of 2 h (Hirayama et al. 2013), and the methane assimilation rate of this methanotroph under the maintenance state for 1-day incubation was used as the positive control: 20 nmol/day/L with a cell density of 1.3×10^4 cells/mL (Table 3). In a similar manner, the autotrophic bicarbonate assimilation of natural deep-sea diffusing hydrothermal fluid was determined to be 2.3 nmol/day/L with a cell density of 1.0×10^2 cells/mL (Table 3). These methane assimilation rates of *M. vadi* and the bicarbonate assimilation of the natural autotrophic microbial community in the discharging fluid are good comparative indexes to estimate the functions of the methanotrophic and autotrophic populations and even the whole microbial community in the CORK fluid.

The methane assimilation rates were determined to be 0.036, 0.049 and 0.039 nmol/day/L in the CORK fluid under in situ, microaerobic and fully anaerobic conditions, respectively (Table 3). These values were significantly higher than those of the negative controls with the same tracers and are attributed to activity from the microbial populations in the CORK fluid. In addition, these values correspond to the methane assimilation activity by $2.3–3.2 \times 10^1$ cells/mL through aerobic methanotrophy in the maintenance state. Carbon monoxide assimilation was not detected, while heterotrophic carbon assimilation was found to be the highest with formate (0.06 nmol/day/L) and the lowest with acetate (0.004 nmol/day/L). These values are similar to or lower than the methane assimilation activity. In the CORK fluid obtained 5 months after the valve opening, autotrophic bicarbonate assimilation activity was the highest (1.4 nmol/day/L), and methanotrophic and heterotrophic (with formate and acetate) assimilations were lower at 0.2 and 0.05 nmol/day/L, respectively (Table 3). The methanotrophic assimilation was higher than any of the methane assimilation activities using the CORK fluid obtained 1 to 6 days after valve opening and the heterotrophic activity was similar to values reported previously from the CORK fluid, although the cell density in the CORK fluid decreased from $4.4–6.6 \times 10^2$ to 1.0×10^2 cells/mL during the 5 months period. Compared to the values of positive controls, these autotrophic and methanotrophic assimilation activities correspond to 7.0×10^1 cells/mL of a natural autotrophic microbial

Table 3 Radio-isotope-tracer carbon assimilation of microbial community in the CORK fluids and positive controls

Cruise ID	HPD dive#	Tracer and incubation condition	Incubation time (days)	Net assimilation (nmol day^{-1} L^{-1})
NT09-01	943	$^{14}C-CH_4$ (0.25 MBq per bottle and 0.4 mM)	54	0.036
	943	$^{14}C-CH_4$ (0.25 MBq per bottle and 0.4 mM) with 40 μM O_2	54	0.049
NT09-01	947	^{14}C-Leucine (0.25 MBq per bottle and 0.1 mM)	52	0.014
	947	^{14}C-Formate (0.25 MBq per bottle and 0.1 mM)	52	0.06
	947	^{14}C-Acetate (0.25 MBq per bottle and 0.1 mM)	52	0.004
NT09-01	948	$^{14}C-CH_4$ (0.25 MBq per bottle and 0.4 mM) with 10 mM $NaSO_4$ and 0.5 mM Na_2S	49	0.039
	948	^{14}C-CO (0.25 MBq per bottle and 0.4 mM)	49	n.d.[a]
NT09-07	1007	$^{14}C-CH_4$ (0.25 MBq per bottle and 0.4 mM)	123	0.2
	1007	$^{14}C-HCO_3^-$ (0.25 MBq per bottle and 0.1 mM) with 0.4 mM H_2	123	1.4
	1009	^{14}C-Formate and ^{14}C-acetate (each of 0.25 MBq per bottle and 0.1 mM)	122	0.05
An aerobic methanotroph culture		$^{14}C-CH_4$ (0.25 MBq per bottle and 0.4 mM)	1	20
An autotrophic community in diffusing hydrothermal fluid		$^{14}C-HCO_3^-$ (0.25 MBq per bottle and 0.1 mM)	5	2.3

[a]*n.d.* not detected

community and 1.3×10^2 cells/mL of relatively inactive aerobic methanotrophs. These results indicate that the small number of microbial cells present in the CORK fluid obtained 5 months post-valve opening may be functionally active and capable of methanotrophic and autotrophic metabolisms. Nevertheless, radio-isotope-tracer experiments clearly indicate that microbial metabolisms and activities in the CORK fluids are detectable.

Discussion

Post-drilling fluid alteration

Sulfate reduction coupled with CH_4 oxidation

Except for sulfate, the consistency in the major ion fluid chemistry between the CORK fluid and the DPW suggests that CORK fluids originate from the deep SCS formation fluid. Sulfate depletions in the CORK fluids, as opposed to the sulfate enrichment seen in the DPW, likely resulted from perturbation of the subseafloor geochemical environment during drilling activity. To estimate the abundance of reactive components (particularly CH_4 and H_2) in the pristine SCS formation fluid, it is necessary to consider processes associated with the drilling operation that could alter the fluid chemistry at ~ 4 °C.

A previous study (Wheat et al. 2008) pointed out that the methane oxidation can cause SO_4 depletion in the CORK fluid, as follows:

$$CH_4 + OH^- + SO_4^{2-} \rightarrow CO_3^{2-} + HS^- + 2H_2O \text{ (R1)}.$$

R1 produces 1 mol of dissolved sulfide and two equivalents of alkalinity for each mole of OH^-, CH_4, and SO_4^{2-} consumed. This stoichiometry is consistent with the stoichiometry exhibited by the differences between the DPW and CORK fluids (Figs. 3 and 4). Assuming that R1 fully accounts for the ~ 27 mM-equivalent sulfate consumption in the CORK fluids, the pristine SCS formation fluids would have contained CH_4 at concentrations of 35–65 mM (Fig. 4c). The measured pH values of the CORK fluids (12.1–12.3) were higher than would be expected given the consumption of ~ 27 mM OH^- and a DPW with a pH = 12.5 ± 0.1 that is representative of the SCS formation fluid (Fig. 4d). In addition, some of the measured total DIC contents of the CORK fluids (17–51 mM) were also inconsistent with a DIC concentration of ~ 50 mM expected by a combination of the ~ 27 mM DIC production and the 23 mM DIC being originally present in the pristine SCS formation fluid (Mottl et al. 2003). Differences in the OH^- and DIC values between the measured and predicted fluid compositions may be due to brucite-magnesite buffer through the following reaction:

$$Mg(OH)_2 + CO_3^{2-} \Leftrightarrow MgCO_3 + 2OH^- \text{(R2)}.$$

The maximum CH_4 concentration estimated (65 mM) seems too high but is not unrealistic because it is comparable to the solubility of CH_4 clathrate hydrate under relevant concentration-pressure-temperature-salinity condition within the SCS (~ 60 mM at 30 MPa, 2 °C and 3.5%) (Tishchenko et al. 2005). No geophysical data, however, support the presence of layered hydrate boundary, such as bottom simulating reflector (BSR).

Possible H_2 consumption coupled with sulfate reduction

The low concentrations of H_2 in the CORK fluid (< 0.1 mM) suggest that it may be consumed by sulfate reduction in addition to CH_4. H_2 oxidation coupled with sulfate reduction progresses as follows:

$$4H_2 + SO_4^{2-} \rightarrow HS^- + OH^- + 3H_2O \text{ (R3)}.$$

Because R3 produces OH^-, H_2 consumption during sulfate reduction may contribute to measured pH values that are higher than predicted by R1 (Fig. 4d). Assuming the pH balance is made by R1 and R3, the native SCS formation fluid may contain as much as ~ 40 mM of H_2. This H_2 level may be realistic because 27 mM of H_2 has been observed in a seafloor serpentinization-associated hydrothermal system at > 296 °C (Konn et al. 2015) and > 100 mM of H_2 can be attained through serpentinization at 200–300 °C from thermodynamic calculation (McCollom and Bach 2009). If the cool SCS formation fluid contains such abundant H_2, the reduction of CO_2 to form CH_4 is thermodynamically favorable, according to the following reaction:

$$4H_2 + CO_3^{2-} \rightarrow CH_4 + H_2O + 2OH^- \text{(R4)}.$$

However, abiotic CH_4 formation represented by R4 is unlikely due to kinetic barriers at the low temperatures within the SCS body (4 °C). In addition to CH_4 and H_2, native iron and other metals in the casing pipe may function as reductant and cation for the dissolved sulfate removal.

Drilling induced sulfate reduction

The discussions above suggest that abundant CH_4 and SO_4 coexist in the native SCS formation fluid and is consistent with their presence in the DPW (Mottl et al. 2003). This coexistence is unusual in the subseafloor environment because sulfate reduction coupled with methane oxidation is thermodynamically favorable and usually promoted by ubiquitous microbial consortia of methanotrophic archaea and sulfate reducing bacteria, known as AOM (Orphan et al. 2001). In other words, both abundant CH_4 and SO_4 in the native SCS formation fluid point to unusually slow kinetics of the sulfate reduction (R1) within the native SCS body. In fact, SO_4 reduction coupled with CH_4 oxidation (R1) likely occurs in the post-drilling CORK fluids, derived from subseafloor regions affected by the Hole 1200C drilling, casing, and/or CORK installation.

Drilling-induced introduction of external microbial communities into the subsurface environment may account for the post-drilling progression of R1 due to AOM. While the relatively impermeable serpentinite matrix prevented access of the over-pressured aquifer to the microbe-bearing seawater, the drilling operation and the hole made the deep aquifer exposed to deep-sea water and shallow subseafloor sediment containing exogenous microbial communities, including AOM consortia (Mottl et al. 2003). However, the post-drilling microbial AOM function is likely insignificant, as suggested by multiple lines of microbiological evidence such as small biomass, a lack of evidence for 16S rRNA gene phylotype signatures of previously known AOM consortia, and the negligible CH_4 assimilation during the ^{14}C experiment (see the "Results" section). Assuming that the dissimilatory AOM activity of consortia does not exceed 10^2 times higher than the anoxic methane assimilation rate, as the highest methane assimilation activity estimated in this study is 0.0002 µmol/day/L, it can be calculated that a 27 mmol/L SO_4 consumption requires $> 10^6$ days. This amount of time is much longer than the time elapsed since drilling ceased ($< 10^4$ days). This calculation, together with other microbiological results, suggests that the microbial AOM function is not significantly involved in the post-drilling CH_4-SO_4 reactions, based on experimental conditions at 2 °C. Although faster microbial reaction rates may occur deeper within the seamount where the environment is warmer, the extent of their involvement would be still negligible. It is noted that microbial activity at the pristine interior of the SCS body is more stricly limited than the rate calculated above due to their impermeability and over-pressured aquifer, and probably less biomass.

There are no reliable data for the kinetics of R1 without the involvement of microbial AOM. However, the sudden and rapid progression of R1 after drilling would require some catalytic effects associated with the drilling, casing, and CORK installation operations, other than the introduction of AOM populations. A possible catalyst introduced by drilling and casing the borehole is native iron and other metals used in the casing pipe and the CORK structure. Transition metals such as iron and nickel generally exhibit catalytic function for redox reactions due to their multiple valences. In addition to catalytic reactions occurring locally on the metal surface, high electron conductivity of the elongate casing pipe allows remotely located reducing and oxidizing molecules to be reacted through electron transfer, similar to how a battery functions. Nevertheless, the extent to which drilling and casing operations promoted abiotic sulfate reduction coupled with CH_4 oxidation is still not entirely clear.

Origins of CH_4 and SO_4 in the native serpentine formation fluid of the SCS

Abiotic hydrocarbon synthesis associated with serpentinization—environmental factors

Although it is difficult in natural environments of the modern Earth to distinguish between abiotic synthesis of CH_4 from other CH_4 sources such as microbial methanogenesis (biogenic CH_4) and thermal degradation of sedimentary organic matter (thermogenic CH_4) (see reviews, Etiope and Sherwood Lollar, 2013 for example), multiple lines of geochemical and microbiological evidence suggest that contributions of biogenic and thermogenic CH_4 in the SCS system are not signficant. During the early stage of subduction, where the temperatures are below the known 122 °C upper limit for the growth of microbes (Takai et al. 2008), microbial methanogenesis is probably absent or insignificant due to the abundantly dissolved sulfate in the sediment pore water in the Pacific plate eastward from the Mariana Trench (ODP sites 800–802) (France-Lanord et al. 1992). During the later stages of subduction, temperatures exceed 122 °C, which is too high for microbial life to survive. The same is also true for regions of extensive serpentinization within the overlying mantle wedge. In the SCS formation fluid regime, where temperatures again become cool and habitable, there is little potential to produce abundant CH_4 due to low biomass and microbial function, revealed by the 45 months long observation of the CORK fluids. In addition, the highest known pH limit for the growth of methanogens is < 10 (*Methanosalsum zhilinae*: Mathrani et al. 1988), significantly lower than the pH of > 12 in the SCS fluids. The extremely low organic contents in the SCS serpentine mud (Mottl et al. 2003) indicate that there is little potential for thermogenic hydrocarbon generation beneath the SCS.

Possibility of abiotic synthesis from thermodynamic and kinetic viewpoints Abiotic organic synthesis, including the generation of CH_4 from CO_2, is generally more favorable at lower temperatures from a thermodynamic viewpoint despite unfavorable kinetic constraints (e.g., McCollom, 2008). In the case of the SCS, thermodynamically favorable conditions for the abiotic synthesis of CH_4 is expected through the entire formation fluid regime due to sufficiently reducing conditions resulting from serpentinization occur at 200–300 °C (McCollom and Bach, 2009) and subsequent cooled as fluids ascend. The ascending fluid is quenched sufficiently for microbial activity (122 °C: Takai et al. 2008) at 13 km below the seafloor, estimated from the thermal gradient observed (9.4 °C/km). Since formation fluids probably ascend on a geological time span, continuous abiotic synthesis and accumulation of organic compounds is plausible, even at extremely low reaction rates.

The rate of abiotic hydrocarbon synthesis depends on the carbon substrate (McCollom et al. 2010). McCollom et al. (2010) demonstrated that CO-based synthesis is feasible within an experimental timeframe. Even when the CO-based abiotic synthesis is feasible under conditions of the SCS formation fluid, CO production from CO_2 represents a bottle-neck in the process due to the several orders of magnitude less CO compared to CO_2 in the CORK fluid. Abiotic methanogenesis based on solid-phase carbon, such as graphite derived from metamorphism and carbonate mineral deposits (Etiope and Sherwood Lollar, 2013), cannot be precluded, although the possibility cannot be evaluated here due to limited information about the rates of such processes.

Suggestions from stable isotope signatures Stable isotope signatures should be interpreted in the context of isotope ratios of individual molecules (Fig. 5a, b) and isotope fractionation between relevant molecules (Fig. 5c) (Etiope and Sherwood Lollar, 2013). The CH_4-H_2O hydrogen isotopic fractionation factor of the CORK fluid ($\alpha^{H}_{CH4\text{-}H2O}$) is 0.87, which is inconsistent with typical biogenic values of ~0.70 (Okumura et al. 2016b) (Fig. 5c). Thermogenic CH_4 reservoirs generally have $\alpha^{H}_{CH4\text{-}H2O}$ values corresponding to the CH_4-H_2O isotope equilibrium values at the formation/reservoir temperature (e.g., Stolper et al. 2014; Wang et al. 2015). The $\alpha^{H}_{CH4\text{-}H2O}$ and $\alpha^{H}_{C2H6\text{-}H2O}$ values of the CORK fluids (> 0.86) correspond to those of the isotope equilibrium values at > 200 °C (Fig. 5b, c) (Horibe and Craig 1995; Reeves et al. 2012), consistent with the uppermost temperature expected at the plate boundary and mantle wedge beneath the SCS (~350 °C) (Mottl et al. 2004; Hulme et al. 2010). This suggests that the hydrogen isotope signature of hydrocarbons recorded at > 200 °C has not been altered after these fluids have been cooled to their discharge temperatures. The lack of hydrogen isotope re-equilibrium between CH_4 and H_2O seems consistent with the half-life for equilibrium exchange, estimated to be approximately one million years at 200 °C (Sessions et al. 2004; Wang et al. 2015). In addition, the hydrogen isotope signature shows no evidence for a contribution of biogenic CH_4 produced in the cool formation fluid regime of the SCS. The H_2-H_2O thermometer value, indicating ~ 2 °C, is inconsistent with the CH_4-H_2O thermometer but reasonable because of the more feasible H_2-H_2O equilibrium, even at lower temperatures (Valentine et al. 2004; Proskurowski et al. 2008; Kawagucci et al. 2010; 2016; Pester et al. 2018).

The $\delta^{13}C$-CH_4 values of about − 37‰ and the isotope fractionation between CH_4-DIC ($\alpha^{C}_{CH4\text{-}DIC}$) of ~0.963 seem to be an equivocal signature for thermogenic, biogenic, or abiotic origins (Fig. 5c) (Kawagucci et al. 2013 and references therein). Hydrothermal experiments simulating hydrothermal hydrocarbons generation from the seafloor sediment at 325–400 °C show an $\alpha^{C}_{CH4\text{-}CO2}$ as high as 0.985 (Seewald et al. 1994) (Fig. 5c). Carbon isotope fractionation produced by microbial hydrogenotrophic methanogenesis in the seafloor sediment typically displays $\alpha^{C}_{CH4\text{-}DIC}$ of ~0.930 (Okumura et al. 2016b and references therein). It is also known, however, that microbial metabolisms sometimes cause unusual isotope effects under extreme growth and survival conditions (e.g., Valentine et al. 2004; Penning et al. 2005; Takai et al. 2008). Microbial methanogenesis in the SCS formation fluid regime may cause an unusual carbon isotope effect as CO_3^{2-} is the dominant DIC speciation in highly alkaline solutions.

Carbon and hydrogen isotope systematics of CORK fluids suggest that abiotic methanogenesis is occurring at > ~200 °C. Hydrothermal experiments, demonstrating CO- and CO_2-based abiotic hydrocarbon syntheses, provide consistent $\alpha^{C}_{CH4\text{-}CO2}$ values of 0.97–0.98 (Fu et al. 2007; McCollom et al. 2010), slightly higher than that observed in the CORK fluids (0.963). Experimental determination of CH_4-CO_2 isotope equilibrium at 200 °C exhibits a $\alpha^{C}_{CH4\text{-}CO2}$ of 0.966 (Horita, 2001), very close to the value observed from the CORK fluid (Fig. 5c). Although carbon isotope exchange between CH_4-CO_2 below ~400 °C is generally thought to become kinetically sluggish without a catalyst, particularly in the CH_4-generating direction (e.g., McCollom, 2008), the CH_4-CO_2 isotope equilibrium could be attained at ~200 °C beneath the SCS because of thermodynamically favorable condition for CH_4 generation and geological time span. For the hydrogen isotope equilibrium between hydrocarbons and H_2O, terminal locality of hydrogen atoms would be more feasible for the exchange with the surrounding hydrogen-bearing molecules, namely, H_2O (Reeves et al. 2012; Wang et al. 2015). The H exchange would erase the extremely D-depleted signature ($\alpha^{H}_{CH4\text{-}H2O}$ = ~0.45: McCollom et al. 2010) exhibited by the (experimental) abiotic CH_4 generation. Carbon and hydrogen isotopic thermometers both indicating > ~200 °C left in the 2 °C CORK fluid (Fig. 5c), imply that the reactions for isotope exchange are limited at lower temperatures. Thus, the rate of abiotic synthesis would decease after cooling the SCS formation fluid below 200 °C, yet it must continue within the serpentinite matrix at lower temperatures.

Slab-derived SO_4 conservation through serpentinization

The abundant dissolved sulfate in the SCS formation fluid was thought to originate from sulfate in the subducting slab (Mottl et al. 2003). This conclusion seems inconsistent with abiotic organic synthesis via serpentinization, in which sulfate should be reduced from a thermodynamic viewpoint. On the other hand, closed-system experiments simulating serpentinization at 200–300 °C (Janecky and Seyfried Jr., 1986; Seyfried 2007) confirmed barriers to sulfate reduction during serpentinization due to rapid sulfate precipitation as anhydrite upon initial heating of the experiment and its subsequent metastable persistence, even in a highly reducing fluid. Accordingly, anhydrite precipitation in the downgoing slab may facilitate sulfate conservation through high-temperature regions of subduction and diapir. In addition, a recent quadruple sulfur isotope study of SCS DPW and CORK fluid suggests that the mantle wedge represents one of the possible sources of sulfate in the SCS formation fluid (Aoyama et al. 2018).

Hypothetical sequence of geochemical processes in the SCS system

We propose a hypothesis for the possible sequence of (bio) geochemical processes in the SCS serpentinite formation fluid regime, based on CORK fluid characteristics and observations from previous studies (Fig. 7). The oligotrophic western North Pacific water provides little sedimentary organic matter, resulting in pelagic sediments of the Pacific plate eastward from the Mariana Trench (ODP sites 800–802) that contain sulfate-rich porewater (France-Lanord et al. 1992; Alt and Shanks III, 2006) and low sedimentary biomass (Kellmeyer et al., 2012). Global seafloor drilling that penetrates into the upper basaltic crust (e.g., Gieskes et al. 1990, 1993; Hensen and Wallmann, 2005) have indeed revealed dissolved sulfate that is ubiquitously abundant in the uppermost part of subducting plate, despite their great depth and distance from the trench seafloor. At an early stage of subduction, some or all of the porewater sulfate is precipitated as anhydrite with requisite Ca supplied by dissolution of sedimentary calcite or reactions with involving basaltic crust. At the deep plate boundary region at ~ 0.6 GPa and ≤ ~ 350 °C (Hulme et al. 2010), metamorphic processes such as compactive dewatering, expulsion of interlayer water, dehydration, and decarbonation forms the DIC-rich initial upwelling fluid (Mottl et al. 2004), which harbors no microbial population due to its high temperature. The initial fluid accompanied with metamorphic minerals such as blueschist and anhydrite penetrates into the overlying mantle wedge and induces serpentinization. Extensive serpentinization produces highly alkaline and strongly reducing H_2-rich fluid

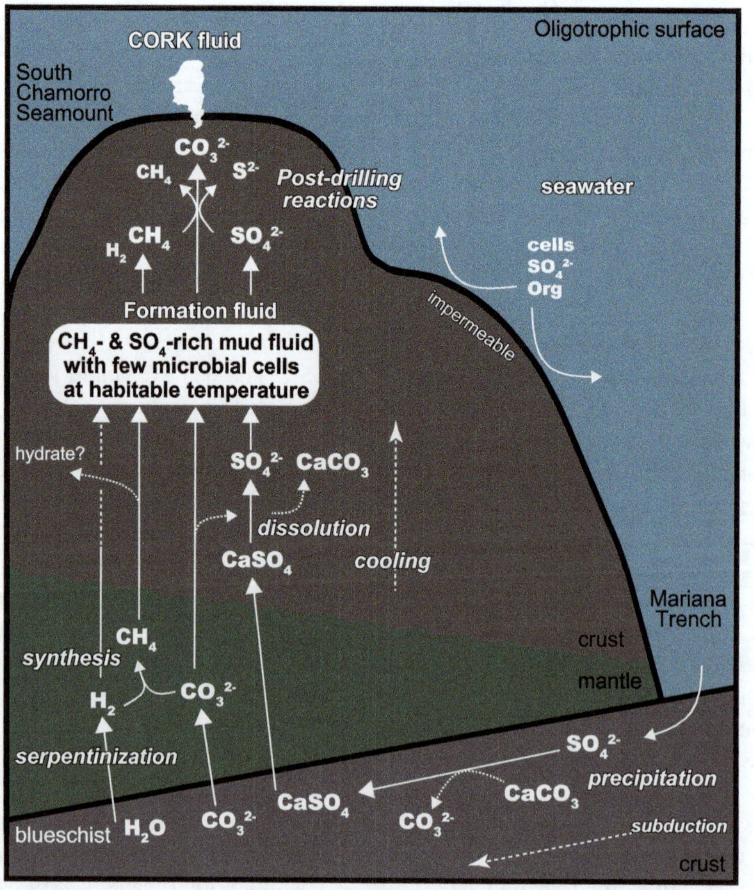

Fig. 7 Illustration for possible C-S-H reactions beneath the SCS. "Cells" represents microbial cells and "Org" represents organic matter in the seawater

(McCollom and Bach 2009), leading to the thermodynamic conditions favorable for the abiotic organic synthesis from slab-derived inorganic carbon. Sulfate minerals are not reduced, due to kinetic constraints. The abiotic synthesis of organic molecules occurs within the ascending serpentinite mud until its temperature decreases to ~ 200 °C. Further ascend and cooling decrease the rate of abiotic synthesis while allowing for the dissolution of anhydrite (anhydrite-derived calcium is completely removed by abundant DIC). This sequence results in the co-existence of abiotic CH_4 (and possibly H_2, C_{2+} hydrocarbons, and some organic matter) and slab-derived dissolved SO_4 in the SCS formation fluid. Although the abundant CH_4 and SO_4 exhibit sufficient bioavailable energy potentials (Nakamura and Takai 2014) within the habitable temperature range (< ~ 122 °C: Takai et al. 2008), limited microbial biomass and activity are present because the relatively impermeable SCS prevents penetration of microbial community from ambient seawater. As the major conduit for the mud/fluid ascending path consists of 60–100% serpentinite muds (Shipboard Sci. Party 2002), the fluid-serpentinite interaction rather than fluid-peridotite interaction (serpentinization) dominates the current fluid chemistry of the SCS formation fluid. The highly alkaline porewater pH of ~ 12.5 likely prevents the development of subseafloor microbial community due to the CO_3^{2-}-dominated DIC speciation being disadvantageous for autotrophy (e.g., Schrenk et al. 2013) whereas the known upper pH limit for chemoorganotrophic microorganisms is 12.4 (Takai et al. 2001). Ongoing reactions between the CH_4 and SO_4, as observed in this study, would have been triggered by the very recent ODP drilling and casing operations carried out in 2001 (Shipboard Sci. Party 2002; Wheat et al. 2008).

Conclusions

The present study investigated the chemical and microbiological characteristics of formation fluids from a serpentinite mud volcano, the SCS. The gas chemistry characterized by abundant CH_4 and low H_2 is the first of its kind from deep-sea highly alkaline (serpentinization-associated) geofluid systems. The SCS formation fluid regime may also represent the first reported example on Earth where abiotic processes are conspicuous with little biotic processes, despite conditions with sufficient bioavailable energy potentials and temperatures within the habitable range. However, there are still many uncertainties associated with the hypothesis presented here. It is not sufficient to clearly distinguish the geochemical processes in each of the spatio-temporally different stages of fluid-rock interactions beneath the SCS, such as metamorphism associated with subduction, fluid released from the plate boundary zone, extensive serpentinization at the wedge mantle, fluid maturation and cooling in the ascending pathway, episodic mud volcanism, and mass wasting of the SCS body. The behavior of sulfate in subduction systems is a key factor for understanding the (bio) geochemical cycle at forearc geofluid systems. Constraint in keeping the co-existence of the (probably) slab-derived sulfate and abiotic CH_4 in the SCS formation, unexpected due to the strong thermodynamic drive for their reaction, remains unidentified. Broadly speaking, abiotic organic synthesis associated with subseafloor serpentinization has been considered to progress in many of the natural serpentinization-associated geofluid systems including the SCS, but has never been confirmed by multiple lines of solid evidences that serve to distinguish it from organic matter degradation and biological carbon fixation. Future studies of the SCS and other serpentinite mud volcanoes in the Mariana forearc are needed to refine conceptual models of reactions within these interesting features that link seafloor materials with the descending slab.

Abbreviations

CORK: Circulation Obviation Retrofit Kit; DPW: Deep porewater; IC: Ion chromatography; ICP-AES: Inductively coupled plasma atomic emission spectrometry; ICP-MS: Inductively coupled plasma mass spectrometry; ODP: Ocean Drilling Program; PVC: Polyvinyl chloride; ROV: Remotely operated vehicle; SCS: South Chamorro Seamount

Acknowledgements

We thank the operation team of the ROV Hyper-Dolphin and the crews of R/V Natsushima for their skillful support during cruises NT09-1, NT09-7, NT12-04, and NT12-23. We also thank Dr. Uta Konno for gas and isotope analyses. Drs. Kentaro Nakamura, Takazo Shibuya, and Chong Chen are gratefully acknowledged for useful and fruitful discussions. SK thanks Prof. Gretchen Bernasconi-Green for providing space and opportunity to draft the manuscript as an academic visitor of ETH Zürich.

Funding

This work was partly supported by JSPS KAKENHI Grant Number 25701004 (SK).

Authors' contributions

SK, JM, CGW, and KT proposed the topic, conceived, and designed the study. All authors joined cruises for sample collection. JM, YM, JSS, and CGW analyzed the samples and helped in their interpretation. JSS and CGW collaborated with the corresponding author (SK) in the construction of manuscript. All authors read and approved the final manuscript.

Competing interests

The authors declare that they have no competing interest.

Author details

[1]Department of Subsurface Geobiological Analysis and Research (D-SUGAR), Japan Agency for Marine-Earth Science and Technology (JAMSTEC), 2-15 Natsushima-cho, Yokosuka 237-0061, Japan. [2]Kochi Institute for Core Sample Research, Japan Agency for Marine-Earth Science and Technology (JAMSTEC), Nankoku, Kochi 783-8502, Japan. [3]Department Marine Chemistry and Geochemistry (MC&G), Woods Hole Oceanographic Institution (WHOI), Woods Hole, MA 02543, USA. [4]Global Undersea Research Unit, University of Alaska Fairbanks, P. O. Box 475, Moss Landing, CA 95039, USA. [5]Institute of Geochemistry and Petrology, ETH Zürich, Zürich, Switzerland.

References

Alt JC, Shanks WC III (2006) Stable isotope compositions of serpentinite seamounts in the Mariana forearc: serpentinization processes, fluid sources and sulfur metasomatism. Earth Planet Sci Lett 242:272–285.

Aoyama S, Nishizawa M, Miyazaki J, Shibuya T, Ueno Y, Takai K (2018) Recycled Archean sulfur in the mantle wedge of the Mariana Forearc and microbial sulfate reduction within an extremely alkaline serpentine seamount. Earth Planet Sci Lett 491:109–120.

Barnes I, LaMarche j VC, Himmelberg G (1967) Geochemical evidence of present-day serpentinization. Science 156:830–832.

Barnes I, O'neil JR, Trescases JJ (1978) Present day serpentinization in New Caledonia, Oman and Yugoslavia. Geochim Cosmochim Acta 42:144–145.

Burruss RC, Laughrey CD (2010) Carbon and hydrogen isotopic reversals in deep basin gas: Evidence for limits to the stability of hydrocarbons. Organic Geochemistry 41:1285–1296. https://doi.org/10.1016/j.orggeochem.2010.09.008.

Charlou JL, Donval JP, Fouquet Y, Jean-Baptiste P, Holm N (2002) Geochemistry of high H2 and CH4 vent fluids issuing from ultramafic rocks at the Rainbow hydrothermal field (36°14'N, MAR). Chem Geol 191:345–359.

de Lange GJ, Cranston RE, Hydes DH, Boust D (1992) Extraction of pore water from marine sediments: a review of possible artifacts with pertinent examples from the North Atlantic. Mar Geol 109:53–76.

Etiope G, Sherwood Lollar B (2013) Abiotic methane on earth. Rev Geophys 51:1–24.

France-Lanord C, Michard A, Karpoff AM (1992) Major element and Sr isotope composition of interstitial waters in sediments from leg 129: the role of diagenetic reactions. In: Larson RL, Lancelot Y et al (eds) Proceedings of teh Ocean Drilling Program Scientific Results, vol 129, pp 267–281.

Früh-Green GL, Kelley DS, Bernasconi SM, Karson JA, Ludwig KA, Butterfield DA, Boschi C, Proskurowski G (2003) 30,000 years of hydrothermal activity at the Lost City vent field. Science 301:495–498.

Fryer P (2011) Serpentinite mud volcanism: observations, processes, and implications. Annu Rev Mar Sci 4:345–373.

Fryer P, Mottl MJ (1997) "Shinkai 6500" investigations of a resurgent mud volcano on the Southeastern Mariana forearc. JAMSTEC-R 13:103–114.

Fu Q, Sherwood Lollar B, Horita J, Lacrampe-Couloume G, Seyfried WE Jr (2007) Abiotic formation of hydrocarbons under hydrothermal conditions: constraints from chemical and isotope data. Geochim Cosmochim Acta 71: 1982–1998.

Gieskes J, Gamo T, Kastner M (1993) Major and minor element geochemistry of interstitial waters of site 808, Nankai trough: an overview. In: Hill IA, Taira A, Firth JV et al (eds) Scientific results, Ocean Drilling Program, Leg 131. Ocean Drilling Program, College Station, Texas, pp 387–396.

Gieskes JM, Vroluk P, Blank G (1990) Hydrogeochemistry of the Northern Barbados accretionary complex transect: Ocean Drilling Project Leg 110. J Geophys Res 95:8809–8818.

Hensen C, Wallmann K (2005) Methane formation at Costa Rica continental margin - constraints for gas hydrate inventories and cross-décollment fluid flow. Earth Planet Sci Lett 236:41–60.

Hirayama H, Fuse H, Abe M, Miyazaki M, Nakamura T, Nunoura T, Furushima Y, Yamamoto H, Takai K (2013) Methylomarinum vadi gen. nov. sp. Nov., a methanotroph isolated from two distinct marine environments. Int J Syst Evol Microbiol 63:1073–1082.

Hirayama H, Sunamura M, Takai K, Nunoura T, Noguchi T, Oida H, Furushima Y, Yamamoto H, Oomori T, Horikoshi K (2007) Culture-dependent and -independent characterization of microbial communities associated with a shallow submarine hydrothermal system occurring within a coral reef off Taketomi Island, Japan. Appl Environ Microbiol 73:7642–7656.

Homma A, Tsukahara H (2008) Chemical characteristics of hot spring water and geological environment in the northernmost area of the Itoigawa Shizuoka tectonic line. Bull Earthw Res Inst Univ Tokyo 83:217–225.

Horibe Y, Craig H (1995) D/H fractionation in the system methane-hydrogen-water. Geochim Cosmochim Acta 59:5209–5217.

Horita J (2001) Carbon isotope exchange in the system CO2-CH4 at elevated temperatures. Geochim Cosmochim Acta 65:1907–1919.

Hulme SM, Wheat CG, Fryer P, Mottl MJ (2010) Pore water chemistry of the Mariana serpentinite mud volcanoes: a window to the seismogenic zone. Geochem Geophys Geosyst 11:Q01X09. https://doi.org/10.1029/2009GC002674.

Janecky DR, Seyfried WE Jr (1986) Hydrothermal serpentinization of peridotite within the oceanic crust: experimental investigations of mineralogy and major element chemistry. Geochim Cosmochim Acta 50:1357–1378.

Kawagucci S, Miyazaki J, Noguchi T, Okamura K, Shibuya T, Watsuji T et al (2016) Fluid chemistry in the solitaire and dodo hydrothermal fields of the central Indian ridge. Geofluids 5:988–1005. https://doi.org/10.1111/gfl.12201.

Kawagucci S, Toki T, Ishibashi J, Takai K, Ito M, Oomori T, Gamo T (2010) Isotopic variation of molecular hydrogen in 20°–375°C hydrothermal fluids as detected by a new analytical method. J Geophys Res 115:G03021–9. https://doi.org/10.1029/2009JG001203.

Kawagucci S, Ueno Y, Takai K, Toki T, Ito M, Inoue K, Makabe A, Yoshida N, Muramatsu Y, Takahata N, Sano Y, Narita T, Teranishi G, Obata H, Nakagawa S, Nunoura T, Gamo T (2013) Geochemical origin of hydrothermal fluid methane in sediment-associated fields and its relevance to the geographical distribution of whole hydrothermal circulation. Chem Geol 339:213–225.

Kelley DS, Karson JA, Blackman DK, Früh-Green GL, Butterfield DA et al (2001) An off-oxis hydrothermal vent field near the Mid-Atlantic Ridge at 30°N. Nature 412:145–149.

Kellmeyer J, Pockalny R, Adhikari RR, Smith DC, D'Hondt S (2012) Global distribution of microbial abundance and biomass in subseafloor sediment. PNAS 109:16213–16216.

Konn C, Charlou JL, Lolm NG, Mousis O (2015) The production of methane, hydrogen, and organic compound in ultramafic-hosted hydrothermal vents of the Mid-Atlantic Ridge. Astrobiology 15:381–399.

Kumagai H, Nakamura N, Toki T, Morishita T, Okino K, Ishibashi J et al (2008) Geological background of the Kairei and Edmond hydrothermal fields along the Central Indian ridge: implications of their vent fluids' distinct chemistry. Geofluids 8:239–251.

Mathrani IM, Boone DR, Mah RA, Fox GE, Lau PP (1988) Methanohalophilus zhilinae sp. Nov., an alkaliphili, halophilic, methylotrophic methanogen. Int J Syst Bacteriol 38:139–142.

McCollom TM (2008) Observational, experimental, and theoretical constraints on carbon cycling in Mid-Ocean Ridge hydrothermal systems. In: Magma to microbe: modeling hydrothermal processes at ocean spreading centers geophysical monograph series, vol 178, pp 193–213.

McCollom TM (2013) Laboratory simulations of abiotic hydrocarbon formation in Earth's deep subsurface. Rev Mineral Geochem 75:467–494.

McCollom TM, Bach W (2009) Thermodynamic constraints on hydrogen generation during serpentinization of ultramafic rocks. Geochim Cosmochim Acta 73:856–875.

McCollom TM, Sherwood Lollar B, Lacrampe-Couloume G, Seewald JS (2010) The influence of carbon source on abiotic organic synthesis and carbon isotope fractionation under hydrothermal conditions. Geochim Cosmochim Acta 74: 2717–2740.

McDermott JM, Ono S, Tivey MK, Seewald JS, Shanks WC III, Solow AR (2015) Identification of sulfur sources and isotopic equilibria in submarine hot-springs using multiple sulfur isotopes. Geochimica et. Cosmochimica Acta 160:169–187. https://doi.org/10.1016/j.gca.2015.02.016.

Miyazaki J, Makabe A, Matsui Y, Ebina N, Tsutsumi S, Ishibashi J, Chen C, Kaneko S, Takai K, Kawagucci S (2017) WHATS-3: an improved flow-through multi-bottle fluid sampler for Deep-Sea Geofluid research. Front Earth Sci 5:45. https://doi.org/10.3389/feart.2017.00045.

Morono Y, Inagaki F (2010) Automatic Slide-Loader Fluorescence Microscope for Discriminative Enumeration of Subseafloor Life. Scientific Drilling 9:32–36. https://doi.org/10.5194/sd-9-32-2010.

Morono Y, Terada T, Kallmeyer J, Inagaki F (2013) An improved cell separation technique for marine subsurface sediments: applications for high-throughput analysis using flow cytometry and cell sorting. Environ Microbiol 2:2841–2849. https://doi.org/10.1111/1462-2920.12153.

Morono Y, Terada T, Masui N, Inagaki F (2009) Discriminative detection and enumeration of microbial life in marine subsurface sediments. The ISME Journal 3:503–511. https://doi.org/10.1038/ismej.2009.1.

Mottl MJ, Komor SC, Fryer P, Moyer CL (2003) Deep-slab fluids fuel extremophilic archaea on a Mariana forearc serpentinite mud volcano: Ocean Drilling Program Leg 195. Geochem Geophys Geosyst 4:9009.

Mottl MJ, Wheat CG, Fryer P, Gharib J, Martin J (2004) Chemistry of springs across the Mariana forearc shows progressive devolatilization of the subduction plate. Geochim Cosmochim Acta 68:4915–4933.

Nakagawa S et al (2006) Microbial community in black rust exposed to hot ridge flank crustal fluids. Appl Environ Microbiol 72:6789–6799.

Nakamura K, Takai K (2014) Theoretical constraints of physical and chemical properties of hydrothermal fluids on variations in chemolithotrophic microbial

communities in seafloor hydrothermal systems. Prog Earth Planet Sci 1:5.
Nigro LM, Harris K, Orcutt BN, Hyde A, Clayton-Luce S, Becker K, Teske A (2012) Microbial communities at the borehole observatory on the Costa Rica Rift flank (Ocean Drilling Program Hole 896A). Front Microbiol 3:232. https://doi.org/10.3389/fmicb.2012.00232.
Nunoura T, Takaki Y, Kazama H, Hirai M, Ashi J, Imachi H, Takai K (2012) Microbial diversity in deep-sea methane seep sediments presented by SSU rRNA gene tag sequencing. Microbes Environ 27:382–390. https://doi.org/10.1264/jsme2.ME12032.
Ohara Y et al (2012) A serpentinite-hosted ecosystem in the southern Mariana Forearc. Proc Natl Acad Sci U S A 109:2831–2835. https://doi.org/10.1073/pnas.1112005109.
Okumura T, Kawagucci S, Saito Y, Matsui Y, Takai K, Imachi H (2016b) Hydrogen and carbon isotope systematics in hydrogenotrophic methanogenesis under H2-limited and H2-enriched conditions: implications for the origin of methane and its isotopic diagnosis. Prog Earth Planet Sci 3:14.
Okumura T et al (2016a) Brucite chimney formation and carbonate alteration at the Shinkai seep field, a serpentinite-hosted vent system in the southern Mariana forearc. Geochem Geophys Geosyst 17. https://doi.org/10.1002/2016GC006449.
Onishi Y, Yamanaka T, Okumura T, Kawagucci S, Watanabe HK, Ohara Y (2018) Evaluation of nutrient and energy sources of the deepest known serpentinite-hosted ecosystem using stable carbon, nitrogen, and sulfur isotopes. PLoS One 13(6):e0199000.
Orphan VJ, House CH, Hinrichs K-U, McKeegan KD, DeLong EF (2001) Methane-consuming archaea revealed by directly coupled isotopic and phylogenetic analysis. Science 293:484–487.
Penning H, Plugge CM, Galand PE, Conrad R (2005) Variation of carbon isotope fractionation in hydrogenotrophic methanogenic microbial cultures and environmental samples at different energy status. Glob Chang Biol 11:2103–2113.
Pester NJ, Conrad ME, Knauss KG, DePaolo DJ (2018) Kinetics of D/H isotope fractionation between molecular hydrogen and water. Geochim Cosmochim Acta 242:191–212.
Proskurowski G, Lilley MD, Seewald JS, Fru h-Green GL, Olson EJ, Lupton JE, Sylva SP, Kelley DS (2008) Abiogenic Hydrocarbon Production at Lost City Hydrothermal Field. Science 319:604–607. https://doi.org/10.1126/science.1151194.
Redfield AC (1958) The biological control of chemical factors in the environment. Am Sci 46:205–221.
Reeves E, Seewald JS, Sylva SP (2012) Hydrogen isotope exchange between n- alkanes and water under hydrothermal conditions. Geochim Cosmochim Acta 77:582–599.
Russell MJ et al (2014) The drive to life on wet and icy worlds. Astrobiology 14: 308–343.
Saegusa S, Tsunogai U, Nakagawa F, Kaneko S (2006) Development of a multibottle gas-tight fluid sampler WHATS II for Japanese submersibles/ROVs. Geofluids 6:234–240.
Sano Y, Marty B (1995) Origin of carbon in fumarolic gas from island arcs. Chem Geol 119:265–274.
Schrenk MO, Brazelton WJ, Lang SQ (2013) Serpentinization, carbon, and deep life. Rev Mineral Geochem 75:575–606.
Seewald JS, Doherty KW, Hammar TR, Liberatore SP (2001) A new gas-tight isobaric sampler for hydrothermal fluids. Deep Sea Res 49:189–196. https://doi.org/10.1016/S0967-0637(01)00046-2.
Seewald JS, Seyfried WE Jr, Shanks WC III (1994) Variations in the chemical and stable isotope composition of carbon and sulfur species during organic-rich sediment alteration: an experimental and theoretical study of hydrothermal activity at Guaymas Basin, Gulf of California. Geochim Cosmochim Acta 58: 5065–5082.
Sessions AL, Sylva SP, Summons RE, Hayes JM (2004) Isotopic exchange of carbon-bound hydrogen over geological timescales. Geochim Cosmochim Acta 68:1545–1559.
Seyfried WE Jr, Foustoukos DI, Fu Q (2007) Redox evolution and mass transfer during serpentinization: an experimental and theoretical study at 200 °C, 500 bar with implications for ultramafic-hosted hydrothermal systems at Mid-Ocean Ridge. Geochim Cosmochim Acta 71:3872–3886.
Seyfried WE Jr., Pester NJ, Ding K, Rough M (2011) Vent fluid chemistry of the Rainbow hydrothermal system (36°N, MAR): Phase equilibria and in situ pH controls on subseafloor alteration processes. Geochimica et Cosmochimica Acta 75:1574–1593. https://doi.org/10.1016/j.gca.2011.01.001.
Shibuya T, Komiya T, Takai K, Maruyama S, Russell MJ (2017) Weak hydrothermal carbonation of the Ongeluk volcanics: evidence for low CO2 concentrations in seawater and atmosphere during the Paleoproterozoic global glaciation. Prog Earth Planet Sci 4:31. https://doi.org/10.1186/s40645-017-0145-6.
Sherwood Lollar B, Westgate TD, Ward JA, Slater GF, Lacrampe-Couloume G (2002) Abiogenic formation of alkanes in the Earth's crust as a minor source for global hydrocarbon reservoirs. Nature 416:522–524. https://doi.org/10.1038/416522a.
Shipboard Sci. Party (2002) In: Peters LL (ed) Site 1200. Proc. Ocean Drill. Prog. Initial Rep., Vol. 195. Ocean Drill. Prog./Texas A&M Univ, College Station. https://doi.org/10.2973/odp.proc.ir.195.103.2002.
Stern RJ, Ren M, Kelley KA, Ohara Y, Martinez F, Bloomer SH (2014) Basaltic volcaniclastics from the Challenger Deep forearc segment, Mariana convergent margin: Implications for tectonics and magmatism of the southernmost Izu-Bonin-Mariana arc. Island Arc 23:368–382. https://doi.org/10.1111/iar.12088.
Stolper DA, Lawson M, Davis CL, Ferreia AA, Santos Neto EV, Ellis GS, Lewan MD, Martini AM, Tang Y, Schoell M, Sessions AL, Eiler JM (2014) Formation temperatures of thermogenic and biogenic methane. Science 344:1500–1503.
Suda K, Ueno Y, Yoshizaki M, Nakamura H, Kurokawa K, Nishiyama E, Yoshino K, Hongoh Y, Kawachi K, Omori S, Yamada K, Yoshida N, Maruyama S (2014) Origin of methane in serpentinite-hosted hydrothermal systems: The CH4–H2–H2O hydrogen isotope systematics of the Hakuba Happo hot spring. Earth and Planetary Science Letters 386:112–125. https://doi.org/10.1016/j.epsl.2013.11.001
Takai K, Gamo T, Tsunogai U, Nakayama N, Hirayama H, Nealson KH, Horikoshi K (2004) Geochemical and microbiological evidence for a hydrogen-based, hyperthermophilic subsurface lithoautotrophic microbial ecosystem (HyperSLiME) beneath as active deep-sea hydrothermal field. Extremophiles 8:269–282.
Takai K, Moser DP, Onstott TC, Spoelstra N, Pfiffner SM, Dohnalkova A, Fredrickson JK (2001) Alkaliphilus transvaalensis gen. nov., sp. Nov., an extremely alkaliphilic bacterium isolated from a deep South African gold mine. Int J Syst Evol Microbiol 51:1245–1256.
Takai K, Moyer CL, Miyazaki M, Nogi Y, Hirayama H, Nealson KN, Horikoshi K (2005) Marinobacter alkaliphilus sp. Nov., a novel alkaliphilic bacterium isolated from subseafloor alkaline serpentine mud from Ocean Drilling Program Site 1200 at South Chamorro Seamount, Mariana Forearc. Extremophiles 9:17–27.
Takai K, Nakamura K, Suzuki K, Inagaki F, Nealson KH, Kumagai H (2006) Ultramafic-Hydrothermalism-Hydrogenesis-HyperSLiME (UltraH3) linkage: a key insight into early microbial ecosystem in the Archean deep-sea hydrothermal systems. Paleontological Res 10:269–282.
Takai K, Nakamura K, Toki T, Tsunogai U, Miyazaki M, Miyazaki J, Hirayama H, Nakagawa S, Nunoura T, Horikoshi K (2008) Cell proliferation at 122 °C and isotopically heavy CH4 production by a hyperthermophilic methanogen under high-pressure cultivation. Proc Natl Acad Sci U S A 105:10949–10954.
Tishchenko P, Hensen C, Wallmann K, Wong CS (2005) Calculation of the stability and solubility of methane hydrate in seawater. Chem Geol 219:37–52.
Ueda H, Shibuya T, Sawaki Y, Saitoh M, Takai K, Maruyama S (2016) Reactions between komatiite and CO2-rich seawater at 250 and 350 °C, 500 bars: implications for hydrogen generation in the Hadean seafloor hydrothermal system. Progress in Earth and. Planet Sci 3:35. https://doi.org/10.1186/s40645-016-0111-8..
Valentine DL, Chidthaisong A, Rice A, Reeburgh WS, Tyler SC (2004) Carbon and hydrogen isotope fractionation by moderately thermophilic methanogens. Geochim Cosmochim Acta 68:1571–1590.
Wang DT, Gruen DS, Lollar BS, Hinrichs KU, Stewart LC, Holden JF, Hristov AN, Pohlman JW, Morrill PL, Konneke M, Delwiche KB, Reeves EP, Sutcliffe CN, Ritter DJ, Seewald JS, McIntosh JC, Hemond HF, Kubo MD, Cardace D, Hoehler TM, Ono S (2015) Nonequilibrium clumped isotope signals in microbial methane. Science 348:428–431. https://doi.org/10.1126/science.aaa4326.
Wheat CG, Fryer P, Fisher AT, Hulme S, Jannasch H, Mottl MJ, Becker K (2008) Borehole observations of fluid flow from South Chamorro Seamount, an active serpentinite mud volcano in the Mariana forearc. Earth Planet Sci Lett 267:401–409.
Wiesenburg DA, Guinasso NL Jr (1979) Equilibrium solubilities of methane, carbon monoxide, and hydrogen in water and sea water. J Chem Eng Data 24:356–560.

3

Sound velocities of bridgmanite from density of states determined by nuclear inelastic scattering and first-principles calculations

Catherine McCammon[1*], Razvan Caracas[2], Konstantin Glazyrin[3], Vasily Potapkin[1], Anastasia Kantor[1,4], Ryosuke Sinmyo[1,6], Clemens Prescher[1,7], Ilya Kupenko[1,4,8], Aleksandr Chumakov[4,5] and Leonid Dubrovinsky[1]

Abstract

Sound velocities of bridgmanite measured in the laboratory are a key to deciphering the composition of the lower mantle. Here, we report Debye sound velocities determined using nuclear inelastic scattering (NIS) for one majorite composition ($Mg_{0.82}Fe_{0.18}SiO_3$) and five bridgmanite compositions ($Mg_{0.82}Fe_{0.18}SiO_3$, $Mg_{0.86}Fe_{0.14}Si_{0.98}Al_{0.02}O_3$, $Mg_{0.88}Fe_{0.12}SiO_3$, $Mg_{0.6}Fe_{0.4}Si_{0.63}Al_{0.37}O_3$, $Mg_{0.83}Fe_{0.15}Si_{0.98}Al_{0.04}O_3$) measured in a diamond anvil cell at pressures up to 89 GPa at room temperature. Debye sound velocities for majorite determined from NIS are consistent with literature data from Brillouin scattering and ultrasonics, while Debye sound velocities for bridgmanite are significantly lower than literature values from the same methods. We calculated partial and total density of states (DOS) for $MgSiO_3$ and $FeSiO_3$ bridgmanite using density functional theory and demonstrate that Debye sound velocities calculated from the reduced DOS using the same approach as for the experimental data (i.e., the limit of $D(E)/E^2$ as energy goes to zero) give the same sound velocities for each phase irrespective of which partial DOS is used. In addition, we show that Debye sound velocities calculated using this approach are consistent with values obtained from the calculation of the full elastic tensor. Comparison of the calculated DOS with the one obtained from NIS indicates that the experimental DOS has enhanced intensity at low energies that leads to a different slope of the DOS and hence a lower sound velocity. This effect is present in all of the bridgmanite samples examined in this study.

Keywords: Elasticity, Perovskite, Nuclear resonance, Sound velocity, Lower mantle, Density functional theory

Background

Insight into the accretion of the Earth and its subsequent differentiation can be gained through knowledge of the present-day Earth composition. The lower mantle in particular has been the focus of much attention as well as controversy, and questions such as how closely it approximates a chondritic composition have occupied geochemists for decades. Comparison of laboratory measurements of elastic wave velocities of lower mantle minerals with seismic data has played a crucial role in the discussion starting with Birch (1952), and experimental techniques such as inelastic X-ray scattering (e.g., Fiquet et al. 2004), Brillouin scattering (e.g., Sinogeikin and Bass 2002; Jackson et al. 2005; Murakami et al. 2007; Murakami et al. 2008), ultrasonics (e.g., Irifune et al. 2008; Gwanmesia et al. 2009; Liu et al. 2015), and impulsive stimulated scattering (e.g., Crowhurst et al. 2008) have provided important results on transition zone and lower mantle minerals. In situ measurements are important, because some transitions (for example, spin transitions) are not quenchable, and may influence the elastic properties of iron-containing minerals.

Nuclear inelastic scattering (NIS) offers the attractive possibility to determine elastic wave velocities of iron-

* Correspondence: catherine.mccammon@uni-bayreuth.de
[1]Bayerisches Geoinstitut, Universität Bayreuth, Bayreuth 95440, Germany
Full list of author information is available at the end of the article

Table 1 Starting materials used for DAC experiments

Sample	Phase	x_{Fe}	x_{Al}	x_{Mg}	x_{Si}	$Fe^{3+}/\Sigma Fe$ precursor	$Fe^{3+}/\Sigma Fe$ bridgmanite	^{57}Fe enrichment	Ref
U1219	Majorite	0.18(1)	0	0.82(1)	1.00(1)	0.09(2)	0.15–0.20	0.9	a, b
SL16	Bridgmanite	0.14(2)	0.02(1)	0.86(3)	0.98(4)	–	0.30(5)	0.6	c
SL18	Pyroxene	0.12(1)	0	0.88(2)	1.00(1)	n.a.	0.15(4)	0.6	d, e
S4850	Bridgmanite	0.40(3)	0.37(2)	0.60(3)	0.63(3)	–	0.80(5)	0.9	f
S4883	Bridgmanite	0.15(2)	0.04(1)	0.83(2)	0.98(1)	–	0.27(7)	0.6	g

a McCammon and Ross (2003); b McCammon et al. (2010); c Lauterbach et al. (2000); d Lauterbach (2000); e McCammon et al. (2008); f Potapkin et al. (2013); g this study
n.a. not analyzed

containing minerals in the laser-heated diamond anvil cell through direct measurement of the partial density of states (e.g., Sturhahn and Jackson 2007), and results for mantle minerals have already been reported for ferropericlase (Lin et al. 2006; Wicks et al. 2010; Chen et al. 2012; Sinmyo et al. 2014), orthoenstatite (Jackson et al. 2009) and post-perovskite (Mao et al. 2006). So far, however, no investigations using NIS have been carried out on the Earth's most abundant mineral that constitutes the bulk of the lower mantle, namely bridgmanite.

Computational studies provide an important complement to experimental measurements and allow a more extensive probe of parameter space. Methods have advanced substantially in recent years to enable densities and elastic wave velocities of candidate lower mantle assemblages to be calculated for the appropriate compositions at relevant pressure and temperature compositions (e.g., Wang et al. 2015 and references therein). Electronic transitions in iron present a challenge, however, and the apparent inconsistency between experimental and computational results on the stability of the intermediate spin state of Fe^{2+} in bridgmanite remains controversial (e.g., Stackhouse 2008; McCammon et al. 2013; Shukla et al. 2015). Nevertheless, a strong advantage of computations is the capability to calculate a wide range of properties including experimental observables based on vibrational density of states (DOS), for example infrared spectra of hydrous post-perovskite (Townsend et al. 2015). The combination of theoretical DOS calculations with experimental determinations using NIS provides a powerful new tool to understand the vibrational properties of iron-containing phases.

In this work, we present an experimental study using NIS to determine sound velocities of five different compositions of $(Mg,Fe)(Si,Al)O_3$ bridgmanite at pressures between 0 and 89 GPa at room temperature, including a comparison with results for $(Mg,Fe)(Si,Al)O_3$ majorite with the same composition. We present a parallel theoretical study to calculate the DOS of two bridgmanite end-members ($MgSiO_3$ and $FeSiO_3$) at comparable pressures that provides insight into the anomalous experimental results.

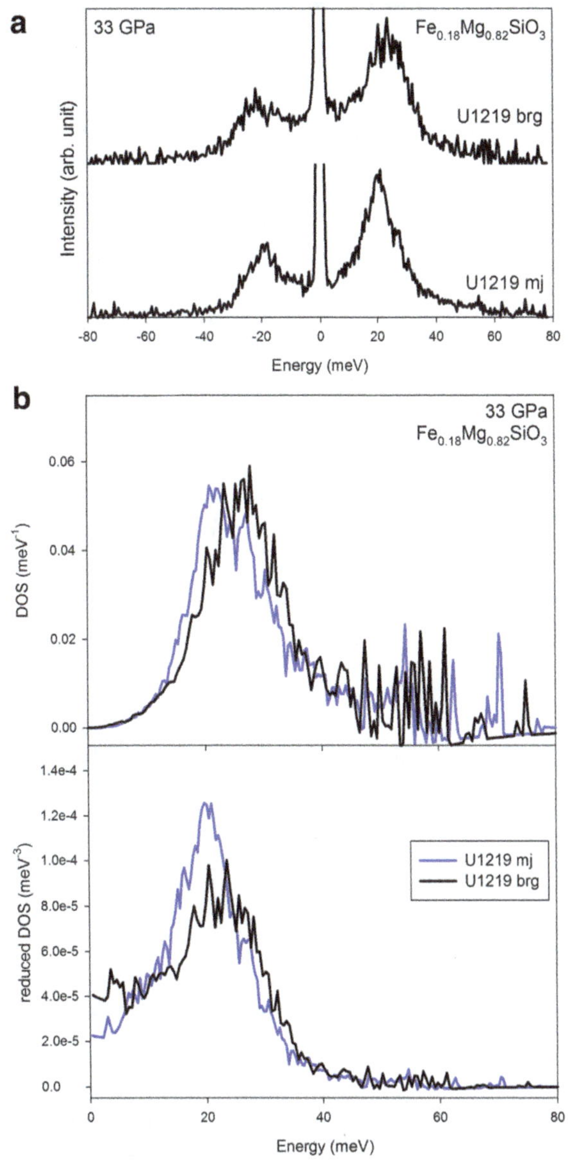

Fig. 1 Comparison of U1219 sample data for the majorite and bridgmanite phases at 33 GPa. **a** NIS spectra. **b** pDOS (*upper*) and reduced pDOS (*lower*). Majorite was measured first and then transformed to bridgmanite by laser heating, after which the NIS data for bridgmanite were collected

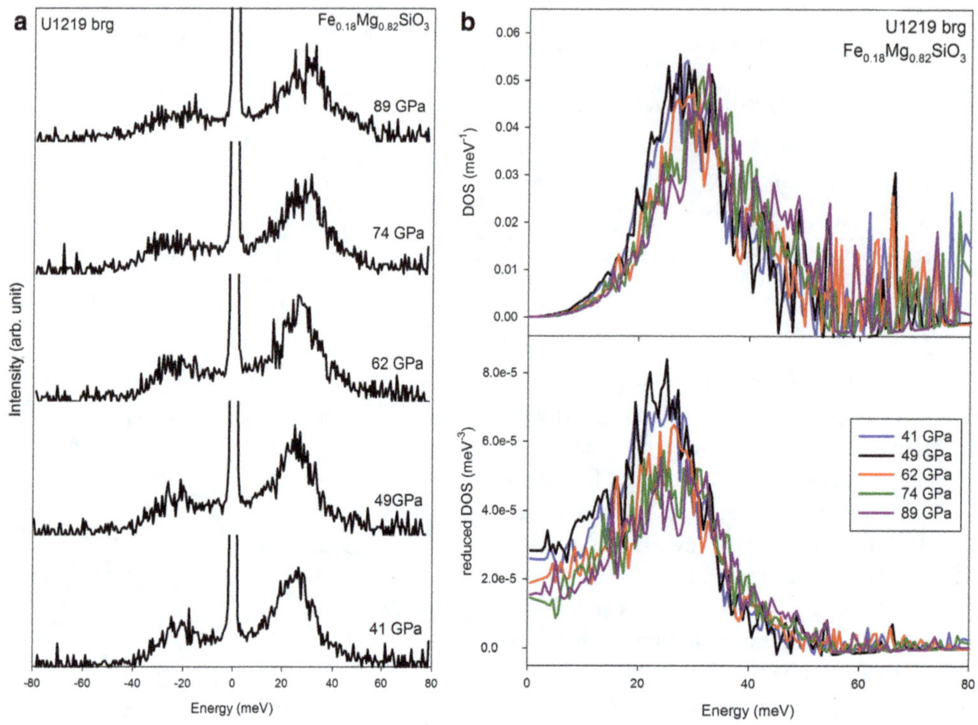

Fig. 2 Influence of pressure on U1219 bridgmanite. **a** NIS spectra. **b** pDOS (*upper*) and reduced pDOS (*lower*). Data were collected on compression during the same beamtime experiment and follow on from the spectra shown in Fig. 1

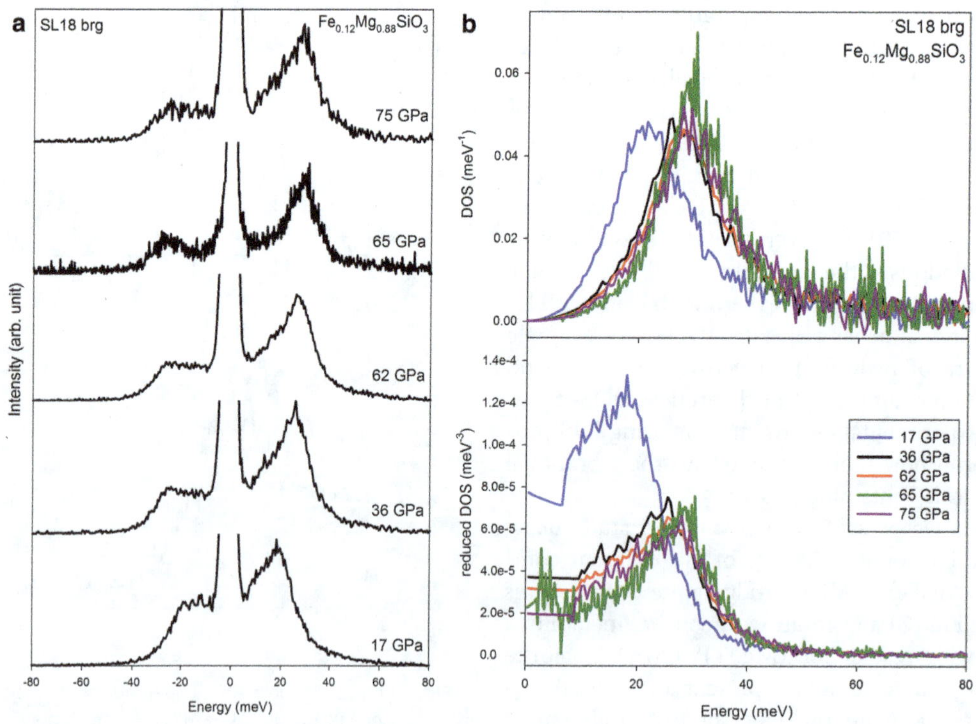

Fig. 3 Influence of pressure on SL18 bridgmanite. **a** Selected NIS spectra. **b** Selected pDOS (*upper*) and reduced pDOS (*lower*). The SL18 pyroxene starting material was transformed to bridgmanite by laser heating in the DAC, and data were collected during three different beamtime experiments: #1—17, 36, 62, and 75 GPa; #2—50 GPa (not shown); and #3—65 GPa

Methods

The starting materials for the DAC experiments are listed in Table 1. Sample S4883 was synthesized from orthopyroxene loaded into a Re capsule in a Kawai-type multianvil press with a $LaCrO_3$ heater assembly and $W_{75}Re_{25}/W_{97}Re_{03}$ thermocouple at 26 GPa and 2000 °C for 40 min (Keppler and Frost 2005). The compositions of the starting materials were determined using the electron microprobe and conventional (radioactive point source) Mössbauer spectroscopy.

NIS measurements were made using panoramic DACs that were designed and constructed at Bayerisches Geoinstitut. For each experiment, a starting material in the form of powder was loaded into a Be gasket together with small ruby chips for pressure determination using the ruby scale of Mao et al. (1986). Generally, LiF or NaCl was used as a pressure medium. The diameters of the diamond culets and opening holes in the gaskets varied depending on the pressure range of the experiment. The samples were laser-heated in the DAC using a portable system mounted on the beamline incorporating either single-sided (Dubrovinsky et al. 2009) or double-sided (Kupenko et al. 2012) laser heating.

The NIS data were collected at ambient conditions on the nuclear resonance beamline ID18 at the European Synchrotron Radiation Facility (Rüffer and Chumakov 1996) between 2008 and 2010. Further details of the NIS setup and data analysis are given in Glazyrin et al. (2013) and references therein. Nuclear forward scattering (NFS) data were collected at the same time as the NIS data to monitor the electronic state of iron, and X-ray diffraction was used to confirm the identity of bridgmanite off-line after laser heating. The NIS data were collected typically over a range of −80 to 80 meV around the ^{57}Fe nuclear resonance energy of 14.4 keV in steps of 0.2 meV. The energy resolution of the X-ray beam was 1 meV. Debye sound velocities (V_D) were determined from the reduced partial density of states (pDOS) using the "homogeneous model" described in Sinmyo et al. (2014). The densities required for the calculation of V_D of bridgmanite were determined using a third-order Birch-Murnaghan equation of state based on volumes and elastic parameters reported in Boffa Ballaran et al. (2012) (sample S4850) and Glazyrin et al. (2014) (samples U1219, SL16, SL18, S4883), where the generalized model in the latter work was used to calculate values relevant to the different compositions. Equations of state parameters for majorite were taken from McCammon and Ross (2003) (cell volume) and Kavner et al. (2000) (elastic parameters). The adiabatic bulk modulus can be used to calculate longitudinal wave (V_P) and transverse wave (V_S) velocities from V_D and vice versa (e.g., Sturhahn and Jackson 2007).

First-principles calculations were performed based on the local density approximation of DFT to determine the DOS for $MgSiO_3$ bridgmanite at 60 GPa and antiferromagnetic $FeSiO_3$ bridgmanite at 20, 40, and 60 GPa. We computed the dynamical matrices on a regular grid of 4 × 4 × 4 special q points (Monkhorst and Pack 1976). We used Fourier interpolation techniques to obtain the interatomic force constants on a dense grid in the reciprocal space, from which we obtained the phonon DOS (Gonze et al. 2005). We determined both the total and

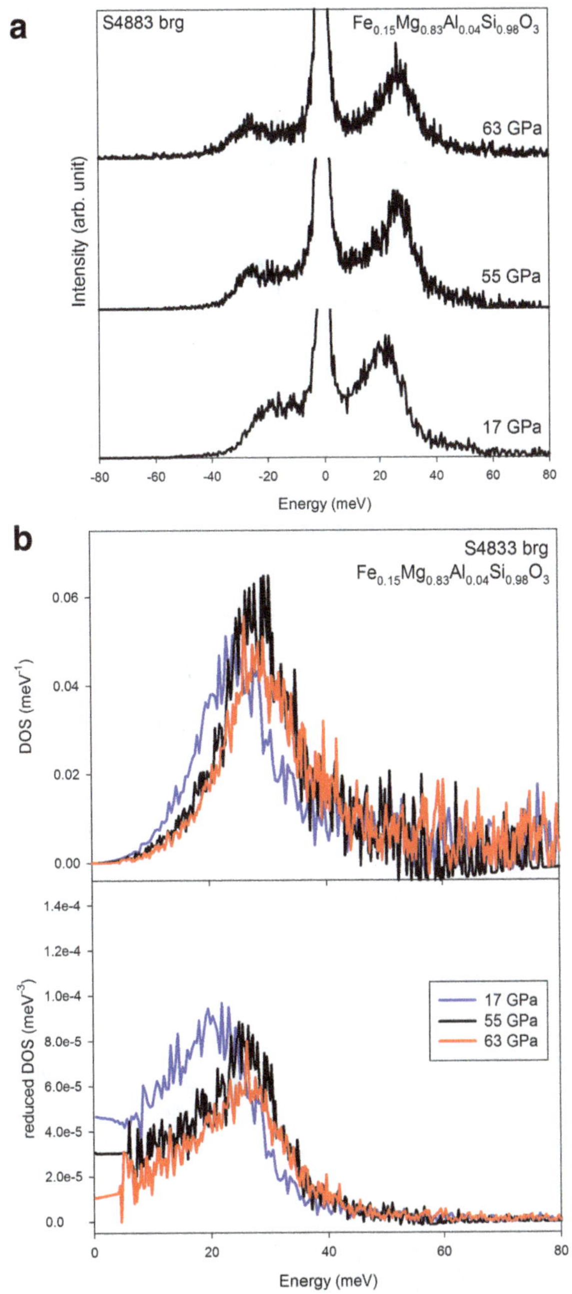

Fig. 4 Influence of pressure on S4883 bridgmanite. **a** NIS spectra. **b** pDOS (*upper*) and reduced pDOS (*lower*). The sample was loaded as bridgmanite into the DAC, and data were collected on compression during the same beamtime experiment

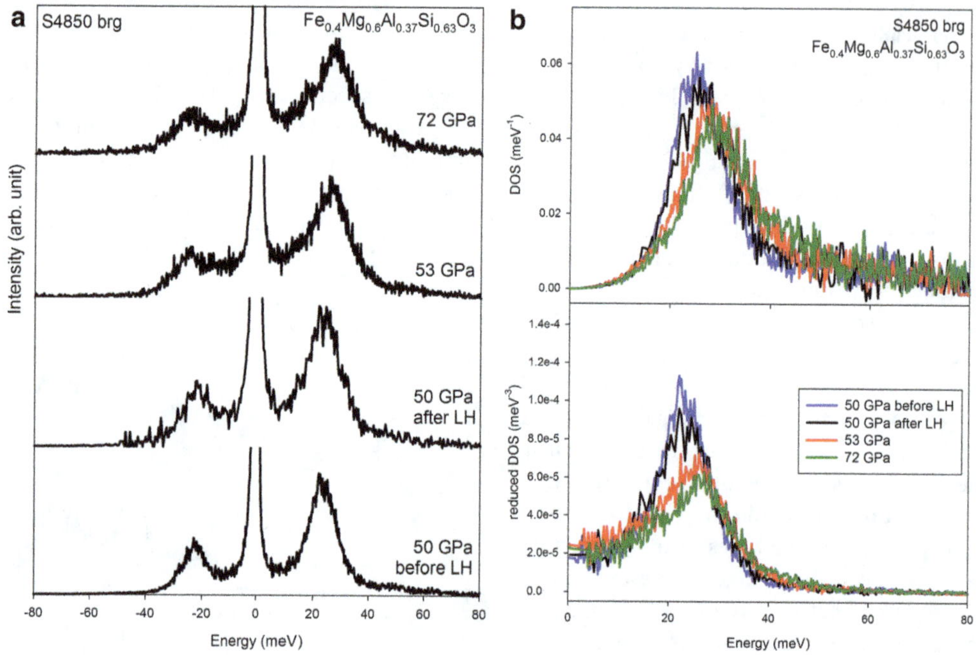

Fig. 5 Influence of pressure on S4850 bridgmanite. **a** NIS spectra. **b** pDOS (*upper*) and reduced pDOS (*lower*). The sample was loaded as bridgmanite into the DAC, and data were collected during two different beamtime experiments: #1—50 GPa and #2—53 and 72 GPa. The sample was laser-heated at 50 GPa, and data were collected both before and after laser heating

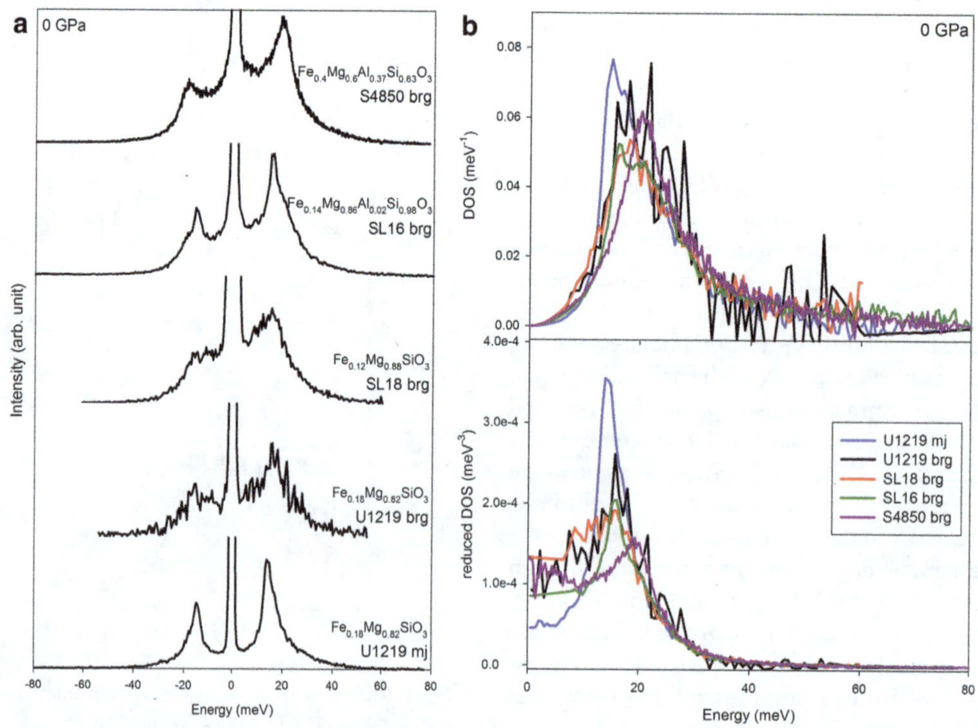

Fig. 6 Comparison of majorite and bridgmanite sample data at 0 GPa. **a** NIS spectra. **b** pDOS (*upper*) and reduced pDOS (*lower*). Majorite was measured as the starting material in air while bridgmanite samples were measured after decompression

Table 2 Debye sound velocities at high pressure determined from NIS at room temperature and DFT at 0 K

Sample	Phase	P	V_D	σ_{VD}	ρ	Method	Remarks
		GPa	km/s	km/s	g/cm^3		
U1219	Majorite	0	5.19	0.27	3.70	NIS	Starting material
U1219	Majorite	33	5.45	0.62	4.29	NIS	
U1219	Bridgmanite	0	3.56	0.46	4.31	NIS	After decompression
U1219	Bridgmanite	33	5.51	0.59	4.78	NIS	
U1219	Bridgmanite	41	5.80	0.66	4.87	NIS	
U1219	Bridgmanite	49	6.27	0.88	4.97	NIS	
U1219	Bridgmanite	62	5.62	0.55	5.11	NIS	
U1219	Bridgmanite	74	6.79	1.15	5.23	NIS	
U1219	Bridgmanite	89	7.23	1.17	5.37	NIS	
SL16	Bridgmanite	0	3.53	0.05	4.25	NIS	Multianvil sample in air
SL18	Bridgmanite	0	3.21	0.12	4.24	NIS	After decompression
SL18	Bridgmanite	0	3.97	0.15	4.24	NIS	After decompression
SL18	Bridgmanite	17	4.45	0.31	4.50	NIS	
SL18	Bridgmanite	36	5.72	0.38	4.74	NIS	
SL18	Bridgmanite	50	6.84	1.62	4.90	NIS	
SL18	Bridgmanite	62	5.95	0.31	5.03	NIS	
SL18	Bridgmanite	65	6.65	1.16	5.06	NIS	
SL18	Bridgmanite	75	6.47	1.15	5.16	NIS	
S4850	Bridgmanite	0	3.81	0.12	4.43	NIS	After decompression
S4850	Bridgmanite	50	6.95	1.26	5.15	NIS	Before laser heating
S4850	Bridgmanite	50	6.35	1.39	5.15	NIS	After laser heating
S4850	Bridgmanite	53	5.54	0.47	5.19	NIS	
S4850	Bridgmanite	72	6.14	0.69	5.40	NIS	
S4883	Bridgmanite	17	5.03	0.72	4.52	NIS	
S4883	Bridgmanite	55	5.11	0.53	4.99	NIS	
S4883	Bridgmanite	63	5.39	0.88	5.07	NIS	
$MgSiO_3$	Bridgmanite	60	8.21	0.32	5.03	DFT	From Mg partial DOS
$FeSiO_3$	Bridgmanite	60	6.73	0.22	6.22	DFT	From Fe partial DOS

partial DOS for all the elements and calculated V_D using the same approach as for the experimental data. The equation of state parameters were taken from Caracas and Cohen (2005). We also determined seismic wave velocities from the full elastic tensor where elastic constants were obtained from the linear relation between stresses and strains. Uniaxial and/or pure shear strains on the order of ±1 % and ±2 % were applied, and then the positions of the atoms were relaxed. The elastic constants were determined from the residual stresses divided by the strains. Voigt-Reuss-Hill schemes were employed to average the elastic tensor, and seismic wave velocities were determined based on homogeneous aggregates.

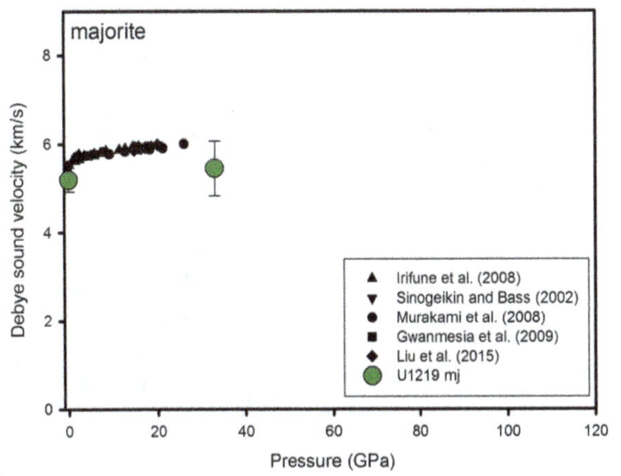

Fig. 7 Influence of pressure on V_D for majorite. Values from the present study are indicated by *large green circles*, while literature data are indicated by *black symbols* according to the legend

Results and discussion

We measured room temperature NIS spectra of one majorite composition and five bridgmanite compositions

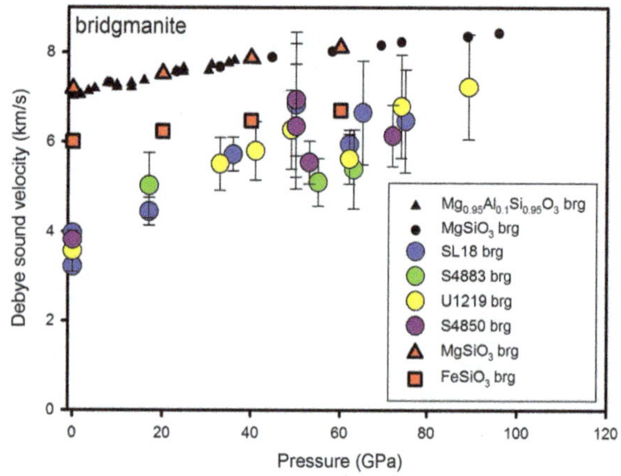

Fig. 8 Influence of pressure on V_D for bridgmanite. Experimental values from the present study are indicated by *large colored circles*, while Brillouin scattering data from the literature are indicated by *small black triangles* (Jackson et al. 2005) and *small black circles* (Murakami et al. 2007). Values from a previous computational determination of the elastic tensor from finite differences (Caracas and Cohen 2005) are indicated as *red triangles* for $MgSiO_3$ bridgmanite and *red squares* for antiferromagnetic $FeSiO_3$ bridgmanite

as a function of pressure. In the case of majorite, sample U1219 was first measured at 33 GPa and then laser-heated to produce bridgmanite and measured again without changing pressure. The NIS spectra show a clear change between majorite and bridgmanite (Fig. 1a) that leads to different pDOS (Fig. 1b). Subsequent compression of the U1219 bridgmanite sample produced systematic changes in the NIS spectra (Fig. 2a) that are reflected in the pDOS (Fig. 2b). A similar behavior on compression was observed in bridgmanite with lower iron content (SL18, Fig. 3) and Al-containing bridgmanite with lower (S4883) and higher (S4850) amounts of Fe^{3+} (Figs. 4 and 5, respectively). Finally, we collected data for the five different samples at ambient conditions for comparison (Fig. 6).

Debye sound velocities (V_D) were calculated from the reduced DOS in the limit as energy goes to zero by averaging $D(E)/E^2$ in the low-energy range as described in Sinmyo et al. (2014). V_D values for majorite (Table 2) are slightly lower than the results obtained for end-member $MgSiO_3$ majorite and solid solutions containing Fe and/or Al that were obtained using ultrasonics ($Mg_{0.59}Fe_{0.04}Ca_{0.18}Na_{0.03}Al_{0.23}Cr_{0.01}Si_{0.90}O_3$: Irifune et al. 2008; $Mg_{0.875}Al_{0.25}Si_{0.875}O_3$ and $Mg_{0.85}Al_{0.3}Si_{0.85}O_3$: Gwanmesia et al. 2009; $Mg_{0.95}Al_{0.1}Si_{0.95}O_3$: Liu et al. 2015) and Brillouin scattering ($MgSiO_3$ and $Mg_{0.875}Al_{0.25}Si_{0.875}O_3$: Sinogeikin and Bass 2002, $Mg_{0.79}Fe_{0.08}Al_{0.30}Si_{0.84}O_3$: Murakami et al. 2008) (Fig. 7). These lower values are consistent with the higher iron content of our majorite ($Mg_{0.82}Fe_{0.18}SiO_3$). In contrast, the V_D values for bridgmanite (Table 2) are unrealistically low compared to the results from Brillouin scattering (Fig. 8). We can rule out the influence of local clustering as a significant factor that lowers velocities

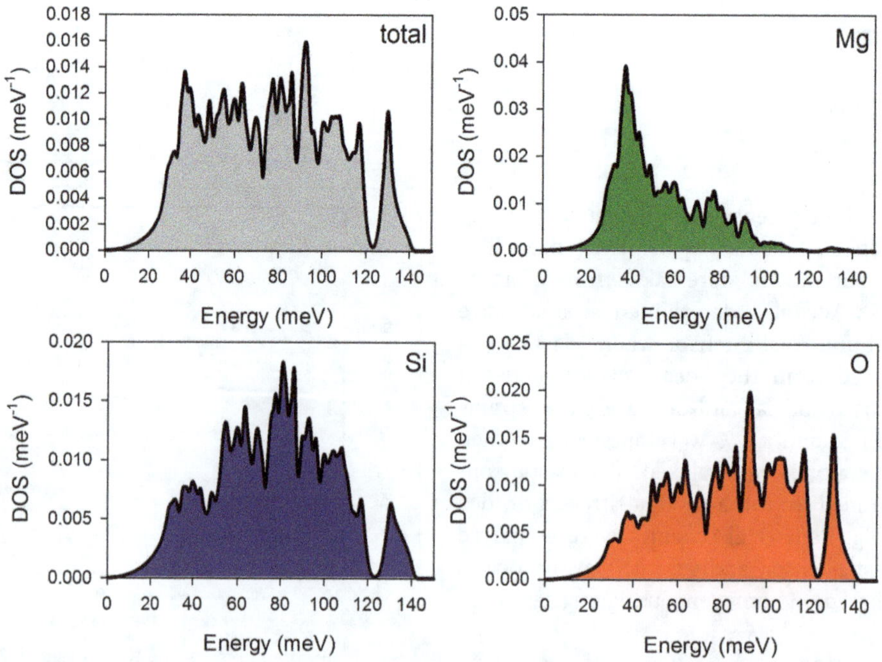

Fig. 9 DOS for $MgSiO_3$ bridgmanite at 60 GPa calculated using DFT. The total DOS is shown in the *upper left panel*, while the pDOS are shown in the *other panels* for the indicated element

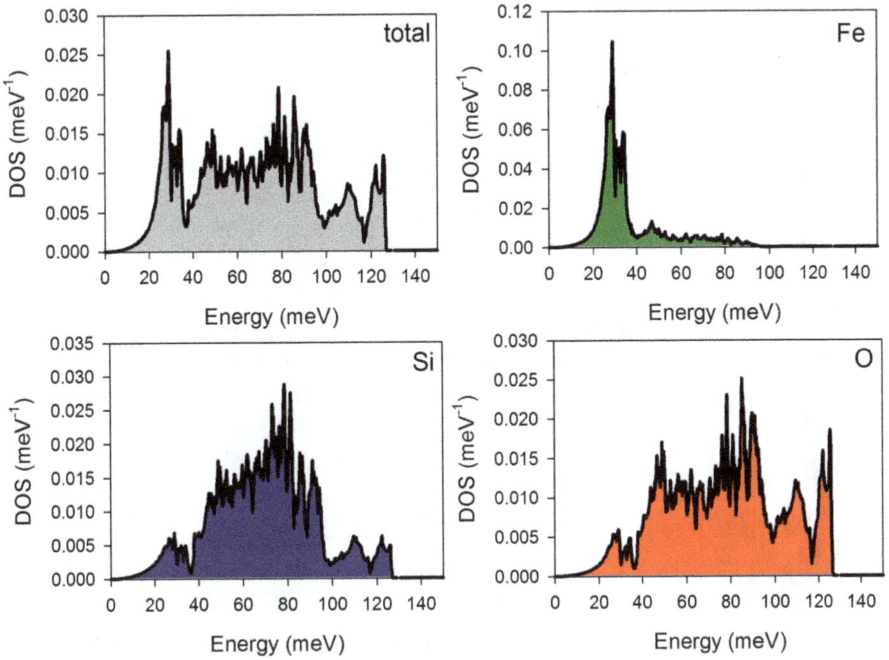

Fig. 10 DOS for antiferromagnetic $FeSiO_3$ bridgmanite at 60 GPa calculated using DFT. The total DOS is shown in the *upper left panel*, while the pDOS are shown in the *other panels* for the indicated element

since the effect is too small to account for the large discrepancy (e.g., Sinmyo et al. 2014).

For a better understanding of the DOS for end-member bridgmanite, we calculated the total DOS for $MgSiO_3$ bridgmanite and antiferromagnetic $FeSiO_3$ bridgmanite as well as the pDOS for each element at 60 GPa (Figs. 9 and 10). For consistency with previous finite difference computations (Caracas and Cohen 2005), we calculated the full elastic tensor to confirm that we obtained the same elastic velocities within error. We then applied the same approach used to obtain V_D from the experimental DOS obtained from NIS data to the theoretical DOS shown in Figs. 9 and 10. We found that the V_D values calculated from each pDOS as well as the total DOS for each phase were identical within error, i.e., the same result is obtained irrespective of which element pDOS is used. In addition, we found that the V_D values calculated using this approach (Table 2) are consistent with values that we obtained from the full elastic tensor (8.23 and 6.78 km/s for $MgSiO_3$ and $FeSiO_3$, respectively).

A Birch plot for bridgmanite data from both calculations and experiments highlights the anomalous behavior of sound velocities determined using NIS (Fig. 11). To obtain further insight into the origin of the low velocities of bridgmanite, we compare the pDOS for experiments and calculations. The breadth of vibrational states is smaller for bridgmanite calculated at 0 K compared to the experimental data at room temperature, but it is similar for the experimental data for both bridgmanite and majorite at room temperature (Fig. 12, upper). In the reduced pDOS, the enhanced intensity at low energies in the experimental bridgmanite data leads to a significantly higher intercept and hence lower velocity compared to both calculated bridgmanite and experimental majorite (Fig. 12, lower). While particle size (e.g., Marquardt et al. 2011) or anisotropy (Chumakov et al. 2009) could be a factor, we note that this effect is present in all the bridgmanite compositions that we measured using NIS, irrespective of sample synthesis method or history (i.e., whether they were

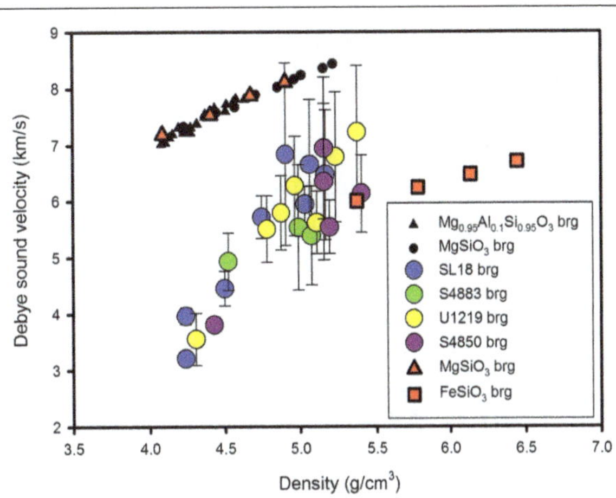

Fig. 11 Birch plot of V_D versus density for bridgmanite. *Symbols* are the same as for Fig. 8

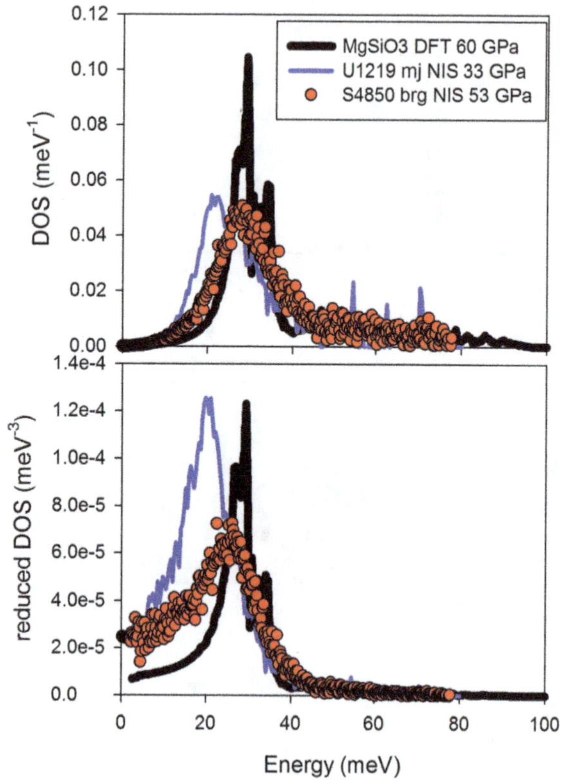

Fig. 12 Comparison of pDOS (*upper*) and reduced pDOS (*lower*) determined by experiment and calculation. Experimental data are from the room temperature NIS spectra of S4850 bridgmanite at 53 GPa (*red circles*) and U1219 majorite at 33 GPa (*blue line*) while the *black line* indicates DFT calculations of antiferromagnetic $FeSiO_3$ bridgmanite at 60 GPa and 0 K

synthesized in the multianvil press or by laser heating in the DAC). While it is clear from our results that sound velocities for bridgmanite determined using NIS cannot be used for comparison with seismic data, there is clearly scope for further work to identify the contributions to the vibrational DOS derived from NIS data in order to more fully understand the lattice vibrational properties of bridgmanite.

Conclusions

Our work has presented an extensive dataset for bridgmanite with different compositions based on NIS measurements in a DAC as well as first-principles calculations:

(1) NIS measurements of majorite give sound velocities that are consistent with literature data from other methods, while NIS measurements of bridgmanite give sound velocities that are significantly lower than literature data.

(2) Sound velocities of bridgmanite determined from NIS data are consistently low, irrespective of composition or sample history.

(3) Treatment of the DOS calculated using DFT for $MgSiO_3$ and $FeSiO_3$ bridgmanite using the same approach as for the NIS data gives sound velocities that are consistent with each other for each phase irrespective of which partial DOS is used.

(4) Treatment of the DOS calculated using DFT for $MgSiO_3$ and $FeSiO_3$ bridgmanite using the same approach as for the NIS data gives sound velocities that are consistent with the values obtained from the calculation of the full elastic tensor.

(5) Comparison of the calculated and experimental DOS shows enhanced intensity at low energies in the latter that leads to the lower sound velocities.

Abbreviations

DAC: diamond anvil cell; DFT: density functional theory; DOS: density of states; NFS: nuclear forward scattering; NIS: nuclear inelastic scattering; pDOS: partial density of states.

Competing interests

The authors declare that they have no competing interests.

Authors' contributions

CM, RC, and LD proposed the topic and conceived and designed the study. The experimental data were collected by CM, KG, VP, AK, RS, CP, IK, AC, and LD. CM, KG, CP, and AC analyzed the experimental data, and RC carried out the ab initio calculations. All authors provided input to the interpretation of the data and the writing of the manuscript. All authors read and approved the final manuscript.

Acknowledgements

We acknowledge the European Synchrotron Radiation Facility for the provision of synchrotron radiation facilities (ID18), and we would like to thank R Rüffer and J-P Celse for additional assistance. The project was supported by funds from the German Science Foundation (DFG) in their normal funding program and Priority Program SPP1236, the PROCOPE exchange program, and the German Federal Ministry for Education and Research (BMBF). RS was supported by a Research Fellowship for Postdoctoral Researchers awarded by the Alexander von Humboldt Foundation.

Author details

Bayerisches Geoinstitut, Universität Bayreuth, Bayreuth 95440, Germany. [2]Laboratoire de Géologie de Lyon UMR CNRS 5276, Ecole Normale Supérieure de Lyon, Lyon Cedex 07 69364, France. [3]Deutsches Elektronen-Synchrotron, Hamburg 22607, Germany. [4]European Synchrotron Radiation Facility, Grenoble Cedex 9 38043, France. [5]National Research Center "Kurchatov Institute", Moscow 123182, Russia. [6]ELSI, Tokyo Institute of Technology, Tokyo 152-8550, Japan. [7]Institut für Geologie und Mineralogie, Universität zu Köln, Köln 50969, Germany. [8]Institut für Mineralogie, Universität Münster, Münster 48149, Germany.

References

Birch F (1952) Elasticity and constitution of the Earth's interior. J Geophys Res 57: 227–286

Boffa Ballaran T, Kurnosov A, Glazyrin K, Frost DJ, Merlini M, Hanfland M, Caracas R (2012) Effect of chemistry on the compressibility of silicate perovskite in the lower mantle. Earth Planet Sci Lett 333–334:181–190

Caracas R, Cohen RE (2005) Effect of chemistry on the stability and elasticity of the perovskite and post-perovskite phases in the $MgSiO_3$-$FeSiO_3$-Al_2O_3 system and implications for the lowermost mantle. Geophys Res Lett 32: L16310. doi:10.1029/2005GL023164

Chen B, Jackson JM, Sturhahn W, Zhang D, Zhao J, Wicks JK, Murphy CA (2012) Spin crossover equation of state and sound velocities of $(Mg_{0.65}Fe_{0.35})O$ ferropericlase to 140 GPa. J Geophys Res 117(B08208):2012. doi:10.1029/2012JB009162

Chumakov AI, Bosak A, Rüffer R (2009) Contribution of acoustic modes to the density of vibrational states measured by inelastic scattering techniques. Phys Rev B 90(094303):1–8

Crowhurst JC, Brown JM, Goncharov AF, Jacobsen SD (2008) Elasticity of (Mg, Fe)O through the spin transition of iron in the lower mantle. Science 319: 451–453

Dubrovinsky L, Glazyrin K, McCammon C, Narygina O, Greenberg E, Übelhack S, Chumakov AI, Pascarelli S, Prakapenka V, Bock J, Dubrovinskaia N (2009) Portable laser-heating system for diamond anvil cells. J Synchrotron Radiat 16:737–741

Fiquet G, Badro J, Guyot F, Bellin C, Krisch M, Antonangeli D, Requardt H, Mermet A, Farber D, Aracne-Ruddle C, Zhang J (2004) Application of inelastic X-ray scattering to the measurements of acoustic wave velocities in geophysical materials at very high pressure. Phys Earth Planet Int 143–144:5–18

Glazyrin K, Pourovskii LV, Dubrovinsky L, Narygina O, McCammon C, Hewener B, Schünemann V, Wolny J, Muffler K, Chumakov AI, Crichton W, Hanfland M, Prakapenka V, Tasnádi F, Ekholm M, Aichhorn M, Vildosola V, Ruban AV, Katsnelson MI, Abrikosov IA (2013) Importance of correlation effects in hcp iron revealed by a pressure-induced electronic topological transition. Phys Rev Lett 110:117206

Glazyrin K, Boffa Ballaran T, Frost DJ, McCammon C, Kantor A, Merlini M, Hanfland M, Dubrovinsky L (2014) Effect of iron oxidation state on the bulk sound velocity of the Earth's lower mantle. Earth Planet Sci Lett 393:182–186

Gonze X, Rignanese G-M, Caracas R (2005) First-principles studies of the lattice dynamics of crystals, and related properties. Z Krist 220:458–472

Gwanmesia GD, Wang L, Triplett R, Liebermann RC (2009) Pressure and temperature dependence of the elasticity of pyrope–majorite [Py60Mj40 and Py50Mj50] garnets solid solution measured by ultrasonic interferometry technique. Phys Earth Planet In 174:105–112

Irifune T, Higo Y, Inoue T, Kono Y, Ohfuji H, Funakoshi K (2008) Sound velocities of majorite garnet and the composition of the mantle transition region. Nature 451:814–817

Jackson JM, Zhang J, Shu J, Sinogeikin SV, Bass JD (2005) High-pressure sound velocities and elasticity of aluminous $MgSiO_3$ perovskite to 45 GPa: implications for lateral heterogeneity in Earth's lower mantle. Geophys Res Lett 32, doi:10.1029/2005GL023522, 022005

Jackson JM, Hamecher EA, Sturhahn W (2009) Nuclear resonant X-ray spectroscopy of (Mg, Fe)SiO3 orthoenstatites. Eur J Mineral 21:551–560

Kavner A, Sinogeikin SV, Jeanloz R, Bass JD (2000) Equation of state and strength of natural majorite. J Geophys Res 105(B3):5963–5971

Keppler H, Frost DJ (2005) Introduction to minerals under extreme conditions. In: Miletich R (ed) Mineral behaviour at extreme conditions, vol 7, EMU notes in mineralogy. Eötvös University Press, Budapest, pp 1–30

Kupenko I, Dubrovinsky L, Dubrovinkskaya N, McCammon C, Glazyrin K, Bykova E, Boffa Ballaran T, Sinmyo R, Chumakov AI, Potapkin V, Kantor A, Rüffer R, Hanfland M, Crichton W, Merlini M (2012) Portable double-sided laser-heating system for energy-domain Mössbauer spectroscopy and single crystal X-ray diffraction experiments at synchrotron facilities with diamond anvil cells. Rev Sci Instrum 83:124501

Lauterbach S (2000) Der Oxidationsgrad des Eisens im unteren Mantel: Eine Studie des Fe^{3+}-Gehaltes des Silikat-Perowskits in Abhängigkeit seines Al-Gehaltes mit Mössbauer-Spektroskopie und Elektronen-Energie-Verlust-Spektroskopie (EELS). University of Bayreuth, Dissertation

Lauterbach S, McCammon CA, van Aken P, Langenhorst F, Seifert F (2000) Mössbauer and ELNES spectroscopy of (Mg, Fe)(Si, Al)O3 perovskite: a highly oxidised component of the lower mantle. Contrib Mineral Petrol 138:17–26

Lin J-F, Jacobsen SD, Sturhahn W, Jackson JM, Zhao J, Yoo C-S (2006) Sound velocities of ferropericlase in the Earth's lower mantle. Geophys Res Lett 33, doi:10.1029/2006GL028099, 022006

Liu Z, Irifune T, Gréaux S, Arimoto T, Shinmei T, Higo Y (2015) Elastic wave velocity of polycrystalline Mj80Py20 garnet to 21 GPa and 2,000 K. Phys Chem Miner 42:213–222

Mao H-K, Xu J-A, Bell PM (1986) Calibration of the ruby pressure gauge to 800 kbar under quasi-hydrostatic conditions. J Geophys Res 91:4673–4676

Mao WL, Mao H-K, Sturhahn W, Zhao J, Prakapenka VB, Meng Y, Shu J, Fei Y, Hemley RJ (2006) Iron-rich post-perovskite and the origin of ultralow-velocity zones. Science 312:564–565

Marquardt H, Gleason A, Marquardt K, Speziale S, Miyagi L, Neusser G, Wenk H-R, Jeanloz R (2011) Elastic properties of MgO nanocrystals and grain boundaries at high pressures by Brillouin scattering. Phys Rev B 84(064131):1–9

McCammon CA, Ross NL (2003) Crystal chemistry of ferric iron in (Mg, Fe)(Si, Al)O3 majorite with implications for the transition zone. Phys Chem Minerals 30:206–216

McCammon C, Kantor I, Narygina O, Rouquette J, Ponkratz U, Sergueev I, Mezouar M, Prakapenka V, Dubrovinsky L (2008) Stable intermediate-spin ferrous iron in lower mantle perovskite. Nat Geosci 1:684–687

McCammon C, Dubrovinsky L, Narygina O, Kantor I, Wu X, Glazyrin K, Sergueev I, Chumakov A (2010) Low-spin Fe^{2+} in silicate perovskite and a possible layer at the base of the lower mantle. Phys Earth Planet Int 180:215–221

McCammon C, Glazyrin K, Kantor A, Kantor I, Kupenko I, Narygina O, Potapkin V, Prescher C, Sinmyo R, Chumakov A, Rüffer R, Sergueev I, Smirnov G, Dubrovinsky L (2013) Iron spin state in silicate perovskite at conditions of the Earth's deep interior. High Pressure Res 33:663–672

Monkhorst HJ, Pack JD (1976) Special points for Brillouin-zone integrations. Phys Rev B 13:5188–5192

Murakami M, Sinogeikin SV, Hellwig H, Bass JD, Li J (2007) Sound velocity of $MgSiO_3$ perovskite to Mbar pressure. Earth Planet Sci Lett 256:47–54

Murakami M, Sinogeikin SV, Litasov K, Ohtani E, Bass JD (2008) Single-crystal elasticity of iron-bearing majorite to 26 GPa: implications for seismic velocity structure of the mantle transition zone. Earth Planet Sci Lett 274:339–345

Potapkin V, McCammon C, Glazyrin K, Kantor A, Kupenko I, Prescher C, Sinmyo R, Smirnov GV, Chumakov AI, Rüffer R, Dubrovinsky L (2013) Effect of iron oxidation state on the electrical conductivity of the Earth's lower mantle. Nat Commun 4(1427):1–6

Rüffer R, Chumakov AI (1996) Nuclear resonance beamline at ESRF. Hyperfine Interact 97(98):589–604

Shukla G, Wu Z, Hsu H, Floris A, Cococcioni M, Wentzcovitch RM (2015) Thermoelasticity of Fe^{2+}-bearing bridgmanite. Geophys Res Lett 42: 1741–1749

Sinmyo R, Glazyrin K, McCammon C, Kupenko I, Kantor A, Potapkin V, Chumakov AI, Rüffer R, Dubrovinsky L (2014) The influence of solid solution on elastic wave velocity determination in (Mg, Fe)O using nuclear inelastic scattering. Phys Earth Planet Int 229:16–23

Sinogeikin SV, Bass JD (2002) Elasticity of majorite and a majorite-pyrope solid solution to high pressure: implications for the transition zone. Geophys Res Lett 29:10.1029/2001GL013937

Stackhouse S (2008) The spin deep within. Nat Geosci 1:648–650

Sturhahn W, Jackson JM (2007) Geophysical applications of nuclear resonant spectroscopy. In: Ohtani E (ed) Advances in high-pressure mineralogy, GSA special paper 421. Geological Society of America, Boulder, pp 157–174

Townsend JP, Tsuchiya J, Bina CR, Jacobsen SD (2015) First-principles investigation of hydrous post-perovskite. Phys Earth Planet In 244:42–48

Wang X, Tsuchiya T, Hase A (2015) Computational support for a pyrolitic lower mantle containing ferric iron. Nat Geosci 8:556–559

Wicks JK, Jackson JM, Sturhahn W (2010) Very low sound velocities in iron-rich (Mg,Fe)O: implications for the core-mantle boundary region. Geophys Res Lett 37:L15304. doi:10.1029/2010GL043689

4

Weak hydrothermal carbonation of the Ongeluk volcanics: evidence for low CO_2 concentrations in seawater and atmosphere during the Paleoproterozoic global glaciation

Takazo Shibuya[1,2,3*], Tsuyoshi Komiya[4], Ken Takai[1,2,3], Shigenori Maruyama[5,7] and Michael J. Russell[6]

Abstract

It was previously revealed that the total CO_2 concentration in seawater decreased during the Late Archean. In this paper, to assess the secular change of total CO_2 concentration in seawater, we focused on the Paleoproterozoic era when the Earth experienced its first recorded global glaciation. The 2.4 Ga Ongeluk Formation outcrops in the Kaapvaal Craton, South Africa. The formation consists mainly of submarine volcanic rocks that have erupted during the global glaciation. The undeformed lavas are mostly carbonate-free but contain rare disseminated calcites. The carbon isotope ratio of the disseminated calcite ($\delta^{13}C_{cc}$ vs. VPDB) ranges from − 31.9 to − 13.2 ‰. The relatively low $\delta^{13}C_{cc}$ values clearly indicate that the carbonation was partially contributed by ^{13}C-depleted CO_2 derived from decomposition of organic matter beneath the seafloor. The absence of $\delta^{13}C_{cc}$ higher than − 13.2‰ is consistent with the exceptionally ^{13}C-depleted CO_2 in the Ongeluk seawater during glaciation. The results suggest that carbonation occurred during subseafloor hydrothermal circulation just after the eruption of the lavas. Previously, it was reported that the carbonate content in the uppermost subseafloor crust decreased from 3.2 to 2.6 Ga, indicating a decrease in total CO_2 concentration in seawater during that time. However, the average CO_2 (as carbonate) content in the Ongeluk lavas (< 0.001 wt%) is much lower than those of 2.6 Ga representatives and even of modern equivalents. This finding suggests that the total CO_2 concentration in seawater further decreased during the period between 2.6 and 2.4 Ga. Thus, the very low content of carbonate in the Ongeluk lavas is probable evidence for the extremely low CO_2 concentration in seawater during the global glaciation. Considering that the carbonate content of the subseafloor crusts also shows a good correlation with independently estimated atmospheric pCO_2 levels through the Earth history, it seem highly likely that the low carbonate content in the Ongeluk lavas reflects the low atmospheric pCO_2 at that time. We conclude that the continuous decrease in CO_2 concentration of seawater/atm. from 3.2 Ga was one of the contributing factors to the Paleoproterozoic global glaciation.

Keywords: Paleoproterozoic global glaciation, Ongeluk volcanics, Carbonation, Carbon and oxygen isotopes, Seawater/ atmosphere CO_2 level

* Correspondence: takazos@jamstec.go.jp
[1]Department of Subsurface Geobiological Analysis and Research (D-SUGAR), Japan Agency for Marine-Earth Science and Technology (JAMSTEC), 2-15 Natsushima-cho, Yokosuka 237-0061, Japan
[2]Research and Development Center for Submarine Resources, Japan Agency for Marine-Earth Science and Technology (JAMSTEC), 2-15 Natsushima-cho, Yokosuka 237-0061, Japan
Full list of author information is available at the end of the article

Introduction

Because the Sun was approximately 30% less luminous at 4.6 Ga (Gough 1981), the mean temperature of the Earth's surface would have been below the freezing point of seawater if the albedo and atmospheric composition of the early Earth had been the same as at the present (Sagan and Mullen 1972). However, the occurrence of pillow lavas and sedimentary rocks clearly indicate the presence of liquid water on the early Earth (Sagan and Mullen 1972). This "faint young Sun paradox" has been explained by positing a greenhouse effect to counteract the lower solar luminosity (e.g., Kasting 1987, 1993; Sleep and Zahnle 2001). Yet, at present, this paradox has not yet been entirely resolved because the geologic record suggests that CO_2 levels in the Archean atmosphere potentially exerted enough of a greenhouse effect to avoid Earth freezing for contemporary albedos, whereas the presence of other greenhouse gases such as CH_4 and OCS was also reported (e.g., Pavlov et al. 2003; Ueno et al. 2009). Furthermore, the changes in albedos of the early Earth also had the potential to compensate the faint young Sun (Rosing et al. 2010). However, comprehensive geological studies provide evidence suggesting that atmospheric CO_2 levels in the early Earth were substantially higher than at present: e.g., > 100 PAL at 3.8–1.8 Ga (Ohmoto et al. 2004), > 100 PAL at 3.5–3.2 Ga (Lowe and Tice 2004), at > 7 PAL bar (25 °C) at 3.2 Ga (Hessler et al. 2004), 23×/÷3 PAL at 2.2 Ga (Sheldon 2006), 30–190 PAL at 2.15 Ga, and various other estimates for 2.77–1.85 Ga (Kanzaki and Murakami 2015).

The putative CO_2-rich atmosphere in the early Earth would also have kept oceans acidic and CO_2- and metal-rich (Konhauser et al. 2007). Such CO_2-rich seawater provided seafloor hydrothermal vent environments that would sustain the emergence and early evolution of life (Takai et al. 2006; Russell et al. 2010, 2014; Shibuya et al. 2015, 2016). Furthermore, chemical exchange reactions would have had particular characteristics as the then CO_2-rich seawater interfaced oceanic lithosphere (e.g., intense CO_2 fixation into the crust) (Nakamura and Kato 2004; Rouchon and Orberger 2008) resulting in the generation of novel high-temperature hydrothermal fluids (e.g.., alkaline, silica-rich, metal-poor fluid, and Mg-rich fluid) (Shibuya et al. 2010; Ueda et al. 2016). These processes would have played a significant role in controlling the chemistry of early oceans. Therefore, the secular change of CO_2 levels in the atmosphere and seawater is important for understanding the evolution of surface environments and life on Earth.

In this context, the degree of carbonation of greenstones has been considered to qualitatively reflect CO_2 concentrations in ancient seawater (Kitajima et al. 2001; Nakamura and Kato 2004; Hofmann and Harris 2008). The 3.5 and 3.2 Ga greenstones in extensional geological settings (mid-ocean ridge or rift basin) underwent strong carbonation caused by intense interaction with ambient CO_2-rich seawater (Nakamura and Kato 2004; Shibuya et al. 2007a, 2012). Furthermore, the lower carbonate content in the 2.6 Ga greenstones compared to those in the 3.5 and 3.2 Ga greenstones indicated that the CO_2 concentration in seawater decreased during Late Archean period—possibly caused by the formation and breakdown of the supercontinent at around 2.7 Ga (Shibuya et al. 2013a). However, there are no data for the carbonation of subseafloor crust after 2.6 Ga, although the data from the DSDP/ODP drill cores have been obtained from the Phanerozoic crusts (Alt and Teagle 1999; Gillis and Coogan 2011).

In this paper, we focus on the Paleoproterozoic era when global climate drastically cooled to produce an ice age, though it reverted thereafter (Evans et al. 1997; Kirschvink et al. 2000; Harada et al. 2015). In this era, the atmospheric CO_2 level was temporarily insufficient to keep the Earth's surface above the freezing point of water. Previously, a number of triggers for the Paleoproterozoic global glaciation have been proposed: (1) the collapse of a methane greenhouse caused by the global rise of oxygen over the period ~2.5–2.2 Ga (Kopp et al. 2005), (2) the removal of atmospheric CO_2 via increased silicate weathering of the amalgamated supercontinent at ~2.5 Ga (Young 2013) and/or of the newly created voluminous continental flood basalts at 2.45 Ga (Melezhik 2006), and (3) the shutdown of volcanic CO_2 emissions resulting from a global magmatic lull that lasted from 2.45 to 2.2 Ga (Condie et al. 2009). All these models are premised upon low atmospheric CO_2 levels. To test this supposed prerequisite and to reveal the secular change of CO_2 concentration in seawater after 2.6 Ga, we investigated the carbonation of the Paleoproterozoic submarine volcanics in the Ongeluk Formation, Transvaal Supergroup, South Africa. Based on the lines of geological evidence, the Ongeluk volcanics are considered to have erupted during the global glaciation (Kirschvink et al. 2000; Kopp et al. 2005). Therefore, it is expected that the degree of carbonation of the Ongeluk volcanics qualitatively reflects the CO_2 concentration in seawater and atmosphere during that glaciation.

Methods

The Ongeluk formation

The Griqualand West Sequence in South Africa mainly comprises Late Archean to Paleoproterozoic sedimentary rocks and intercalating volcanic rocks deposited on the Kaapvaal Craton (Beukes and Smit 1987; Cornell et al. 1996). In the Griqualand West Sequence, the Transvaal Supergroup is subdivided into three groups by major unconformities; Ghaap Group, Postmasburg Group and Olifantshoek Group in stratigraphicaclly ascending order. The Paleoproterozoic Ongeluk Formation belongs to the Postmasburg Group that mainly consists of diamictite, volcanic rocks, terrigenous clastic rocks, banded iron formations, manganese formations, and shallow marine dolostones (Beukes and Smit 1987). The Postmasburg Group unconformably overlies the Ghaap Group, including Koegas and Asbestos Hill Subgroups composed of iron formation, shale, and quartzite (Fig. 1).

The Ongeluk Formation consists ~ 500–1000 m of basaltic-andesitic lavas comprising pillow lavas, massive sheet flows, and hyaloclastites, intercalated with jaspilite beds (Grobler and Botha 1976) (Fig. 2), indicating the subaqueous eruption of the Ongeluk lavas. The formation conformably overlies the glaciogenic deposits (Makganyene diamictite). The diamictite/lava contacts appear conformable without paleosols in borehole cores (de Villiers and Visser 1977; Evans et al. 1997). Additionally, volcanic shards and fragments are abundant in the upper part of the Makganyene Formation (de Villiers and Visser 1977; Evans et al. 1997), and the diamictite and the lava are interdigitated (Kirschvink et al. 2000; Polteau et al. 2006). These lines of geological evidence indicate that the Ongeluk volcanism and Makganyene glaciation overlapped in time. Therefore, the depositional paleolatitude of the Ongeluk Formation (11 ± 6°) provides strong evidence for low-latitude glaciation in a maritime setting (Evans et al. 1997; Kirschvink et al. 2000; Hoffman 2013; Gumsley et al. 2017). Furthermore, the Hotazel Formation, consisting predominantly of manganese formation/banded iron formation which includes dropstones at its base (Kirschvink et al. 2000; Polteau et al. 2006), conformably overlies the Ongeluk Formation. The contact between these two formations is characterized by parallel bedding planes in both, ferruginous alteration at the top of the Ongeluk Formation, and the absence of a basal conglomerate. These relationships indicate that the Ongeluk lavas erupted during the global glacial period (Kirschvink et al. 2000; Kopp et al. 2005) (Fig. 2). Metamorphism of the Ongeluk Formation does not exceed prehnite-pumpellyite facies, so that the effects of ancient passive subseafloor hydrothermal alteration are still plain to see (Gutzmer et al. 2001). The age of the Ongeluk Formation had long been under debate because two possible depositional ages (2.2 and 2.4 Ga) were constrained by other formations stratigraphically close to the Ongeluk Formation (Bau et al. 1999; Cornell et al. 1996; Dorland 2004). However, a recent geochronogical work on the Ongeluk Formation itself strongly suggested that the exact age of the volcanism of the Ongeluk lavas was 2426 ± 3 Ma (Gumsley et al. 2017). Thus, 2.4 Ga is tentatively used here as the age of the Ongeluk Formation. The Ongeluk lavas are considered to have erupted in a shallow marine environment during the development of

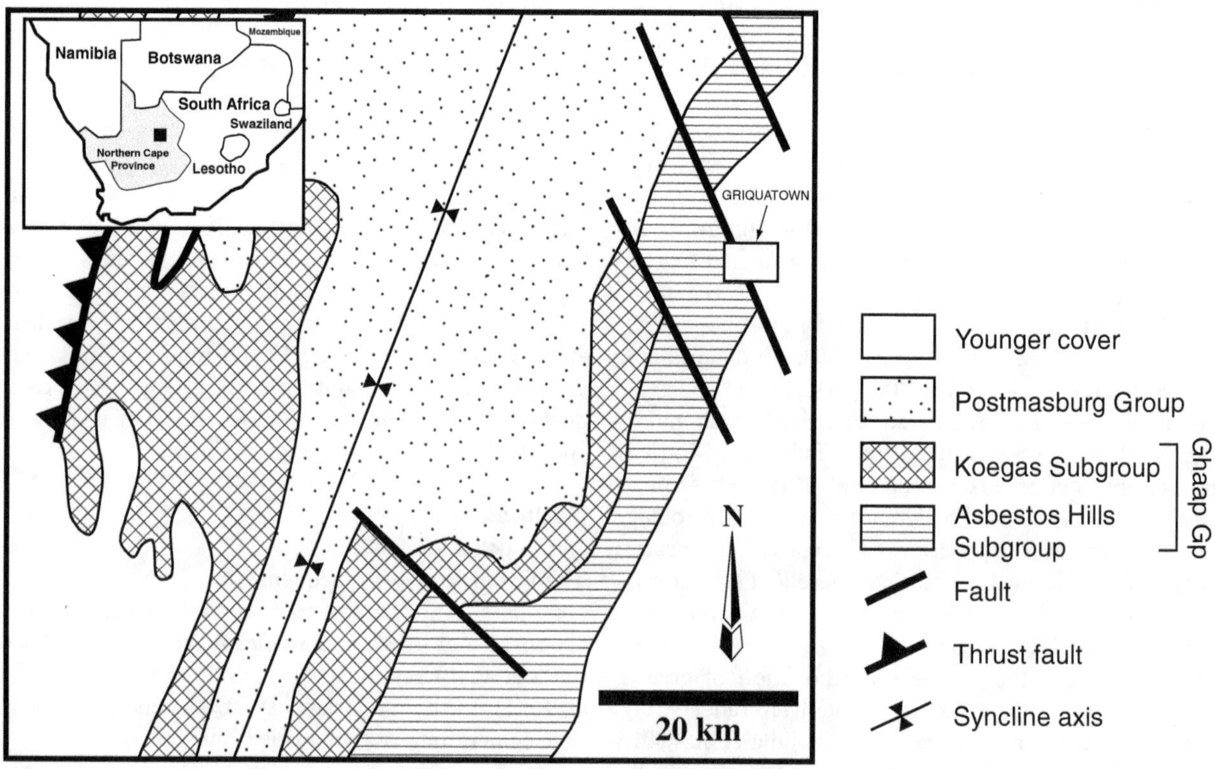

Fig. 1 Simplified geological map of the Transvaal Supergroup, Northern Cape Province, South Africa (modified after Gutzmer et al. 2003)

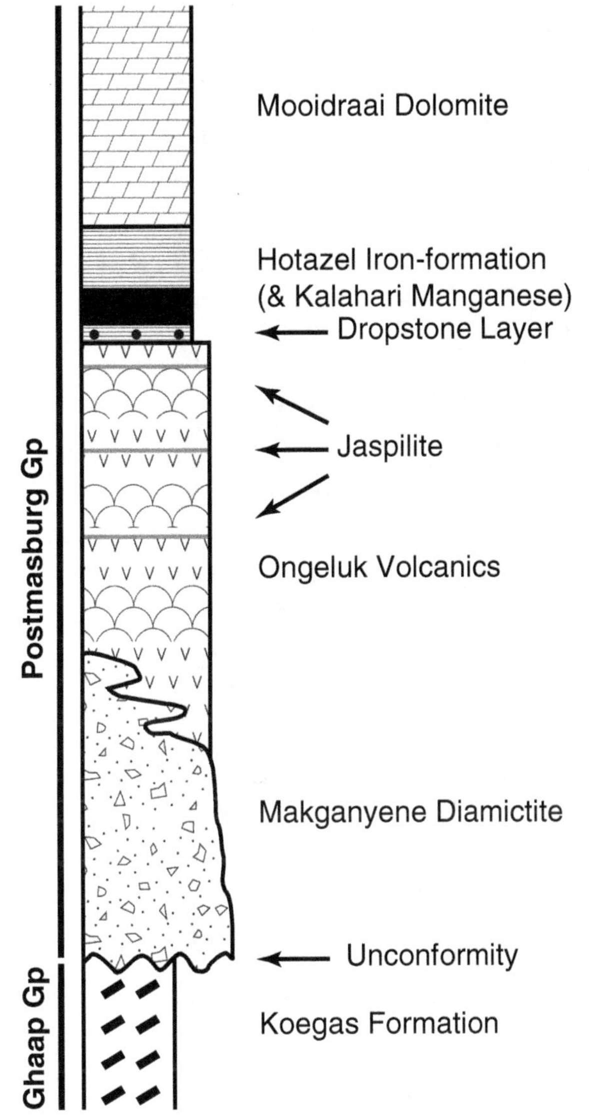

Fig. 2 Stratigraphy of the Transvaal Supergroup in the Griqualand West region, South Africa. The relationship among the Ongeluk Formation, the Makganyene diamictite, and the overlying Hotazel Formation are shown (after Kirschvink et al. 2000). The Ongeluk Formation (~ 500–1000 m thick) conformably overlies the Makganyene Diamictite (~ 3–70 m thick) and underlies a dropstone layer at the bottom of the Hotazel Formation (~ 100–250 m thick) (Hoffman 2013)

an extensional rifting basin on the Kaapvaal Craton (Cornell et al. 1996).

Lavas and hydrothermal jaspilites in the Ongeluk formation

The Ongeluk basaltic andesites are mostly undeformed, and igneous textures, such as intergranular, intersertal, and subophitic textures, are well preserved (Cornell et al. 1996; Gutzmer et al. 2001, 2003) (see also Fig. 3a). Igneous plagioclase and glass are mainly replaced by chlorite, albite, quartz, and minor pumpellyite and calcite, but igneous clinopyroxene survives. Although low-temperature clay minerals, often observed for modern altered seafloor rocks, are minor in the Ongeluk lavas, such textures are similar to those of modern hydrothermally altered seafloor volcanic rocks (e.g., Alt et al. 2010). The degree of carbonation of the Ongeluk basaltic andesites is weak; 17 of the 115 samples were identified to contain calcite under a microscope. These samples mainly contain disseminated calcites replacing igneous phases, but a few samples have vein-filling calcites (Fig. 3b, c).

The Ongeluk pillow lavas still retain open drainage cavities and interpillow voids. These are filled with quartz and/or jaspilite, which was likely precipitated penecontemporaneously with the interbedded jaspilite (Grobler and Botha 1976; Gutzmer et al. 2001, 2003). The precipitates in the drainage cavities have smooth, oval tubular shapes and are oriented parallel to the pillow tubes themselves or to the direction of lava flow. The cavities are isolated from one another and never cross-cut pillow rims nor do the precipitates in the voids ever cut the chilled margins of these rims. In the absence of such precipitates, the open spaces in volcanic rocks are obliterated by compaction during burial—strong evidence that these quartz precipitates formed relatively soon after eruption of the host volcanic rocks (Gutzmer et al. 2003). The jaspilite precipitates are composed mainly of microquartz and hematite and frequently form banded textures of red and white layers. The red layer consists of hematite and minor amounts of microquartz, magnetite, and pyrite, whereas the white layer comprises microquartz and larger subhedral andradite with subordinate hematite and disseminated calcite (Fig. 3d, e), as previously reported (Gutzmer et al. 2001). As observed for the lavas, minor jaspilite samples exhibit deformation textures such as brecciation and/or calcite veining (Fig. 3f). The formation temperature of these hydrothermal precipitates was estimated to be 64–160 °C from the homogenization temperature of fluid inclusions in the quartz (Gutzmer et al. 2001, 2003; Saito et al. 2016).

Analytical methods

Powdered samples were prepared with a tungsten mill from slab samples from which the weathered surface had been removed. In sealed vessels, CO_2 was extracted from the powder samples by using anhydrous phosphoric acid. The stable carbon and oxygen isotope ratios were analyzed using an isotope ratio mass spectrometer equipped with an automatic sampler (DELTA V plus and GasBench II at SI Science Co., Ltd., Japan) (Révész and Landwehr 2002). The isotopic ratios are presented in conventional δ notation. The carbon and oxygen isotope ratios were calibrated relative to the Vienna Pee Dee

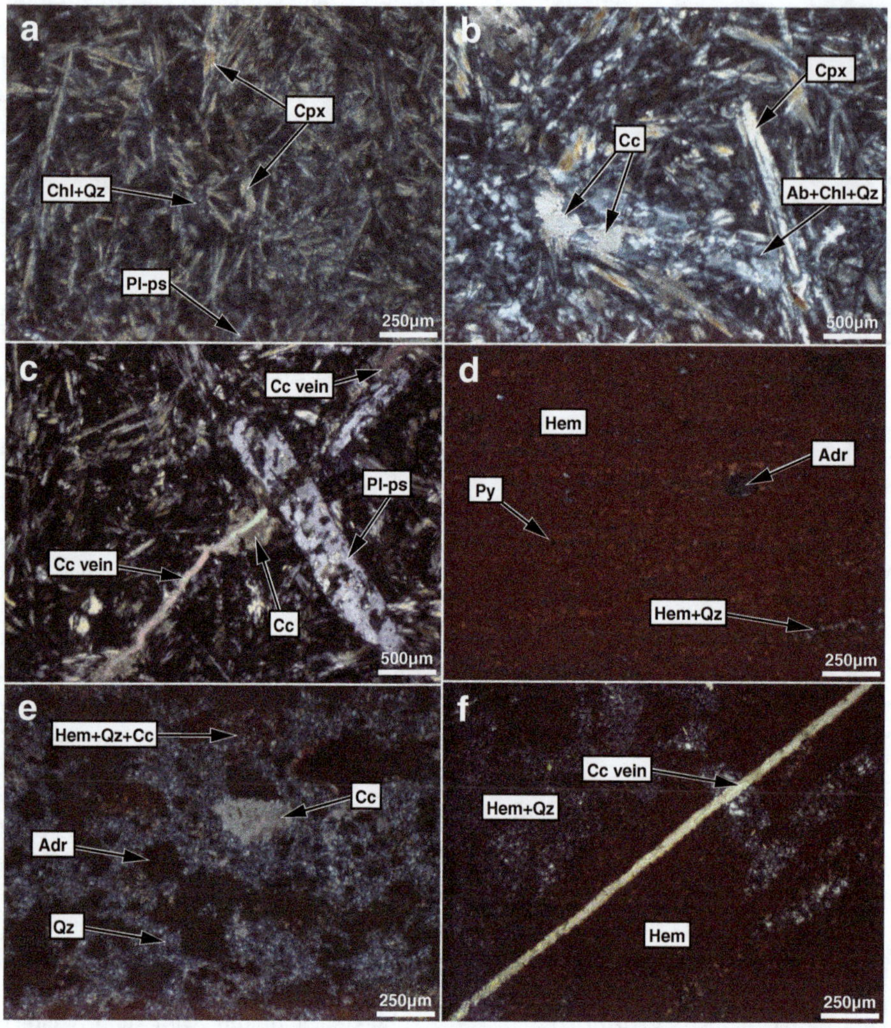

Fig. 3 Photomicrographs of basaltic andesites and hydrothermal jaspilites in the Ongeluk Formation. **a** A typical basaltic andesite in which the igneous minerals and glass have been replaced by secondary minerals but the igneous texture is well preserved. **b** Disseminated calcite-bearing basaltic andesite in which igneous plagioclase has been replaced by secondary albite, chlorite, quartz, and calcite. **c** A calcite vein cutting all alteration minerals and igneous texture. **d** A hematite-rich layer of interpillow hydrothermal jaspilite. **e** A microquartz-rich layer of the interpillow hydrothermal jaspilite that includes calcite and andradite. **f** A calcite vein cutting jaspilite

Belemnite (VPDB). Based on replicate analyses of laboratory standard, the analytical reproducibility is estimated to be better than ± 0.2‰ for carbon and better than ± 0.5‰ for oxygen isotope ratios. In the analysis of the CO_2 content in the samples, the relative error is 5–10% of the reported values, depending on the amount of extracted CO_2.

Results and discussion

Based on the occurrence and mineral texture of the Ongeluk lavas and jaspilites, the disseminated calcites are considered to have formed during subseafloor hydrothermal circulation at that time, as previously reported (Gutzmer et al. 2001). In contrast, the calcite veins must have formed at least after the formation of the disseminated calcites, which might have occurred during a later stage of subseafloor hydrothermal circulation, though they could also relate to a later metamorphic or deformation event. The oxygen isotope ratios ($\delta^{18}O$) of the disseminated calcites in the lavas and hydrothermal jaspilites range from − 17.9 to − 10.2 ‰ while their carbon isotope ratio ($\delta^{13}C$) shows relatively low values ranging from − 31.9 to − 13.2 ‰ (Table 1 and Fig 4). Most of the calcites in hydrothermal jaspilites have lower $\delta^{18}O$ values than those in the Ongeluk lavas, but there is no obvious correlation to be made between the $\delta^{18}O$ and $\delta^{13}C$ values for all samples, indicating that the isotopic values have not been greatly modified since their formation. The CO_2

Table 1 Isotopic compositions of calcites and CO_2 content in the Ongeluk basaltic andesites and hydrothermal jaspilites

Sample no.	Rock type	Classification of calcite	CO_2 (wt%)	$\delta^{13}C$ (VPDB, ‰)	$\delta^{18}O$ (VPDB, ‰)	Temperature (°C)[a]
GU033	Basaltic andesite	Disseminated	0	− 14.2	− 15.2	123
GU034	Basaltic andesite	Disseminated	0	− 15.6	− 13.8	109
GU035	Basaltic andesite	Disseminated	0	− 13.2	− 14.6	117
GU036	Basaltic andesite	Disseminated	0.01	− 15.8	− 15.8	130
GU037	Basaltic andesite	Disseminated	0.01	− 16.1	− 14.3	114
GU038	Basaltic andesite	Disseminated	0	− 15.9	− 12.9	100
GU041	Basaltic andesite	Disseminated	0	− 18.9	− 14.1	112
GU045	Basaltic andesite	Disseminated	0	− 18.3	− 13.3	104
GU048	Basaltic andesite	Disseminated	0	− 15.7	− 13.2	103
GU049	Basaltic andesite	Disseminated	0	− 14.9	− 14.6	117
GU054	Basaltic andesite	Disseminated	0	− 23.5	− 14.3	114
GU055	Basaltic andesite	Disseminated	0.01	− 24.9	− 15	121
GU063	Basaltic andesite	Disseminated	0	− 22.3	− 12.3	94
GU110	Pillowed basaltic andesite (core)	Disseminated	0	− 15.6	− 10.2	77
GU172	Basaltic andesite	Disseminated	0	− 18.0	− 10.8	82
GU174	Basaltic andesite with calcite veins	Disseminated/vein	0.060	− 16.1	− 17.1	147
GU144	Interpillow jaspilite	–	0	− 17.0	− 17.6	153
GU149	Interpillow jaspilite	–	0	− 21.3	− 13.4	105
GU151	Interpillow jaspilite	–	0.01	− 31.9	− 17.9	158
GU150	Interpillow jaspilite with calcite/quartz veins	–	0.1	− 15.4	− 17.7	155
GU158	Brecciated interpillow jaspilite	–	0.05	− 12.5	− 17.8	156

[a]Formation temperature was estimated from the reported oxygen fractionation factor (O'Neil et al. 1969; Friedman and O'Neil 1977; Chacko et al. 2001) assuming a fluid with δ18O = 0‰ (SMOW)

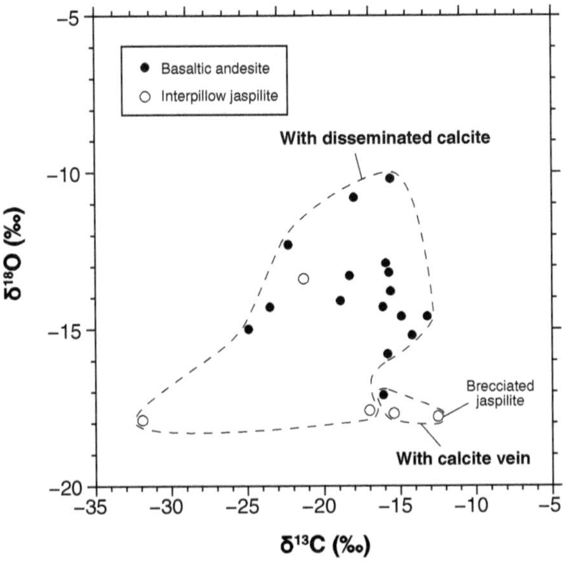

Fig. 4 Relationship between $\delta^{18}O$ and $\delta^{13}C$ values of the hydrothermal calcites from the Ongeluk Formation

contents of the undeformed samples (with disseminated calcites only) are lower than 0.007 wt% (almost equivalent to 0.016 wt% as $CaCO_3$) whereas those of the deformed samples (with calcite vein) have clearly higher CO_2 contents reaching 0.099 wt% (0.23 wt% as $CaCO_3$) (Table 1 and Fig 5). Based on these data, we discuss the carbonation of lavas and the formation temperature of jaspilites in the following sections.

Oxygen isotopes of the calcites in the Ongeluk lavas and jaspilites

The formation temperatures of calcites in the Ongeluk lavas and hydrothermal jaspilite should—if the calcites were formed due to the seafloor hydrothermal circulation—ideally be consistent with the temperatures of seafloor hydrothermal alteration independently estimated from fluid inclusions in the Ongeluk hydrothermal quartz (Gutzmer et al. 2003; Saito et al. 2016). To estimate the formation temperatures of calcites from their $\delta^{18}O$ values, a $\delta^{18}O$ value of seawater must be assumed. However, there have long been two different models on the evolution of the $\delta^{18}O$ value of seawater through the Earth's history; the one is a constant $\delta^{18}O$ value (e.g., Muehlenbachs and Clayton 1976; Blake et al.

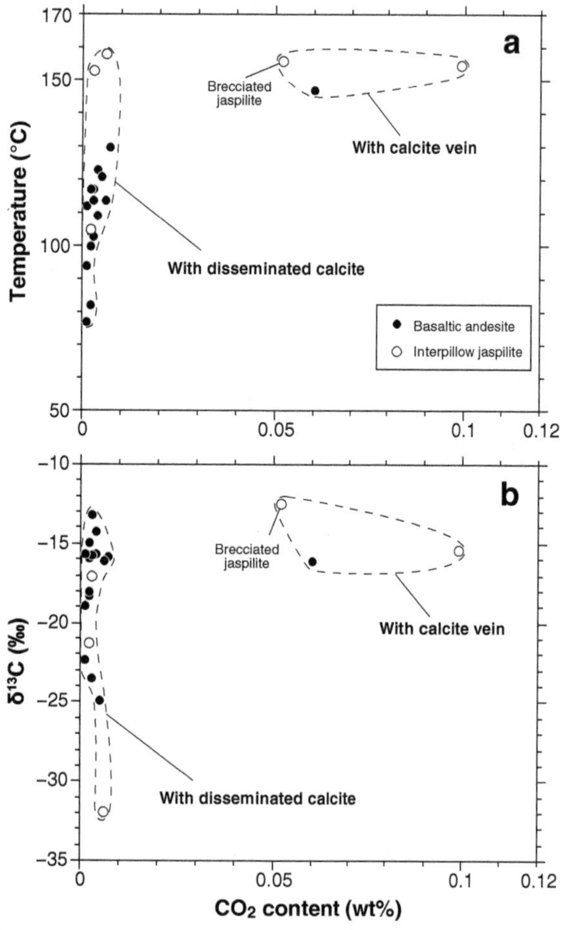

Fig. 5 Content of CO_2 fixed as carbonate vs. **a** temperature obtained from $\delta^{18}O$ values and **b** $\delta^{13}C$ values (‰, VPDB) of calcites from the Ongeluk basaltic andesites and hydrothermal jaspilites. The deformed samples (brecciated or with calcite vein) have higher CO_2 content than the undeformed samples

2010; Tartèse et al. 2017) and the other is a gradually increasing $\delta^{18}O$ value (e.g., Kasting et al. 2006).

In the former case, using the oxygen isotope fractionation factors (O'Neil et al. 1969; Friedman and O'Neil 1977; Chacko et al. 2001), the formation temperatures of disseminated calcites in the lavas are calculated to be 77–130 °C (Table 1 and Fig. 5a). The $\delta^{18}O$ values of the calcites in the undeformed jaspilites provide a slightly higher formation temperature range (105–158 °C). The difference in the formation temperatures indicates that the disseminated calcites in the lavas were formed mainly as a result of interaction with relatively low-temperature downwelling seawater from the seafloor, whereas the precipitation of jaspilites took place with a larger contribution of high-temperature iron-rich hydrothermal fluids derived from the lower part of the hydrothermal circulation cell. Be that as it may, the temperatures obtained for both calcites in the undeformed lavas and the hydrothermal jaspilites fall within the range of homogenization temperatures of the fluid inclusions (64–160 °C) in the Ongeluk hydrothermal quartz precipitates (Gutzmer et al. 2001, 2003; Saito et al. 2016). The consistency of the formation temperatures suggests that the generation of disseminated calcites in the lavas and jaspilites coincided with the precipitation of the hydrothermal quartz during subseafloor hydrothermal circulation. On the other hand, the calcites in the deformed samples show only high formation temperatures (147–156 °C), which are consistent with a metamorphic temperature not exceeding prehnite-pumpellyite facies (Gutzmer et al. 2001).

In the latter case that $\delta^{18}O$ value of seawater has increased from − 9‰ (vs. SMOW) at 4.6 Ga to the present value (e.g., Kasting et al. 2006), and the $\delta^{18}O$ value of seawater is approximately − 8‰ at 2.4 Ga. Using this value, the formation temperature of hydrothermal calcites and jaspilite was estimated to be 28–75 °C, which is clearly much lower than the formation temperature (64–160 °C) of hydrothermal quartz estimated from fluid inclusions therein (Gutzmer et al. 2003; Saito et al. 2016) and the alteration temperature estimated from the assemblage of secondary minerals in the lavas (Gutzmer et al. 2001). This discrepancy also forces the conclusion that the $\delta^{18}O$ value of seawater has remained almost constant through the Earth history (e.g., Holmden and Muehlenbachs 1993).

Carbon isotopes of the calcites in the Ongeluk lavas and jaspilites

Variation of calcite $\delta^{13}C$ values ($\delta^{13}C_{cc}$) in seafloor volcanic rocks tends to reflect the geological setting of hydrothermal alteration and the carbon sources (Fig. 6). In general, modern and Archean unsedimented ridge-derived oceanic crusts have $\delta^{13}C_{cc}$ values close to those of seawater (mostly − 4 to 4‰), because subseafloor volcanic sequences are generally dominated by seawater-derived fluid due to their high permeability (Nakamura and Kato 2004; Gillis and Coogan 2011; Shibuya et al. 2012). On the other hand, the $\delta^{13}C_{cc}$ values of some rift lake volcanics in the Afar Rift (− 7 to − 1‰) are explained as a mixture of meteoric water (~− 15‰) and seawater, with a possible contribution of magmatic CO_2 (~− 5‰) (Fouillac et al. 1989), whereas hydrothermal calcite chimneys in Lake Abhé in the Afar Rift have $\delta^{13}C_{cc}$ values (~ 2 to 4‰) consistent with that of the alkaline lake water (0.6‰) equilibrated with atmospheric CO_2 (~− 7‰) (Dekov et al. 2014). In contrast, all sediment-related settings, such as modern sedimented ridge-derived oceanic crusts and modern/Archean oceanic rift basin basalts, bare relatively low $\delta^{13}C_{cc}$ values (<− 15‰) as well as the values near the $\delta^{13}C$ of CO_2 ($\delta^{13}C_{CO2}$) in seawater which would represent a significant input of CO_2 derived from the decomposition of

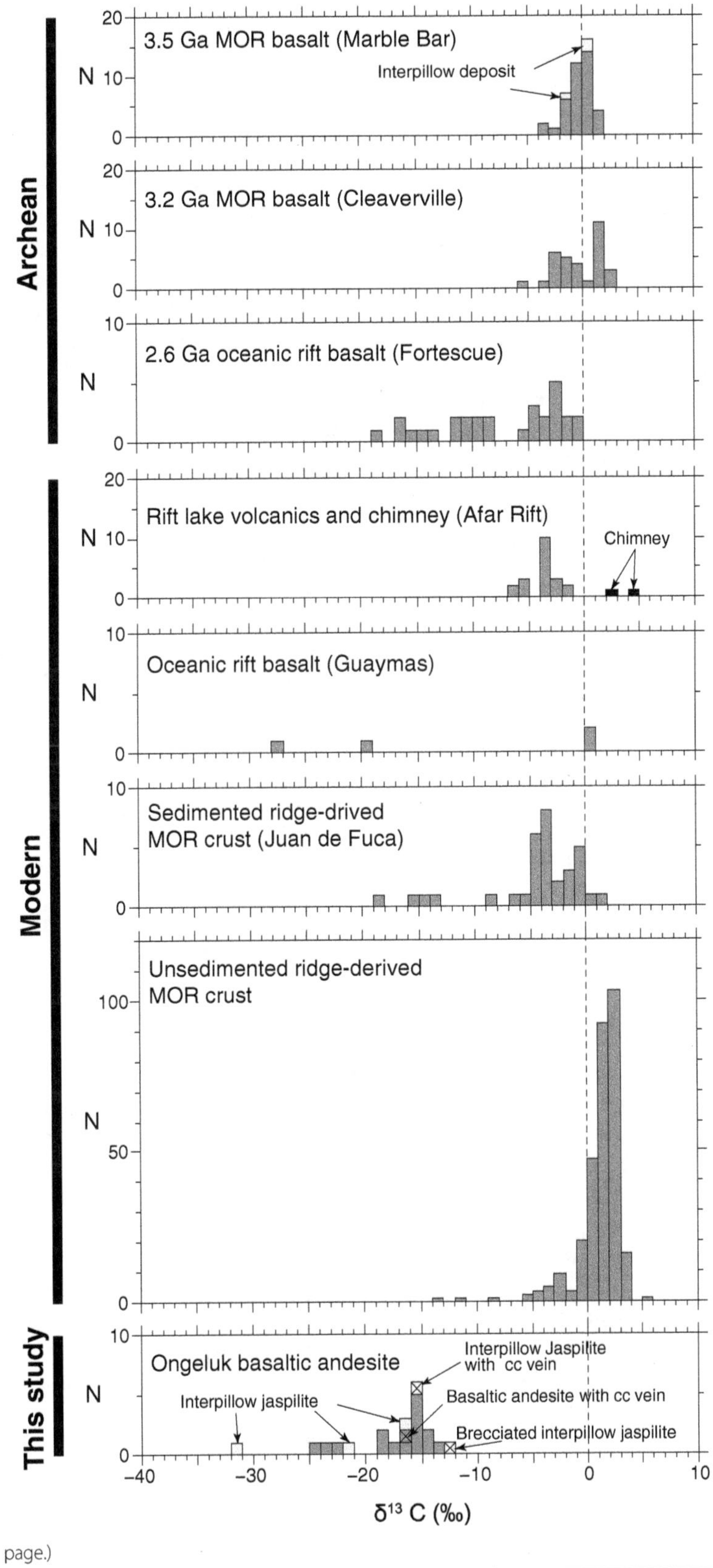

Fig. 6 (See legend on next page.)

(See figure on previous page.)
Fig. 6 $\delta^{13}C$ values (‰, VPDB) of hydrothermal calcites from the Ongeluk Formation. For comparison, data from modern unsedimented ridge-derived mid-ocean ridge (MOR) crusts (Gillis and Coogan 2011), sedimented ridge-derived MOR crusts (Coggon et al. 2004), rift-zone seafloor basalts (Gieskes et al. 1982), rift lake basement volcanics (Fouillac et al. 1989), and hydrothermal chimney (Dekov et al. 2014) from Afar Rift, Archean MOR basalts (Nakamura and Kato 2004; Shibuya et al. 2012), and rift-zone seafloor basalts (Shibuya et al. 2013a) are also shown

organic matter (Gieskes et al. 1982; Coggon et al. 2004; Shibuya et al. 2013a).

In this regard, the relatively low $\delta^{13}C_{cc}$ values down to −31.9‰ in the Ongeluk Formation is consistent with its geological setting (a sedimented rifting basin) where decomposition of organic matter in sedimentary rocks (e.g., underlying Makganyene Formation) could occur during the hydrothermal circulation. Another characteristic point is that $\delta^{13}C_{cc}$ values higher than − 13.2‰ were not identified in the undeformed lavas. This is consistent with the exceptionally ^{13}C–depleted CO_2 in the Ongeluk seawater at that time; the Paleoproterozoic long-term seawater $\delta^{13}C_{CO2}$ generally decreased and reached its lowest value (− 14 to − 8‰) at the time of the glaciation and then recovered thereafter (Polteau et al. 2006). Therefore, the wide variation of $\delta^{13}C_{cc}$ values of the Ongeluk lavas can be well explained by varying the mixing ratios between the CO_2 derived from the decomposition of subseafloor sedimentary organic matter and the ^{13}C-depleted CO_2 in seawater. For this reason, the measured CO_2 (as carbonate) content in the lavas likely corresponds to the upper limit of seawater-derived CO_2.

Secular change of carbonate content in the altered subseafloor crust

Because permeability decreases while temperature increases with increasing depth below the seafloor, carbonate minerals tend to become unstable, thereby leading to a decrease in fixed CO_2 in the basaltic crust and ophiolite (Gillis and Robinson 1988; Gillis and Coogan 2011; Alt and Teagle 1999; Shibuya et al. 2007a, 2007b, 2012). However, the carbonation of highly permeable, low- to moderate-temperature (<c. 200 °C) uppermost basaltic sequences presumably tends to reflect seawater CO_2 concentration. Furthermore, carbonation of seafloor basalts would be less significantly influenced by the seawater pH governing the speciation of CO_{2aq}, HCO^{3-}, and CO_3^{2-}. Under constant pressure-temperature conditions, calcite precipitation from a solution is governed by activities of CO_3^{2-} and Ca^{2+} ions. In other words, the saturation state of calcite mainly depends on the total CO_2 concentration (CO_{2aq} + HCO^{3-} + CO_3^{2-}) and pH value in seawater, at least when the activity of Ca^{2+} is constant. In seafloor hydrothermal systems, however, pH variations at low-moderate temperatures are much smaller than in solution because alteration minerals substantially buffer the pH of a coexisting fluid. For example, thermodynamic calculations indicated that fluid pH at 150 °C is potentially buffered to be 7.7 and 7.6 even after reactions with the Hadean weakly acidic, CO_2-rich seawater (total CO_2 = 200 mmol/kg) and with the present-day seawater (total CO_2 = 2 mmol/kg), respectively (Macleod et al. 1994). Such behavior of pH at low-moderate temperatures is also distinct from high-temperature fluids (> 300 °C) where pH values are influenced by an input of magmatic volatiles (e.g., Butterfield et al. 2003) and the presence or absence of carbonate minerals (Shibuya et al. 2010, 2013b). Accordingly, it is suggested that the content of fixed CO_2 as carbonate in the uppermost subseafloor crust qualitatively reflects total CO_2 concentration in seawater under a constant pressure-temperature condition. On the other hand, it was pointed out that high-temperature bottom seawater elevates temperature of subseafloor fluids and kinetically facilitates the formation reactions of carbonate minerals in the uppermost crust (Gillis and Coogan 2011), although carbonate minerals in altered mafic rocks become unstable with increasing temperature under equilibrium condition (e.g., Shibuya et al. 2013b). In this section, therefore, we discuss the secular change of total CO_2 concentration, pH, temperature, and partial pressure of CO_2 (pCO_2) in seawater based on the CO_2 content in the subseafloor uppermost crust (< 1 km deep from the seafloor) through geologic time.

In the Phanerozoic, the content of CO_2 as carbonate in the Jurassic/Cretaceous crusts is generally higher than that of younger crusts (Fig. 7a). This trend is more clearly evidenced in crusts with greater penetration depth. The CO_2 contents in the young crusts (< 13.6 Ma) with penetration depth of > 200 m (Holes 369B, 896A, and 504B) are lower than 1 wt% (Gillis and Coogan 2011; Alt and Teagle 1999) (Fig. 7a). These values probably reflect the total CO_2 concentration in seawater because the temperature of bottom seawater is presumably almost constant (e.g., 2 °C) and not high enough to effect carbonate formation in the uppermost crust during this period. In contrast, Jurassic and Cretaceous crusts with penetration depths of > 200 m (Holes 801C, 417A, 417D, and 418A) have higher CO_2 contents, up to 4.2 wt.%, than the younger crusts. It was previously suggested that the enhanced CO_2 uptake by the crust in the Cretaceous period can be explained only

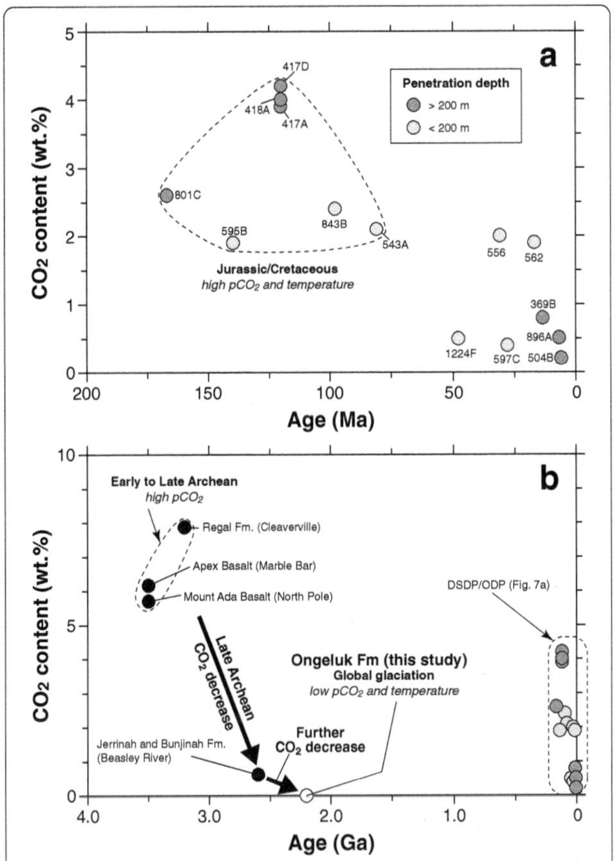

Fig. 7 Secular change of CO_2 (as carbonate) content in the altered subseafloor crust. **a** DSDP/ODP drill cores that penetrated basaltic basement (Alt and Teagle 1999; Gillis and Coogan 2011). The CO_2 content in the Jurassic/Cretaceous crusts is generally higher than that of younger crusts. Because the crusts with shallower penetration depth have large variation of CO_2 content, the crusts with penetration depth > 200 m is mainly discussed in the text. **b** The CO_2 content in Precambrian greenstones with geologically reconstructed depth > 1000 m from the ancient seafloor. Data from 3.5 Ga Apex Basalt in the Marble Bar area (Nakamura and Kato 2004), 3.5 Ga Mount Ada Basalt in the North Pole area (Kitajima et al. 2001; Terabayashi et al. 2003), 3.2 Ga Regal Formation in the Cleaverville area (Shibuya et al. 2012), 2.6 Ga Jeerinah and Bunjinah Formations in Beasley River area (Shibuya et al. 2013a) in the Pilbara Craton (Western Australia), and this study. Note that the CO_2 contents in the 3.5 Ga Mount Ada Basalt and the 3.2 Ga Regal Formation were calculated from the modal abundance of carbonate minerals in greenstones

by higher seawater temperatures (Brady and Gíslason 1997; Gillis and Coogan 2011). In this case, the high temperature of seawater was likely maintained by the strong greenhouse effect induced by high atmospheric pCO_2 in this era (Berner and Kothavala 2001), as suggested by other geologic records (e.g., paleosols; Ekart et al. 1999). High atmospheric pCO_2 might also elevate the total CO_2 concentration in seawater, in which case, the high total CO_2 concentration along with high temperature of seawater potentially accounts for the high carbonate content in the Jurassic/Cretaceous uppermost crust.

In the Archean, the carbonate contents in the 3.5–3.2 Ga subseafloor crusts (5.7–7.9 wt% as CO_2) are obviously much higher than modern equivalents (Kitajima et al. 2001; Terabayashi et al. 2003; Nakamura and Kato 2004; Shibuya et al. 2012, 2013a) (Fig. 7b). Although the Archean seawater temperature has not been quantitatively constrained (Knauth and Lowe 2003; Robert and Chaussidon 2006; Blake et al. 2010), it is highly possible that the strong carbonation of Archean greenstones reflects high total CO_2 concentration in seawater. Furthermore, many geological and theoretical investigations indicate that the Early to Middle Archean atmosphere was extremely enriched in CO_2 to compensate the faint young Sun (Sagan and Mullen 1972; Kasting 1987; Ohmoto et al. 2004; Lowe and Tice 2004). Therefore, the high pCO_2 in the Early to Middle Archean seawater/atm. would be the primary factor elevating the total CO_2 concentration in seawater and the carbonate content in the subseafloor crust. In contrast, the 2.6 Ga subseafloor crust has relatively low CO_2 content (0.6 wt.%) that is comparable to modern equivalents, which suggests that total CO_2 concentration in seawater decreased during the period between 3.2 and 2.6 Ga (Shibuya et al. 2013a). The trigger for this decrease in seawater CO_2 is possibly the formation and breakdown of the supercontinent in the late Archean, the removal of precipitated sedimentary carbonate onto land during the amalgamation of continents, and the precipitation of platform carbonate on newly created passive margins during/after the continental breakdown (Shibuya et al. 2013a).

Compared with the contents of fixed CO_2 as carbonate of the Archean and Phanerozoic subseafloor uppermost crusts, the average CO_2 content in the Ongeluk lavas (< 0.001 wt% including calcite veins) is clearly lower than those of Late Archean and even the modern equivalents (e.g., 0.21 wt% in 504B), even though the CO_2 in the Ongeluk lavas includes organic matter-derived CO_2 in addition to the original seawater CO_2. These results carry the implication that total CO_2 concentration in seawater further decreased from Late Archean to the Paleoproterozoic because the temperature of the Ongeluk bottom seawater would not have been much lower than the modern temperature (e.g., 2 °C). Thus, the total CO_2 concentration in 2.4 Ga seawater may have been lower than the present level. Such decrease in the total CO_2 concentration in seawater from 3.2 to 2.4 Ga is also consistent with the decrease in atmospheric pCO_2 from the late Archean to Paleoproterozoic estimated from the paleosols (Kanzaki and Murakami 2015).

In principle, carbonate formation in the uppermost subseafloor crust should be influenced by not only temperature and total CO_2 concentration of seawater, but also other factors such as pressure dependence of carbonate stability. Nevertheless, as discussed above, the degree of hydrothermal carbonation of the uppermost subseafloor crust shows a good correlation with independently estimated atmospheric pCO_2 levels throughout Earth history. This overall trend implies that the secular change of atmospheric pCO_2 relates more closely to that of total CO_2 concentration in seawater rather than pH change. Thus, the exceptionally low CO_2 contents in the Ongeluk lavas would also reflect extremely low CO_2 levels in the seawater/atm. during the Paleoproterozoic global glaciation. Our conclusion supports the low atmospheric CO_2 levels at that time—a premise of the previously proposed hypotheses for the trigger of the Paleoproterozoic global glaciation (Kopp et al. 2005; Melezhik 2006; Condie et al. 2009; Young 2013).

Conclusions

The carbonate content in the Ongeluk lavas and its comparison with other equivalents of various ages revealed that the total CO_2 concentration in 2.4 Ga seawater was extremely low, which potentially accounts for the assumed low atmospheric CO_2 levels during the Paleoproterozoic global glaciation. More importantly, the present study indicates the possibility that the degree of carbonation of subseafloor crusts reflects the qualitative secular change of CO_2 concentration in seawater and atmosphere. Therefore, the carbonate content in post-2.4 Ga subseafloor crusts should be investigated to decode secular changes of CO_2 concentrations in seawater/atm., especially during the Proterozoic. Moreover, it is suggested that more quantitative compositions of ancient seawater can be estimated by analysis of subseafloor crust materials such as hydrothermally precipitated minerals and their fluid inclusions.

Abbreviations

Adr: Andradite; Cc: Calcite; Chl: Chlorite; Cpx: Clinopyroxene; Hem: Hematite; MOR: Mid-ocean ridge; PAL: Present atmospheric level; Pl-ps: Plagioclase pseudomorph; Py: Pyrite; Qz: Quartz; VPDB: Vienna Pee Dee Belemnite

Acknowledgements

We thank for M. Cloete (Council for Geoscience, South Africa), K. Kitajima, S. Yamamoto, N. Enomoto, and A. Nakajima for assistance during the fieldwork. We are grateful to two anonymous reviewers for constructive comments and H. Kawahata for editorial handling and suggestions.

Funding

This research was partially supported by MEXT/JSPS KAKENHI Grant Numbers JP25707038, JP17H04580, and JP17H06455, and the Ministry of Education and Science of the Russian Federation, Project no. 14.Y26.31.0018.

Authors' contributions

TS and SM proposed the topic and conceived and designed the study. TK and TS carried out the geological investigation. TS conducted the petrological description and obtained the geochemical data. TK, KT, and MJR helped in their interpretation and collaborated with the corresponding author in the construction of the manuscript. All authors read and approved the final manuscript.

Authors' information

MJR's research was carried out at the Jet Propulsion Laboratory, California Institute of Technology, under a contract with the National Aeronautics and Space Administration, through the NASA Astrobiology Institute under cooperative agreement issued through the Science Mission directorate, No. NNH13ZDA017C (Icy Worlds), at the Jet Propulsion Laboratory. US Government sponsorship acknowledged.

Competing interests

The authors declare that they have no competing interests.

Author details

[1]Department of Subsurface Geobiological Analysis and Research (D-SUGAR), Japan Agency for Marine-Earth Science and Technology (JAMSTEC), 2-15 Natsushima-cho, Yokosuka 237-0061, Japan. [2]Research and Development Center for Submarine Resources, Japan Agency for Marine-Earth Science and Technology (JAMSTEC), 2-15 Natsushima-cho, Yokosuka 237-0061, Japan. [3]Project Team for Development of New-Generation Research Protocol for Submarine Resources, Japan Agency for Marine-Earth Science and Technology (JAMSTEC), 2-15 Natsushima-cho, Yokosuka 237-0061, Japan. [4]Department of Earth Science and Astronomy, The University of Tokyo, 3-8-1 Komaba, Meguro-ku, Tokyo 153-8902, Japan. [5]Earth-Life Science Institute, Tokyo Institute of Technology, 2-12-1 Ookayama, Meguro-ku, Tokyo 152-8551, Japan. [6]Jet Propulsion Laboratory, California Institute of Technology, 4800 Oak Grove Drive, Pasadena, CA 91108, USA. [7]Novosibirsk State University, Pirogova 1 630090, Russia.

References

Alt JC, Laverne C, Coggon RM, Teagle DAH, Banerjee NR, Morgan S, Smith-Duque CE, Harris M, Galli L (2010) Subsurface structure of a submarine hydrothermal system in ocean crust formed at the East Pacific Rise, ODP/IODP Site 1256. Geochem Geophy Geosy 11:Q10010. https://doi.org/10.1029/2010gc003144

Alt JC, Teagle DAH (1999) The uptake of carbon during alteration of ocean crust. Geochim Cosmochim Acta 63:1527–1535

Bau M, Romer RL, Lüders V, Beukes NJ (1999) Pb, O, and C isotopes in silici¢ed Mooidraai dolomite (Transvaal Supergroup, South Africa): implications for the composition of Paleoproterozoic seawater and 'dating' the increase of oxygen in the Precambrian atmosphere. Earth Planet Sci Lett 174:43–57

Berner RA, Kothavala Z (2001) GEOCARB III: a revised model of atmospheric CO_2 over Phanerozoic time. Am J Sci 301:182–204

Beukes NJ, Smit CA (1987) New evidence for thrust faulting in Griqualand West, South Africa: implications for stratigraphy and the age of red beds. S Afr J Geol 90:378–394

Blake RE, Chang SJ, Lepland A (2010) Phosphate oxygen isotopic evidence for a temperate and biologically active Archaean ocean. Nature 464:1029–1032

Brady PV, Gíslason SR (1997) Seafloor weathering controls on atmospheric CO_2 and global climate. Geochim Cosmochim Acta 61:965–973

Butterfield DA, Seyfried WE Jr, Lilley MD (2003) Composition and evolution of hydrothermal fluids. In: Halbach PE, Tunnicliffe V, Hein JR (eds) Energy and mass transfer in marine hydrothermal systems. Dahlem University Press, Berlin, pp 123–161

Chacko T, Cole DR, Horita J (2001) Equilibrium oxygen, hydrogen and carbon isotope fractionation factors applicable to geologic systems. In: Valley JW, Cole DR (eds) Stable isotope geochemistry. Reviews in Mineralogy and Geochemistry. Mineralogical Society of America, Washington, D. C, pp 1–81

Coggon RM, Teagle DAH, Cooper MJ, Vanko DA (2004) Linking basement carbonate vein compositions to porewater geochemistry across the eastern flank of the Juan de Fuca Ridge, ODP Leg 168. Earth Planet Sci Lett 219:111–128

Condie KC, O'Neill C, Aster RC (2009) Evidence and implications for a widespread magmatic shutdown for 250 My on Earth. Earth Planet Sci Lett 282:294–298

Cornell DH, Schütte SS, Eglington BL (1996) The Ongeluk basaltic andesite formation in Griqualnd West, South Africa: submarine alteration in a 2222 Ma Proterozoic sea. Precambr Res 79:101–123

de Villiers PR, Visser JNJ (1977) The glacial beds of the Griqualand West Supergroup as revealed by four deep boreholes between Postmasburg and Sishen. Trans Geol Soc S Afr 80:1–8

Dekov VM, Egueh NM, Kamenov GD, Bayon G, Lalonde SV, Schmidt M, Liebetrau V, Munnik F, Fouquet Y, Tanimizu M, Awaleh MO, Guirreh I, Le Gall B (2014) Hydrothermal carbonate chimneys from a continental rift (Afar Rift): mineralogy, geochemistry, and mode of formation. Chem Geol 387:87–100

Dorland HC (2004) Provenance ages and timing of sedimentation of selected Neoarchean and Paleoproterozoic successions on the Kaapvaal Craton. Rand Afrikaans University, Johannesburg

Ekart DD, Cerling TE, Montanez IP, Tabor NJ (1999) A 400 million year carbon isotope record of pedogenic carbonate: Implications for paleoatmospheric carbon dioxide. Am J Sci 299:805–827

Evans DA, Beukes NJ, Kirschvink JL (1997) Low-latitude glaciation in the Palaeoproterooic era. Nature 386:262–266

Fouillac AM, Fouillac C, Cesbron F, Pillard F, Legendre O (1989) Water-rock interaction between basalt and high-salinity fluilds in the Asal Rift, Republic of Djibouti. Chem Geol 76:271–289

Friedman I, O'Neil JR (1977) Compilation of stable isotope fractionation factors of geochemical interest. U.S. Geological Survey, Professional Paper:440–KK

Gieskes J, Elderfield H, Lawrence JR, Johnson J, Meyers B, Campbell A (1982) Geochemistry of interstitial waters and sediments, Leg-64, Gulf of California. Initial Rep Deep Sea 64:675–694

Gillis KM, Coogan LA (2011) Secular variation in carbon uptake into the ocean crust. Earth Planet Sci Lett 302:385–392

Gillis KM, Robinson PT (1988) Distribution of alteration zones in the upper oceanic crust. Geology 16:262–266

Gough DO (1981) Solar interior structure and luminosity variations. Sol Phys 74: 21–34

Grobler NJ, Botha BJV (1976) Pillow lavas and hyaloclastites in the Ongeluk andesite formation in a road-cut west of Griquatown, South Africa. Trans Geol Soc S Afr 79:53–57

Gumsley AP, Chamberlain KR, Bleeker W, Söderlund U, de Kock MO, Larsson ER, Bekker A (2017) Timing and tempo of the Great Oxidation Event. Proc Natl Acad Sci U S A 114:1811–1816

Gutzmer J, Banks DA, Luders V, Hoefs J, Beukes NJ, von Bezing KL (2003) Ancient sub-seafloor alteration of basaltic andesites of the Ongeluk Formation, South Africa: implications for the chemistry of Paleoproterozoic seawater. Chem Geol 201:37–53

Gutzmer J, Pack A, Lüders V, Wilkinson JJ, Beukes NJ, van Niekerk HS (2001) Formation of jasper and andradite during low-temperature hydrothermal seafloor metamorphism, Ongeluk Formation, South Africa. Contrib Mineral Petrol 142:27–42

Harada M, Tajika E, Sekine Y (2015) Transition to an oxygen-rich atmosphere with an extensive overshoot triggered by the Paleoproterozoic snowball Earth. Earth Planet Sci Lett 419:178–186

Hessler AM, Lowe DR, Jones RL, Bird DK (2004) A lower limit for atmospheric carbon dioxide levels 3.2 billion years ago. Nature 428:736–738

Hoffman PF (2013) The Great Oxidation and a Siderian snowball Earth: MIF-S based correlation of Paleoproterozoic glacial epochs. Chem Geol 362: 143–156

Hofmann A, Harris C (2008) Silica alteration zones in the Barberton greenstone belt: a window into subseafloor processes 3.5-3.3 Ga ago. Chem Geol 257: 221–239

Holmden C, Muehlenbachs K (1993) The $^{18}O/^{16}O$ ratio of 2-billion-year-old seawater inferred from ancient oceanic crust. Science 259:1733–1736

Kanzaki Y, Murakami T (2015) Estimates of atmospheric CO_2 in the Neoarchean–Paleoproterozoic from paleosols. Geochim Cosmochim Acta 159:190–219

Kasting JF (1987) Theoretical constraints on oxygen and carbon dioxide concentrations in the Precambrian atmosphere. Precambr Res 34:205–229

Kasting JF (1993) Earth's early atmosphere. Science 259:920–926

Kasting JF, Howard MT, Wallmann K, Veizer J, Shields G, Jaffrés J (2006) Paleoclimates, ocean depth, and the oxygen isotopic composition of seawater. Earth Planet Sci Lett 252:82–93

Kirschvink JL, Gaidos EJ, Bertani LE, Beukes NJ, Gutzmer J, Maepa LN, Steinberger RE (2000) Paleoproterozoic snowball Earth: extreme climatic and geochemical global change and its biological consequences. Proc Natl Acad Sci U S A 97:1400–1405

Kitajima K, Maruyama S, Utsunomiya S, Liou JG (2001) Seafloor hydrothermal alteration at an Archaean mid-ocean ridge. J Metamorph Geol 19:583–599

Knauth LP, Lowe DR (2003) High Archean climatic temperature inferred from oxygen isotope geochemistry of cherts in the 3.5 Ga Swaziland Supergroup, South Africa. Geol Soc Am Bull 115:566–580

Konhauser KO, Amskold L, Lalonde SV, Posth NR, Kappler A, Anbar A (2007) Decoupling photochemical Fe(II) oxidation from shallow-water BIF deposition. Earth Planet Sci Lett 258:87–100

Kopp RE, Kirschvink JL, Hilburn IA, Nash CZ (2005) The Paleoproterozoic snowball Earth: a climate disaster triggered by the evolution of oxygenic photosynthesis. Proc Natl Acad Sci U S A 102:11131–11136

Lowe DR, Tice MM (2004) Geologic evidence for Archean atmospheric and climatic evolution: fluctuating levels of CO_2, CH_4, and O_2 with an overriding tectonic control. Geology 32:493–496

Macleod G, McKeown C, Hall AJ, Russell MJ (1994) Hydrothermal and oceanic pH conditions of possible relevance to the origin of life. Orig Life Evol Biosph 24: 19–41

Melezhik VA (2006) Multiple causes of Earth's earliest global glaciation. Terra Nova 18:130–137

Muehlenbachs K, Clayton RN (1976) Oxygen isotope composition of the oceanic crust and its bearing on seawater. J Geophys Res 81:4365–4369

Nakamura K, Kato Y (2004) Carbonatization of oceanic crust by the seafloor hydrothermal activity and its significance as a CO_2 sink in the Early Archean. Geochim Cosmochim Acta 68:4595–4618

Ohmoto H, Watanabe Y, Kumazawa K (2004) Evidence from massive siderite beds for a CO_2-rich atmosphere before ~1.8 billion years ago. Nature 429:395–399

O'Neil JR, Clayton RN, Mayeda TK (1969) Oxygen isotope fractionation in divalent metal carbonates. J Chem Phys 51:5547–5558

Pavlov AA, Hurtgen MT, Kasting JF, Arthur MA (2003) Methane-rich Proterozoic atmosphere? Geology 31:87–90

Polteau P, Moore JM, Tsikos H (2006) The geology and geochemistry of the Palaeoproterozoic Makganyene diamictite. Precambr Res 148:257–274

Révész KM, Landwehr JM (2002) $\delta^{13}C$ and $\delta^{18}O$ isotopic composition of $CaCO_3$ measured by continuous flow isotope ratio mass spectrometry: statistical evaluation and verification by application to Devils Hole core DH-11 calcite. Rapid Commun Mass Sp 16:2102–2114

Robert F, Chaussidon M (2006) A palaeotemperature curve for the Precambrian oceans based on silicon isotopes in cherts. Nature 443:969–972

Rosing MT, Bird DK, Sleep NH, Bjerrum CJ (2010) No climate paradox under the faint early Sun. Nature 464:744–747

Rouchon V, Orberger B (2008) Origin and mechanisms of K-Si-metasomatism of ca. 3.4-3.3 Ga volcaniclastic deposits and implications for Archean seawater evolution: examples from cherts of Kittys Gap (Pilbara craton, Australia) and Msauli (Barberton Greenstone Belt, South Africa). Precambr Res 165:169–189

Russell MJ, Barge L, Bhartia R, Bocanegra D, Bracher P, Branscomb E, Hand K, Kidd R, McGlynn S, Meier DH, Nitschke W, Shibuya T, Vance S, White L, Kanik I (2014) The drive to life on wet and icy worlds. Astrobiology 14:308–343

Russell MJ, Hall AJ, Martin W (2010) Serpentinization as a source of energy at the origin of life. Geobiology 8:355–371. https://doi.org/10.1111/j.1472-4669.2010.00249.x

Sagan C, Mullen G (1972) Earth and Mars: evolution of atmospheres and surface temperatures. Science 177:52–56

Saito T, Shibuya T, Komiya T, Kitajima K, Yamamoto S, Nishizawa M, Ueno Y, Kurosawa M, Maruyama S (2016) PIXE and microthermometric analyses of fluid inclusions in hydrothermal quartz from the 2.2 Ga Ongeluk Formation, South Africa: implications for ancient seawater salinity. Precambr Res 286: 337–351

Sheldon ND (2006) Precambrian paleosols and atmospheric CO_2 levels. Precambr Res 147:148–155

Shibuya T, Kitajima K, Komiya T, Terabayashi M, Maruyama S (2007a) Middle Archaean ocean ridge hydrothermal metamorphism and alteration recorded in the Cleaverville area, Pilbara Craton, Western Australia. J Metamorph Geol 25:751–767

Shibuya T, Komiya T, Anma R, Ota T, Omori S, Kon Y, Yamamoto S, Maruyama S (2007b) Progressive metamorphism of the Taitao ophiolite; evidence for axial and off-axis hydrothermal alterations. Lithos 98:233–260

Shibuya T, Komiya T, Nakamura K, Takai K, Maruyama S (2010) Highly alkaline, high-temperature hydrothermal fluids in the early Archean ocean. Precambr Res 182:230–238

Shibuya T, Russell MJ, Takai K (2016) Free energy distribution and hydrothermal mineral precipitation in Hadean submarine alkaline vent systems: Importance of iron redox reactions under anoxic conditions. Geochim Cosmochim Acta 175:1–19

Shibuya T, Tahata M, Kitajima K, Ueno Y, Komiya T, Yamamoto S, Igisu M, Terabayashi M, Sawaki Y, Takai K, Yoshida N, Maruyama S (2012) Depth variation of carbon and oxygen isotopes of calcites in Archean altered upper oceanic crust: implications for the CO_2 flux from ocean to oceanic crust in the Archean. Earth Planet Sci Lett 321–322:64–73. https://doi.org/10.1016/j.epsl.2011.12.034

Shibuya T, Tahata M, Ueno Y, Komiya T, Takai K, Yoshida N, Maruyama S, Russell MJ (2013a) Decrease of seawater CO_2 concentration in the Late Archean: an implication from 2.6 Ga seafloor hydrothermal alteration. Precambr Res 236: 59–64

Shibuya T, Yoshizaki M, Masaki Y, Suzuki K, Takai K (2013b) Reactions between basalt and CO_2-rich seawater at 250 and 350 °C, 500 bars: implications for the CO_2 sequestration into the modern oceanic crust and the composition of hydrothermal vent fluid in the CO_2-rich early ocean. Chem Geol 359:1–9

Shibuya T, Yoshizaki M, Satoh M, Shimizu K, Nakamura K, Omori S, Suzuki K, Takai K, Hideo T, Maruyama S (2015) Hydrogen-rich hydrothermal environments in the Hadean ocean inferred from serpentinization of komatiites at 300 °C and 500 bar. Prog Earth Planet Sci 2:46. https://doi.org/10.1186/s40645-015-0076-z

Sleep NH, Zahnle K (2001) Carbon dioxide cycling and implications for climate on acient Earth. J Geophys Res 106:1373–1399

Takai K, Nakamura K, Suzuki K, Inagaki F, Nealson KH, Kumagai H (2006) Ultramafics-Hydrothermalism-Hydrogenesis-HyperSLiME (UltraH^3) linkage: a key insight into early microbial ecosystem in the Archean deep-sea hydrothermal systems. Paleontol Res 10:269–282

Tartèse R, Chaussidon M, Gurenko A, Delarue F, Robert F (2017) Warm Archean oceans reconstructed from oxygen isotope composition of early-life remnants. Geochemical Perspectives Letters 3:55–65. https://doi.org/10.7185/geochemlet.1706

Terabayashi M, Masuda Y, Ozawa H (2003) Archean ocean-floor metamorphism in the North Pole area, Pilbara Craton, Western Australia. Precambr Res 123:167–180

Ueda H, Shibuya T, Sawaki Y, Saitoh M, Takai K, Maruyama S (2016) Reactions between komatiite and CO_2-rich seawater at 250 °C and 350 °C, 500 bars: implications for hydrogen generation in the Hadean seafloor hydrothermal system. Prog Earth Planet Sci 3:35. DOI: https://doi.org/10.1186/s40645-016-0111-8

Ueno Y, Johnson MS, Danielache SO, Eskebjerg C, Pandey A, Yoshida N (2009) Geological sulfur isotopes indicate elevated OCS in the Archean atmosphere, solving faint young sun paradox. Proc Natl Acad Sci U S A 106:14784–14789

Young GM (2013) Precambrian supercontinents, glaciations, atmospheric oxygenation, metazoan evolution and an impact that may have changed the second half of Earth history. Geosci Front 4:247–261

5

Mass transfer in the Earth's interior: fluid-melt interaction in aluminosilicate–C–O–H–N systems at high pressure and temperature under oxidizing conditions

Bjorn Mysen

Abstract

Understanding what governs the speciation in the C–O–H–N system aids our knowledge of how volatiles affect mass transfer processes in the Earth's interior. Experiments with aluminosilicate melt + C–O–H–N volatiles were, therefore, carried out with Raman and infrared spectroscopy to 800 °C and near 700 MPa in situ in hydrothermal diamond anvil cells. The measurements were conducted in situ with the samples at the desired temperatures and pressures in order to avoid possible structural and compositional changes resulting from quenching to ambient conditions prior to analysis. Experiments were conducted without any reducing agent and with volatiles added as H_2O, CO_2, and N_2 because both carbon and nitrogen can occur in different oxidation states.

Volatiles dissolved in melt comprise H_2O, CO_3^{2-}, HCO_3^-, and molecular N_2, whereas in the coexisting fluid, the species are H_2O, CO_2, CO_3^{2-}, and N_2. The HCO_3^-/CO_3^{2-} equilibrium in melts shift toward CO_3^{2-} groups with increasing temperature with $\Delta H = 114 \pm 22$ kJ/mol. In fluids, the CO_2 abundance is essentially invariant with temperature and pressure. For fluid/melt partitioning, those of H_2O and N_2 are greater than 1 with temperature-dependence that yields ΔH values of -6.5 ± 1.5 and -19.6 ± 3.7 kJ/mol, respectively. Carbonate groups, CO_3^{2-} are favored by melt over fluid. Where redox conditions in the Earth's interior exceed that near the QFM oxygen buffer (between NNO and MW buffers), N_2 is the stable nitrogen species and as such acts as a diluent of both fluids and melts. For fluids, this lower silicate solubility, in turn, enhances alkalinity. This means that in such environments, the transport of components such as high field strength cations, will be enhanced. Effects of dissolved N_2 on melt structure are considerably less than on fluid structure.

Keywords: Volatiles, Spectroscopy, Melt structure, Fluid structure, Temperature, Pressure

Introduction

Characterization of the processes that govern the budget and recycling of volatile components in the Earth's interior is central to our understanding of the formation and evolution of the solid Earth, its oceans, and atmosphere (e.g., Kasting et al. 1993; Jambon 1994; Tolstikhin and Marty 1998; Deines 2002). However, our knowledge of the nitrogen abundance in the Earth's mantle and in mantle-derived magma is not well established with model-dependent mantle abundance suggested between about 1 and 35 ppm (Marty, 1995; Javoy, 1997). The nitrogen abundance likely also is heterogeneously distributed with considerably greater nitrogen contents in subduction zone environments where abundances above 0.1% N_2 have been reported (Fisher et al. 2002). This latter environment likely also is where most of the exchange of volatiles (including nitrogen) occurs between the solid Earth and the atmosphere (Schmidt and Poli, 2003; Bebout, 2007).

Transfer of nitrogen from the Earth's interior to the surface and atmosphere is accomplished via partitioning of nitrogen between solid mantle, partial melts and perhaps multicomponent fluids with melt/mineral and fluid/mineral N_2 partition coefficients exceeding 1. Mass transport is at the core of these processes.

Correspondence: bmysen@carnegiescience.edu
Geophysical Laboratory, Carnegie Instn. Washington, Washington DC, USA

Characterization of transport processes requires data and models that describe physico-chemical properties of silicate melts and C–O–H–N fluids at high temperature and pressure under controlled redox conditions. Redox conditions are important because even within the redox range of natural magmatic processes in the Earth's interior (Carmichael and Ghiorso 1990; Wood et al. 1990), carbon and nitrogen, in particular, may occur in different oxidation states (Libourel et al. 2003; Kadik et al. 2004 2010; Mysen et al. 2008 Mysen et al. 2011; Li and Keppler 2014). The nitrogen solubility and solution mechanism(s) in silicate melts are sensitive to the redox state of nitrogen (Libourel et al., 2003; Mysen and Yamashita, 2010; Kadik et al. 2015) and needs to be controlled, therefore, during experiments in order to describe the behavior of nitrogen during melting and crystallization in the Earth's interior.

Nitrogen is but one of several volatile components in the Earth's interior with the system C–O–H–N comprising the main components (Jambon, 1994). Solubility and solution mechanisms of C–O–H–N components in silicate melts and of silicate components in C–O–H–N fluid are, therefore, necessary, and may be carried out experimentally (e.g., Franz 1998; Newton and Manning 2005 2006; Li and Keppler 2014; Kadik et al. 2015). However, most of that information is from analysis of glasses formed by temperature-quenching of volatile-containing melts to form a glass at ambient temperature at high pressure followed by decompression and, finally, examination of these glasses. Application of such data to characterize high-temperature/pressure melts and fluids relies on the assumption that there are no changes of these materials as they are brought from high temperature and high pressure to ambient conditions. The structure of a glass formed is this way is not that of the melt. Rather, it is the structure frozen in at the glass transition (e.g., Moynihan et al. 1976; Richet and Bottinga 1986; Dingwell and Webb 1990; Richet et al. 1992) because melt structure is temperature (and pressure) dependent (e.g., Stebbins 1988; Mysen et al. 2003; Agee 2008; Yamada et al. 2011; Wu et al. 2014). Moreover, in melts that contain volatiles, their speciation can be temperature- and pressure-dependent (Nowak and Behrens, 1995; Mysen and Yamashita 2010; Wu et al. 2014). In addition, during quenching to ambient temperature and pressure, some or all of the volatiles may exsolve from the cooling melt because the solubility of volatiles in melts varies with both temperature and pressure. Therefore, neither the volatile content nor the speciation of a glass is that which existed in its molten state at high temperature and pressure.

An additional problem associated with examination of quenched material as a proxy of high-temperature/pressure is that the structure of C–O–H–N fluids in equilibrium with silicate at high temperature and pressure cannot be quenched for quantitative structure study at ambient conditions. This is partly because silicate and other oxides dissolved in fluids at such high temperatures and pressures often precipitate during quenching (e.g., Zhang and Frantz 2000; Mysen and Armstrong 2002; Kessel et al. 2004), and partly because the C–O–H–N fluid structure may change as pressure and temperature are reduced from the experimental to ambient conditions (e.g., Mysen and Yamashita 2010; Wu et al. 2014). In summary, it is necessary, therefore, to study melts and fluids at the temperature and pressure of interest in order to attain quantitative understanding of their structure and their structure-property relations.

Recent experimental studies have addressed structural and compositional aspects of silicate–COH fluid + melt systems while samples were at high temperature and pressure (Ni and Keppler 2013; Mysen 2015a, b). Less is known about chemically more complex situation in which the C–O–H–N system of volatiles in equilibrium with aluminosilicate melts. The objective of the present study is to address this latter type of system in order to provide experimental data with which to further our understanding the structure of fluids and melts of aluminosilicate–C–O–H–N systems relevant in the temperature and pressure range of the Earth's crust and upper mantle under oxidizing conditions.

The experiments were conducted under conditions where nitrogen exists in its oxidized state, N_2. We note, though, that nitrogen may be reduced to amide and ammine forms at and above hydrogen fugacities between those defined by the NNO–H_2O (nickel–nickel oxide) and MW–H_2O (magnetite–wüstite) redox conditions (Mysen et al. 2008), which are redox conditions within the range encountered in the Earth's interior (Wood et al. 1990; Frost and McCammon, 2008). This information is, therefore, relevant only to the most oxidized part of this range of redox conditions.

Methods/Experimental

Optical and vibrational spectroscopic data from the samples were recorded while these were at the desired high pressure and temperature while contained in externally heated, hydrothermal diamond anvil cells (HDAC) of the type originally described by Bassett et al. (1996). In this cell, the upper and lower diamonds are surrounded by Mo wrap-around wire heaters. Two thermocouples in contact with the upper and lower diamonds, respectively, are located within less than 1 mm from the sample. The samples, themselves, were contained in the cell within 125 μm thick Ir gaskets with a 500 μm diameter sample hole. Iridium was used as gasket material because it does not react with the sample (melt and fluid) at the temperature and pressure conditions of these experiments.

The temperature was measured with K-type thermocouples that were calibrated against the melting temperature of NaCl (800.5 °C). Temperature error was about ± 3 °C with the thermal gradient across the sample volume less than 2 °C. Pressure was generated by the fluid at high temperature in the cell. Pressure is, therefore, a dependent variable in these experiments. It was determined from the one-phonon Raman shift of ^{13}C synthetic diamond, which was used to avoid interference with the Raman signals of ^{12}C from the diamond cell itself (Bassett et al. 1996; Schiferl et al. 1997; Mysen and Yamashita 2010). Ten to 30 μm across diamond pieces were grown from $^{13}CH_4$ by using CVD techniques (Liang et al. 2009). Pressure uncertainty at high temperature is ± 110 MPa and precision ± 40 MPa determined from the precision of the Raman spectrometer (± 0.1 cm^{-1} by using Ne lines as internal spectrometer calibrant) and the uncertainty in the calibration by Mysen and Yamashita (2010).

The samples in the diamond cell were examined through optical microscopy and via confocal micro-Raman and microFTIR methods while at the temperature and pressure of interest. In the Raman spectroscopic studies, the samples (melt and fluid) were excited with the 490 nm line of a solid state laser in a JASCO model NRS-3100 confocal micro-Raman spectrometer. The spectrometer is equipped with a single monochromator, a holographic notch filter, and holographic gratings with 600, 1200, and 2400 grooves/mm. The laser was focused to about 1 μm diameter at the sample with power at the sample typically near 30 mW. Spectra were recorded through a × 50 magnification/0.42 N.A. long-working distance Mitutoyo objective lens. An Andor Model DV401-F1 1024 × 128 pixel (25 μm pixel size) Peltier-cooled CCD was used for signal detection. By using the confocal optics, > 85% of the signal was recorded from within 30 μm of a sample surface (Mysen 2015a). For infrared spectroscopy, a JASCO model IMV4000 FTIR microscope system operating in transmission mode with 10X objective and condenser cassegrain lenses, an InSb detector, CaF_2 beam splitter, and a halogen light source was employed.

Staring materials were aluminosilicate glass chips of composition $(Na_2Si_4O_9)_{90}(Na(NaAl)_4O_9)_{10}$ (denoted NA10) together with solid $Ag_2C_2O_4$ and AgN_3 as sources of CO_2 and N_2, respectively. No reducing agent was added to the starting materials so that all components, including N_2 and CO_2, remain in their oxidized form. The $(Na_2Si_4O_9)_{90}(Na(NaAl)_4O_9)_{10}$ glass was from the batch originally synthesized by Mysen (2007), but remelted and quenched to glass to be used for starting materials here in order to eliminate incipient crystals and adsorbed and water that may contributed to the sample during its storage. The $Ag_2C_2O_4$ breaks down to metallic Ag and CO_2 gas in between 150° and 200 °C and AgN_3 to metallic Ag and N_2 gas under similar temperature conditions (Boettcher et al. 1973; Roskosz Mysen and Cody, 2006). A few experiments were also conducted with $Ag_2C_2O_4 + H_2O$ only in order to generate reference spectra of volatiles in the COH fluid system.

The sample volume defined by the central hole in the Ir gaskets is about 0.025 mm^3, which poses a major challenge for quantitative addition of individual components to the starting material. To overcome some of the problems posed by these very small samples, a mixture of NA10 + $Ag_2C_2O_4$ + AgN_3 with 8.7 wt% CO_2 and 8.0 wt% N_2 was prepared prior to sample loading. Double-distilled and deionized H_2O was added with the aid of a micro-syringe, but the proportion of H_2O relative to solid materials could not be determined quantitatively because the sample masses are on the order of 0.0x μg. It was, nevertheless attempted to add total solid (NA10 + $Ag_2C_2O_4$ + AgN_3) and liquid H_2O visually estimated to about 1:1 volume proportions.

For both Raman and infrared spectroscopic measurements, the sample was first heated to the highest planned temperature and was kept at this condition for ~ 60 min. This duration is more than sufficient to reach equilibrium for which only several minutes are needed (Horita 1988 2014). Spectroscopic measurements of the sample (fluid, melt, and/or supercritical fluid) together with the ^{13}C diamond were carried out at this highest temperature and pressure condition. The temperature was then decreased at 1 °C/s to the next temperature and pressure condition with the sample remaining in the HADC, and the measurement protocol was repeated. There was some deformation of the Ir gaskets at the highest temperatures of these experiments, which led to sample volume changes. This is why pressure does not decrease monotonously with decreasing temperature (Table 1). Ambient conditions Raman measurements of the ^{13}C diamond were carried before and after a series of high-temperature and high-pressure experiments.

The FTIR measurements were conducted with the same sample contained in the diamond cell as that used for Raman measurements. This sample was subjected to the same temperatures and pressures by using the same protocol as the Raman measurements and was carried out after the Raman measurements were completed. Pressure could not be measured with the diamond shift in the FTIR system. It was assumed, therefore, that for given temperature, the pressure recorded with the ^{13}C diamond in the Raman measurement series was the same during the subsequent FTIR measurements. The validity of this assumption was assessed by monitoring the temperature of gas exsolution from the fluid phase during the Raman and FTIR experiments. Gas exsolution at the same temperature at constant sample volume implies the same pressure at each temperature. In the present experiments, this exsolution temperature was

Table 1 Experimental data

NA10–C–O–H–N		
Temperature, °C*	Pressure, MPa*	Phases
475	63	Melt + fluid
550	262	Melt + fluid
625	361	Melt + fluid
700	489	Melt + fluid
750	679	Melt + fluid
800	643	Single-phase fluid
25	0.1	Glass + gas bubbles
C–O–H		
Temperature, °C*	Pressure, MPa*	Phases
800	865	Fluid
700	267	Fluid
600	132	Fluid
25	0.1	Gas

*Pressure and temperature uncertainties and precision are described in the text

within 10°–15 °C in the Raman and FTIR measurement series. This uncertainty corresponds to a ± 0.02 g/cm^3 density uncertainty for pure H_2O, which for the H_2O density of the experiments reported here corresponds to ± 20 MPa. Equation-of-state data for the fluid mixture are not available, but given that H_2O is a dominant volatile component, ± 20 MPa uncertainty seems reasonable. This conclusion implied a temperature/pressure evolution in the FTIR experiments within error the same as in the Raman measurement series.

The Raman and infrared transmission spectra include contributions from the diamond anvils. This background signal was eliminated in the FTIR spectra by normalizing to those of an empty diamond cell at the same temperature after the sample was removed. For the Raman spectra, the procedure is more complex, and the diamond background was subtracted by using the procedure of Dalou et al. (2015). Briefly, this required recording Raman spectra of the diamond cell away from the sample hole in the Ir gasket but within the same focal plane as the sample measurement at the same temperature and pressure. This spectrum was then subtracted from the sample spectra (melts, fluids, and supercritical fluids), which results in a spectrum without signal contribution from the second-order Raman signal intensity from the diamond cell itself.

The subtraction method used for the Raman spectra could not be used over more than several hundred cm^{-1} frequency range in part because the sensitivity of the CCD detector varies with frequency. Subtraction over wider ranges could, therefore, result in artificial intensity variations. Moreover, in frequency regions with rapid intensity variations, the spectrometer positioning uncertainty (2–3 cm^{-1}) could result in artificial sharp intensity variations. Care was, therefore, exercised to narrow the frequency ranges of Raman spectra to within those of particular interest when conducting the background correction. The infrared spectra suffered from complications resulting from major variations in intensity across the spectra so that in order to obtain reasonable signals from CO_2 and from the overtone/combination bands of water species (see also below) portions of the spectra could not be recorded as the signals commonly saturated the detector. Sample thickness is governed by the gasket thickness and could not, therefore, be varied greatly. It is for this reason that only portions of the infrared spectra are shown.

Results

Experimental conditions are shown in Table 1. In the temperature range from 475 to 750 °C, the samples comprise 30–40 μm diameter melt spheres in a matrix of fluid (Fig. 1). At 800 °C, there is a single fluid phase only. The melt spheres and the fluid matrix could be probed quantitatively with the confocal micro-Raman system because with the ~ 1 μm diameter laser beam the sampling depth is within the range of the sphere diameter or less. In the FTIR spectrometer, however, the minimum aperture is > 30 μm, which means that with the 30–40 μm diameter melt spheres in these experiments, quantitative infrared absorption spectra of melts were not possible because a significant portion of the infrared beam will pass through neighboring fluid. Only infrared absorption spectra of fluid spectra are, therefore, reported.

Infrared absorption spectra

Infrared absorption spectra of fluid in the frequency range of the combination and overtone bands of OH and H_2O (4000–7500 cm^{-1}) can be used to infer changes in total H_2O content as well as proportion of molecular H_2O and OH groups that form bonding with Si^{4+} and Al^{3+} in silicate (e.g., Scholze 1959; Malfait and Xue 2014). The presence of molecular H_2O in fluids (and melts) characteristically results in absorption centered near 5200 cm^{-1}, which has been assigned to a combination band of the H–O–H bending vibration near 1600 cm^{-1} and the OH fundamentals near 3600 cm^{-1} (Fig. 2) (Scholze 1959). The 7200 cm^{-1} band is the first overtone of the 3600 cm^{-1} OH stretch vibrations and has been used to obtain total water content of the sample (Scholze 1959; Dixon et al. 1995; Behrens and Nowak 2003). The 5200 and 7200 cm^{-1} bands broaden significantly with increasing temperature with the broadening of the 7200 cm^{-1} including development of an

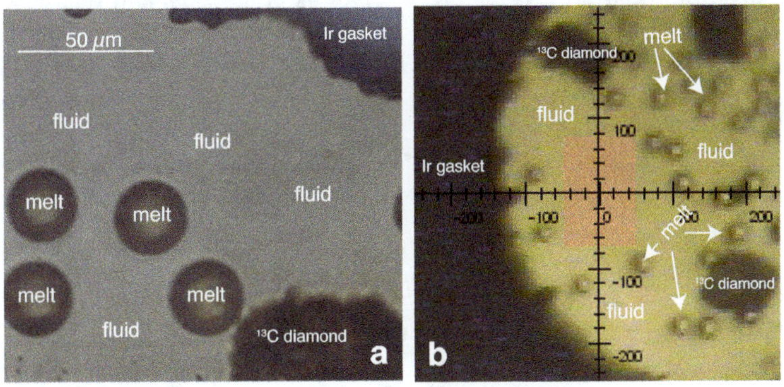

Fig. 1 Microphotograph of samples in the diamond cell at 550 °C and 262 MPa. **a** As viewed through the optical microscope during micro-Raman measurements. The width of the field of view is 200 μm. **b** View of the sample through the optical microscope in the microFTIR system. The numbering on the crosshair indicates μm. The shaded area centered in the cross-hear indicates the position and dimension of the aperture (100 × 150 μm) for fluid examination in this particular experiment

asymmetry toward higher frequency at the highest temperatures.

There is also minor intensity centered near 4500 cm^{-1} at the higher temperatures in the infrared spectra of fluids (see insert in Fig. 2). The 4500 cm^{-1} band is a combination band of the OH fundamentals at 3600 cm^{-1} and vibrations in OH groups attached to Si and Al to form (Si,Al)–OH bond with its main vibrational band between 900 and 950 cm^{-1} (Malfait and Xue 2014). Its integrated intensity increases approximately linearly with increasing temperature to a maximum relative intensity of few percent relative to the 5200 cm^{-1} intensity (Fig. 3). Conversion of this information to concentration requires molar absorption coefficients, which are not available for these fluids. However, if we assume that the 10% difference in these coefficients between the

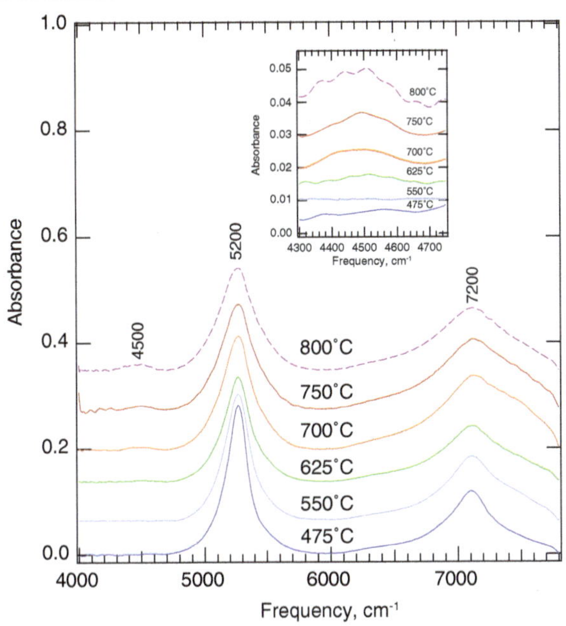

Fig. 2 FTIR spectra of fluid at temperatures indicated. Corresponding pressures can be found in Table 1. The dashed line is the spectrum from a single-phase fluid. Insert shows expanded view of frequency region centered at 4500 cm^{-1}. Note that all spectra are offset for clarity. The 0-intensity line passes through the background at 600 cm^{-1}. In the insert, the 0 line is approximately that near 4300 cm^{-1}

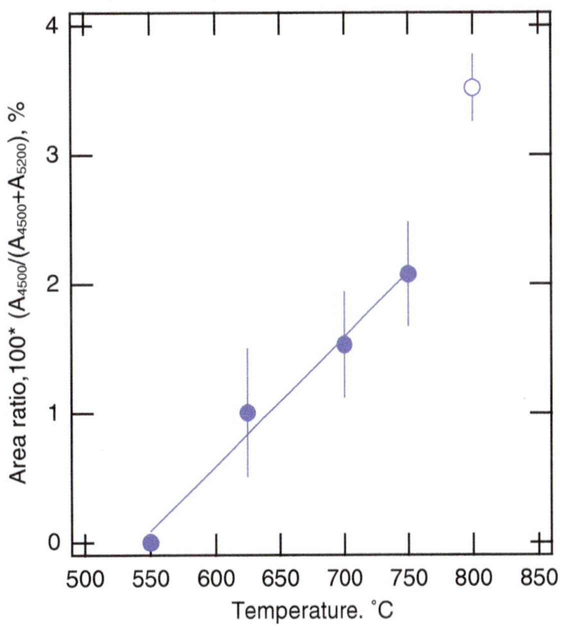

Fig. 3 Relationship between the area of the band centered near 4500 cm^{-1} and that near 5200 cm^{-1} (expressed in percent) from FTIR absorption spectra of fluid. The data point from 800 °C (open circle) is from single-phase fluid. The straight line is a simple linear fit to the data point (not including that from single-phase fluid at 800 °C); area ratio = 0.0100 ± 0.0009–5.4 ± 0.5 T(°C)

4500 and 5200 cm^{-1} absorption from $Na_2O–SiO_2$ glasses (Yamashita et al. 2008), from the relationship;

$$C_{OH}/C_{H2O} = A_{4500}/\varepsilon_{4500}/(A_{5200}/\varepsilon_{5200}), \quad (1)$$

where C is concentration, A is integrated area, and ε is molar absorption coefficient. The ε_{5200} is about 10% larger than ε_{4500}, which means that the area ratio in Fig. 3 at most underestimates the proportion of OH by about 10%.

Infrared spectroscopy also was employed to probe for species that comprise carbon or nitrogen (CO_2, CO, N_2, N_2O, etc.) or both (e.g., carboxyl groups or CN groups). Among these species, molecular N_2 is infrared inactive, but can be identified, if present, with Raman spectroscopy (see below).

In the spectra of fluids, the only absorption observed is that near 2350 cm^{-1} (Fig. 4), which is assigned as the ν_3 antisymmetric stretching in the CO_2 molecule (see Nowak et al. 2003, and references therein). Its integrated intensity is essentially independent of temperature and pressure (see insert A in Fig. 4). Notably, though, the width of this absorption broadens with increasing temperature. It also is asymmetric, wherein a second band could be fitted near 2320 cm^{-1} (see insert B, Fig. 4). This second absorption might exist because there may be variations in the geometry of the CO_2 molecule when dissolved in silicate-saturated C–O–H–N fluid.

Raman spectra

Raman spectra provide information on a range of potentially volatile species as well as silicate species in fluids and melts.

H_2O component

The 3000–4000 cm^{-1} frequency region of the Raman spectra comprises that where signals from the fundamental OH vibrations will occur when water is present in a sample (Walrafen 1967; Holtz et al. 1996). This Raman intensity

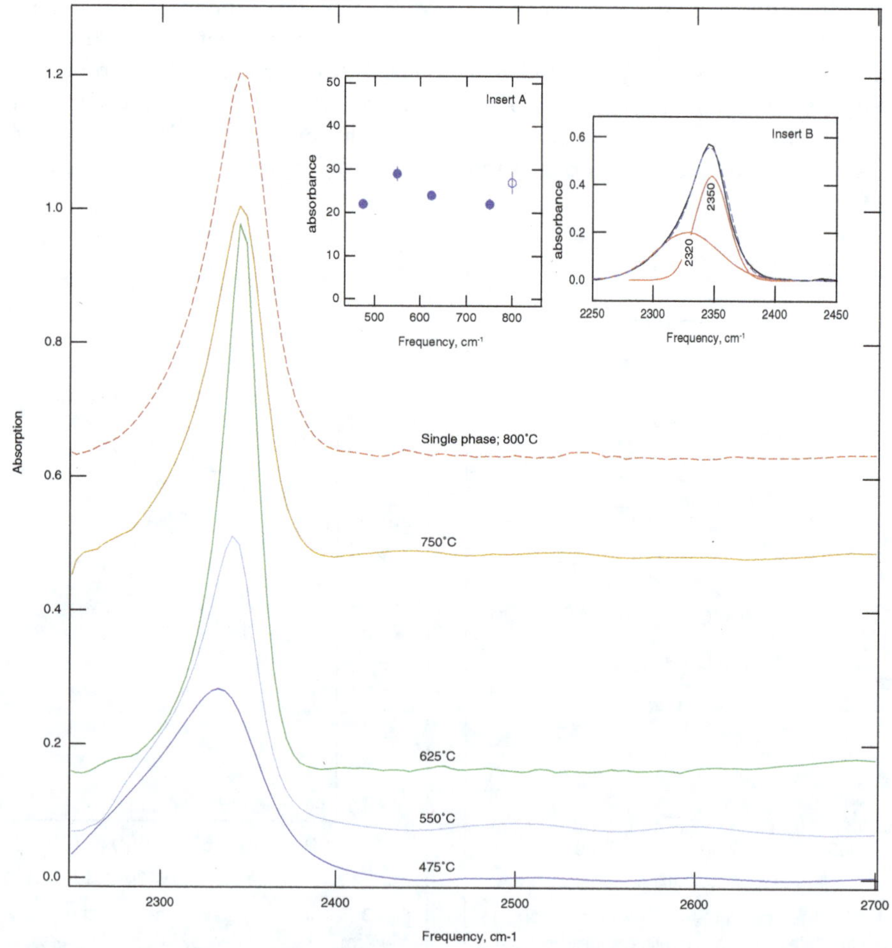

Fig. 4 Infrared absorption spectra of fluids in the spectral region of C–O stretching in CO_2 molecules as a function of temperature and pressure (see Table 1 for pressures). The spectra are offset for clarity. The vertical *line* is the scale bar for the Absorption intensity of the spectra. Inserts A and B show the integrated intensities of the absorption and an example of 2-line curve-fitted infrared spectrum by using Gaussian *line* shape. The spectrum shown was recorded while the sample was at 750 °C. *Dashed line* is from 800 °C where a single fluid phase only exists

includes contributions from both molecular H_2O and structurally bound OH groups.

Sharp, nearly symmetric bands with intensity maxima slightly below 3600 cm^{-1} were recorded in the spectra of all fluids and all melts (Fig. 5a, b). By normalizing the integrated intensities to that in the spectrum of single-phase fluid at 800 °C (diamond in Fig. 5c), we find that this intensity passes through at maximum near 625 °C before decreasing with further temperature increase. The intensities in the spectra of fluid always exceed those of coexisting melt.

The width of the 3600 cm^{-1} Raman bands decreases with increasing temperature (Fig. 5d), which is qualitatively similar to that reported for pure H_2O as well as that of water in silicate-rich hydrous fluids and melts in silicate–H_2O systems (Frantz et al. 1993; Walrafen et al. 1999; Foustoukos and Mysen 2015). This decrease is associated with diminishing asymmetry on the low-frequency-side of the spectrum, a feature that has been ascribed to less influence of hydrogen bonding among water species with increasing temperature (Walrafen et al. 1996, 1999; Foustoukos and Mysen 2012). At all temperatures and pressures, the width of this band in spectra of hydrous melts is greater than in the spectra of silicate-bearing aqueous fluids. This difference reflects either different hydrogen bonding environment in fluids and melts or a broader range of different forms of OH groups might exist in hydrous melts, or both, in hydrous melts as compared with the silicate-bearing C–O–H–N fluids. It is likely that both features contribute. It has been shown, for example, that the enthalpy of hydrogen bond formation of water in silicate melts is less than 50% of the value in aqueous fluids (Mysen 2014). Given the much more silica-rich nature of hydrous melts compared with silicate-saturated C–O–H–N fluids in the temperature and pressure range under a study, it is not

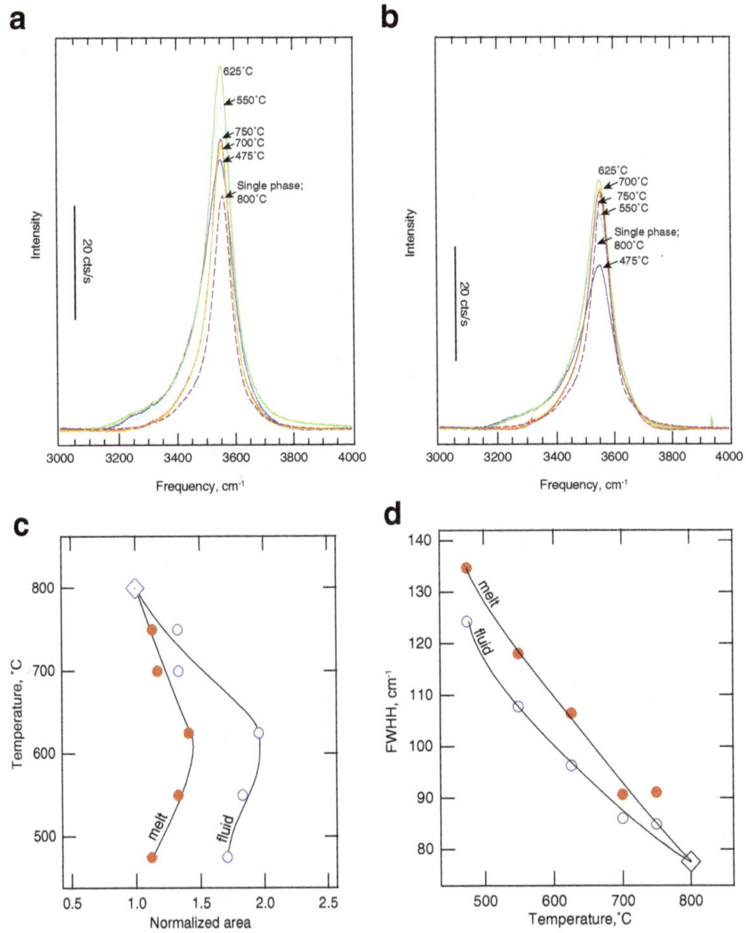

Fig. 5 Raman spectra of coexisting fluid and melt in the high-frequency region of fundamental O–H vibrations. **a** Spectra of fluid at temperatures indicated (corresponding pressures can be found in Table 1). **b** Spectra of coexisting melt (corresponding pressures can be found in Table 1). The spectra are normalized to the same acquisition time with identical focal plane and with otherwise identical instrumental parameters. **c** Normalized areas of coexisting melt and fluid as indicated. **d** Full width at half height, FWHH, from the spectra in Figs. A and B

surprising that a broader range of types of OH formation would be found in water-bearing silicate melts.

Nitrogen component

The most common N-bearing species in the C–O–H–N system could be N_2, N_2O, (N..H)-species, nitride and perhaps CN complexes. Their individual stability depends on bulk composition, temperature, pressure, and redox conditions. There is no evidence in the present spectra for (N..H) complexes (in the 3000–3400 cm^{-1} range; Yeo and Ford, 1994), nitride substitution in SiO_4 tetrahedra (> 1200 cm^{-1}; Sato et al. 1990), and CN-complexes (2000–2200 cm^{-1} frequency range; see, for example, Nakamoto 2009). The lack of Raman signals in the frequency ranges of these possible compounds is consistent with the fact that nitrogen was added as N_2 (from AgN_3 decomposition) and no reducing agent was added to the sample.

Absent a reducing agent, nitrogen in the C–O–H–N system may exist as N_2O, azide (N^{3-}) or as N_2, which are all detectable with Raman spectroscopy. For example, in an experimental study of N-bearing silicate glasses in which Raman and ^{15}N MAS NMR spectroscopies were the structural tools, Roskosz Mysen and Cody (2006) reported that the principal Raman signals from nitrosyl (N_2O) and azide (N≡N) are near 2100 and 2200 cm^1. No Raman signals were detected in this frequency region thus ruling out detectable amounts of such species in either fluid or melt. Whether or not NO_3 groups are formed will be addressed below in the presentation of the low-frequency range comprising vibrations of the silicate network because N–O vibrations occur in the same frequency range (between 1100 and 1200 cm^{-1}; Fung et al. 1994). There is no Raman signal in this frequency range in spectra of fluids. The signal intensity in spectra of melts can be satisfied without resorting to the presence of Raman bands assignable to N–O vibrations in NO_3 groups. This leaves molecular N_2 as the only likely nitrogen species in fluids and melts in the silicate–C–O–H–N system under study here.

The N=N stretch vibrations from molecular N_2 dissolved in silicate melts at high pressure result in Raman bands slightly above 2300 cm^{-1} (Roskosz Mysen and Cody, 2006). The spectra of fluids and melts in the silicate–C–O–H–N system exhibit a single sharp line near 2320 cm^{-1} after the background from second-order Raman peaks in this region from the diamonds in the diamond cell, itself, was subtracted (Fig. 6a, b). The peaks are Gaussian in shape, which is similar to that reported by Roskosz Mysen and Cody (2006) in Raman spectra of N_2-saturated silicate glasses quenched from melts at pressures ≥ 1 GPa. In contrast, gaseous N_2 spectra have sharp Lorentzian peaks. This difference likely reflects some interaction between the N_2 molecules and the hydrous silicate melt matrix in which they are dissolved.

The integrated Raman intensities of these bands, normalized to the same acquisition time and retaining the same focal plane for all measurements show distinctly different abundance behavior with temperature and pressure relative to the value in the spectra of single phase fluid (Fig. 6c). Whereas the normalized abundance in fluid passes through a maximum before decreasing rapidly with increasing temperature and pressure, the normalized abundance of N_2 in the coexisting melt is insensitive to temperature (and pressure).

Carbon component

Carbon speciation in fluids and melts is redox sensitive. Under reducing conditions such as those near the magnetite–wüstite buffer and more reducing, the most common species in silicate–COH system is CH_4 with the dominant Raman signals near 2900–3000 cm^{-1} (Kadik et al. 2004; Mysen et al. 2011; Armstrong et al. 2015). In the absence of hydrogen and under reducing conditions, oxycarbide complexes have been proposed (Libourel et al. 2003; Renlund et al. 1991; Soraru et al. 1995) with the two dominant Raman bands near 1350 and slightly above 1500 cm^{-1}, respectively (Soraru et al. 1995). There is no evidence for any of these species in the Raman spectra of fluids and melts, which is, of course, understandable given the conditions of no reducing agent in samples were carbon was added to the starting materials as CO_2.

Molecular CO_2, CO_3^{2-}, and HCO_3^- groups are the three main C-bearing species under oxidizing conditions. Molecular CO_2 was detected in the infrared absorption spectra of fluid as discussed above (Fig. 4). However, the Fermi diad of molecular CO_2, which at ambient conditions occurs at 1280 and 1380 cm^{-1}, respectively (Rosso and Bodnar 1995; Lamadric et al. 2017), cannot be identified in Raman spectra of the samples contained in diamond cells at high temperatures because the one-phonon Raman shift from the natural diamonds used as anvils in the cell results in a Raman band with intensity that covers those frequency regions. As noted above, infrared spectra could not be obtained from the melt spheres because of their small size. It could not be established, therefore, whether molecular CO_2 also exists in the hydrous silicate melts at high temperature and pressure.

The Raman spectra of fluids and melts may, contain information relevant to possibility of CO_3^{2-} and HCO_3^- groups because of their dominant C–O stretch vibrations near 1070 and 1020 cm^{-1} at ambient temperatures and pressures. The C–O stretch vibration in HCO_3^- is particularly sensitive to temperature and is located near

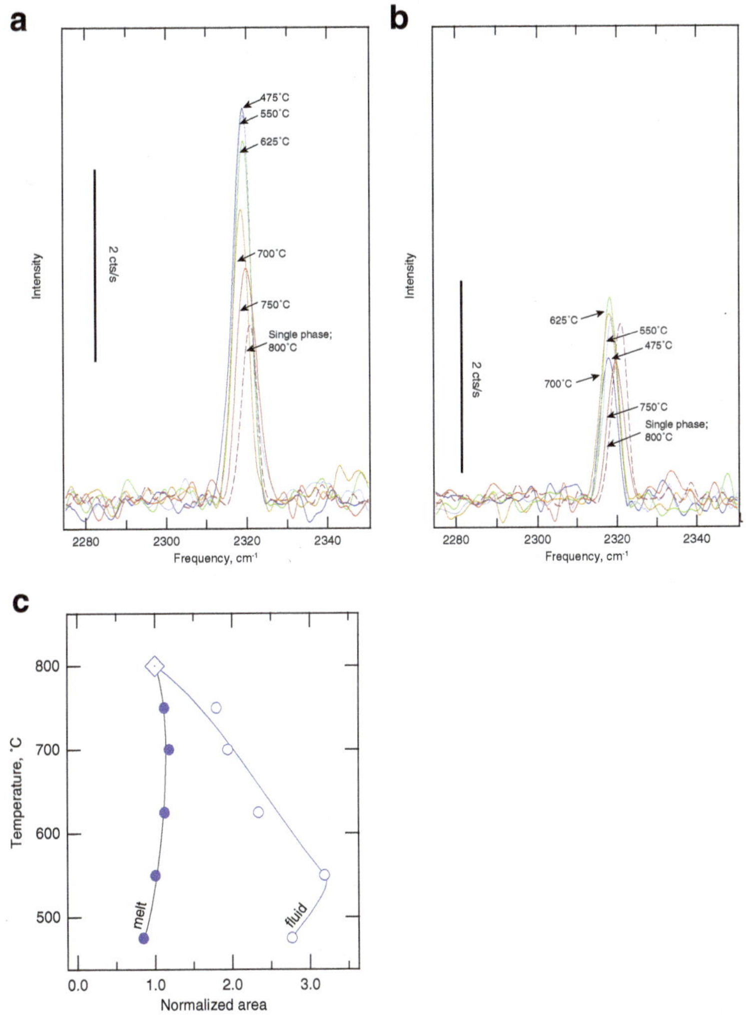

Fig. 6 Raman spectra of coexisting fluid and melt in the high-frequency region of fundamental N=N vibrations. **a** Spectra of fluid at temperatures indicated (corresponding pressures can be found in Table 1). **b** Spectra of coexisting melt (corresponding pressures can be found in Table 1). The spectra are normalized to the same acquisition time with identical focal plane and with otherwise identical instrumental parameters. **c** Areas of coexisting melt and fluid normalized to the area of single phase fluid at 800 °C as a function of temperature and pressure

950 cm^{-1} near 500 °C, for example (Frantz 1998). In simple COH fluids equilibrated at 700 °C and above (without nitrogen and without silicate components), there is evidence for CO_3^{2-} groups from Raman intensity near 1070 cm^{-1} (Fig. 7), which is assigned to C–O stretch vibrations in CO_3^{2-} groups (White 1974; Frantz 1998). A peak that occurs slightly below 1100 cm^{-1} in the spectra of silicate–C–O–H–N melt from 550 °C to the highest temperatures (Fig. 8). The same peak can be detected in spectra of fluids from 550 °C to the highest temperatures. It is assigned to C–O vibrations in CO_3^{2-} groups. Within the scatter of the data, there are no other peaks in this frequency region in the spectra of fluid (Fig. 8a). In the spectra of melts, a second band appears near 1000 cm^{-1} (Fig. 8b), which is assigned to C–OH vibrations in HCO_3^- groups (Frantz 1998). We cannot detect HCO_3^- groups in spectra of fluids. Neither is there any signal above 1100 cm^{-1}, which could be due to N–O vibrations in NO_3 groups in the fluids (Fung et al. 1994).

Absolute concentration of CO_3^{2-} and HCO_3^- groups cannot be obtained from the spectra without assuming that the Raman cross-sections for the two vibrations are similar. It is clear, nevertheless, that in both fluid and coexisting melt, the CO_3^{2-} abundance increase with temperature and pressure until the univariant line that separates melt + fluid from a single fluid is crossed (near 800 °C) (Fig. 9). In this regard, the evolution of the CO_3^{2-} abundance with temperature and pressure resembles those of N_2 and H_2O (Figs. 5c and 6c). There is, however, a difference in that for N_2 and H_2O, these

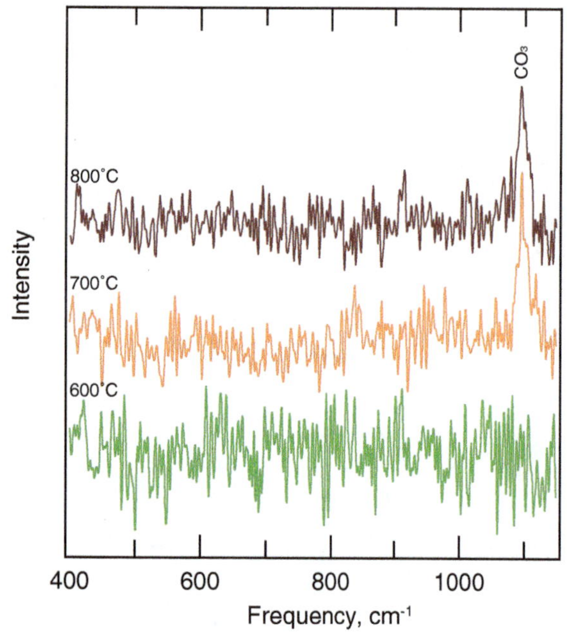

Fig. 7 Raman spectra of COH fluids at 600°, 700°, and 800 °C and corresponding pressures (see Table 1). The peak near 1070 cm^{-1} is assigned to C–O stretch vibrations in CO_3 groups (Frantz et al. 1993) and marked CO_3

components are favored by the fluid, whereas for the CO_3^{2-} groups, these are favored by melt.

Silicate components

Silicate components and their species exist in coexisting silicate melts and C–O–H–N fluid. There are, however, significant differences between the silicate-saturated C–O–H–N fluid and the C–O–H–N-saturated silicate melts as seen, for example, in their significantly different Raman spectra (Fig. 10). In spectra of fluid, the dominant peak is near 780 cm^{-1}, and is assigned to Si–O vibrations in Q^0 species (Zotov and Keppler 2002; Mibe et al. 2008). The spectra of fluid characteristically also comprise a broad band centered near 600 cm^{-1}, with two weaker bands near 810 and 900 cm^{-1}, respectively. The latter two Raman bands are assigned Si–O vibrations in Q^1 and Q^2 species, respectively, whereas that centered near 600 cm^{-1} is an Si–O–Si bending vibration, which requires, therefore, bridging oxygen such as, for example, in Q^1 and Q^2 species (Zotov and Keppler 2002). At frequencies higher than 1000 cm^{-1} in spectra of fluids, there is a weak band near 1070 cm^{-1}, assigned to C–O stretching in CO_3 groups, as discussed under C-bearing species above (see also Fig. 8).

The spectra of coexisting melts differ significantly from those of fluids. They do comprise the 780, 810, and 900 cm^{-1} bands also seen in spectra of fluids, but in addition, there is a band near 1100 cm^{-1} not seen in the fluid spectra. There is also much more intensity at frequencies below 700 cm^{-1} (Fig. 10). The 1100 cm^{-1} is assigned to Si–O vibrations in Q^3 species, whereas the extensive intensity below 700 cm^{-1} is due to bending and rocking motions in polymerized silicate species (McMillan 1984; Neuville et al. 2006; Mibe et al. 2008). The intensity near 500 cm^{-1}, coupled with the shoulder near 550 cm^{-1}, in the spectra of melts suggests that Q^4 species may exist in the C–O–H–N-saturated silicate melts. The 1070 and 1000 cm^{-1} bands, assigned to C–O stretching in CO_3^{2-} and HCO_3^- species, were discussed above (Fig. 8).

The abundance of Q-species and degree of polymerization, NBO/T, of silicate in fluids and melts were derived from the Raman spectra. This was accomplished by assuming that the Raman cross sections of Si–O stretch vibrations in Q^0, Q^1, Q^2, and Q^3 species were the same. The NBO/T can then be calculated from the simple relationship;

$$\mathrm{NBO/T} = \sum_{n=1}^{n=4} (\mathrm{NBO/T})_{Q^n} \bullet X_{Q^n}, \tag{2}$$

where $(\mathrm{NBO/T})_{Q_n}$ is the degree of polymerization of a specific Q-species and X_{Q_n} its mol fraction.

The C–O–H–N-saturated melts become increasingly depolymerized (NBO/T increases) with increasing temperature and pressure, whereas that of silicate in the C–O–H–N fluid follows the opposite trend (Fig. 11). These trends qualitatively are the same as those observed in the simpler silicate–H_2O systems (Mysen 2009) and reflect how the H_2O component in silicate–C–O–H–N melts in breaks up the silicate network. The effect becomes more pronounced with increasing temperature and pressure because such increases also most likely are associated with increasing water concentration in the melt. In the C–O–H–N fluids, on the other hand, the silicate solubility increases with increasing temperature and pressure. Increasing silicate concentration in water-bearing fluids results in increasing silicate polymerization (Mysen et al. 2013).

The NBO/T evolution with temperature and pressure reflects the abundance evolution of the Q-species (Fig. 11). For melts, the Q^0 abundance increases with temperature and pressure, that of Q^1 is insensitive, whereas the abundance of more polymerized species, Q^2 and Q^3, decreases with increasing temperature and pressure and, therefore, increasing water content. In other words, the C–O–H–N-saturated melts behave as simpler H_2O-saturated melts, which become increasingly depolymerized with increasing water content (Zotov and Keppler 1998). This behavior contrasts with the silicate dissolved in C–O–H–N fluid, where the abundance of

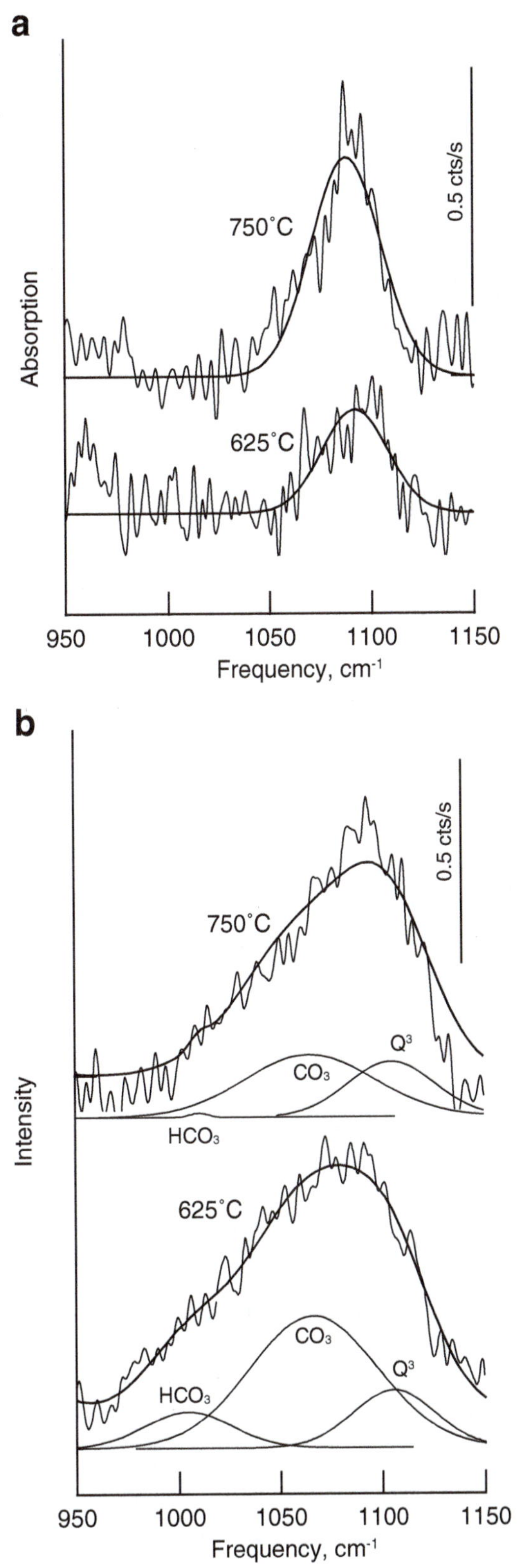

Fig. 8 Example of Raman spectra of silicate-C-O-H-N fluid and melt at 625° and 750 °C, respectively, fitted to Gaussian lines in the 950–1150 cm^{-1} frequency range. **a** Spectra from fluid with a single line assigned to C-O vibrations in CO_3 groups. **b** Spectra from melts where three lines are needed. These are assigned to C-O stretching in HCO_3^- and CO_3^{2-} groups and to Si-O stretching in Q^3 silicate species (see text for discussion on these assignments)

Q^0 species decreases and that of Q^1 increases with increasing temperature and pressure and, therefore, likely total water content (Fig. 11).

Discussion

The solubility mechanisms of volatile species in silicate–C–O–H–N melts qualitatively resemble those in the simpler systems such as silicate–H_2O and silicate–COH. Addition of nitrogen to the system under oxidizing conditions results in formation of molecular N_2, which does not interact chemically with the other volatile components.

Water in the form of molecular H_2O dominates in C–O–H–N fluids with OH groups forming less than 3% of the total water concentration even at the highest temperatures and pressures (Fig. 3). This is less than the proportion of OH groups in pure silicate–H_2O systems at similar temperatures and pressures, where this proportion often exceeds 10% (Mysen 2010). In fact, by using the relationship between proportion of OH in fluid, expressed

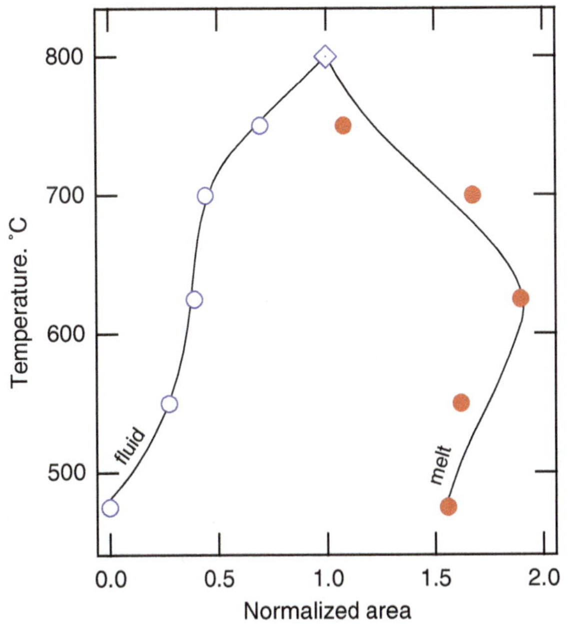

Fig. 9 Evolution of CO_3^{2} abundance in coexisting fluid and melt as a function of temperature and pressure with concentration, expressed as area of the ~ 1070 cm^{-1} band normalized to the value for single phase fluid at 800 °C (diamond symbol)

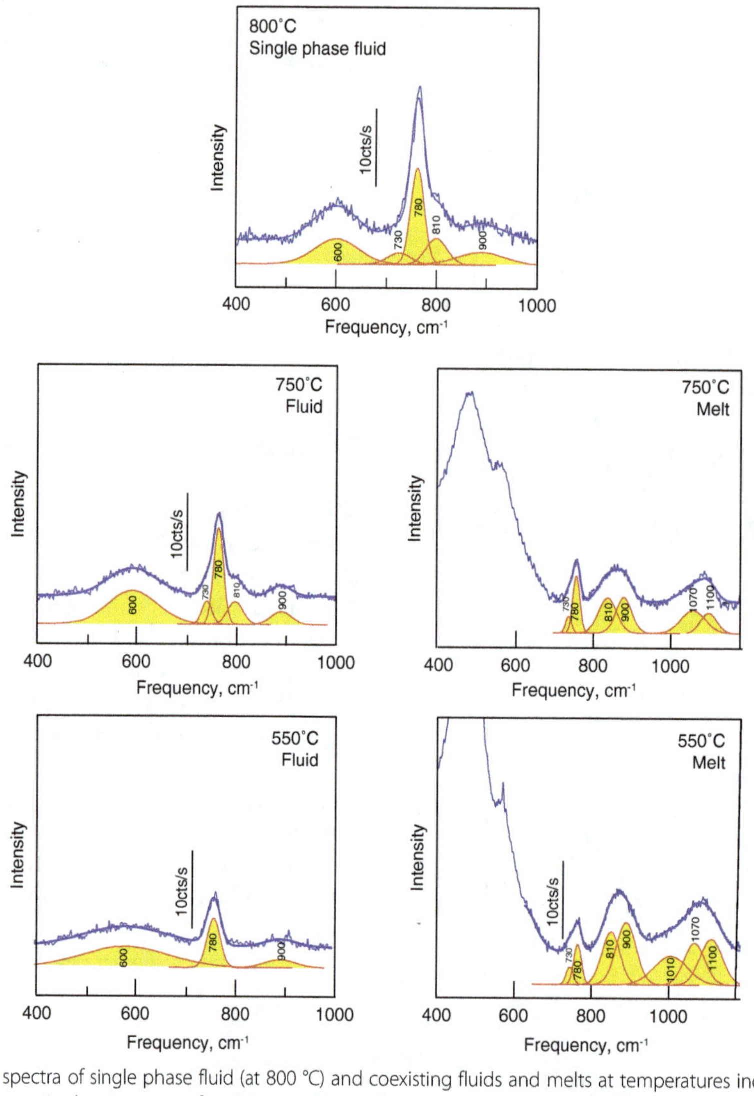

Fig. 10 Examples of Raman spectra of single phase fluid (at 800 °C) and coexisting fluids and melts at temperatures indicated (corresponding pressures in Table 1). Also shown in the spectra are fitted Gaussian lines relevant to structural assignments of silicate and carbonate species. The intensity scale in each spectrum is identical

as $A_{4500}/(A_{4500} + A_{5200})$ from the FTIR data, and fugacity of H_2O for the same silicate composition and pure H_2O (Mysen 2010) (see Fig. 12a), the much lower proportion of OH groups in the C–O–H–N–silicate fluid implies very considerably dilution of water because the f_{H2O} of pure H_2O at the pressures and temperatures of these experiments is in the 200–700 MPa range.

Equivalent OH/H_2O for melt were not obtained for reasons discussed above. However, the overall degree of silicate polymerization as a function of water fugacity, using the NA10–H_2O data from Mysen (2010) for calibration, results in an equivalent f_{H2O} quite similar to that of pure H_2O (Fig. 12b). This observation may suggest that the presence of C and N components in the melt has a much smaller influence on activity composition relations in the (C–O–H–N)-saturated melt than in fluid. This difference reflects the much smaller solubility if silicate components in the fluid than in the melt.

Fluid/melt partition coefficients can be obtained from the ratio of integrated intensities of bands in spectra of coexisting melts and fluids assigned to specific vibrations provided that the instrumental parameters such as acquisition time, slit settings, and focal depth are same during acquisition of both spectra. This can be done because in a ratio of integrated intensities from bands assigned to the same vibration, the remaining variables affecting Raman intensity is temperature, which, of course, is the same for coexisting fluid and melt. Then, the partition coefficient of component, *i*, between fluid and melt, f/m, is;

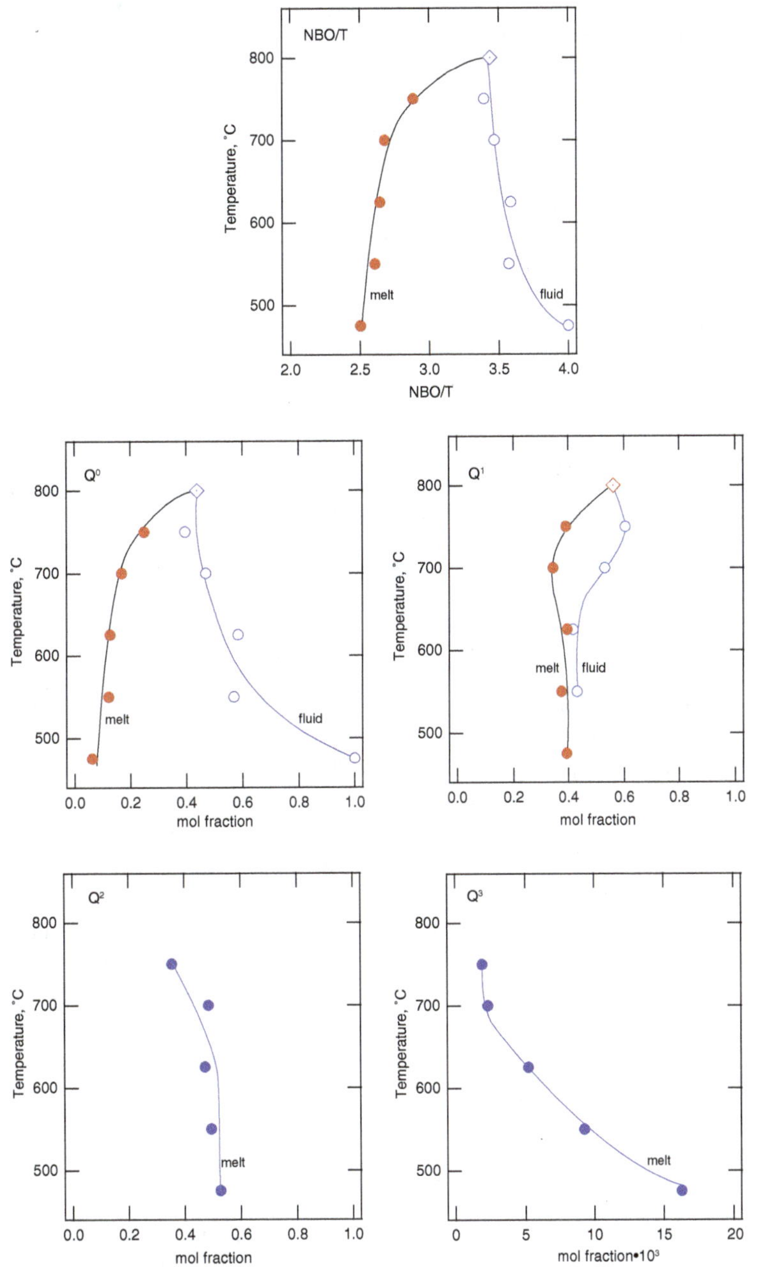

Fig. 11 Evolution of Q-species abundance and bulk NBO/T of melt and coexisting fluid as a function of temperature (and pressure, see Table 1). Note that diamond symbol represents single phase

$$K_i^{f/m} = A_i^{\mathrm{f}}/A_i^{\mathrm{m}}, \tag{3}$$

where A_i^{f} and A_i^{m} are the integrated areas from the Raman spectra of a band assigned to a specific type of vibration in component, *i*.

The partition coefficient for total H_2O (which includes all OH-bearing species) can be derived with Eq. (3) from the 3600 cm^{-1} band intensity in spectra of fluids and melts (Fig. 5). This partition coefficient is always greater than unity, which is similar to the partition coefficient in silicate–H_2O systems (Mysen 2013; Webster et al. 2014). From the linear fit to the ln $K_{\mathrm{H2O}}^{\mathrm{f/m}}$ vs/ 1/T (T is temperature in kelvin), the enthalpy from this partioning is -6.5 ± 1.5 kJ/mol (Fig. 13), which is quite similar to the results in the simpler silicate–H_2O (Mysen 2013) where $\Delta H = -7.6 \pm 0.7$ kJ/mol. That fact that this ΔH value differs from 0 reflects, most likely, the different extent and type of speciation of H_2O in the coexisting melt and fluid. We should also note that the decreasing partition

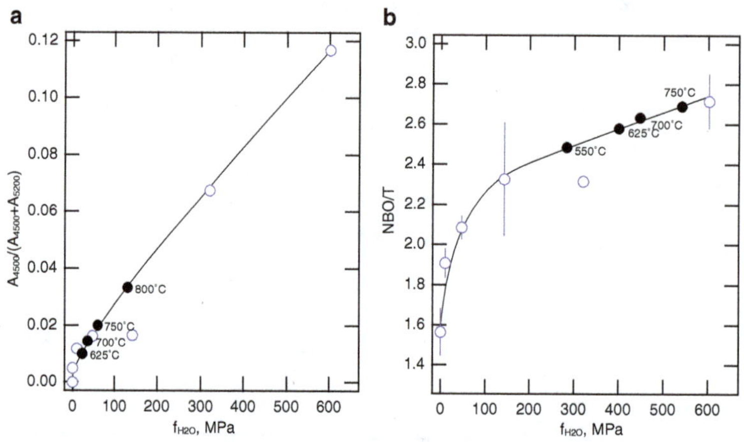

Fig. 12 Relationship between fugacity of H_2O, f_{H2O}, and species abundance. **a** Proportion of OH and molecular H_2O as a function of f_{H2O} in NA10–H_2O fluid from Mysen (2010) (open symbols). Solid symbols plotted onto the line are the date from the current silicate–C–O–H–N fluid. In both cases, the silicate starting composition is NA10. See text for discussion of approximation of OH/(OH + H_2O) ratio with the FTIR absorbance ratio, $A_{4500}/(A_{4500} + A_{5200})$. Silicate polymerization of melt, NBO/T, as a function of f_{H2O} in NA10–H_2O fluid from Mysen (2010) (open symbols). **b** Evolution of NBO/T of the silicate component in the fluid as a function of temperature. Solid symbols plotted onto the line are the data from the current silicate–C–O–H–N fluid. In both cases, the silicate starting composition is NA10

coefficient with increasing temperature actually reflects an increase in pressure as well, which results in increased silicate solubility in fluid and decreased silicate content of hydrous melt (its water content increases). In the isobaric (0.2 GPa) experiments by Webster et al. (2014), no temperature-dependence was observed. That notwithstanding, their fluid/melt partitions fall naturally on the extension of the present data and those of Mysen (2013).

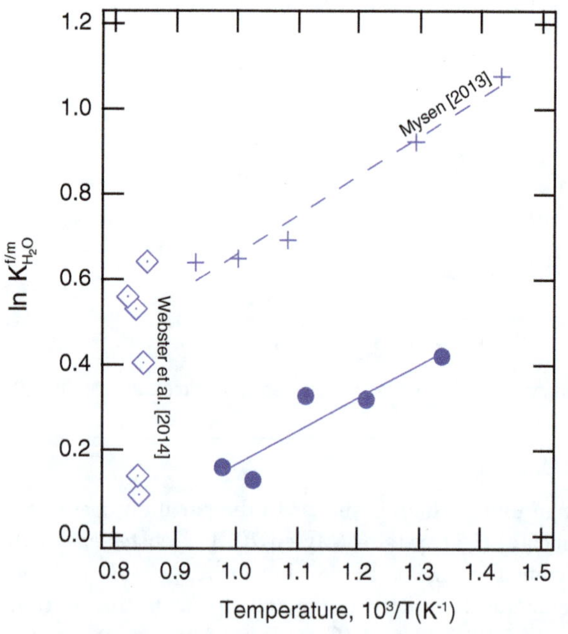

Fig. 13 Partition coefficient between fluid and melt of total H_2O, $K^{f/m}_{H2O}$, as a function of temperature. The least squares straight line fit is, $K^{f/m}_{H2O} = -0.6 \pm 0.2 + 782 \pm 189/T$, $r^2 = 0.85$. Also shown are data from natural samples by Webster et al. (2014) and from experimental data in silicate–H_2O from Mysen (2014). The least squares fit through the data of Mysen (2013) yields the expression, $K^{f/m}_{H2O} = -0.3 \pm 0.1 + 782 \pm 920/T$, $r^2 = 0.97$

For carbon species in melts and fluids in the silicate–C–O–H–N system, there are also similarities between the data and those in the simpler silicate–COH system (Mysen 2015a). However, there also are differences. For example, in the silicate-saturated C–O–H–N fluid, we have evidence only of molecular CO_2 and CO_3^{2} groups, whereas in the silicate–COH system, Mysen (2015a) also observed evidence of HCO_3^{-} groups in the fluid. This difference implies that in the C–O–H–N fluids studied here, the simple carbonate equilibrium;

$$2HCO_3^{-} = 2CO_3^{2} + 2H^+, \tag{4}$$

is shifted further to the left than was the case in the COH fluids studied by Mysen (2015a). The implication is that the C–O–H–N fluids examined here are more alkaline that those examined by Mysen (2015a).

For C–O–H–N-saturated melts, on the other hand, there is spectroscopic evidence for both CO_3^{2} and HCO_3^{-} groups. No evidence could be found for molecular CO_2 in the spectra of melt. A simple solution mechanism involving an equilibrium between CO_3^{2} and HCO_3^{-} is (Mysen 2015a):

$$2CO_3^{2-} + H_2O + 2Q^n = 2HCO_3^{-} + 2Q^{n-1}. \tag{5}$$

In Eq. (5), Q^n, and Q^{n-1}, are silicate species with, respectively, n and n−1 bridging oxygen. The CO_3^{2-}/HCO_3^{-} abundance ratio in melts increases with increasing temperature with ΔH = 114 ± 22 kJ/mol (Fig. 14). The

HCO_3^- abundance in melt diminishes rapidly with increasing temperature and can barely be observed at 750 °C and cannot be seen in spectra of single phase fluids at 800 °C. In comparison, in the N-free system silicate–COH, also examined in the hydrothermal diamond anvil cell with similar protocol, the CO_3/HCO_3^- ratio was determined for the fluid to be considerably less sensitive to temperature with a ΔH = 17 ± 3 kJ/mol (Fig. 14).

The fluid/melt partition coefficients for CO_3^{2-} groups are in the range of ~ 0.2 to ~ 0.7 as a function of increasing temperature and pressure with an enthalpy of 52 ± 14 kJ/mol (Fig. 15). These values fall in the range reported by Webster et al. (2014) and Mysen (2015b) even though the compositions in those experiments differed significantly from those of the present compositions.

Even though nitrogen exists in coexisting fluid and melt only as molecular N_2, there are differences in the interaction between the C–O–H–N fluid matrix in fluid and N_2 and silicate matrix in melts and N_2. This difference results in distinctly temperature-dependent N_2 partition coefficients with an enthalpy change (– 20 ± 4 kJ/mol; see Fig. 16) that is considerably greater than the – 6.5 ± 1.5 kJ/mol observed for H_2O partitioning between fluid and melt, for example (Fig. 13). The nitrogen partition coefficients also are more than 20% greater (and increasing with decreasing temperature) than for H_2O. It would seem, therefore, that the activity-composition relations of N_2 in melts, and perhaps fluids, differ from those of H_2O. This difference likely is because H_2O may dissolve both as molecular H_2O and as OH groups forming bonding the silicate network, whereas there is no evidence that N_2 behaves in such a manner.

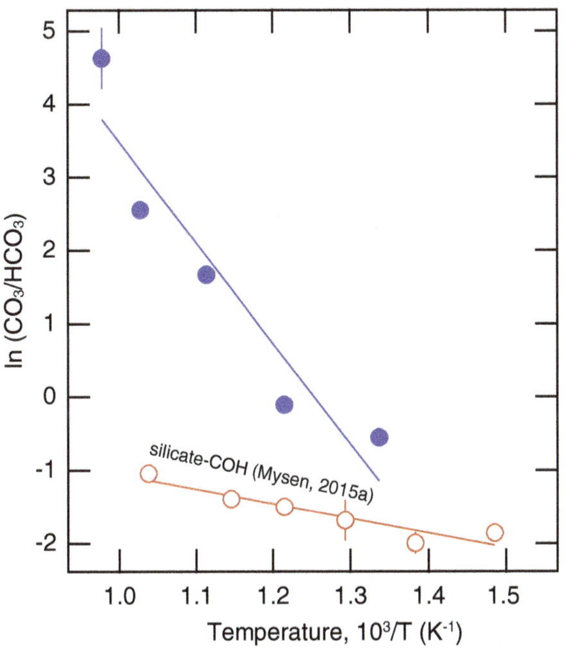

Fig. 14 Abundance ratio, CO_3/HCO_3, as a function of temperature (and pressure, see Table 1) in silicate–C–O–H–N melts. The straight line is least squares fitted to the expression, $\ln(CO_3/HCO_3) = -13{,}738 \pm 2680/T + 17 \pm 3$, $r^2 = 0.95$. Also shown are the experimental data from an aluminosilicate–COH system under similar conditions from Mysen (2015a). Note that the label on the *y*-axis in the original plot in Mysen (2015a) was inverted. This is corrected for the present figure. The regression line for that regression is $\ln(CO_3^{2}/HCO_3^-) = -1999 \pm 367/T + 0.9 \pm 0.4$, $r^2 = 0.94$

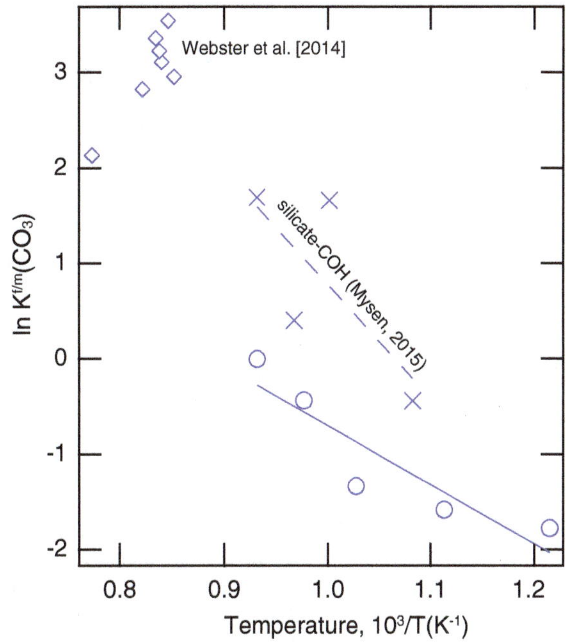

Fig. 15 Partition coefficient of CO_3^{2} groups between coexisting fluid and melt as a function of temperature. Partition coefficients were determined as discussed under Eq. (1) in the text. The straight line fit by least squares regression yields; $\ln K = 6 \pm 2 - 6203 \pm 1620/T$, $r^2 = 0.83$. Also shown are data from Mysen (2015a) from a silicate–COH system, also under oxidizing conditions. The dashed lines represents a least squares linear fit through those data, with the expression $\ln(CO_3^{2-}/HCO_3^-) = -11{,}893 \pm 7610/T + 13 \pm 8$, $r^2 = 0.74$ as well as data from Webster et al. (2014) using natural silicate melts and COH fluids that also contains halogens and sulfur

Conclusions

It follows from the observations here and in the literature (Mysen et al. 2008; Li and Keppler 2014) that where redox conditions in the Earth are more oxidizing than those defined by the NNO–H_2O and FMQ–H_2O buffer nitrogen released into fluid or melt will most likely exist as molecular N_2. In other words, there is no evidence to indicate chemical interaction between N_2 and other volatile and silicate species under such circumstances.

Redox conditions similar to these are those observed in subduction zones, for example (Wood et al. 1990; Cottrell and Kelley 2011). Under these conditions,

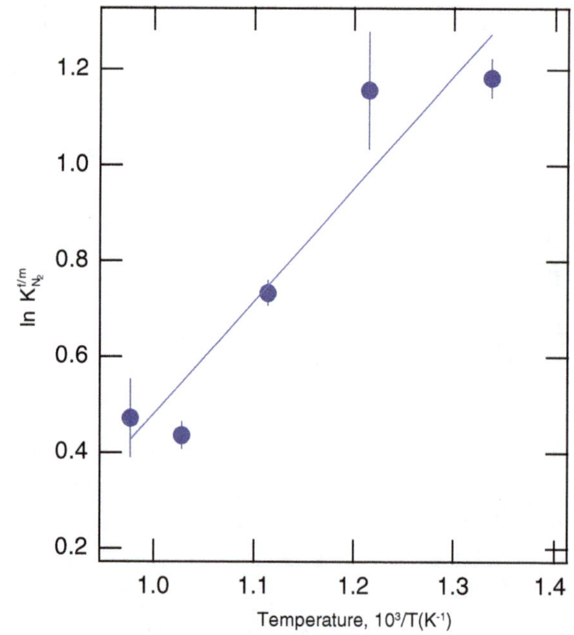

Fig. 16 Partition coefficient between fluid and melt of N_2, $K^{f/m}_{N2}$, as a function of temperature. The least squares straight line fit is, $K^{f/m}_{H2O} = -1.9 \pm 0.5 + 2356 \pm 450/T$, $r^2 = 0.90$

nitrogen effectively serves as a diluent of the volatile species. Dilution is indicated, for example, by the low OH/H_2O ratio of the C–O–H–N fluid (Fig. 12), which indicates lesser water activity than in silicate–H_2O fluids. This much lower OH/H_2O ratio probably this because silicate solubility in C–O–H–N fluids must be considerably less than in the absence of N_2. This follows from simple relationships of the form;

$$Si^{4+} + H_2O = Si(OH)^{3+} + H^+, \qquad (6)$$

Where Si^{4+} is a proxy for silicate components and $Si(OH)^{3+}$ a proxy for hydroxylated silicate. This means that as the activity of silicate decreases, the equation shifts to the left thus reducing the proportion of OH groups in the system,

It also appears, therefore, that a decrease of silicate activity in solution in the fluid resulting from the dilution of H_2O by other fluid components enhances fluid alkalinity. This feature, in turn, may alter the solubility of pH-dependent minor and trace elements in the fluid (e.g., high field strength cations; see Ayers et al. 2012) and, therefore result in different element patterns resulting from C–O–H–N fluid release than from release of fluids without N_2, where the silicate solubility is greater and the fluid likely is less alkaline. These features are, of course, also dependent on redox conditions because under reducing conditions nitrogen may form ammonia or ammine-type complexes (Mysen and Fogel, 2010; Li and Keppler 2014), which results in significant changes the behavior of nitrogen in both fluids and melt.

It appears that C–O–H–N-saturated melts are less sensitive to the presence of N_2 than C–O–H–N fluids as both the degree of melt polymerization, NBO/T, and silicate species abundance, appear insensitive to the existence of molecular N_2. However, as is the case for fluids, when magmatic activity occurs where conditions are sufficiently reducing to nitrogen to ammonia and ammine groups (somewhere near the hydrogen fugacity defined by the magnetitie–wüstite–H_2O redox buffer; see Mysen et al. 2008), this may change because the solubility of reduced nitrogen in silicate melts is much greater than oxidized nitrogen. Furthermore, the solution mechanism of reduced nitrogen as NH-complexes resembles that of dissolved H_2O (Mysen et al. 2014). Thus, nitrogen in magmatic liquids under such reducing conditions would tend to enhance melt properties that are sensitive to melt polymerization.

Acknowledgements
Instrument, electronics and library support was provided by Geophysical Laboratory technical staff. Their support is gratefully acknowledged.

Funding
This research was supported in by grant EAR1212854 from the National Science Foundation.

Author's contributions
All research reported in this report was carried out by the author.

Competing interests
The author declares that he has no competing interests.

References

Agee CR (2008) Static compression of silicate melt and the effect of water on planetary differentiation. Earth Planet Sci Lett 265,641–654.

Armstrong LS, Hirschmann MM, Stanley BD, Falksen EG, Jacobsen SD (2015) Speciation and solubility of reduced C-O-H-N volatiles in mafic melt: implications for volcanism atmospheric evolution and deep volatile cycles in the terrestrial planets. Geochim Cosmochim Acta 171:283–302

Ayers JC, Zhang L, Luo Y, Peters TJ (2012) Zircon solubility in alkaline aqueous fluids at upper crustal conditions. Geochim Cosmochim Acta 96:18–28

Bassett WA, Wu T-C, Cho I-M, Haselton T, Frantz JD, Mysen BO, Schiferl D (1996) The hydrothermal diamond anvil cell (DAC) and its applications. In: Dyar MD, McCammon CA, Schaefer M (eds) Mineral spectroscopy: a tribute to Roger G burns. The Geochemical Society, Houston, pp 261–272

Bebout GE (2007) Metamorphic chemical geodynamics of subduction zones. Earth Planet Sci Lett 260:373–393

Behrens H, Nowak M (2003) Quantification of H_2O speciation in silicate glasses and melts by IR spectroscopy - in situ versus quench techniques. Phase Transit 76:45–61

Boettcher AL, Mysen BO, Allen JC (1973) Techniques for the control of water fugacity and oxygen fugacity for experimentation in solid-media high-pressure apparatus. J Geophys Res 78:5898–5902

Carmichael ISE, Ghiorso MS (1990) Controls on oxidation-reduction relations in magmas. In: Nicholls J, Russell JK (eds) Modern methods of igneous petrology: understanding magmatic processes. The Mineralogical Society of America, Washington DC, pp 191–212

Cottrell E, Kelley KA (2011) The oxidation state of Fe in MORB glasses and the oxygen fugacity of the upper mantle. Earth Planet Sci Lett 308:1–13

Dalou C, Le Losq C, Mysen BO (2015) In situ study of the fractionation of hydrogen isotopes between aluminosilicate melts and coexisting aqueous fluids at high pressure and high temperature-implications for the delta D in magmatic processes. Earth Planet Sci Lett 426:158–166

Deines P (2002) The carbon isotope geochemistry of mantle xenoliths. Earth-Sci Rev 58:247–278.

Dingwell DB, Webb SL (1990) Relaxation in silicate melts. Eur J Mneral 2:427–449

Dixon JE, Stolper EM, Holloway JR (1995) An experimental study of water and carbon dioxide solubilities in mid-ocean ridge basaltic liquids part 1 calibration and solubility models. J Petrol 36:1607–1631

Fisher TP, Hilton DR, Zimmer M, Shaw AM, Sharp ZD, Walker JA (2002) Subduction and recycling if nitrogen along the central American margin. Science 297:1154–1157

Foustoukos D, Mysen BO (2015) The structure of water-saturated carbonate melts. Am Mineral 100:35–46

Foustoukos DI, Mysen BO (2012) D/H isotopic fractionation in the H_2-H_2O system at supercritical water conditions: composition and hydrogen bonding effects. Geochim Cosmochim Acta 86:88–102

Frantz G (1998) Raman spectra of potassium carbonate and bicarbonate aqueous fluids at elevated temperatures and pressures: comparison with theoretical simulations. Chem Geol 152:211–225

Frantz JD, Dubessy J, Mysen B (1993) An optical cell for Raman spectroscopic studies of supercritical fluids and its applications to the study of water to 500°Cand 2000 bar. Chem Geol 106:9–26

Frost DJ, McCammon CA (2008) The redox state of the Earth's mantle. Annu Rev Earth Planet Sci 36:389–384

Fung KH, Imre DG, Tang IN (1994) Detection limits for sulfates and nitrates in aerosol particles by Raman spectroscopy. J Aerosol Sci 25:479–485

Holtz F, Beny JM, Mysen BO, Pichavant M (1996) High-temperature Raman spectroscopy of silicate and aluminosilicate hydrous glasses: implications for water speciation. Chem Geol 128:25–39

Horita J (1988) Hydrogen isotope analyses of natural waters using an H_2-water equilibration method: a special implication to brines. Chem Geol 72:89–94

Horita J (2014) Oxygen and carbon isotope fractionation in the system dolomite–water–CO_2 to elevated temperatures. Geochim Cosmochim Acta 129:111–124

Jambon A (1994) Earth degassing and large-scale geochemical cycling of volatile elements. In: Carroll MR, Holloway JR (eds) Volatiles in Magmas. Mineralogical Society of America, Washington DC, pp 479–517

Javoy M (1997) The major volatile elements of the earth: their origin, behaviour and fate. Geophys Res Lett 24:177–180

Kadik A, Pineau F, Litvin Y, Jendrzejewski N, Martinez I, Javoy M (2004) Formation of carbon and hydrogen species in magmas at low oxygen fugacity. J Petrol 45:1297–1310

Kadik AA, Koltashev VV, Kryukova EB, Plotnichenko VG, Tsekhonya TI, Kononkova NN (2015) Solubility of nitrogen carbon and hydrogen in FeO-Na_2O-Al_2O_3-SiO_2 melt and liquid iron alloy: influence of oxygen fugacity. Geochem Int 53:849–868

Kadik AA, Kurovskaya NA, Ignat'ev YA, Kononkova NN, Koltashev VV, Plotnichenko VG (2010) Influence of oxygen fugacity on the solubility of carbon and hydrogen in FeO-Na_2O-SiO_2-Al_2O_3 melts in equilibrium with liquid iron at 15 GPa and 1400 degrees. Geochem Int 48:953–960

Kasting JF, Eggler DH, Raeburn SP (1993) Mantle redox evolution and the oxidation state of the Archean atmosphere. J Gol 101:245–257

Kessel R, Ulmer P, Pettke T, Schmidt MW, Thompson AB (2004) A novel approach to determine high-pressure high-temperature fluid and melt compositions using diamond-trap experiments. Am Mineral 89:1078–1086

Lamadric HM, More LR, Moncada D, Rimstidt RJ, Burruss RC, Bodnar RJ (2017) Reassessment of the Raman CO_2 densimeter. Chem Geol 450:210–222

Li Y, Keppler H (2014) Nitrogen speciation in mantle and crustal fluids. Geochim Cosmochim Acta 129:13–32

Liang Q, Yan C-S, Meng Y, Lai J, Krasnicki A, Mao H-K, Hemley R (2009) Recent advances in high-growth rate single-crystal CVD diamonds. Diamonds Rel Mater 18:698–703

Libourel G, Marty B, Humbert F (2003) Nitrogen solubility in basaltic melt part I effect of oxygen fugacity. Geochim Cosmochim Acta 67:4123–4136

Malfait WJ, Xue X (2014) Hydroxyl speciation in felsic magmas. Geochim Cosmochim Acta 140:606–620

Marty B (1995) Nitrogen content of the mantle inferred from N_2-Ar correlation in oceanic basalt. Nature 377:326–329

McMillan P (1984) Structural studies of silicate glasses and melts-applications and limitations of Raman spectroscopy. Am Mineral 69:622–644

Mibe K, Chou I.-M., Bassett W A (2008) In situ Raman spectroscopic investigation of the structure of subduction-zone fluids. J Geophys Res 113. https://doi.org/10.1029/2007JB005179.

Moynihan CT, Macedo PB, Montrose CJ, Gupta PK, De Bolt MA, Dill JF, Wilder JA (1976) Structural relaxation in vitreous materials. Ann N Y Acad Sci 279:15–35

Mysen B (2010) Structure of H_2O-saturated peralkaline aluminosilicate melt and coexisting aluminosilicate-saturated aqueous fluid determined in-situ to 800 C and 800 MPa. Geochim Cosmochim Acta 74:4123–4139

Mysen B (2013) Effects of fluid and melt density and structure on high-pressure and high-temperature experimental studies of hydrogen isotope partitioning between coexisting melt and aqueous fluid. Am Mineral 98:1754–1764

Mysen B (2015a) Carbon speciation in silicate-C-O-H melt and fluid as a function of redox conditions: an experimental study in situ to 1.7 GPa and 900 degrees. Am Mineral 100:872–882

Mysen B (2015b) Hydrogen isotope fractionation and redox-controlled solution mechanisms in silicate-COH melt plus fluid systems. J Geophys Res 120:7440–7459

Mysen BO (2007) The solution behavior of H_2O in peralkaline aluminosilicate melts at high pressure with implications for properties of hydrous melts. Geochim Cosmochim Acta 71:1820–1834

Mysen BO (2009) Solution mechanisms of silicate in aqueous fluid and H_2O in coexisting silicate melts determined in-situ at high pressure and high temperature. Geochim Cosmochim Acta 73:5748–5763

Mysen BO (2014) Water-melt interaction in hydrous magmatic systems at high temperature and pressure. Prog Earth Planet Sci 1. doi:https://doi.org/10.1186/2197-4284-1-4

Mysen BO, Armstrong L (2002) Solubility behavior of alkali aluminosilicate components in aqueous fluids and silicate melts at high pressure and temperature. Geochim Cosmochim Acta 66:2287–2298

Mysen BO, Fogel ML (2010a) Nitrogen and hydrogen isotope compositions and solubility in silicate melts in equilibrium with reduced (N+H)-bearing fluids at high pressure and temperature: effects of melt structure. Am Mineral 95:987–999

Mysen BO, Kumamoto K, Cody GD, Fogel ML (2011) Solubility and solution mechanisms of C-O-H volatiles in silicate melt with variable redox conditions and melt composition at upper mantle temperatures and pressures. Geochim Cosmochim Acta 75:6183–6199

Mysen BO, Lucier A, Cody GD (2003) The structural behavior of Al^{3+} in peralkaline melts and glasses in the system Na_2O-Al_2O_3-SiO_2. Am Mineral 88:1668–1678

Mysen BO, Mibe K, Chou I-M, Bassett WA (2013) Structure and equilibria among silicate species in aqueous fluids in the upper mantle: experimental SiO_2-H_2O and MgO-SiO_2-H_2O data recorded in-situ to 900°C and 54 GPa. J Geophys Res 118:6076–6085

Mysen BO, Yamashita S (2010) Speciation of reduced C-O-H volatiles in coexisting fluids and silicate melts determined in-situ to similar to 1.4 GPa and 800 degrees. Geochim Cosmochim Acta 74:4577–4588

Mysen BO, Yamashita S, Chertkova N (2008) Solubility and solution mechanisms of NOH volatiles in silicate melts at high pressure and temperature-amine groups and hydrogen fugacity. Am Mineral 93:1760–1770

Nakamoto K (2009) Infrared and Raman spectra of inorganic and organic coordination compounds part B: applications in coordination Crganometallic and Cioorganic chemistry chapter 1. John Wiley, sons, Hoboken NJ

Neuville DR, Cormier L, Massiot D (2006) Al coordination and speciation in calcium aluminosilicate glasses: effects of composition determined by ^{27}Al MQ-MAS NMR and Raman spectroscopy. Chem Geol 229:173–185

Newton RC, Manning CE (2005) Solubility of anhydrite CaSO4 in NaCl–H_2O solutions at high pressures and temperatures: applications to fluid–rock interaction. J Petrol 46:701–716

Newton RC, Manning CE (2006) Solubilities of corundum wollastonite and quartz in H_2O-NaCl solutions at 800°C and 10 kbar: interaction of simple minerals with brines at high pressure and temperature. Geochim Cosmochim Acta 70:5571–5582

Ni HW, Keppler H (2013) Carbon in silicate melts. Rev Mineral Geochem Carbon in Earth 75:251–287

Nowak M, Behrens H (1995) The speciation of water in haplogranitic glasses and melts by in-situ near-infrared spectroscopy. Geochim Cosmochim Acta 59:3445–3450

Nowak M, Porbatzki D, Spickenbom K, Diedrich O (2003) Carbon dioxide speciation in silicate melts: a restart. Earth Planet Sci Lett 207:131–139

Renlund GM, Prochazka S, Doremus RH (1991) Silicon oxycarbide glasses 2 structure and properties. J Mater Res 6:2723–2734
Richet P, Bottinga Y (1986) Thermochemical properties of silicate glasses and liquids: a review. Rev Geophys 24:1–25
Richet P, Neuville DR, Courtial P, Tequi C (1992) Thermodynamic and rheological properties of silicate melts. In: Anonymous (ed) 29th international geological congress; abstracts, vol 29, p 540
Roskosz Mysen BO, Cody GD (2006) Dual speciation of nitrogen in silicate melts at high pressure and temperature: an experimental study. Geochim Cosmochim Acta 70:2902–2918
Rosso KM, Bodnar RJ (1995) Microthermometric and Raman spectroscopic detection limits of CO_2 in fluid inclusions and the Raman spectroscopic characterization of CO_2. Geochim Cosmochim Acta 59:3961–3975
Sato R, Bolvin J, McMillan PF (1990) Synthesis and characterization of a SiAlON glass. J Amer Ceram Soc 73:2494–2497
Schiferl D, Nicol M, Zaug JM, Sharma SK, Cooney TF, Wang S-Y, Fleischer JF (1997) The diamond $^{13}C/^{12}C$ isotope Raman pressure sensor system for high temperature/pressure diamond-anvil cells with reactive samples. J Appl Phys 82:3256–3265
Schmidt MW, Poli S (2003) Generation of mobile components during subduction of oceanic crust. Treatise on Geochemistry: Elsevier:567–591
Scholze H (1959) Über die quantitative IR-spektroskopische Wasser-Bestimmung in Silicaten. Angew Chem 71:678
Soraru GD, D'Andrea G, Compostrini R (1995) Structural characterization and high-temperature behavior of silicon oxycarbide glasses prepared from sol-gel precursors containing Si-H bonds. J Amer Ceram Soc 78:379–387
Stebbins JF (1988) Effects of temperature and composition on silicate glass structure and dynamics: Si-29 NMR results. J Non-Cryst Solids 106:359–369
Tolstikhin IN, Marty B (1998) The evolution of terrestrial volatiles: a view from helium argon and nitrogen isotope modeling. Chem Geol 147:27–52
Walrafen GE (1967) Raman spectral studies of HDO in H_2O. J Chem Phys 48:244–251
Walrafen GE, Yang W-H, Chu YC (1999) Raman spectra from saturated water vapor to the supercritical fluid. J Phys Chem B 103:1332–1338
Walrafen GE, Yang WH, Chu YC, Hokmabadi MS (1996) Raman OD-stretching overtone spectra from liquid D_2O between 22 and 152 degrees C. J Phys Chem 100:1381–1139
Webster JD, Goldoff B, Sintoni MF, Shimizu N, De Vivo B (2014) C-O-H-Cl-S-F volatile Solubilities partitioning and mixing in Phonolitic-Trachytic melts and aqueous-carbonic vapor +/− saline liquid at 200 MPa. J Petrol 55:2217–2247
White WB (1974) The carbonate minerals. In: Farmer VC (ed) Infrared spectra of minerals. Mineralogical Society of London, London, p Ch 12
Wood BJ, Bryndzia LT, Johnson KE (1990) Mantle oxidation state and its relationship to tectonic environment and fluid speciation. Science 248:337–345
Wu L, Yang D-B, Xie H-S, Li F-F, Hu B, Yu Y, Gao C-X (2014) Pressure-induced elastic and structural changes in hydrous basalt glasses: the effect of H_2O on the gravitational stability of basalt melts at the base of the upper mantle. Earth Planet Sci Lett 406:165–173
Yamada A, Inoue T, Urakawa S, Funakoshi K-I, Nobumasa F, Kikegawa T, Irifune T (2011) In situ X-ray diffraction study on pressure-induced structural changes in hydrous forsterite and enstatite melts. Earth Planet Sci Lett 308:115–123
Yamashita S, Behrens H, Schmidt BC, Dupree R (2008) Water speciation in sodium silicate glasses based on NIR and NMR spectroscopy. Chem Geol 256:231–241
Yeo GA, Ford TA (1994) Ab initio molecular orbital calculations of the energetic structural vibrational and electronic properties of some hydrogen bonded complexes of water ammonia and hydroxylamine. Spectrochim Acta 50A:5–18
Zhang Y-G, Frantz JD (2000) Enstatite-forsterite-water equilibria at elevated temperatures and pressures. Am Mineral 85:918–925
Zotov N, Keppler H (1998) The influence of water on the structure of hydrous sodium tetrasilicate glasses. Am Mineral 83:823–834
Zotov N, Keppler H (2002) Silica speciation in aqueous fluids at high pressures and high temperatures. Chem Geol 184:71–82

6

Incorporation of Mg^{2+} in surface Ca^{2+} sites of aragonite: an *ab initio* study

Jun Kawano[1,2*], Hiroshi Sakuma[3] and Takaya Nagai[2]

Abstract

First-principles calculations of Mg^{2+}-containing aragonite surfaces are important because Mg^{2+} can affect the growth of calcium carbonate polymorphs. New calculations that incorporate Mg^{2+} substitution for Ca^{2+} in the aragonite {001} and {110} surfaces clarify the stability of Mg^{2+} near the aragonite surface and the structure of the Mg^{2+}-containing aragonite surface. The results suggest that the Mg^{2+} substitution energy for Ca^{2+} at surface sites is lower than that in the bulk structure and that Mg^{2+} can be easily incorporated into the surface sites; however, when Mg^{2+} is substituted for Ca^{2+} in sites deeper than the second Ca^{2+} layer, the substitution energy approaches the value of the bulk structure. Furthermore, Mg^{2+} at the aragonite surface has a significant effect on the surface structure. In particular, CO_3 groups rotate to achieve six-coordinate geometry when Mg^{2+} is substituted for Ca^{2+} in the top layer of the {001} surface or even in the deeper layers of the {110} surface. The rotation may relax the atomic structure around Mg^{2+} and reduces the substitution energy. The structural rearrangements observed in this study of the aragonite surface induced by Mg^{2+} likely change the stability of aragonite and affect the polymorph selection of $CaCO_3$.

Keywords: Aragonite; Impurity; Surface structure; First-principles calculation

Background

The formation of calcium carbonate ($CaCO_3$) polymorphs, calcite, aragonite, and vaterite has been extensively investigated due to their importance in geological and biological environments. To account for the formation of a particular polymorph, the role of impurities has been proposed as the controlling factors in many studies (e.g., Kitano 1962; Davis et al. 2000); however, the mechanism for the incorporation of impurities during crystal growth is poorly understood. In this study, we focus on the incorporation of Mg^{2+} in the aragonite surface and analyze its behavior using first-principles calculations.

Many researchers have previously reported that alkaline-earth cations other than Ca^{2+} affect the growth kinetics of $CaCO_3$ (e.g., De Yoreo and Vekilov 2003; Astilleros et al. 2010; Nielsen et al. 2013). In particular, Mg^{2+} has been considered important to the formation of $CaCO_3$ polymorphs. For example, Kitano (1962) indicated that the addition of Mg^{2+} to a solution promoted the metastable formation of aragonite. Recently, detailed atomic force microscopy (AFM) observations of the growth surface of calcite suggested that Mg^{2+} inhibits the crystal growth of calcite by blocking the propagation of kink sites (Nielsen et al. 2013) or by increasing the mineral solubility (Davis et al. 2000). To analyze this phenomenon on calcite surfaces at the atomic level, atomistic simulations were also conducted using static lattice energy minimization (Titiloye et al. 1998), molecular dynamics (MD) (de Leeuw and Parker 2001), and electronic structure calculations based on density functional theory (DFT) (Sakuma et al. 2014).

However, there are relatively few studies focusing on the aragonite surface. To discuss the mechanism for the formation of $CaCO_3$ polymorphs, not only should the atomic behavior on the calcite surface be understood but also that on the aragonite surface. Moreover, the Mg content in coral fossils comprising aragonite has been used to reconstruct the past climatic record (e.g., Mitsuguchi et al. 1996); however, the location of Mg^{2+} in the coral skeleton is strongly debated (Finch and Allison 2008). Therefore, understanding the mechanism for the incorporation of

* Correspondence: j-kawano@mail.sci.hokudai.ac.jp
[1]Creative Research Institution (CRIS), Hokkaido University, N21 N10, Kita-ku, Sapporo, Hokkaido 001-0021, Japan
[2]Faculty of Science, Hokkaido University, N10 W8, Kita-ku, Sapporo, Hokkaido 060-0810, Japan
Full list of author information is available at the end of the article

Mg^{2+} into the aragonite surface is important not only for the mineral and material sciences but also for the biological and environmental sciences.

Divalent cations smaller than Ca^{2+}, such as Mg^{2+}, do not generally enter the aragonite structure, whereas larger cations, such as Ba^{2+}, cannot be incorporated in the calcite structure. However, the structure near a crystal surface differs from the bulk crystal because of its flexibility. Thus, a crystal surface can incorporate ions that are unstable in the bulk structure and play an important role during the formation and subsequent crystal growth of calcium carbonate polymorphs. We investigated the substitution of Mg^{2+} ions at the Ca^{2+} sites of aragonite surfaces. Mg^{2+} is unstable in ninefold coordination in aragonite and does not readily enter into the bulk aragonite structure; however, Mg^{2+} is expected to be substitutable for Ca^{2+} at sites near the surface. Recently, Ruiz-Hernandez et al. (2012) performed MD calculations regarding Mg^{2+} substitution at the aragonite surface. However, they analyzed only the Mg^{2+} substitution into Ca^{2+} sites at the top surface. To discuss the incorporation of an ion into a specific surface, the ion substitution energy for Ca^{2+} sites should be estimated at the top surface and deeper. Furthermore, the substitution of Mg^{2+} for Ca^{2+} may change the surface structure. This could affect the stability relations among polymorphs, as surface energy differences among polymorphs have been proposed to account for their stability field (Navrotsky 2004; Kawano et al. 2009). Hence, an in-depth analysis of these surface structures and their incorporation of ions is important; however, details regarding the surface structural changes when a Mg^{2+} ion is incorporated at the surface are presently lacking. Therefore, in this study, the stability of Mg^{2+} near the aragonite surface and the structure of Mg^{2+}-containing aragonite surface were investigated using first-principles calculations, and the effect of Mg^{2+} on the formation of polymorphs was examined.

Methods

The optimized geometries and total energies of the surfaces were obtained using DFT with the Vienna *ab initio* simulation package (VASP) code (Kresse and Hafner 1993, 1994; Kresse and Furthmüller 1996a, b; Kresse and Joubert 1999) and the Perdew-Burke-Ernzerhof version of the generalized gradient approximation (GGA-PBE) (Perdew et al. 1996). The energy cutoff of the plane-wave basis set was 900 eV, which was tested for energy convergence. The valence states for Ca, Mg, C, and O are $3s^23p^64s^2$, $2p^63s^2$, $2s^22p^2$, and $2s^22p^4$, respectively, following previous DFT calculations for $CaCO_3$ (Hossain et al. 2009) and $MgCO_3$ (Hossain et al. 2010).

Prior to calculation of the aragonite surface, the structural parameters of aragonite were simulated. The calculated lattice parameters are $a = 5.022$ Å, $b = 8.042$ Å, and $c = 5.816$ Å, whereas the experimental values are $a = 4.962$ Å, $b = 7.969$ Å, and $c = 5.743$ Å (Balmain et al. 1999). The calculated C-O bond lengths are 1.291 to 1.301 Å, which are comparable to previous experimentally and theoretically obtained values (Balmain et al. 1999; Akiyama et al. 2011).

Aragonite surfaces were simulated as repeated slabs. The unit supercell contained 80 atoms and 4 or 5 Ca and CO_3 layers of the unit (2 × 1) surface structure with a 15 Å thick vacuum layer. The macroscopic dipole was removed, and the neutrality of the supercell was ensured by the two equivalent surfaces on opposite sides of the slab. To calculate the substitution energy, the total energy of the slab was simulated by relaxing the atoms except those on the bottom layer. In the calculation, the supercell parameters were fixed using the calculated aragonite unit cell parameters, because Mg ions are not supposed to be substituted into the bulk Ca sites but only into surface sites; hence, substitution does not affect the lattice constants. Optimization was performed with a convergence threshold of 5.0×10^{-6} eV/atom for the maximum energy change and 0.05 eV/Å for the maximum force. The atomic structures of the slab were drawn with the VESTA software (Momma and Izumi 2008). $CaCO_3$ crystals generally grow in an aqueous solution with H_2O molecules just above the surface; however, the presence of the vacuum layer above the surface was considered here. The validity of this setting will be discussed in the next section.

Results and discussion

Substitution energy of Mg^{2+} for Ca^{2+} near the aragonite surface

We calculated the surface energies of the aragonite {001}, {010}, and {110} surfaces, which are generally expressed in experimental morphology. On the {001} surface, the CO_3 groups (triangles) are arranged parallel to the surface, whereas these groups are perpendicular to the {010} and {110} surfaces. Recently, Akiyama et al. (2011) determined the stable surface structure of calcite and aragonite via DFT calculations. They reported that the most stable configuration of {001} is a CO_3-terminated surface, wherein the coverage of CO_3 ions is 0.5. In contrast, the most stable configurations for {010} and {110} are Ca-terminated planes, wherein the coverage of Ca^{2+} ions is 0.5. The surface energy of these stable configurations were calculated to be 0.49, 0.57, and 0.49 J/m^2 for the {001}, {010}, and {110} surfaces, respectively (Table 1). In the calculation, the positions of all the slab atoms were relaxed, and the surface energies E_{surf} were obtained by dividing the difference between the total energy of the slab E_{slab} and that of the bulk crystal E_{bulk}, including the same number of atoms, by the total area of the calculated surface A, which includes the top and bottom surfaces:

Table 1 Estimated surface energies (J/m²) of the aragonite surfaces

	This work	Akiyama et al.	de Leeuw and Parker	
Surface	DFT	DFT	Empirical potential	
	Pure	Pure	Pure	Hydrated
{001} $\theta_{CO_3} = 0.5$	0.49	0.58	0.85	0.90
{010} $\theta_{Ca} = 0.5$	0.57	0.73	0.96	0.24
{110} $\theta_{Ca} = 0.5$	0.49	0.64	0.88	0.56

$$E_{surf} = (E_{slab} - E_{bulk})/A. \quad (1)$$

The results suggest that the {001} and {110} surfaces are slightly more stable than the {010} surface, which is consistent with previous DFT and empirical potential calculations (Akiyama et al. 2011; De Leeuw and Parker 1998). Therefore, we focused on the {001} and {110} aragonite surfaces and analyzed the substitution energy of Mg^{2+} for Ca^{2+}. The structures considered in this study are shown in Figure 1. In the first Ca layer of the {001} surface, there are two types of Ca sites because the coverage of CO_3 ions is 0.5 in the most stable structure of this surface. The site labeled A in Figure 1 (denoted as site A hereafter) is the site above which no CO_3 groups are located, whereas the site labeled B in the same figure (site B) is similar to the site in the bulk with respect to the arrangement of surrounding CO_3^{2-} ions, except for the lack of one CO_3^{2-}. In contrast, there are two Ca layers in the {110} surface with different depths between the CO_3 layers.

De Leeuw and Parker (1998) estimated the energies of pure and hydrated surfaces and found that hydration does not stabilize the carbonate-terminated {001} surface and less so the calcium-terminated {110} surface (Table 1). In contrast, the calcium-terminated {010} surface was significantly stabilized by hydration. The surface energies calculated in this study show almost the same trend. Therefore, the substitution energies obtained for nonhydrated and hydrated {001} and {110} surfaces can be considered to have similar features. It is thus reasonable to analyze non-hydrated {001} and {110} surfaces to discuss the exchange energy in aqueous solutions, whereas Nada (2014) recently reported the importance of water layers on the calcite {104} surface.

In an aqueous solution, Mg^{2+} ions are surrounded by H_2O molecules and exist as hydration shells (e.g., Marcus

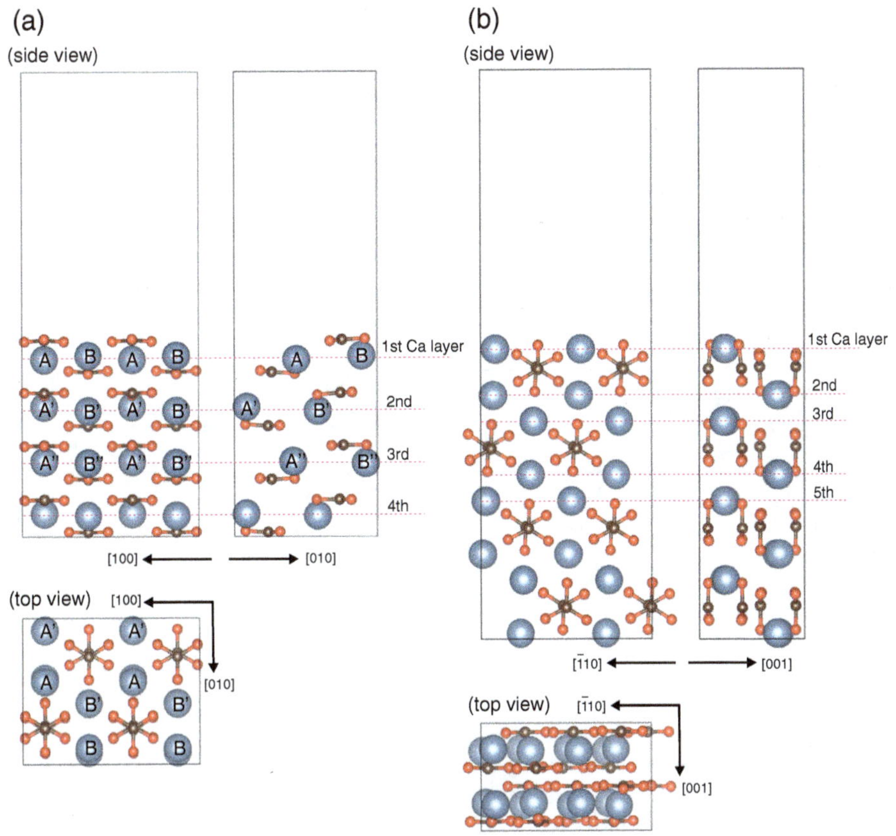

Figure 1 Relaxed atomic arrangement of (a) {001} and (b) {110} surfaces. Red, brown, and blue circles represent O, C, and Ca, respectively. Ca layers are numbered from the top to deeper layers - see text regarding [A] and [B] for the top Ca layer on the {001} surface. In addition, there are two types of sites in the second and third layers of the {001} surface, which are labeled [A′], [B′] and [A″], [B″], respectively.

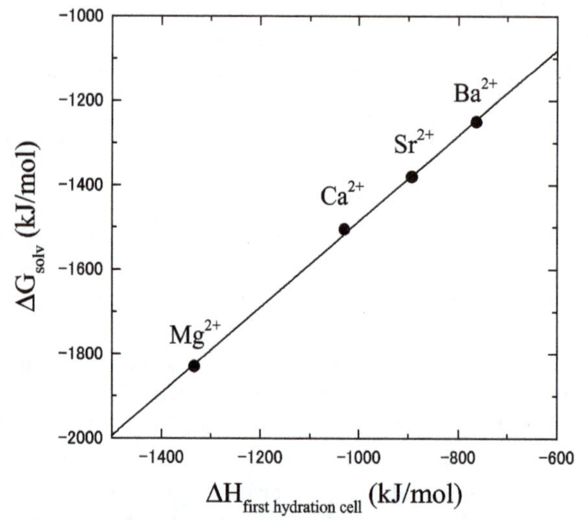

Figure 2 Relation between experimental free energy of solvation and binding enthalpy of the first hydration shell. The binding energies of the hydration shell with six water molecules were calculated.

1987; Rodriguez-Cruz et al. 1999; Bock et al. 2006). Therefore, to substitute into a surface Ca^{2+} site, a Mg atom must be released from the hydration shell. In this study, a six-coordinated complex was assumed for the divalent cations in the hydration shells, and the cohesive energies of the primary hydration shells of Ca^{2+} and Mg^{2+} ions were calculated using the DFT method as references. The calculated and experimental Gibbs free energies for hydration (Marcus 1991) have an excellent 1:1 correspondence, as shown in Figure 2. Following the discussion by Sakuma et al. (2014), such correspondence suggests that the relative binding enthalpy of the first hydration shell with six water molecules can describe the relative free energies of divalent cations in water.

Using these results, the energy ΔE of the Mg^{2+} substitution in surface Ca^{2+} sites of aragonite can be expressed as follows:

$$\Delta E = \left[E_{\mathbf{surface(Mg)}} + E_{\mathbf{hydration\ shell(Ca)}}\right] - \left[E_{\mathbf{surface(Ca)}} + E_{\mathbf{hydration\ shell(Mg)}}\right], \quad (2)$$

where $E_{\text{surface(Ca)}}$ and $E_{\text{surface(Mg)}}$ represent the total energies of the aragonite slab with and without Mg, and $E_{\text{hydration shell(Ca)}}$ and $E_{\text{hydration shell(Mg)}}$ are the energies of the hydration shells of Ca^{2+} and Mg^{2+} with $6H_2O$, respectively. The calculated cohesive energy of the primary hydration shell of Mg^{2+} with $6H_2O$ is approximately −13.8 eV (1,330 kJ/mol) and that for Ca^{2+} is −10.8 eV (1,030 kJ/mol), which indicates that the hydration shell of the smaller cation is more stable than that of the larger cation.

For Mg^{2+} substitution in Ca^{2+} sites of the bulk aragonite structure, the substitution energy obtained from Equation 2 using the energy of the bulk instead of the energy of the slab was calculated as 93.1 kJ/mol. However, the substitution energy for the surface site is much smaller than that for the bulk. Whether Mg ions will actually substitute for Ca depends on the chemical potential difference including entropy, while the entropy of an ion in aqueous solutions is difficult to estimate. Therefore, the negative or positive sign of the substitution energy ΔE does not directly mean that Mg ions enter the crystal or not. However, these energies can provide information regarding the relative stability of Mg in the site around the surface. The results of the present calculations suggest that Mg^{2+} can substitute more easily at surface sites than in the bulk aragonite structure. Figure 3 shows the Mg^{2+} substitution energy ΔE for a Ca^{2+} site as a function of depth for the interface of the {001} and {110} surfaces.

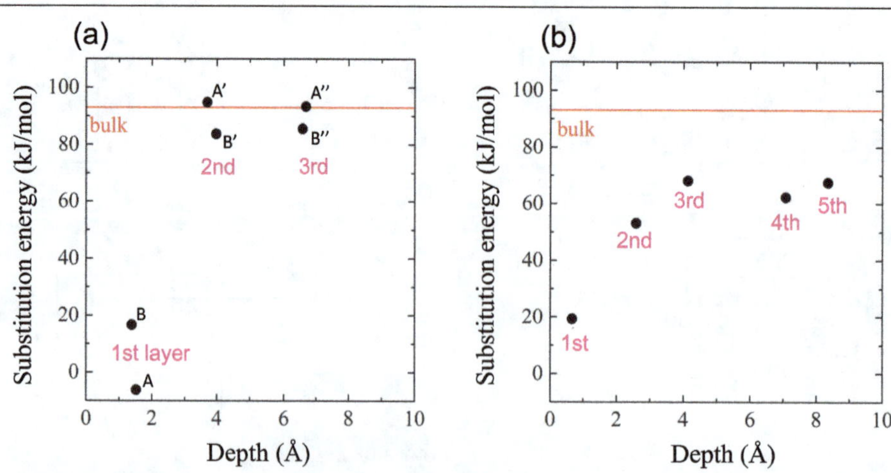

Figure 3 Energies for Mg^{2+} substitution at Ca^{2+} in aragonite surface sites. The substitution energies are shown as a function of the depth from the interface of **(a)** {001} and **(b)** {110} surfaces. Red lines are drawn at 93.1 kJ/mol, which is the energy for Mg^{2+} substitution into bulk aragonite. Layer numbers, [A], [A′], [A″], [B], [B′], and [B″] correspond to the substitution energies for the site shown in Figure 1.

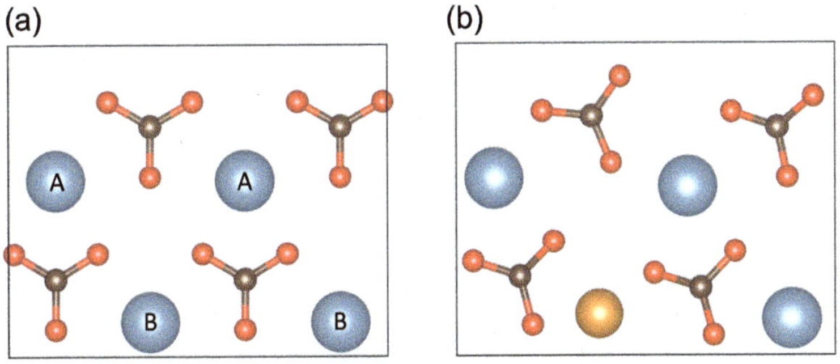

Figure 4 Top view of the relaxed aragonite {001} surface wherein Mg^{2+} substitutes for Ca^{2+}. Comparison of **(a)** surface structure without Mg^{2+} ions and **(b)** with Mg^{2+} substituting for Ca^{2+}. Red, brown, blue, and orange circles represent O, C, Ca, and Mg, respectively. [A] and [B] are labeled as in Figure 1.

We first discuss the {001} surface (Figure 3a). When Mg^{2+} substitutes into site A in the first Ca layer, above which no CO_3 groups are located, the substitution energy is almost zero and Mg^{2+} is easily incorporated into site A, which agrees with the MD calculations (Ruiz-Hernandez et al. 2012). The substitution energy increases when Mg^{2+} is substituted into the B site but is still much lower than that when it enters the Ca site in the bulk aragonite structure. However, for substitution within the deeper layers, the substitution energies increase significantly and reach almost that of the bulk aragonite structure. Thus, Mg^{2+} readily attaches to the first layer of the {001} surface but less so within the deeper layers.

In the {110} surface, the substitution energy of Mg^{2+} into the first Ca layer is almost the same as that for the B site in the {001} surface, and it rapidly increases with substitution in the deeper layers (Figure 3b). However, for this surface, even the energy for the substitution at Ca sites in the fifth layer is smaller than that for the bulk. This suggests that for {110} faces, a slightly higher energy would be required for Mg^{2+} ions to enter the Ca site of the top layer, whereas Mg^{2+} ions would enter the deeper layers more easily than the {001} face.

Structure of aragonite surfaces with Mg^{2+} ions at the Ca^{2+} sites

The structures of aragonite surfaces that contain Mg^{2+} ions were examined next. In the aragonite {001} surface, the CO_3^{2-} ions are arranged parallel to the surface; therefore, the flexibility of the movement and rotation of ions initially seem higher than those on the other surfaces. Figure 4b shows the relaxed structure of the {001} surface containing Mg^{2+} ions in B sites of the first Ca layer, where Mg^{2+} is surrounded by CO_3 ions. Figure 4 shows that the CO_3 groups move from the original positions and rotate 30° in the same direction. Notably, not only the CO_3^{2-} ions above the Mg^{2+} ions, which exist on the surface, but also the CO_3 groups below Mg^{2+}, which make up the second CO_3 layer, rotate in the same manner. The arrangement of CO_3 ions around a Mg atom resembles the structure of calcite. To clarify this, the arrangement of the CO_3 groups around the Mg ions is shown in Figure 5. This figure shows that CO_3 groups around a Mg^{2+} ion move and rotate to assume six-coordinate geometry, which is the arrangement of the MgO_6 octahedron in the calcite structure.

Furthermore, the {110} surface seems to have less flexibility than the {001} surface because the CO_3 groups are arranged perpendicular to the {110} surface. However,

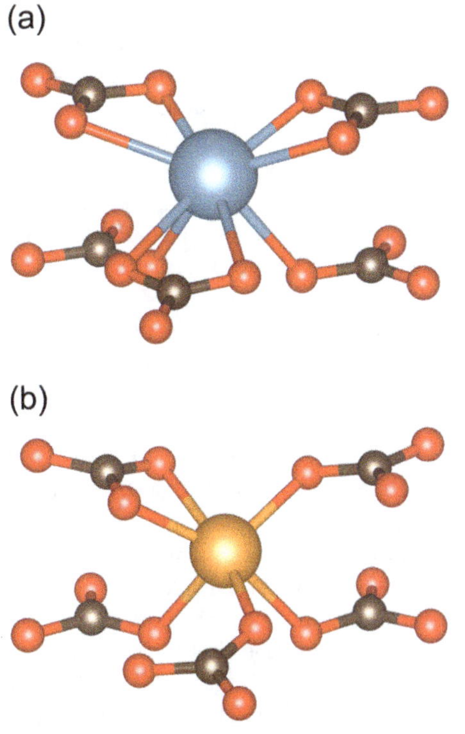

Figure 5 Details of the arrangement of CO_3 groups around the cation on the top {001} surface. (a) Ca and **(b)** Mg in the B site on the top layer of the {001} surface. Mg presumably prefers six-coordinate (octahedral) geometry as in the calcite structure.

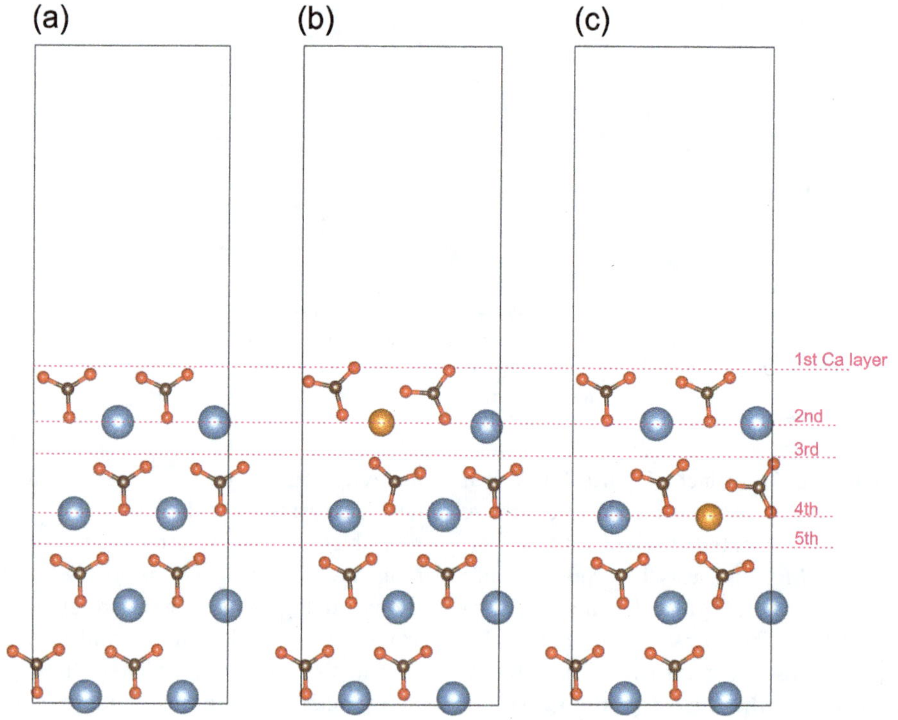

Figure 6 Side views of the relaxed aragonite {110} surface in which Mg^{2+} substitutes for Ca^{2+}. **(a)** No substitution, **(b)** substitution into the second Ca layer, and **(c)** substitution into the fourth Ca layer. Red, brown, blue, and orange circles represent O, C, Ca, and Mg, respectively.

even in this case, the CO_3 groups move and rotate to achieve six-coordinate geometry, as shown in Figure 6. Moreover, for the {110} surface, when Mg^{2+} substitutes at a Ca site, not only in the first but also in the fifth Ca layer, the CO_3 groups also rotate to achieve six-coordinate geometry but do not form MgO_6 octahedra (Figure 7). This is opposite to the {001} surface, wherein the CO_3 groups do not move to accommodate the Mg substitution into the Ca sites near the {001} surface, except in the top layer. This is probably because atoms can move easily perpendicular to the surface, but less easily when parallel to it because the size of the surface is fixed, even though the layer thickness is not fixed. Therefore, contrary to the first expectation, it is easier for CO_3 groups to rotate in the {110} surface where atoms easily move parallel to the CO_3 groups. In contrast, in the {001} surface, atoms do not readily move in the long direction parallel to the CO_3 groups, because the size of this plane is fixed.

The differences between the {001} and {110} surfaces lead to differences in the substitution energies; when Mg^{2+} ions substitute in the deeper Ca layers, the atomic arrangement near the {110} surface is more relaxed than that near the {001} surface. Hence, the substitution energies for {110} are lower than those for the {001} surface. The energy gained by the rotation of CO_3 is estimated to be around 20 eV, by comparison of the substitution energy for the deep layers of the {110} surface and the value for the substitution in the bulk where CO_3 groups are not rotated.

The results suggest that the CO_3 groups near the surface easily move and rotate relative to their original positions. Moreover, Mg^{2+} ions strongly prefer six-coordinate geometry. Therefore, the presence of Mg^{2+} affects the surface stability of aragonite, and it may further affect the

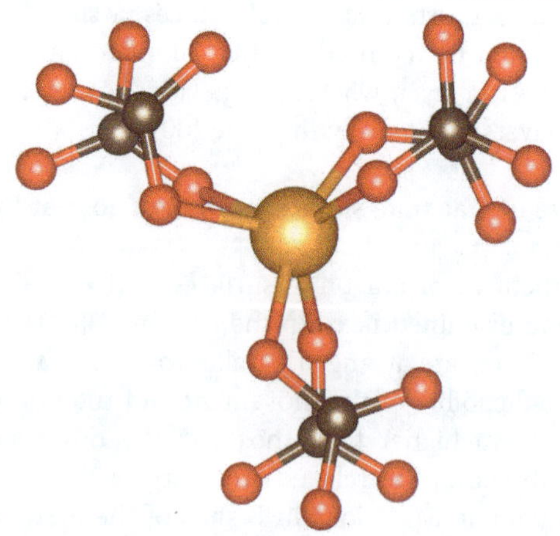

Figure 7 Detailed arrangement of CO_3 groups around Mg on the {110} surface. The arrangement of CO_3 groups around Mg at the Ca site of the fourth layer in the {110} surface is shown, which suggests that Mg achieves six-coordinate geometry but does not form MgO_6 octahedra as in the calcite structure.

structure of the small clusters that appear during the early formation of $CaCO_3$, which has more flexibility than the surface, both of which affect the polymorph selection of $CaCO_3$.

Conclusions

First-principles calculations were performed for Mg^{2+}-containing aragonite surfaces. The results suggest that the substitution energy of Mg^{2+} for Ca^{2+} at the surface is lower than the substitution energy of Mg^{2+} for Ca^{2+} in the bulk structure. However, for the {001} surface, when Mg^{2+} substitutes for Ca^{2+} deeper than the second Ca layer, the substitution energy is almost the same as that for substitution in the bulk aragonite structure. In contrast, for the {110} surface, even when Mg^{2+} ion substitutes into deeper layers, the substitution energy is still lower than the substitution energy in the bulk aragonite structure. Thus, Mg^{2+} ions easily attach onto the {001} surface with lower energy; however, it should be difficult for these ions to move to deeper layers. In contrast, for the {110} surface, a relatively higher energy is required for Mg^{2+} ions to substitute for Ca^{2+} at the top surface sites, whereas they enter more easily to deeper layers than the {001} face. This is probably because the atomic structure of this surface is more relaxed, and the CO_3 groups move and rotate from their original positions even when Mg^{2+} ions are in deeper layers. In contrast, for the {001} surface, the CO_3 groups do not move when Mg^{2+} substitutes for Ca^{2+}, except in the top layer sites where CO_3 groups easily move and rotate to achieve six-coordinate geometry, such as CaO_6 octahedra in the calcite structure. Mg^{2+} generally assumes a preferential six-coordinate geometry, even at the aragonite surface, indicating that it changes the surface stability of aragonite, which may affect the formation of $CaCO_3$ polymorphs.

Abbreviations

AFM: atomic force microscopy; DFT: density functional theory; GGA-PBE: generalized gradient approximation-Perdew-Burke-Ernzerhof; MD: molecular dynamics; VASP: Vienna *ab initio* simulation package.

Competing interests

The authors declare that they have no competing interests.

Authors' contributions

JK proposed the topic, designed the study, conducted the simulations, and interpreted the results. HS contributed to the simulations and data interpretation. TN contributed to the discussions with the corresponding author throughout the study. All authors read and approved the final manuscript.

Acknowledgements

The authors would like to thank Y. Kimura, one of the session conveners of the 2013 JpGU annual meeting, for his recommendation for submission of the present study to PEPS. Thanks are also due to two anonymous reviewers and the editor for their constructive comments. This study was supported by a MEXT grant for the tenure-tracking system and JSPS KAKENHI Grant Number 26870010.

Author details

[1]Creative Research Institution (CRIS), Hokkaido University, N21 N10, Kita-ku, Sapporo, Hokkaido 001-0021, Japan. [2]Faculty of Science, Hokkaido University, N10 W8, Kita-ku, Sapporo, Hokkaido 060-0810, Japan. [3]Environment and Energy Materials Division, National Institute for Materials Science, 1-1 Namiki, Tsukuba, Ibaraki 305-0044, Japan.

References

Akiyama T, Nakamura K, Ito T (2011) Atomic and electronic structures of $CaCO_3$ surfaces. Phys Rev B 84:085248, doi:10.1103/PhysRevB.84.085428

Astilleros JM, Fernández-Díaz L, Putnis A (2010) The role of magnesium in the growth of calcite: an AFM study. Chem Geol 271:52–58

Balmain J, Hannoyer B, Lopez E (1999) Fourier transform infrared spectroscopy (FTIR) and X-ray diffraction analyses of mineral and organic matrix during heating of mother of pearl (nacre) from the shell of the mollusk Pinctada maxima. J Biomed Mater Res 48:749–754

Bock CW, Markham GD, Katz AK, Glusker JP (2006) The arrangement of first- and second-shell water molecules around metal ions: effects of charge and size. Theor Chem Acc 115:100–112

Davis KJ, Dove PM, De Yoreo JJ (2000) The role of Mg^{2+} as an impurity in calcite growth. Science 290:1134–1137, doi:10.1126/science.290.5494.1134

de Leeuw NH, Parker CC (1998) Surface structure and morphology of calcium carbonate polymorphs calcite, aragonite, and vaterite: an atomistic approach. J Phys Chem B 102:2914–2922

de Leeuw NH, Parker CC (2001) Surface-water interactions in the dolomite problem. Phys Chem Chem Phys 3:3217–3221

De Yoreo JJ, Vekilov PG (2003) Principles of crystal nucleation and growth. Rev Mineral Geochem 54(1):57–93

Finch AA, Allison N (2008) Mg structural state in coral aragonite and implications for the paleoenvironmental proxy. Geophys Res Lett 35, L08704, doi:10.1029/2008GL033543

Hossain FM, Murch GE, Belova IV, Turner BD (2009) Electronic, optical and bonding properties of $CaCO_3$ calcite. Solid State Comm 149:1201–1203, doi:10.1016/j.ssc.2009.04.026

Hossain FM, Dlugogorski BZ, Kennedy EM, Belova IV, Murch GE (2010) Electronic, optical and bonding properties of $MgCO_3$. Solid State Comm 150:848–851, doi:10.1016/j.ssc.2010.02.008

Kawano J, Shimobayashi N, Miyake A, Kitamura A (2009) Precipitation diagram of calcium carbonate polymorphs: its construction and significance. J Phys Condens Matter 21:425102, doi:10.1088/0953-8984/21/42/425102

Kitano Y (1962) The behaviour of various inorganic ions in the separation of calcium carbonate from a bicarbonate solution. Bull Chem Soc Japan 35:1973–1980

Kresse G, Furthmüller J (1996a) Efficiency of ab-initio total energy calculations for metals and semiconductors using a plane-wave basis set. Comput Mat Sci 6:15–50

Kresse G, Furthmüller J (1996b) Efficient iterative schemes for ab initio total-energy calculations using a plane-wave basis set. Phys Rev B 54:11169–11186

Kresse G, Hafner J (1993) Ab initio molecular dynamics for liquid metals. Phys Rev B 47:558–561

Kresse G, Hafner J (1994) Ab initio molecular-dynamics simulation of the liquid-metal-amorphous-semiconductor transition in germanium. Phys Rev B 49:14251–14269

Kresse G, Joubert J (1999) From ultrasoft pseudopotentials to the projector augmented-wave method. Phys Rev B 59:1758–1775

Marcus Y (1987) Thermodynamics of ion hydration and its interpretation in terms of a common model. Pure Appl Chem 59:1093–1101

Marcus Y (1991) Thermodynamics of solvation of ions. Part 5.-Gibbs free energy of hydration at 298.15 K. J Chem Soc Faraday Trans 87:2995–2999

Mitsuguchi T, Matsumoto E, Abe O, Uchida T, Isdale PJ (1996) Mg/Ca thermometry in coral skeletons. Science 274:961–963

Momma K, Izumi F (2008) VESTA: a three-dimensional visualization system for electronic and structural analysis. J Appl Crystallogr 41:653–658

Nada H (2014) Difference in the conformation and dynamics of aspartic acid on the flat regions, step edges, and kinks of as calcite surface: a molecular dynamics study. J Phys Chem C 118:14335–14345

Navrotsky A (2004) Energetic clues to pathways to biomineralization: precursors, clusters, and nanoparticles. Proc Natl Acad Sci U S A 101:12096–12101

Nielsen LC, De Yoreo JJ, DePaolo DJ (2013) General model for calcite growth kinetics in the presence of impurity ions. Geochim Cosmochim Acta 115:100–114

Perdew JP, Burke K, Ernzerhof M (1996) Generalized gradient approximation made simple. Phys Rev Lett 77:3865–3868

Rodriguez-Cruz SE, Jockusch RA, Williams ER (1999) Hydration energies and structures of alkaline earth metal ions, $M^{2+}(H_2O)_n$, $n = 5–7$, M = Mg, Ca, Sr, and Ba. J Am Chem Soc 121:8898–8906

Ruiz-Hernandez SE, Grau-Crespo R, Almora-Barrios N, Wolthers M, Ruiz-Salvador AR, Fernandez N, de Leeuw NH (2012) Mg/Ca partitioning between aqueous solution and aragonite mineral: a molecular dynamics study. Chemistry 18:9828–9833, doi:10.1002/chem.201200966

Sakuma H, Andersson MP, Bechgaard K, Stipp SLS (2014) Surface tension alteration on calcite, induced by ion substitution. J Phys Chem C 118:3078–3087, doi:10.1021/jp411151u

Titiloye JO, de Leeuw NH, Parker CC (1998) Atomistic simulation of the differences between calcite and dolomite surfaces. Geochim Cosmochim Acta 62:2637–2641

7

Enhanced flux of extraterrestrial ^{3}He across the Permian–Triassic boundary

Tetsuji Onoue[1*], Naoto Takahata[2], Mitsutaka Miura[1], Honami Sato[3], Akira Ishikawa[4], Katsuhito Soda[1], Yuji Sano[2] and Yukio Isozaki[5]

Abstract

The *ca* 252 Ma Permian–Triassic boundary (PTB) represents the most severe mass extinction event of the Phanerozoic, with the disappearance of ~ 80% of marine invertebrate species. Large-scale eruption of the Siberian Traps is the commonly favored cause, although the link between volcanism and extinction remains debated. Here, we report evidence for an extraterrestrial ^{3}He influx in the PTB section of deep-sea bedded cherts from Japan. This unusual signal indicates a significant increase in the influx of interplanetary dust particles, likely related to an asteroid shower in the inner solar system. High-resolution stratigraphy indicates that the peak flux of dust particles occurred during the final 500 kyr of the Permian, concurrent with a pre-extinction decline in radiolarian diversity.

Keywords: Permian, Triassic, Bedded chert, Mino Belt, Helium, Platinum group element, Interplanetary dust particle, Radiolaria

Introduction

The Permian–Triassic boundary (PTB) mass extinction event (MEE) (Erwin 1994; Wignall 2007) was associated with catastrophic environmental changes including oceanic anoxia, global carbon cycle perturbations, methane hydrate release, and global warming (Isozaki 1997; Wignall 2007; Burgess et al. 2014). The release of volcanic and contact metamorphic carbon and sulfur gases (CO_2, CH_4, and SO_2) from the Siberian Traps is commonly invoked as the trigger for climatic perturbations across the PTB (Wignall 2007; Burgess et al. 2017), which led to the end-Permian MEE. Although extraterrestrial causes have been proposed (e.g., bolide impact, Becker et al. 2001, 2004; Basu et al. 2003), they have not been favored (Koeberl et al. 2004; Farley et al. 2005), largely due to a lack of geochemical and mineralogical evidence such as platinum group element (PGE) anomalies and shocked quartz. The accretion of extraterrestrial noble gases such as He and Ar was discussed by Becker et al. (2001), who reported high concentrations of extraterrestrial He ($^3He_{ET}$) in PTB rocks from Japan and China, with a suggestion that $^3He_{ET}$ is trapped in fullerenes associated with an impact event (Becker et al. 2001). However, the reported $^3He_{ET}$ and fullerene concentrations have not been reproduced in independent analyses of the same samples and others from PTB sections in China, Canada, and Austria (Koeberl et al. 2004; Farley et al. 2005).

The solar system contains abundant sub-millimeter interplanetary dust particles (IDPs) enriched in ^{3}He, and ^{3}He concentrations in ancient deep-sea sediments have been used to constrain the IDP flux to Earth over at least the last 100 Myr (Farley 1995; Farley et al. 2012). However, the use of ^{3}He in detecting IDPs is often compromised by the diffusional loss of ^{3}He from sedimentary rocks (Farley et al. 2005; Mukhopadhyay and Farley 2006), with the exception of some Ordovician samples that record a period of unusually high extraterrestrial ^{3}He flux (Patterson et al. 1998).

This is the first report of preservation of $^3He_{ET}$ in Permian bedded chert and overlying Triassic black claystone from Waidani (Sano et al. 2010, 2012a), Japan, which was deposited in the Panthalassa superocean (Isozaki 1997) and subsequently accreted to the Japan margin during the Jurassic. Biostratigraphic and chemostratigraphic data indicate that this PTB section preserves a complete pelagic PTB transition, without a temporal gap at lithological contacts between bedded chert and

* Correspondence: t_onoue@kumamoto-u.ac.jp
[1]Department of Earth and Environmental Sciences, Kumamoto University, Kumamoto 860-8555, Japan
Full list of author information is available at the end of the article

black claystone (Sano et al. 2010, 2012a; Agematsu et al. 2014). The $^3He_{ET}$ flux across the PTB was calculated on the basis of $^3He_{ET}$ concentrations and sedimentary mass accumulation rates estimated from the astronomically tuned cyclostratigraphy of the bedded chert. The $^3He_{ET}$ record is supported by PGE concentrations measured in bulk rock samples from the section, and their major and trace element compositions.

Methods/Experimental

Sample collection and preparation

Samples for whole-rock He isotopic analyses were collected from the siliceous PTB section of the Mino Belt, Japan, which is a Jurassic subduction-generated accretionary complex (Additional file 1: Figure S1). PTB siliceous rocks of the Mino Belt are considered to be long-lived Panthalassa remnants representing an allochthonous proto-Pacific Plate assemblage (Matsuda and Isozaki 1991; Isozaki 1997). The bedded cherts lack carbonate and coarse terrigenous grains, suggesting that their primary depositional site was deeper than the calcium carbonate compensation depth and beyond the depositional range of terrigenous clastic grains (Matsuda and Isozaki 1991; Isozaki 1997). PTB siliceous rocks in Japan are characterized by a very low sedimentation rate (less than a few millimeters per kyr), an absence of coarse-grained terrigenous clastics, and fine-grained sediment compositions (mainly cryptocrystalline quartz). They contain small extraterrestrial particles, including well-preserved melted and unmelted micrometeorites, and cosmic spinels (Onoue et al. 2011, 2012; Uno et al. 2012).

The studied section crops out along a road cutting near Waidani in the Mt. Funabuseyama area of the Mino Belt, central Japan (Additional file 2: Figure S2). The examined rocks correspond to the Permian–Lower Triassic Hashikadani Formation, which is characterized by pelagic, deep-sea siliceous sediments (Sano et al. 2012a). The section was referred to as NF1212R and NF1212C in previous studies (Sano et al. 2010, 2012a), but the term "Waidani section" is used here.

The Waidani section comprises a 3.5 m thick continuous succession of gray and dark gray bedded chert overlain by black claystone, with a pyrite-rich brown claystone layer at its base (Fig. 1). The bedded chert comprises rhythmically alternating beds of chert and relatively thin beds of claystone. Based on high-resolution radiolarian and conodont biostratigraphy (Sano et al. 2010; Agematsu et al. 2014), the bedded chert in the section is assigned to the Upper Permian (Changhsingian), and the overlying black claystone to the lowermost Triassic (Griesbachian). A negative excursion in organic carbon isotopic ratios ($\delta^{13}C_{org}$) occurs at the top of the dark gray bedded chert (Sano et al. 2012a), coinciding with radiolarian fossil disappearances correlated with the MEE horizon in other PTB sections (Feng and Algeo 2014). The Triassic base of the section is defined by the presence of natural assemblages of the conodont *Hindeodus* (*H. parvus* and *H. typicalis*) in the

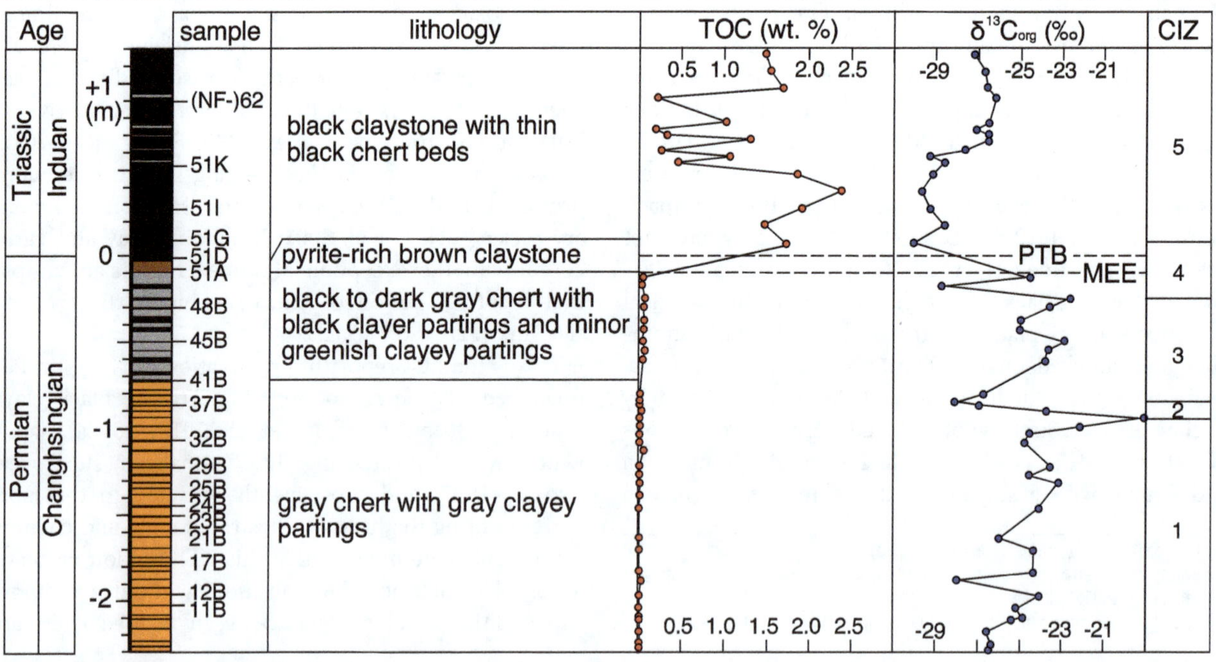

Fig. 1 Lithology and geochemical profiles of the study section. Stratigraphic column showing samples, lithologies, total organic carbon contents (TOC, wt.%), and organic carbon isotopic values ($\delta^{13}C_{org}$, VPDB‰, divided into carbon isotope zones, CIZ). Modified after Sano et al. (2012a)

basal 15 cm thick stratigraphic interval of black claystone (Agematsu et al. 2014).

To avoid variations in chemistry related to lithological changes (Farley et al. 2005), all samples for whole-rock geochemical analysis were collected from claystone beds: 13 Permian claystones and 1 brown and 5 dark gray Triassic claystones. Veins and strongly recrystallized and/or weathered parts of the claystone beds were avoided to minimize effects of diagenetic overprinting on sediment geochemistry. Samples contain brown conodont elements (Agematsu et al. 2014), with a conodont color alteration index (CAI) of 2, indicating that the rocks were exposed to burial temperatures of 60–140 °C.

He isotopic analysis

Prior to He isotopic analysis, the samples were crushed and washed with ultrapure water; after drying, the fragments were carefully hand-picked to avoid contamination by altered and weathered material, and hand-picked fragments (< 5 mm grain size, 10–20 g) were pulverized in an agate mortar.

Both bulk and stepped heating methods were used to determine He concentrations and ^{3}He/^{4}He ratios. The bulk method was applied to 19 samples and involved gas extraction by melting samples at 1800 °C. The stepped heating method (Amari and Ozima 1985; Farley et al. 2005; Mukhopadhyay and Farley 2006) was used to identify the ^{3}He carrier in the most ^{3}He-rich sample (NF-37B): heating began at 550 °C increasing in 100 °C steps, with 30–50 min per step.

Cleaned and homogenized powder samples (0.25 g for the bulk method, ~ 1 g for the stepped method) were wrapped in Al foil and placed in the loading port. Samples were preheated under vacuum at 110 °C for 11 h to drive off adsorbed gases. Samples were then fused under vacuum at a maximum temperature of 1800 °C to extract trapped gases, which were purified using a charcoal trap at liquid nitrogen temperature and with a hot titanium getter. After purification, ^{4}He and ^{20}Ne abundances and ^{4}He/^{20}Ne ratios were determined using a quadrupole mass spectrometer, and He concentrations and ^{3}He/^{4}He ratios using a noble gas mass spectrometer (Helix-SFT; GV Instruments) at the Atmosphere and Ocean Research Institute, University of Tokyo, Tokyo, Japan. Meteorite samples had not previously been analyzed by this measurement system. Air was used as a standard for noble gas isotopic analyses (Sano et al. 2013). He blanks (3×10^{-15} cm^3 STP for ^{3}He; 3×10^{-9} cm^3 STP for ^{4}He) were < 20% of the sample signal.

Acid leaching procedure

Most IDPs found on Earth's surface comprise aggregates of small grains of olivine, pyroxene, Fe–Ni sulfides, and magnetite (Genge et al. 2008). Previous investigations of a Triassic bedded chert unit from the study area identified IDPs preserved with silicate phases (olivine and pyroxene, Onoue et al. 2011). If these minerals act as carrier phases for extraterrestrial ^{3}He, most ^{3}He would be lost during acid leaching (Mukhopadhyay and Farley 2006). Four bulk samples were selected that yielded high (NF-37B, 41B) and low (NF-25B, 51I) ^{3}He concentrations. The samples were treated with a 12M HF–3M HCl mixture for 48 h at room temperature, rinsed four times with ultrapure water, and dried. He was extracted at temperatures of 750 °C, 950 °C, and 1800 °C during heating in a furnace.

Estimation of extraterrestrial ^{3}He concentrations

^{3}He$_{ET}$ concentrations in sediments were calculated using the following equation (Marcantonio et al. 1995):

$$\left[{}^3\mathrm{He}_{\mathrm{ET}}\right] = \left(1 - R_{\mathrm{terrigenous}}/R_{\mathrm{measured}}\right)/\left(1 - R_{\mathrm{terrigenous}}/R_{\mathrm{IDP}}\right) \times \left[{}^3\mathrm{He}_{\mathrm{measured}}\right] \quad (1)$$

where $R_{terrigenous}$ and R_{IDP} are ^{3}He/^{4}He ratios for terrigenous (^{3}He/^{4}He = 0.015 Ra, Sano 2017) and extraterrestrial (^{3}He/^{4}He = 170 Ra, McGee and Mukhopadhyay 2013) end-members, respectively (Ra = atmospheric ^{3}He/^{4}He ratio = 1.382×10^{-6}, Sano 2017). The lack of volcanic material in the analyzed samples precludes the presence of trapped mantle He. The lowest ^{3}He/^{4}He ratio (0.2 Ra) was measured in Triassic beds, with the fraction of extraterrestrial ^{3}He exceeding 90%.

Platinum group element analysis

Platinum group elements (Ir, Ru, Pt, and Pd) and Re were analyzed by isotope dilution mass spectrometry (ID–MS) after quartz tube digestion. Sample digestion, purification, and mass spectrometry were based on procedures described by Ishikawa et al. (2014). Powdered samples (~ 0.1 g) and spike solutions containing ^{185}Re and ^{191}Ir–^{99}Ru–^{196}Pt–^{105}Pd were transferred to a quartz tube. After the addition of 2 mL of inverse *aqua regia*, the tubes were frozen in dry ice–ethanol, sealed with an oxypropane torch, and placed in an oven at 240 °C for 72 h. HF (1 mL, 38%) was added to the residue for desilicification (Ishikawa et al. 2014).

Platinum group element isotopic compositions of purified sample solutions were determined by inductively coupled plasma mass spectrometry (ICP–MS, Thermo-Element XR) at the University of Tokyo, Komaba, Japan. Sample and standard solutions were interspersed throughout analytical sessions to monitor and correct for instrumental fractionation; ^{89}Y, ^{90}Zr, ^{95}Mo, ^{97}Mo, ^{99}Ru, ^{100}Ru, ^{101}Ru, ^{105}Pd, ^{106}Pd, ^{108}Pd, ^{111}Cd ^{178}Hf, ^{185}Re, ^{187}Re, ^{191}Ir, ^{193}Ir, ^{194}Pt, ^{195}Pt, ^{196}Pt, and ^{202}Hg were monitored for possible mass interference. Raw signal intensities were corrected using measured isobaric–oxide interferences, although contributions of interferences to sample signals were insignificant (< 0.1%).

Elemental concentrations were corrected using average total procedural blanks obtained from each batch of analyses (0.7 pg Ir, 2.3 pg Ru, 13 pg Pt, 10 pg Pd, and 0.3 pg Re; $n = 2$). Uncertainties for each sample were estimated by error propagation through ICP–MS measurements (2SE) and blank corrections (< 51%). Blank contributions of Ir, Pt, Pd, and Re were < 12%, < 6%, < 9%, and < 7%, respectively, while the contribution to Ru was slightly higher (5–26%, < 41% for sample NF-21B).

Major and trace element analyses

Major element contents of 21 samples were determined by X-ray fluorescence spectrometry (XRF, PANalytical Epsilon 3XLE with a Mo X-ray tube) on pressed powder pellets, at Kumamoto University, Kumamoto, Japan. Standard rock samples issued by the Geological Survey of Japan were used for calibration purposes. Reproducibility, based on four repeat analyses of standard JSd-1, was better than ± 0.5% for Mg, Al, Si, K, Ca, Ti, and Fe and better than ± 1% for Na and Mn.

Nineteen samples were analyzed for trace and rare earth elements (REE) by ICP–MS after lithium metaborate/tetraborate fusion at Actlabs, Ancaster, Canada. Ten international rock and mineral standards were analyzed for quality control purposes during trace element and REE analyses. Analytical procedures are described in http://www.actlabs.com.

Estimation of sedimentation rate for Changhsingian bedded chert

The average sedimentation rate for Changhsingian bedded chert of the Hashikadani Formation was estimated for a PTB section at Iwaidani (NF195 section; Sano et al. 2012b), cropping out ~ 3 km from the Waidani section (Additional file 2: Figure S2). Lithologically, the NF195 section is very similar to the Waidani section (Sano et al. 2010, 2012b), comprising a lower unit (~ 6.2 m thick) of gray bedded chert, a middle unit (~ 0.7 m thick) of black bedded chert including pyrite nodules near the top, and an upper unit (~ 0.9 m thick) characterized by black claystone with thin black chert beds. Biostratigraphic analysis of radiolarians reveals that the section contains the Wuchiapingian–Changhsingian boundary in the basal ~ 0.3 m of the lower unit and an MEE horizon at the top of the middle unit. The stratigraphic range of the Changhsingian is ~ 6.6 m, and its duration was 2.20 ± 0.16 Myr according to U–Pb ages for the MEE (251.941 ± 0.037 Ma, Burgess et al. 2014) and Wuchiapingian–Changhsingian boundary (254.14 ± 0.12 Ma, Ramezani and Bowring 2017). From these data, the average sedimentation rate for the Changhsingian bedded chert of the Hashikadani Formation is estimated to be ~ 0.3 cm kyr^{-1}.

Time-series analysis

The bedded chert in the studied section comprises rhythmically alternating beds of chert (1–7 cm thick) and relatively thin beds of claystone (1–2 cm thick). Individual chert and claystone thicknesses were measured throughout the section, with "silica ranks" of 0 for claystone and 1 for chert being assigned. Cyclostratigraphic analyses of the Triassic–Jurassic bedded chert sequence were based on bed thicknesses in the Mino Belt (Ikeda and Tada 2014). However, Algeo et al. (2011) suggest that other methods should be used where analyzed intervals are too short for estimating sedimentation rates or to capture astronomical cycles such as the 405 kyr eccentricity cycles (Laskar et al. 2004). Sedimentation rates for the Upper Permian bedded chert sequence were therefore estimated using orbital tuning procedures involving long precession cycles (Yao et al. 2015). Periodicities and modulations in time–frequency space were estimated using REDFIT software (Schulz and Mudelsee 2002) and continuous wavelet analyses with Morlet mother signals (Torrence and Compo 1998). Bandpass filters with Gaussian windows (Paillard et al. 1996) were applied to extract astronomical signals from binary series.

Results

He analyses

Analyzed samples yielded bulk ^{3}He and ^{4}He concentrations of 0.05–0.34 × 10^{-12} cm^3 STP g^{-1} and 0.09–0.47 × 10^{-6} cm^3 STP g^{-1}, respectively (Table 1). ^{3}He concentrations are highest in the topmost ~ 1.5 m of the studied Permian deposits (Fig. 2a), with a maximum ^{3}He concentration of 0.34 × 10^{-12} cm^3 STP g^{-1}, which is 4–5 times greater than that measured in the overlying Triassic unit.

The analyzed samples likely preserve primary fluctuations in ^{3}He concentrations across the PTB because (1) the detected ^{4}He/^{20}Ne ratios are > 100 times greater than the atmospheric value, thereby precluding the existence of significant atmospheric He in the samples and (2) the lack of correlation ($r^2 = 0.3$) between ^{3}He and ^{4}He concentrations suggests that the increase in ^{3}He during the uppermost Permian cannot be explained by He retention alone. ^{3}He/^{4}He ratios follow a similar trend to ^{3}He concentrations, with the maximum ^{3}He/^{4}He ratio of 0.75 Ra being measured in the uppermost Permian bed.

Stepped heating

To identify the source of the high ^{3}He/^{4}He signature, stepwise heating of the sample with the highest ^{3}He concentration (NF-37B) was performed at 550 °C, 650 °C, 750 °C,

Table 1 ^{3}He and ^{4}He concentrations, and $^{3}He/^{4}He$ ratios for samples from the Waidani section

Sample	Height (m)	MAR* ($g\ cm^{-2}\ kyr^{-1}$)	Temperature (°C)	^{3}He ($pcc\ g^{-1}$)	^{4}He ($\mu cc\ g^{-1}$)	$^{3}He/^{4}He$ (R_a)#	min $^{4}He/^{20}Ne$	$^{3}He_{ET}$ ($pcc\ g^{-1}$)	$^{3}He_{ET}$ flux ($pcc\ cm^{-2}\ kyr^{-1}$)
				Bulk					
NF-62	0.88		1800	0.06	0.18	0.25 ± 0.04	40	0.06	
NF-51 K	0.60		1800	0.05	0.17	0.22 ± 0.03	26	0.05	
NF-51I	0.30		1800	0.05	0.20	0.19 ± 0.03	53	0.05	
NF-51G	0.12		1800	0.08	0.32	0.17 ± 0.03	70	0.07	
NF-51D	0.00		1800	0.12	0.47	0.18 ± 0.02	49	0.11	
NF-51A	−0.10		1800	0.08	0.21	0.27 ± 0.03	44	0.08	
NF-48B	−0.33	0.81 ± 0.18	1800	0.23	0.36	0.47 ± 0.02	37	0.23	0.18 ± 0.05
NF-45B	−0.53	0.81 ± 0.18	1800	0.30	0.29	0.75 ± 0.02	82	0.30	0.24 ± 0.05
NF-41B	−0.76	0.81 ± 0.18	1800	0.24	0.32	0.55 ± 0.02	211	0.23	0.19 ± 0.04
NF-37B	−0.90	0.81 ± 0.18	1800	0.34	0.34	0.72 ± 0.02	37	0.34	0.27 ± 0.06
NF-32B	−1.11	0.81 ± 0.18	1800	0.18	0.27	0.47 ± 0.02	49	0.17	0.14 ± 0.03
NF-29B	−1.25	0.81 ± 0.18	1800	0.13	0.15	0.59 ± 0.02	37	0.12	0.10 ± 0.02
NF-25B	−1.42	0.81 ± 0.18	1800	0.13	0.22	0.43 ± 0.02	58	0.13	0.10 ± 0.02
NF-24B	−1.50	0.81 ± 0.18	1800	0.14	0.24	0.42 ± 0.02	60	0.13	0.11 ± 0.02
NF-23B	−1.56	0.81 ± 0.18	1800	0.09	0.14	0.45 ± 0.03	36	0.08	0.07 ± 0.02
NF-21B	−1.63	0.81 ± 0.18	1800	0.07	0.09	0.53 ± 0.03	22	0.06	0.05 ± 0.01
NF-17B	−1.82	0.81 ± 0.18	1800	0.10	0.19	0.35 ± 0.03	34	0.09	0.07 ± 0.02
NF-12B	−2.01	0.81 ± 0.18	1800	0.18	0.27	0.48 ± 0.02	38	0.17	0.14 ± 0.03
NF-11B	−2.08	0.81 ± 0.18	1800	0.09	0.21	0.32 ± 0.03	13	0.09	0.07 ± 0.02
blank				0.01	0.01				
air						1.00	0.32		
				Stepwise heating					
NF-37B	−0.90		550	0.075	0.232	0.23 ± 0.02	194		
NF-37B	−0.90		650	0.071	0.054	0.95 ± 0.02	48		
NF-37B	−0.90		750	0.067	0.009	5.22 ± 0.09	13		
NF-37B	−0.90		850	0.071	0.004	13.5 ± 0.2	6		
NF-37B	−0.90		950	0.019	0.001	13.9 ± 0.5	5		
NF-37B	−0.90		1800	0.006	0.001	3.80 ± 0.22	2		
NF-37B	−0.90		total	0.31	0.30	0.73			
				12 M HF–3 M HCl dissolution					
NF-51I	0.60		750–950	0.002	0.001	1.22 ± 0.07	143		
NF-41B	−0.76		750–950	0.014	< 0.001	107.1 ± 1.9	33		
NF-37B	−0.90		750–950	0.007	< 0.001	149.7 ± 8.8	10		
NF-37B	−0.90		950–1800	0.003	< 0.001	16.6 ± 0.7	34		
NF-25B	−1.42		750–950	< 0.001	< 0.001	2.98 ± 0.27	45		

*Based on the sedimentation rate (mean ± 2SD) derived from an astronomically tuned cyclostratigraphy age model for the Permian section

#Error assigned to the isotopic ratio is 1σ

850 °C, 950 °C, and 1800 °C (Fig. 3), resulting in extraction of He from different mineral phases, which was analyzed for $^{3}He/^{4}He$ ratios at each step. Most ^{4}He was extracted at the lowest step (550 °C), while ~ 50% of ^{3}He was released at intermediate temperatures (750 °C and 850 °C). This decoupled behavior indicates that ^{3}He and ^{4}He are hosted in separate phases within the sample. Previous studies indicate that ^{3}He hosted in IDPs is released at temperatures

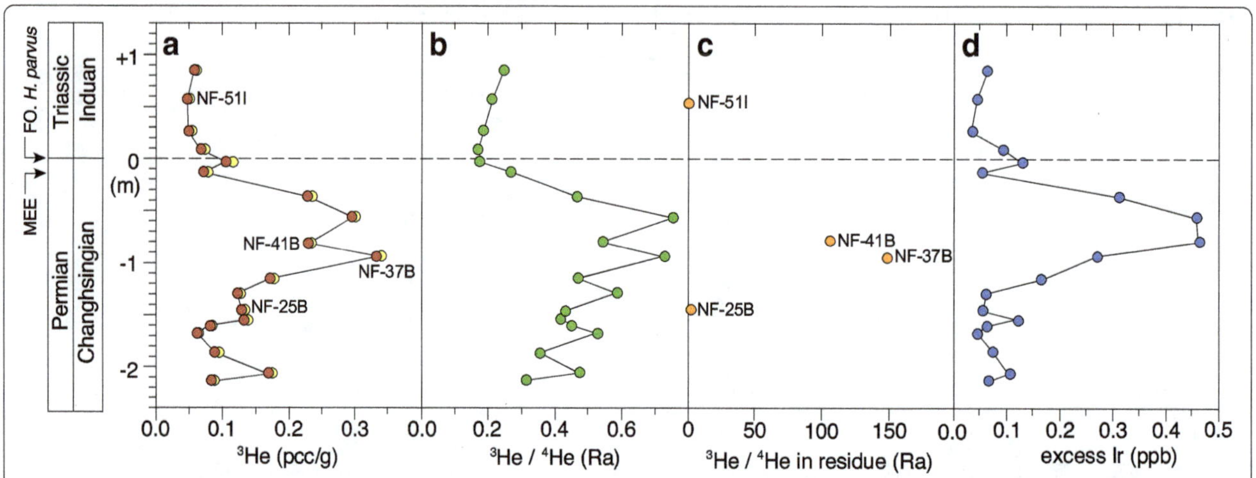

Fig. 2 Stratigraphic He and Ir profiles. Stratigraphic profiles of **a** ^{3}He concentrations, **b** ^{3}He/^{4}He ratios, **c** ^{3}He/^{4}He ratios of acid-insoluble residues produced during stepwise heating at temperatures of 750–950 °C, and **d** excess Ir in bedded cherts of the Waidani section. Red and yellow symbols in the ^{3}He profile represent the extraterrestrial fraction and measured ^{3}He concentrations, respectively. The base of the Triassic and MEE horizons in this section are defined by the occurrence of the conodont *Hindeodus parvus* and disappearance of Permian radiolarian fossils, respectively

of > 800 °C (Farley et al. 2005). Helium extracted at 850 °C from sample NF-37B is similar to that in IDPs in terms of ^{3}He release temperature and high ^{3}He/^{4}He ratios (14 Ra), whereas the component released at 550 °C has a much lower ^{3}He/^{4}He ratio (0.2 Ra). The temperature range over

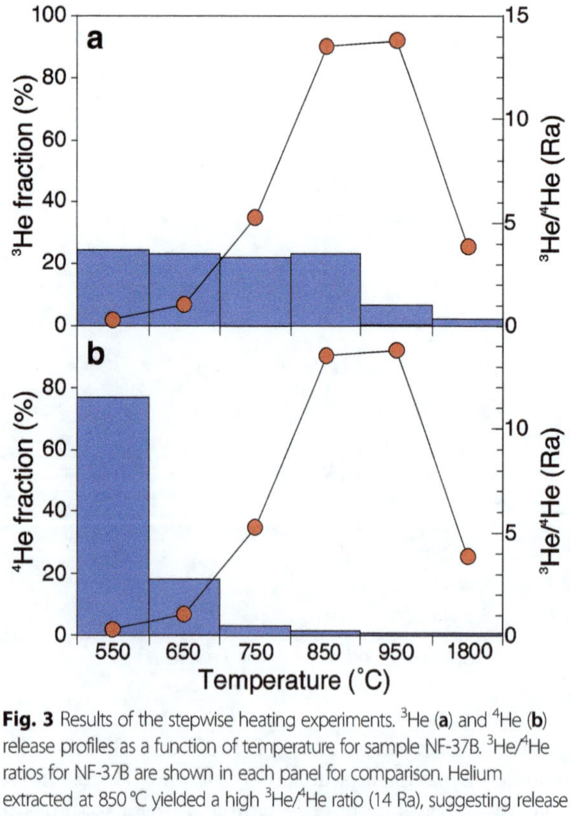

Fig. 3 Results of the stepwise heating experiments. ^{3}He (**a**) and ^{4}He (**b**) release profiles as a function of temperature for sample NF-37B. ^{3}He/^{4}He ratios for NF-37B are shown in each panel for comparison. Helium extracted at 850 °C yielded a high ^{3}He/^{4}He ratio (14 Ra), suggesting release from IDPs. The stratigraphic level of sample NF-37B is shown in Fig. 2

which ^{3}He was released (550–850 °C) is equivalent to that determined for the decomposition of IDPs (Farley et al. 2005). Our results therefore indicate that the unusually high ^{3}He concentration in the studied PTB section is of extraterrestrial origin and is likely hosted in IDPs.

An acid leaching procedure (Mukhopadhyay and Farley 2006) was applied to the PTB samples to confirm the existence of extraterrestrial He. High ^{3}He/^{4}He ratios (up to 150 Ra) were detected during the extraction of He at 750–950 °C from acid-insoluble residues produced from samples with high ^{3}He concentrations (NF-37B, 41B). Low-^{3}He samples (NF-25B, 51I) have low ^{3}He/^{4}He ratios (< 3 Ra, Fig. 2c), with > 96% of ^{3}He being removed during leaching. We therefore interpret the high ^{3}He/^{4}He ratios as recording the existence of extraterrestrial He, hosted mainly in IDPs.

HF–HCl-resistant refractory phases found in meteorites include spinel, corundum, diamond, graphite, SiC, amorphous carbon, and organic material (Patterson et al. 1998). Previous studies excluded such phases (except for spinel) as possible long-term carriers of $^{3}He_{ET}$ (Mukhopadhyay and Farley 2006). A Middle Ordovician (~ 470 Ma) shallow marine limestone in Sweden, which records clear evidence of an asteroid breakup or meteorite shower event, contains $^{3}He_{ET}$ hosted primarily in chromite (Heck et al. 2008). For our samples, we consider that extraterrestrial spinel and/or relict silicate grains may host ^{3}He in acid-insoluble residues.

Stratigraphic distribution of extraterrestrial material

The extraterrestrial fraction of ^{3}He across the PTB (Eq. (1)) indicates that bulk ^{3}He concentrations in the PTB section are highly sensitive to extraterrestrial input.

Furthermore, $^3He_{ET}$ concentrations in the topmost ~ 1.5 m of the uppermost Permian section are at least four times higher those in the pre-event interval. Platinum group elements such as Ir and Pt are often used as tracers of extraterrestrial material. As with concentrations of $^3He_{ET}$, an increase in Pt, Pd, Ru, and Ir concentrations (Table 2) occurs in the topmost ~ 1.5 m of the uppermost Permian section (Fig. 4). Iridium is the least mobile PGE (Evans et al. 1993), and excess Ir concentrations (Ir_{XS}) were calculated using the equation:

$$Ir_{XS} = Ir_{measured} - Al_{measured} \times \left(Ir_{terrigenous}/Al_{terrigenous}\right) \quad (2)$$

where "terrigenous" indicates the average value for the upper continental crust (McLennan 2001). This calculation subtracts the terrigenous Ir (supplied primarily by aeolian dust) from the total Ir content of PTB sediments, allowing the extraterrestrial contribution to be determined. The similar durations and magnitudes (4–6 times) of increases in Ir_{XS} and $^3He_{ET}$ (Fig. 2), and the good positive correlation (r^2 = 0.7) between them (Fig. 5), support our interpretation of enhanced extraterrestrial ^3He in the uppermost Permian and its change across the PTB. Previous studies indicate that Fe–Mn oxyhydroxides contain significant amounts of Ir, which may be remobilized when they are reduced (Colodner et al. 1992). However, reducing conditions were prominent throughout the Upper Permian (Changhsingian) (Isozaki 1997; Sano et al. 2012a; Takahashi et al. 2014) and an oxic-to-anoxic transition was not detected within the studied interval (Sano et al. 2012a) (Additional file 3: Figure S3). Indeed, there is no strong correlation between Ir and redox-sensitive elements such as Fe, Mn, V, Ni, Co, Mo, or U (Additional file 4: Figure S4).

Discussion

Our results provide no evidence of elevated $^3He_{ET}$ levels at the PTB itself, consistent with previous work on other PTB sections (Farley and Mukhopadhyay 2001; Farley et al. 2005). However, concentrations of $^3He_{ET}$ in the PTB section here indicate an abrupt and significant increase in the topmost ~ 1.5 m of the uppermost Permian section. Although He isotopic compositions may be influenced by provenance composition (Winckler and Fischer 2006), the uniform values of terrigenous elemental ratios (e.g., constant Mg/Al, Ti/Al, and Zr/Hf ratios; Additional file 5: Table S1 and Additional file 6: Table S2) indicate no effects on the stratigraphic increase of ^3He by changes of provenance composition of the Permian and Triassic deposits. The contribution of cosmogenic ^3He can also be ignored because the studied rocks were excavated some years previously from a shielded location (2–3 m underground) during road cutting.

Table 2 PGE concentrations in claystone samples from the Waidani section

Sample	Height (m)	Ir (ppb ± 2SE)	Ru (ppb ± 2SE)	Pt (ppb ± 2SE)	Pd (ppb ± 2SE)	Re (ppb ± 2SE)
NF-62	0.88	0.085 ± 0.021	0.117 ± 0.007	2.92 ± 0.14	3.09 ± 0.17	12.5 ± 0.3
NF-51 K	0.60	0.062 ± 0.023	0.114 ± 0.009	3.97 ± 0.19	3.07 ± 0.20	44.9 ± 1.7
NF-51I	0.30	0.052 ± 0.022	0.089 ± 0.007	2.64 ± 0.12	2.05 ± 0.11	18.5 ± 0.4
NF-51G	0.12	0.114 ± 0.021	0.094 ± 0.008	3.15 ± 0.13	1.85 ± 0.35	10.7 ± 0.2
NF-51D	0.00	0.154 ± 0.022	0.151 ± 0.009	2.65 ± 0.13	1.13 ± 0.17	1.01 ± 0.02
NF-51A	− 0.10	0.076 ± 0.024	0.123 ± 0.007	3.84 ± 0.18	1.72 ± 0.43	1.42 ± 0.03
NF-48B	−0.33	0.339 ± 0.042	0.365 ± 0.009	11.3 ± 0.3	7.71 ± 0.28	0.366 ± 0.009
NF-45B	−0.53	0.479 ± 0.031	0.180 ± 0.007	14.7 ± 0.4	6.28 ± 0.25	0.079 ± 0.006
NF-41B	−0.76	0.484 ± 0.035	0.242 ± 0.007	15.4 ± 0.5	6.97 ± 0.32	0.179 ± 0.008
NF-37B	−0.90	0.290 ± 0.033	0.155 ± 0.008	16.9 ± 0.4	7.16 ± 0.25	0.081 ± 0.005
NF-32B	−1.11	0.186 ± 0.023	0.188 ± 0.008	19.8 ± 0.5	10.9 ± 0.5	0.106 ± 0.006
NF-29B	− 1.25	0.082 ± 0.021	0.075 ± 0.007	9.04 ± 0.28	6.47 ± 0.23	0.074 ± 0.005
NF-25B	− 1.42	0.074 ± 0.020	0.093 ± 0.006	11.4 ± 0.4	9.46 ± 0.37	0.072 ± 0.005
NF-24B	−1.50	0.143 ± 0.021	0.159 ± 0.007	5.25 ± 0.18	6.41 ± 0.24	0.077 ± 0.006
NF-23B	−1.56	0.083 ± 0.026	0.085 ± 0.007	4.23 ± 0.18	4.34 ± 0.21	0.098 ± 0.007
NF-21B	−1.63	0.057 ± 0.024	0.032 ± 0.007	1.91 ± 0.11	1.84 ± 0.10	0.048 ± 0.005
NF-17B	− 1.82	0.091 ± 0.020	0.064 ± 0.008	3.54 ± 0.15	2.35 ± 0.18	0.055 ± 0.005
NF-12B	−2.01	0.126 ± 0.022	0.118 ± 0.006	4.61 ± 0.15	3.67 ± 0.14	0.045 ± 0.005
NF-11B	−2.08	0.090 ± 0.019	0.093 ± 0.007	3.69 ± 0.13	3.43 ± 0.14	0.041 ± 0.006

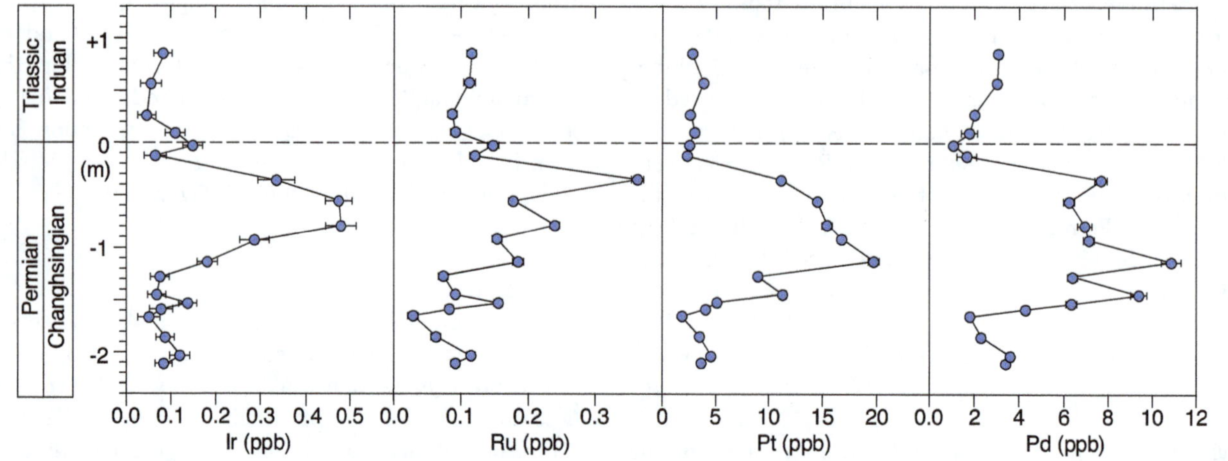

Fig. 4 Stratigraphic profiles of PGE contents of bedded cherts of the Waidani section. See Table 2 for data

There are two possibilities to explain this significant increase in $^3He_{ET}$ concentrations: (1) $^3He_{ET}$ concentrations reflect an approximately fourfold decrease in sedimentation rates or (2) the change reflects a significant increase in $^3He_{ET}$ influx. These possibilities were examined by establishing the floating astronomical timescale of the Upper Permian bedded chert, based on astronomical tuning. The whole-section power spectrum exhibits a dominant peak at 6.0 cm above the 99% confidence level (Fig. 6a). Assuming an average sedimentation rate for Changhsingian bedded chert of ~ 0.3 cm kyr^{-1}, 6.0 cm cycles may represent ~ 20 kyr precession cycles at the PTB. This is consistent with previous studies that demonstrated rhythmic alternations of chert and claystone beds over precession cycles of ~ 20 kyr (Ikeda and Tada 2014). A 20 kyr tuned astrochronology, anchored at the MEE horizon (251.942 Ma, Burgess et al. 2014), allows us to constrain the sedimentation rate for each bed of the Permian section. Results indicate that sedimentation rates varied from 0.2 to 0.5 cm kyr^{-1} (Fig. 6b), with no significant decrease for bedded chert in the topmost ~ 1.5 m of the Permian deposits. Therefore, we conclude that variations of ^3He in the Permian section record changes in the extraterrestrial ^3He flux with time rather than sedimentation rate. Spectral analysis of the precession-tuned time-series also reveals that the interval characterized by elevated $^3He_{ET}$ has a duration of ~ 500 kyr, with such a duration precluding large-impact and meteorite debris origins.

The extraterrestrial ^3He flux was calculated by multiplying the $^3He_{ET}$ concentration by the sedimentary mass accumulation rate (MAR, g cm^{-2} kyr^{-1}):

$$\mathrm{Flux}^3\mathrm{He_{ET}} = \left(\left[^3\mathrm{He_{ET}}\right] \times \mathrm{MAR}\right)/R \quad (3)$$

where R is the fractional retention of ^3He after deposition on the sea floor (Farley 1995), ranging from 1 (total retention) to 0 (total loss). Using an approximate density for shale of 2.6 g cm^{-3}, we estimate an MAR value for the Permian section of 0.81 ± 0.18 g cm^{-2} kyr^{-1} (Table 1). Ignoring possible diffusive loss (R = 1), we estimate a maximum $^3He_{ET}$ flux of 0.27 ± 0.06 pcc cm^{-2} kyr^{-1} (1 pcc = 10^{-12} mL STP) before the MEE (Fig. 7). For a scenario where the MAR remained stable after the MEE, the $^3He_{ET}$ flux is tentatively estimated as 0.04–0.06 pcc cm^{-2} kyr^{-1} in the Triassic section. These values are consistent with those reported for Cenozoic and Cretaceous units (Farley et al. 1998, 2012).

We propose two possible mechanisms to explain the unique IDP signal recorded in the latest Permian samples: a cometary shower or collisions in the asteroid belt. Interplanetary dust particles are generally regarded as having

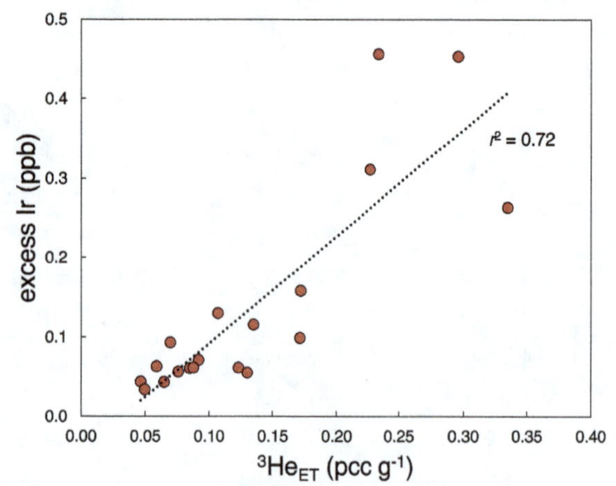

Fig. 5 Bivariate plot of extraterrestrial ^3He ($^3He_{ET}$) versus excess Ir in samples from the Waidani section. Dashed line represents linear regression between the two. See Table 1 for $^3He_{ET}$ and Table 2 for excess Ir data

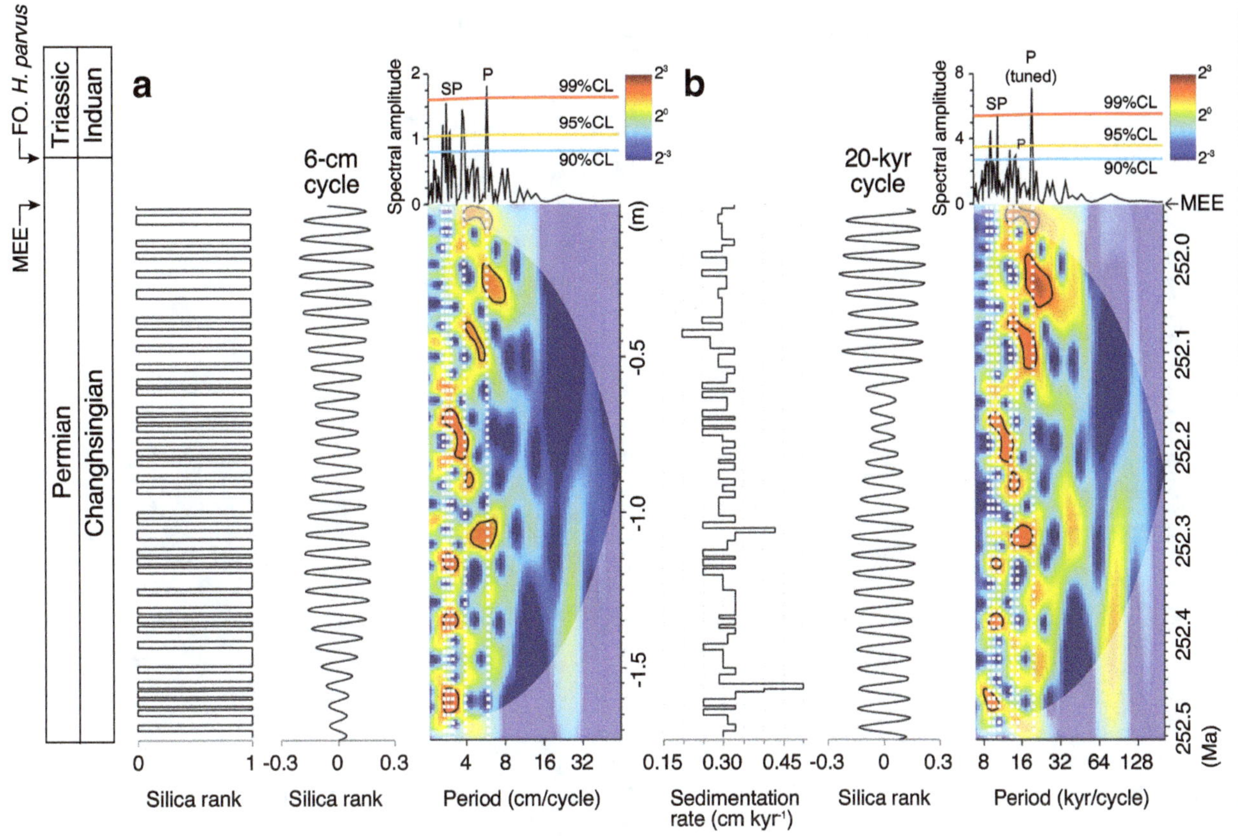

Fig. 6 Cyclostratigraphy and floating astronomical time scale (ATS) of the Waidani section. **a** Silica rank and frequency–amplitude modulations of the Upper Permian (Changhsingian) bedded chert sequence. The average sedimentation rate for the Changhsingian bedded chert is ~ 0.3 cm kyr^{-1}, so the extracted 6 cm cycle was converted to a long precession cycle (~ 20 kyr) for orbital tuning (Yao et al. 2015). **b** Variations in sedimentation rate for each bed, calculated from the Changhsingian ATS and frequency–amplitude modulations of orbital-tuned silica rank. Red, orange, and blue lines in the REDFIT spectra delineate 99%, 95%, and 90% confidence levels (CL) against red-noise background from a first-order autoregressive process, respectively (Schulz and Mudelsee 2002). P and SP at the spectral peaks represent precession and semi-precession (Berger et al. 2006; Soda et al. 2015), respectively. Black contours and shaded regions in the wavelet spectra indicate 95% CL against the red noise and cone of influence due to zero padding, respectively (Torrence and Compo 1998)

originated from dust-producing objects such as comets and asteroids (Kortenkamp and Dermott 1998). The particles migrate toward the Sun, driven by Poynting–Robertson and solar wind drag, and are continuously deposited on Earth. The Poynting–Robertson lifetime for micrometer-sized IDPs is of the order of 10^4–10^5 years (Burns et al. 1979), so most decay products from comets and asteroids are rapidly lost from the solar system. Cometary showers or collisions in the asteroid belt (Farley et al. 1998, 2006) result in episodes of elevated IDP flux to Earth over a few million years or less. Our estimate of the duration of peak $^3He_{ET}$ flux immediately before the PTB is similar in magnitude to those of two well-documented Cenozoic $^3He_{ET}$ events: a major asteroid collision in the late Miocene (Farley et al. 2006) and a dust shower generated from comets or asteroids in the late Eocene (Farley et al. 1998). These events were associated with longer periods (~ 1.5 Myr) of $^3He_{ET}$ flux decay than is observed across the PTB (Farley et al. 2006, 2012). However, the observed short duration of enhanced $^3He_{ET}$ flux (~ 500 kyr) is consistent with the dynamic lifetime (10^4–10^5 years) of fine-grained dust liberated by a single asteroid disruption event. The timing of collision of the Erigone family of asteroids in the inner main asteroid belt is constrained to 220^{+60}_{-80} Ma, close to the PTB (Spoto et al. 2015). Our data are presently insufficient to identify an individual asteroid shower episode.

The present study indicates that an enhanced IDP flux event occurred simultaneously with a marked decline in the diversity of Permian radiolarians from ~ 500 kyr prior to the PTB (Fig. 7). A similar substantial pre-MEE decline of radiolarians has been reported in the Upper Permian (late Changhsingian) in South China (Feng et al. 2007; Feng and Algeo 2014). Furthermore, a recent statistical analysis of Permian ammonoids revealed that high turnover rates and extinction pulses occurred during the last ~ 700 kyr of the Permian (Kiessling et al. 2018). As a first approximation, it seems reasonable to assume that the

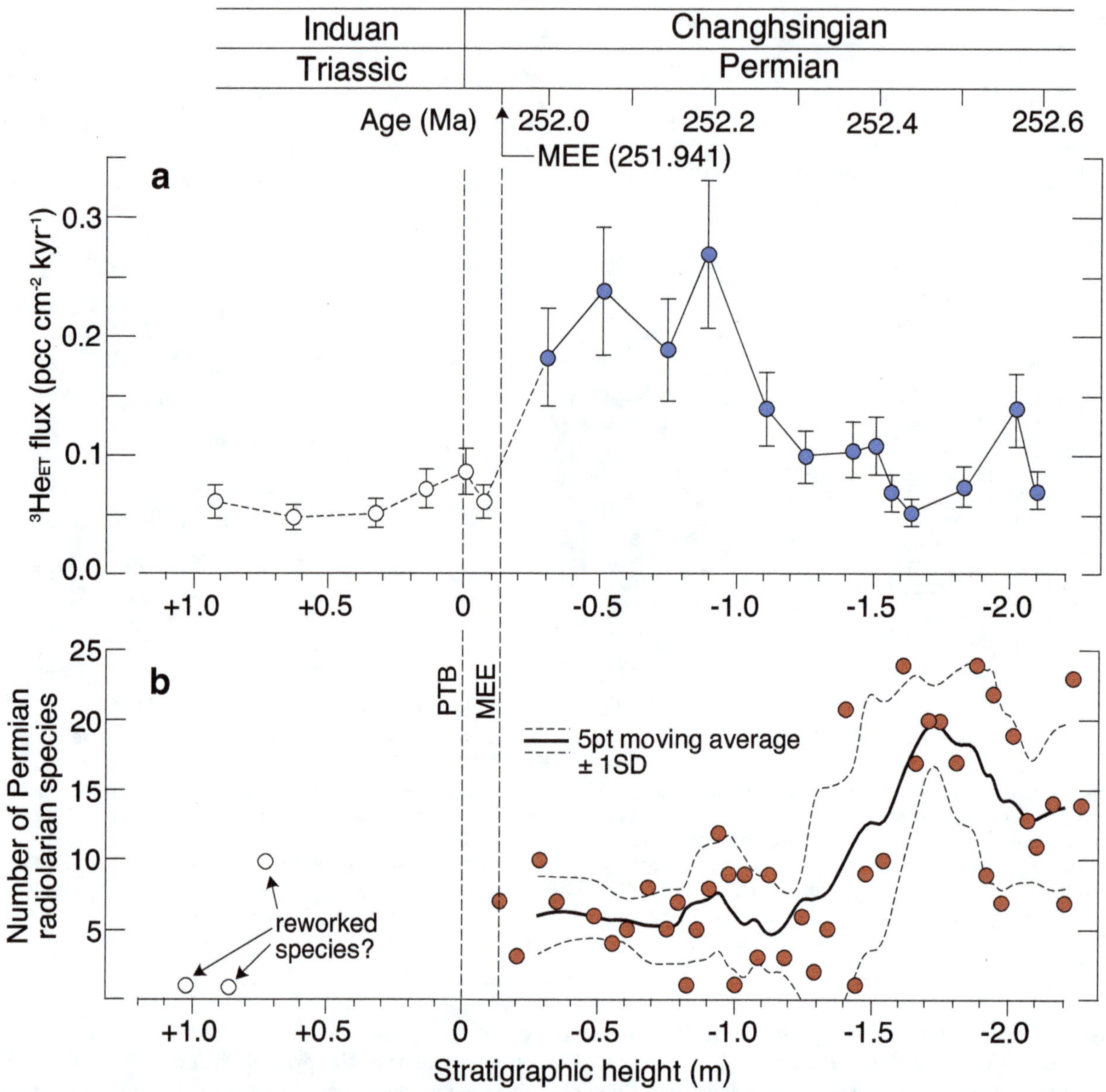

Fig. 7 Extraterrestrial ^{3}He (^{3}He$_{ET}$) flux and radiolarian species diversity across the PTB. **a** ^{3}He$_{ET}$ flux calculated from extraterrestrial ^{3}He concentrations (Table 1), a mean dry bulk density of 2.6 g cm^{-3} for claystone, and mean sedimentation rates calculated using astronomically tuned cyclostratigraphy. Error bars are based on two standard deviations calculated for the MAR of the Permian section. The ^{3}He$_{ET}$ flux after the MEE horizon was tentatively estimated assuming sedimentation rates remained stable after the MEE. See Table 1 for data. **b** Stratigraphic variation of the number of Permian radiolarian species in the study section (Sano et al. 2012a). Marked decline in the diversity of Permian radiolarians began at ~ 500 kyr prior to the PTB. Permian radiolarians in the Triassic section are regarded as reworked species (Sano et al. 2010)

pre-MEE declines of latest Permian radiolarians and ammonoids were related to an extraterrestrial event indicated by the ^{3}He record in the deep-sea PTB section of Japan.

Accretion and ablation of IDPs in Earth's ionosphere may affect the chemistry and aerosol distribution of the upper atmosphere (Grebowsky et al. 2002), providing a possible link between IDP accretion rate and global climate change (Muller and MacDonald 1997; Patterson and Farley 1998). The present study provides the first clear evidence of an extraterrestrial influence immediately prior to the PTB, and a re-evaluation of the cause of the pre-MEE decline is required to confirm the link between extraterrestrial material flux and extinction-related global climate change.

Conclusions

Extraterrestrial ^{3}He is preserved in deep-sea bedded chert from a continuous PTB section at Waidani, Japan, which was deposited in the Panthalassa superocean. High ^{3}He/^{4}He ratios of up to 150 Ra were determined for acid-insoluble residues from the uppermost Permian deposits, indicating the occurrence of extraterrestrial He hosted mainly in IDPs. The extraterrestrial ^{3}He fraction varies across the PTB with $^3He_{ET}$ concentrations being higher in the topmost ~ 1.5 m of the studied Permian deposits (4–5 times higher than in the overlying Triassic unit). The $^3He_{ET}$ flux to Earth across the PTB was estimated from the $^3He_{ET}$ concentration and sedimentary mass accumulation rate of the Waidani PTB section, with the flux increasing over the last 500 kyr of the Permian. This unusual signal suggests a significant increase in the influx of IDPs, likely related to an asteroid shower in the inner solar system. High-resolution stratigraphy indicates that peak IDP flux occurred during the final 500 kyr of the Permian, concurrent with a pre-MEE decline in radiolarian diversity.

Additional files

Additional file 1: Figure S1. Map showing the location of the study area in central Japan. The deep-sea PTB section discovered in a radiolarian chert succession within the Mino Belt (Sano et al. 2012a) represents a Jurassic subduction-generated accretionary complex.

Additional file 2: Figure S2. Geological map of the Mt. Funabuseyama area. Map showing the location of the Waidani section (35°41′11″N, 136° 40′27″E).

Additional file 3: Figure S3. Mo and U enrichment factors and their ratios (Mo_{EF}/U_{EF}) for the studied section. Mo/U ratios are typically moderate (~ 3, similar to modern seawater) under weakly anoxic conditions and high (~ 9, three times that of modern seawater) under highly anoxic–euxinic bottom-water conditions (Takahashi et al. 2014; Algeo and Tribovillard 2009). See Additional file 2: Table S2 for data.

Additional file 4: Figure S4. Correlation coefficients for pairs of the elements/Al ratios determined in this study. Concentrations of the elements were normalized using Al concentrations to compensate the effect of biogenic SiO_2-dilution. Correlation coefficients (r) > 0.75 are indicated by colors according to the ranges of values. ΣREE represents total REE concentrations.

Additional file 5: Table S1. Major element compositions of claystone samples from the Waidani section.

Additional file 6: Table S2. Trace element and REE compositions of claystone samples from the Waidani section.

Abbreviations

CAI: Conodont alteration index; IDP: Interplanetary dust particle; MEE: Mass extinction event; PGE: Platinum group element; PTB: Permian–Triassic boundary

Acknowledgements

We thank S. Maruyama, J. Kirschvink, T. Nakamura, and D. Nakashima for discussions. H. Sano assisted with fieldwork and sample collection, and D. Yamashita assisted with sample processing.

Funding

This work was supported by the Japan Society for the Promotion of Science, grant number 26106005 to TO and NT.

Authors' contributions

TO, NT, YS, and YI designed the study. TO, NT, MM, and YS performed the helium isotopic analyses. HS and AI performed the PGE analyses. TO performed the XRF analyses. KS performed the spectral analysis. All authors participated in the discussions to interpret the analytical results and aided in the writing of the manuscript. All authors read and approved the final manuscript.

Competing interests

The authors declare that they have no competing interests.

Author details

[1]Department of Earth and Environmental Sciences, Kumamoto University, Kumamoto 860-8555, Japan. [2]Atmosphere and Ocean Research Institute, The University of Tokyo, Chiba 277-8564, Japan. [3]Ocean Resources Research Center for Next Generation, Chiba Institute of Technology, Chiba 275-0016, Japan. [4]Department of Earth and Planetary Sciences, Tokyo Institute of Technology, Tokyo 152-8550, Japan. [5]Department of Earth Science and Astronomy, The University of Tokyo, Tokyo 153-8902, Japan.

References

Agematsu S, Sano H, Sashida K (2014) Natural assemblages of Hindeodus conodonts from a Permian-Triassic boundary sequence, Japan. Palaeontology 57:1277–1289. https://doi.org/10.1111/pala.12114.

Algeo TJ, Kuwahara K, Sano H, Bates S, Lyons T, Elswick E, Hinnov L, Ellwood B, Moser J, Maynard JB (2011) Spatial variation in sediment fluxes, redox conditions, and productivity in the Permian-Triassic Panthalassic Ocean. Palaeogeogr Palaeoclimatol Palaeoecol 308:65–83. https://doi.org/10.1016/j.palaeo.2010.07.007.

Algeo TJ, Tribovillard N (2009) Environmental analysis of paleoceanographic systems based on molybdenum–uranium covariation. Chem Geol 268:211–225. https://doi.org/10.1016/j.chemgeo.2009.09.001.

Amari S, Ozima M (1985) Search for the origin of exotic helium in deep-sea sediments. Nature 317:520–522. https://doi.org/10.1038/317520a0.

Basu AR, Petaev MI, Poreda RJ, Jacobsen SB, Becker L (2003) Chondritic meteorite fragments associated with the Permian-Triassic boundary in Antarctica. Science 302:1388–1392. https://doi.org/10.1126/science.1090852.

Becker L, Poreda RJ, Basu AR, Pope KO, Harrison TM, Nicholson C, Iasky R (2004) Bedout: a possible end-Permian impact crater offshore of northwestern Australia. Science 304:1469–1476. https://doi.org/10.1126/science.1093925..

Becker L, Poreda RJ, Hunt AG, Bunch TE, Rampino M (2001) Impact event at the Permian-Triassic boundary: evidence from extraterrestrial noble gases in fullerenes. Science 291:1530–1533. https://doi.org/10.1126/science.1057243..

Berger A, Loutre MF, Mélice JL (2006) Equatorial insolation: from precession harmonics to eccentricity frequencies. Clim Past 2:131–136. https://doi.org/10.5194/cp-2-131-2006.

Burgess SD, Bowring S, Shen SZ (2014) High-precision timeline for Earth's most severe extinction. Proc Natl Acad Sci U S A 111:3316–3321. https://doi.org/10.1073/pnas.1317692111.

Burgess SD, Muirhead JD, Bowring SA (2017) Initial pulse of Siberian Traps sills as the trigger of the end-Permian mass extinction. Nat Commun 8:164. https://doi.org/10.1038/s41467-017-00083-9.

Burns JA, Lamy PL, Soter S (1979) Radiation forces on small particles in the solar system. Icarus 40:1–48. https://doi.org/10.1016/0019-1035(79)90050-2..

Colodner DC, Boyle EA, Edmond JM, Thomson J (1992) Post-depositional mobility of platinum, iridium and rhenium in marine sediments. Nature 358:402–404. https://doi.org/10.1038/358402a0.

Erwin DH (1994) The Permo-Triassic extinction. Nature 367:231–236. https://doi.org/10.1038/367231a0.

Evans NJ, Gregoire DC, Grieve RAF, Goodfellow WD, Veizer J (1993) Use of platinum-group elements for impactor identification: terrestrial impact craters and Cretaceous-Tertiary boundary. Geochim Cosmochim Acta 57:3737–3748. https://doi.org/10.1016/0016-7037(93)90152-M.

Farley KA (1995) Cenozoic variations in the flux of interplanetary dust recorded by ^{3}He in a deep-sea sediment. Nature 376:153–156. https://doi.org/10.1038/376153a0.

Farley KA, Montanari A, Coccioni R (2012) A record of the extraterrestrial ^{3}He flux through the Late Cretaceous. Geochim Cosmochim Acta 84:314–328. https://doi.org/10.1016/j.gca.2012.01.015.

Farley KA, Montanari A, Shoemaker EM, Shoemaker CS (1998) Geochemical evidence for a comet shower in the Late Eocene. Science 280:1250–1253. https://doi.org/10.1126/science.280.5367.1250..

Farley KA, Mukhopadhyay S (2001) An extraterrestrial impact at the Permian-Triassic boundary? Science 293:2343. https://doi.org/10.1126/science.293.5539.2343a.

Farley KA, Vokrouhlicky D, Bottke WF, Nesvorny D (2006) A late Miocene dust shower from the break-up of an asteroid in the main belt. Nature 439:295–297. https://doi.org/10.1038/nature04391.

Farley KA, Ward P, Garrison G, Mukhopadhyay S (2005) Absence of extraterrestrial ^{3}He in Permian-Triassic age sedimentary rocks. Earth Planet Sci Lett 240:265–275. https://doi.org/10.1016/j.epsl.2005.09.054.

Feng Q, He W, Gu S, Meng Y, Jin Y, Zhang F (2007) Radiolarian evolution during the latest Permian in South China. Glob Planet Chang 55:177–192. https://doi.org/10.1016/j.gloplacha.2006.06.012.

Feng QL, Algeo TJ (2014) Evolution of oceanic redox conditions during the Permo-Triassic transition: evidence from deepwater radiolarian facies. Earth-Sci Rev 137:34–51. https://doi.org/10.1016/j.earscirev.2013.12.003.

Genge MJ, Engrand C, Gounelle M, Taylor S (2008) The classification of micrometeorites. Meteorit Planet Sci 43:497–515. https://doi.org/10.1111/j.1945-5100.2008.tb00668.x.

Grebowsky JM, Moses JI, Pesnell WD (2002) Meteoric material—an important component of planetary atmospheres. In: Mendillo M, Nagy A, Waite JH (eds) Atmospheres in the solar system: comparative aeronomy. Geophysical monograph series, vol 130 American Geophysical Union, 235–244. doi: https://doi.org/10.1029/130GM15.

Heck PR, Schmitz B, Baur H, Wieler R (2008) Noble gases in fossil micrometeorites and meteorites from 470 Myr old sediments from southern Sweden, and new evidence for the L-chondrite parent body breakup event. Meteorit Planet Sci 43:517–528. https://doi.org/10.1111/j.1945-5100.2008.tb00669.x.

Ikeda M, Tada R (2014) A 70 million year astronomical time scale for the deep-sea bedded chert sequence (Inuyama, Japan): implications for Triassic–Jurassic geochronology. Earth Planet Sci Lett 399:30–43. https://doi.org/10.1016/j.epsl.2014.04.031.

Ishikawa A, Senda R, Suzuki K, Dale CW, Meisel T (2014) Re-evaluating digestion methods for highly siderophile element and ^{187}Os isotope analysis: evidence from geological reference materials. Chem Geol 384:27–46. https://doi.org/10.1016/j.chemgeo.2014.06.013.

Isozaki Y (1997) Permo-Triassic boundary superanoxia and stratified superocean: records from lost deep-sea. Science 276:235–238. https://doi.org/10.1126/science.276.5310.235.

Kiessling W, Schobben M, Ghaderi A, Hairapetian V, Leda L, Korn D (2018) Pre–mass extinction decline of latest Permian ammonoids. Geology 46:283–286. https://doi.org/10.1130/G39866.1.

Koeberl C, Farley KA, Peucker-Ehrenbrink B, Sephton MA (2004) Geochemistry of the end-Permian extinction event in Austria and Italy: no evidence for an extraterrestrial component. Geology 32:1053–1056. https://doi.org/10.1130/G20907.1.

Kortenkamp SJ, Dermott SF (1998) Accretion of interplanetary dust particles by the Earth. Icarus 135:469–495. https://doi.org/10.1006/icar.1998.5994.

Laskar J, Robutel P, Joutel F, Gastineau M, Correia ACM, Levrard B (2004) A long-term numerical solution for the insolation quantities of the Earth. Astron Astrophys 428:261–285. https://doi.org/10.1051/0004-6361:20041335.

Marcantonio F, Kumar N, Stute M, Anderson RF, Seidl MA, Schlosser P, Mix A (1995) A comparative study of accumulation rates derived by He and Th isotope analysis of marine sediments. Earth Planet Sci Lett 133:549–555. https://doi.org/10.1016/0012-821X(95)00079-R.

Matsuda T, Isozaki Y (1991) Well-documented travel history of Mesozoic pelagic chert in Japan: from remote ocean to subduction zone. Tectonics 10:475–499. https://doi.org/10.1029/90TC02134.

McGee D, Mukhopadhyay S (2013) Extraterrestrial He in sediments: from recorder of asteroid collisions to timekeeper of global environmental changes. In: Burnard P (ed) The noble gases as geochemical tracers. Springer, Berlin, pp 155–176. https://doi.org/10.1007/978-3-642-28836-4_7..

McLennan SM (2001) Relationships between the trace element composition of sedimentary rocks and upper continental crust. Geochem Geophys Geosyst 2:2000GC000109. https://doi.org/10.1029/2000GC000109..

Mukhopadhyay S, Farley KA (2006) New insights into the carrier phase(s) of extraterrestrial ^{3}He in geologically old sediments. Geochim Cosmochim Acta 70:5061–5073. https://doi.org/10.1016/j.gca.2006.06.1566.

Muller RA, MacDonald GJ (1997) Glacial cycles and astronomical forcing. Science 277:215–218. https://doi.org/10.1126/science.277.5323.215..

Onoue T, Nakamura T, Haranosono T, Yasuda C (2011) Composition and accretion rate of fossil micrometeorites recovered in Middle Triassic deep-sea deposits. Geology 39:567–570. https://doi.org/10.1130/G31866.1.

Onoue T, Sato H, Nakamura T, Noguchi T, Hidaka Y, Shirai N, Ebihara M, Osawa T, Hatsukawa Y, Toh Y, Koizumi M, Harada H, Orchard MJ, Nedachi M (2012) Deep-sea record of impact apparently unrelated to mass extinction in the Late Triassic. Proc Natl Acad Sci U S A 109:19134–19139. https://doi.org/10.1073/pnas.1209486109.

Paillard D, Labeyrie L, Yiou P (1996) Macintosh program performs time-series analysis. EOS 77:379. https://doi.org/10.1029/96EO00259.

Patterson DB, Farley KA (1998) Extraterrestrial ^{3}He in seafloor sediments: evidence for correlated 100 kyr periodicity in the accretion rate of interplanetary dust, orbital parameters, and Quaternary climate. Geochim Cosmochim Acta 62: 3669–3682. https://doi.org/10.1016/S0016-7037(98)00263-4.

Patterson DB, Farley KA, Schmitz B (1998) Preservation of extraterrestrial ^{3}He in 480-Ma-old marine limestones. Earth Planet Sci Lett 163:315–325. https://doi.org/10.1016/S0012-821X(98)00197-6.

Ramezani J, Bowring SA (2017) Advances in numerical calibration of the Permian timescale based on radioisotopic geochronology. Geol Soc London Spec Publ 450(1):51–60. https://doi.org/10.1144/SP450.17.

Sano H, Kuwahara K, Yao A, Agematsu S (2010) Panthalassan seamount-associated Permian-Triassic boundary siliceous rocks, Mino terrane, Central Japan. Paleont Res 14:293–314. https://doi.org/10.2517/1342-8144-14.4.293.

Sano H, Kuwahara K, Yao A, Agematsu S (2012b) Stratigraphy and age of the Permian-Triassic boundary siliceous rocks of the Mino terrane in the Mt. Funabuseyama area, Central Japan. Paleont Res 16:124–145. https://doi.org/10.2517/1342-8144-16.2.124.

Sano H, Wada T, Naraoka H (2012a) Late Permian to Early Triassic environmental changes in the Panthalassic Ocean: record from the seamount-associated deep-marine siliceous rocks, Central Japan. Palaeogeogr Palaeoclimatol Palaeoecol 363:1–10. https://doi.org/10.1016/j.palaeo.2012.07.018.

Sano Y (2017) Helium Isotopes. In: White WM (ed) Encyclopedia of geochemistry: a comprehensive reference source on the chemistry of the Earth. Encyclopedia of Earth Sciences Series doi:https://doi.org/10.1007/978-3-319-39193-9_205-1.

Sano Y, Marty B, Burnard P (2013) Noble gases in the atmosphere. In: Burnard P (ed) The noble gases as geochemical tracers. Springer, Berlin, pp 17–31. https://doi.org/10.1007/978-3-642-28836-4_2..

Schulz M, Mudelsee M (2002) REDFIT: estimating red-noise spectra directly from unevenly spaced paleoclimatic time series. Comput Geosci 28:421–426. https://doi.org/10.1016/S0098-3004(01)00044-9.

Soda K, Onoue T, Ikeda M (2015) Cyclostratigraphic examination of Middle Triassic (Anisian) bedded chert in the Chichibu Belt from Tsukumi area, eastern Kyushu, Japan. J Geol Soc Jpn 121:147–152. https://doi.org/10.5575/geosoc.2015.0008.

Spoto F, Milani A, Knezevic Z (2015) Asteroid family ages. Icarus 257:275–289. https://doi.org/10.1016/j.icarus.2015.04.041.

Takahashi S, Yamasaki S, Ogawa Y, Kimura K, Kaiho K, Yoshida T, Tsuchiya N (2014) Bioessential element-depleted ocean following the euxinic maximum of the end-Permian mass extinction. Earth Planet Sci Lett 393:94–104. https://doi.org/10.1016/j.epsl.2014.02.041.

Torrence C, Compo GP (1998) A practical guide to wavelet analysis. Bull Am Meteor Soc 79:61–78. https://doi.org/10.1175/1520-0477(1998)079<0061:APGTWA>2.0.CO;2.

Uno K, Onoue T, Hamada K, Hamami S (2012) Palaeomagnetism of Middle Triassic red bedded cherts from southwest Japan: equatorial palaeolatitude of primary magnetization and widespread secondary magnetization. Geophys J Int 189: 1383–1398. https://doi.org/10.1111/j.1365-246X.2012.05462.x.

Wignall PB (2007) The End-Permian mass extinction - how bad did it get? Geobiology 5:303–309. https://doi.org/10.1111/j.1472-4669.2007.00130.x.

Winckler G, Fischer H (2006) 30,000 years of cosmic dust in Antarctic ice. Science 313:491. https://doi.org/10.1126/science.1127469..

Yao X, Zhou Y, Hinnov LA (2015) Astronomical forcing of a Middle Permian chert sequence in Chaohu, South China. Earth Planet Sci Lett 422:206–221. https://doi.org/10.1016/j.epsl.2015.04.017.

8

Melting phase relations in the $MgSiO_3$–$CaSiO_3$ system at 24 GPa

Ryuichi Nomura[1*], Youmo Zhou[1*] and Tetsuo Irifune[1,2]

Abstract

The Earth's lower mantle is composed of bridgmanite, ferropericlase, and $CaSiO_3$-rich perovskite. The melting phase relations between each component are key to understanding the melting of the Earth's lower mantle and the crystallization of the deep magma ocean. In this study, melting phase relations in the $MgSiO_3$–$CaSiO_3$ system were investigated at 24 GPa using a multi-anvil apparatus. The eutectic composition is (Mg,Ca)SiO_3 with 81–86 mol% $MgSiO_3$. The solidus temperature is 2600–2620 K. The solubility of $CaSiO_3$ component into bridgmanite increases with temperature, reaching a maximum of 3–6 mol% at the solidus, and then decreases with temperature. The same trend was observed for the solubility of $MgSiO_3$ component into $CaSiO_3$-rich perovskite, with a maximum of 14–16 mol% at the solidus. The asymmetric regular solutions between bridgmanite and $CaSiO_3$-rich perovskite and between $MgSiO_3$ and $CaSiO_3$ liquid components well reproduce the melting phase relations constrained experimentally.

Keywords: Bridgmanite, $CaSiO_3$-rich perovskite, Melting phase relations, Multi-anvil experiments, High pressure and high temperature, Thermodynamics

Introduction

Determining melting phase relations in natural rock-forming, multi-component systems at high pressures is essential to understand the origin and consequences of the deep mantle melting (e.g., Williams and Garnero 1996) and the crystallization of the deep magma ocean of the early Earth (e.g., Ito et al. 2004). Although numerous melting experiments have been performed for these purposes, the range of parameter space (pressure (*P*), temperature (*T*), and compositions (*X*)) studied is limited for tracing the crystallization of the magma ocean and for a comprehensive investigation into the origin of the deep mantle melting. Indeed, it is not practical to cover the vast P–T–X space relevant to deep Earth events based on experiments alone. In contrast, experiments in simpler systems combined with thermodynamic models provide a powerful tool to understand the melting chemistry in the multi-component system. Such comprehensive datasets and thermodynamic models have been constructed for pressures up to those of the mantle transition zone (Gasparik 2003).

In addition, the occurrence of a certain degree of temperature gradient is inevitable in melting experiments using laser-heated diamond anvil cells or multi-anvil apparatuses, especially under the high pressures of the lower mantle. The temperature gradient obscures the melting chemistry in multi-component systems because phase segregation changes the bulk compositions of interest. In these cases, only limited information, such as phase equilibria and element partitioning between the phases in contact, can be extracted by assuming local chemical equilibrium. This problem could be overcome by experiments in simple binary systems.

The lower mantle is composed of bridgmanite, ferropericlase, and $CaSiO_3$-rich perovskite (Irifune 1994). Normal mid-ocean ridge basalt has a mineral assemblage of bridgmanite, $CaSiO_3$-rich perovskite, a SiO_2 phase, and an aluminous phase in the lower mantle (Hirose et al. 1999). Therefore, the binary melting phase relations between each mineral-constituting component are important in understanding the melting behaviors of the lower mantle. Previously, the MgO–$MgSiO_3$ system was investigated at the lower-mantle pressures using a multi-anvil press (Liebske and Frost 2012), but little is known about the other systems.

* Correspondence: nomura@sci.ehime-u.ac.jp; youmo@sci.ehime-u.ac.jp
[1]Geodynamics Research Center, Ehime University, 2-5 Bunkyo, Matsuyama, Ehime 790-8577, Japan
Full list of author information is available at the end of the article

This paper is aimed at establishing the melting phase relations in the $MgSiO_3$–$CaSiO_3$ system at 24 GPa, avoiding ambiguous implications for Earth science that require other melting phase relations such as MgO–$CaSiO_3$. The eutectic composition, solidus temperature, and solubility between the two endmembers were investigated experimentally at 24 GPa. Although Gasparik (1996) has performed a melting experiment on the same system at 22.4 GPa, it is difficult to confirm melting from the backscattered electron image presented in his paper. In the present work, we re-investigate the system to find the interaction parameters between the $MgSiO_3$ and $CaSiO_3$ components by thermodynamic analysis in the framework for the endmembers (de Koker and Stixrude 2009; Stixrude and Lithgow-Bertelloni 2011; Bajgain et al. 2015) consistent with a previous work in the MgO-$MgSiO_3$ system (Liebske and Frost 2012).

Methods/experimental

Starting materials

We used three starting compositions, i.e., $Mg_{0.90}Ca_{0.10}SiO_3$ (Mg90), $Mg_{0.75}Ca_{0.25}SiO_3$ (Mg75), and $Mg_{0.50}Ca_{0.50}SiO_3$ (Mg50). Bubble-free, chemically homogenous glasses were synthesized for $Mg_{0.75}Ca_{0.25}SiO_3$ and $Mg_{0.50}Ca_{0.50}SiO_3$ in Pt crucibles at 2023 K using chemical mixtures of SiO_2 (99.9%, Kojundo Chemical Laboratory Co., Ltd.), MgO (> 99.99%, Kojundo Chemical Laboratory Co., Ltd.), and $CaCO_3$ (> 98.0%, Kanto Chemical Co., Inc.), after decarbonation at 1273 K for 16 h. The starting material with a bulk composition of $Mg_{0.90}Ca_{0.10}SiO_3$ was not vitrified completely, which is likely because of its high Mg content. We crushed and ground it to fine particles. The bulk composition of the latter starting material was confirmed via electron microprobe analysis after a subsolidus high-pressure and high-temperature experiment. Table 1 lists the bulk compositions of the starting materials. The starting materials were stored in a furnace at 383 K before use.

High-pressure melting experiments

High-pressure and high-temperature experiments were performed at 24 GPa and 2600–3000 K in a Kawai-type multi-anvil apparatus (Orange 2000, Geodynamics Research Center, Ehime University). A 10/4 (octahedron edge length/truncation edge length) cell assembly was chosen for pressure generation. The pressure of 24 GPa was calibrated based on the coexistence and dissociation of wadsleyite into bridgmanite and periclase at 2273 K in the Mg_2SiO_4 system (Fei et al. 2004), in addition to other phase transitions at room temperature, 1873 and 2273 K, as described in Zhou et al. (2017). The pressure calibration (Additional file 1: Figure S1) suggests that pressure does not vary with temperature above 1873 K. This is also evidenced by the recovery of bridgmanite from 2620 K (OS3099) since the pressure drop of less than 2 GPa is needed to be within the stability field of bridgmanite at that temperature (Gasparik 1996; Fei et al. 2004). A $LaCrO_3$ heater was chosen for temperature generation. The temperature was monitored with a $W_{97}Re_3$-$W_{75}Re_{25}$ thermocouple. No correction was applied for the pressure effect on the electromotive force of the thermocouple. The junction of the thermocouple was positioned at the center of the heater and between two Re capsules. The cell assembly used in this study is identical to that of Zhang et al. (1993) with respect to the octahedron dimension (edge length, 10 mm), the heater material ($LaCrO_3$), and the capsule material (Re). However, we used capsules with much smaller dimensions in both length and diameter and adopted a longer sleeve heater with a much smaller diameter to reduce the thermal gradient. Zhang et al. (1993) reported a temperature drop of ~ 50 K at a distance of ~ 0.5 mm from the hot spot axially. In this study, the center of the sample (~ 0.2 mm thick) is ~ 0.5 mm away axially from the thermocouple junction. Therefore, the uncertainty of the sample temperature caused by the temperature gradient is much less than 50 K. Indeed, we confirmed drastic changes of the melting phase relations within 100 K in a series of experiments. In addition, the obtained electron microprobe images of the recovered samples suggest that the thermal gradient in the axial direction was negligible and it was not the dominant factor of temperature uncertainty (i.e., no systematic error). Therefore, small, non-systematic changes in the geometry and temperature profile caused by the differences in the geometry of components (heater shape and size, the degree of cell deformation during heating, and so on) in each run lead to uncertainties in temperature. We adopted the variations in the heating power-temperature relationships (Additional file 1: Figure S2) as temperature uncertainties, although they represent the variations not only in the thermocouple-sample geometry and temperature profile but also in other factors not related to the temperature uncertainty but to the heating efficiency. Therefore, uncertainties would be overestimated. However, this is the best way to express the uncertainties in experimental temperatures (Table 2). The high-pressure samples were quenched by cutting off the power supply at the target temperatures after heating for 5–20 min. The experimental conditions and results are summarized in Table 2.

Phase identification and chemical analysis

The recovered samples were embedded in epoxy resin and finely polished for subsequent analyses. Phase identification was conducted with a microfocus X-ray diffraction (XRD) apparatus (Rigaku MicroMax-007HF) using Cu Kα radiation. The sample surface was coated with carbon and analyzed using a field emission-type scanning electron microscope (FE-SEM; JEOL JSM7000F).

Table 1 Chemical compositions of starting materials

Name	Phase[a]	MgO	CaO	SiO_2	Total (wt.%)	Mg	Ca	Si	ΣCation[b]	Mg/(Mg + Ca)
Mg90	Mixture	35.63(22)	5.30(45)	58.87(54)	99.79(33)	0.903(1)	0.097(9)	1.001(4)	2.000(4)	0.904(8)
Mg75	Glass (29)	28.99(22)	13.67(10)	57.12(42)	99.78(50)	0.753(5)	0.255(2)	1.004(2)	2.004(3)	0.747(2)
Mg50	Glass (26)	18.53(16)	25.65(15)	54.89(36)	99.07(59)	0.503(3)	0.500(2)	0.999(2)	2.001(2)	0.501(2)

[a]Number of electron microprobe analyses is shown in parentheses. Mg90 is a mixture of glass and oxides; see the main text for details
[b]Number of oxygen is 3

Chemical compositions were measured using an energy-dispersive X-ray spectrometer with a silicon-drift detector (EDS; Oxford Instruments X-Max^N) attached to the FE-SEM at 15 kV and 1 nA with collection times of ~ 30 s. Mg_2SiO_4 forsterite was used as the compositional standard for Mg and Si, while $CaSiO_3$ wollastonite was used for Ca. The grains of bridgmanite, $CaSiO_3$-rich perovskite were analyzed in a focused mode with a beam size of less than 1 μm. The quenched melt was analyzed in a scanning mode larger than the spatial scale of heterogeneity.

Thermodynamic modeling

Among many available thermodynamic approaches that can describe the melting behavior in a multi-component system, we followed that of Liebske and Frost (2012) to maintain consistency with the results for the MgO–$MgSiO_3$ system, which has been studied extensively at high pressures, including lower-mantle pressures. Liebske and Frost (2012) used the chemical potentials given by Lithgow-Bertelloni and Stixrude (2005) and Stixrude and Lithgow-Bertelloni (2011) for the solid phases and by de Koker and Stixrude (2009) for the liquid phases; these papers constructed an internally consistent thermodynamic model, although the equations, parameterization, and parameter set differ slightly from those of Liebske and Frost (2012). For solids, we used the thermodynamic descriptions and parameter sets of Lithgow-Bertelloni and Stixrude (2005) and Stixrude and Lithgow-Bertelloni (2011) based on usability and consistency with other components. For liquids, we followed Liebske and Frost (2012) to maintain consistency with the MgO–$MgSiO_3$ system. We describe all of these equations below for clarification.

Chemical equilibrium is attained when the chemical potentials (μ) of the components become equal between the liquid and solid phases:

$$\mu_Y^{\text{solid}} = \mu_Y^{\text{liquid}}, \tag{1}$$

where μ_Y^{phase} represents the chemical potential of component Y, i.e., $MgSiO_3$ or $CaSiO_3$, in the assigned phase. The chemical potentials of solids and liquid are expressed as follows:

$$\begin{aligned}\mu_Y^{\text{phase}} &= \mu_Y^{o,\text{phase}} + RTln\left(a_Y^{\text{phase}}\right),\\ \mu_Y^{\text{phase}} &= \mu_Y^{o,\text{phase}} + RTln\left(X_Y^{\text{phase}}\right) + RTln\left(\gamma_Y^{\text{phase}}\right),\end{aligned} \tag{2}$$

where μ_Y^o, a_Y, X_Y, and γ_Y are the chemical potentials of the endmember, activity, molar concentration, and activity coefficient of component Y, respectively. The activity coefficients of solid and liquid are described by the asymmetric regular solution expressions

$$RTln\left(\gamma_{\text{MgSiO}_3}^{\text{phase}}\right) = \left[A_o - A_1\left(1-4X_{\text{MgSiO}_3}^{\text{phase}}\right)\right]\left(1-X_{\text{MgSiO}_3}^{\text{phase}}\right)^2, \tag{3}$$

$$RTln\left(\gamma_{\text{CaSiO}_3}^{\text{phase}}\right) = \left[A_o + A_1\left(1-4X_{\text{CaSiO}_3}^{\text{phase}}\right)\right]\left(1-X_{\text{CaSiO}_3}^{\text{phase}}\right)^2, \tag{4}$$

where A_0 and A_1 are expressed as follows:

Table 2 Experimental summary

Run no.	T (K)	Duration (min)	SM[a]	Phase assemblage
P = 24 GPa				
OS3097	2600(36)	20	Mg90	Bdg + CaPv
			Mg75	Bdg + CaPv
OS3099	2620(38)	15	Mg90	[HT] Melt + Bdg, [LT] Bdg + CaPv
			Mg75	[HT] CaPv, [LT] Bdg + CaPv
OS3095	2700(41)	10	Mg90	Melt
			Mg75	Melt + CaPv
OS3150	3000(46)	5	Mg50	Melt + CaPv

SM starting material, *Bdg*, bridgmanite, *CaPv* $CaSiO_3$-rich perovskite, *HT* high-temperature region, *LT* low-temperature region
[a]See Table 1 for details

$$A_0 = \left(W^{\text{phase}}_{\text{CaSiO}_3\text{-MgSiO}_3} + W^{\text{phase}}_{\text{MgSiO}_3\text{-CaSiO}_3}\right)/2, \quad (5)$$

$$A_1 = \left(W^{\text{phase}}_{\text{CaSiO}_3\text{-MgSiO}_3} - W^{\text{phase}}_{\text{MgSiO}_3\text{-CaSiO}_3}\right)/2, \quad (6)$$

where W_{Y-Z} is an interaction parameter between components Y and Z.

Once the chemical potentials of the liquid and solid endmembers are evaluated, we can determine the interaction parameter W by fitting the experimental data to the above equations.

The Gibbs free energy (G) of solid endmembers were determined from the Helmholtz free energy (F) as given by Lithgow-Bertelloni and Stixrude (2005) and Stixrude and Lithgow-Bertelloni (2011):

$$G(P,T) = F(V,T) + PV, \quad (7)$$

$$F(V,T) = F_0(V_0,T_0) + F_{cmp}(V,T_0) + [F_{th}(V,T) - F_{th}(V,T_0)], \quad (8)$$

$$P(V,T) = P_{cmp}(V,T_0) + P_{th}(V,T), \quad (9)$$

where F is composed of a standard-state term (F_0), a compression term (F_{cmp}), and a thermal term (F_{th}). Pressure (P) is composed of a compression term (P_{cmp}) and a thermal term (P_{th}). The compression term is expressed by the following third-order finite-strain (f) model:

$$F_{cmp}(V,T_0) = \frac{9}{2}V_0 K_{T_0} f^2 + \frac{9}{2}V_0 K_{T_0}(K_{T_0}{}' - 4) f^3 \quad (10)$$

$$P_{cmp}(V,T_0) = (1+2f)^{5/2}\left[3K_{T_0} f + \frac{9}{2}K_{T_0}(K_{T0}{}' - 4) f^2\right], \quad (11)$$

$$f = \frac{1}{2}\left[\left(\frac{V_0}{V}\right)^{2/3} - 1\right], \quad (12)$$

where V_0, K_{T0}, and K_{T0}' are the volume at a reference state, the bulk modulus, and the first derivative of the bulk modulus, respectively. The thermal term is derived from the Debye model:

$$F_{th}(V,T) = 9nRTD\left(\frac{\theta}{T}\right), \quad (13)$$

$$D(x) = x^{-3}\int_0^x \ln(1-\exp(-t))t^2 dt, \quad (14)$$

$$P_{th}(V,T) = \frac{\gamma}{V}(E_{th}(V,T) - E_{th}(V,T_0)), \quad (15)$$

$$E_{th}(V,T) = 9nRT\left(\frac{\theta}{T}\right)^{-3}\int_0^{\frac{\theta}{T}} \frac{t^3}{e^t - 1} dt, \quad (16)$$

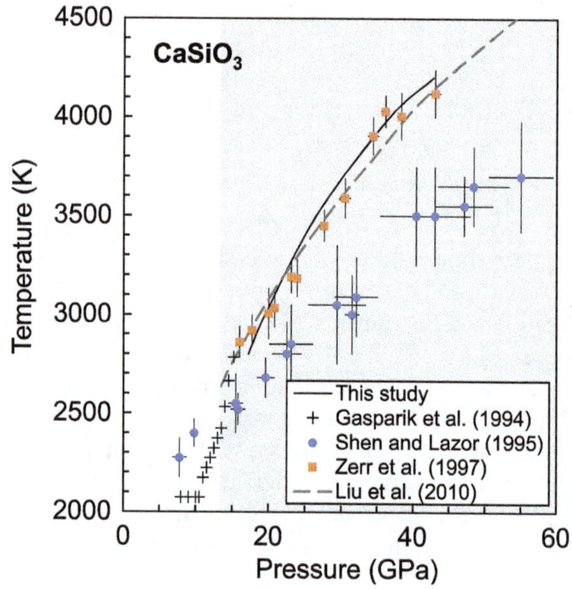

Fig. 1 Melting temperatures of $CaSiO_3$. The black line represents the melting temperatures calculated in this study, the crosses represent the experimental data by Gasparik et al. (1994), the blue circles by Shen and Lazor (1995), the orange squares by Zerr et al. (1997), and the dark gray dashed line represents the molecular dynamics simulation by Liu et al. (2010). The gray region is the stability region of $CaSiO_3$ perovskite at the melting temperatures (Gasparik et al. 1994)

$$\theta = \theta_0\left[1 + 6\gamma_0 f + \frac{1}{2}(-12\gamma_0 + 36\gamma_0{}^2 - 18q_0\gamma_0)f^2\right]^{1/2}, \quad (17)$$

$$\gamma = \gamma_0\frac{(1+2f)}{(1+6\gamma_0 f)}, \quad (18)$$

where the parameters $F(V_0, T_0)$, V_0, K_{T0}, K_{T0}', θ_0, γ_0, and q_0 were given by Stixrude and Lithgow-Bertelloni (2011).

Table 3 Thermodynamic properties of liquids

	$MgSiO_3$	$CaSiO_3$	MgO
T_0 (K)	1773	3000	3000
V_0 (cm^3/mol)	37.20(6)	45.1(12)	16.46(1)
K_{T0} (GPa)	27.3(2)	22.2(30)	34.0(1)
K_{T0}'	5.7(5)	4.60(17)	4.50(5)
γ_0	0.60(2)	0.73	0.96(2)
γ'	− 1.24(5)	− 1.18(5)	− 0.37(5)
C_v (J/K/mol)	173(2)	146(6)	56(3)
S_0 (J/K/mol)	333(3)	*425(3)*	173.5(20)
F_0 (kJ/mol)	*− 1709.09*	*2282.25*	*− 845.22*
Reference	LF2012	B2015	LF2012

Italicized values were constrained in this study. The others are from literature: LF2012, Liebske and Frost (2012); B2015, Bajgain et al. (2015)

Table 4 Thermodynamic properties of solids

	Bdg	CaPv	Pc
T_0 (K)	300	300	300
V_0 (cm^3/mol)	24.45	27.45	11.24
K_{T0} (GPa)	251(3)	236(4)	161(3)
K_{T0}'	4.1(1)	3.9(2)	3.8(2)
γ_0	1.57(5)	1.89(7)	1.36(5)
q_0	1.1(3)	0.9(16)	1.7(2)
θ_0 (K)	905(5)	796(44)	767(9)
F_0 (kJ/mol)	− 1368(1)	− 1463(8)	− 569

The parameters are from Stixrude and Lithgow-Bertelloni (2011)
Bdg bridgmanite, *CaPv* $CaSiO_3$-rich perovskite, *Pc* periclase

For liquids, we followed Liebske and Frost (2012) to maintain consistency. The expressions of the solid compression terms F_{cmp} and P_{cmp} were also applied to liquids, F_{cmp} and P_{cmp}. The thermal part is expressed as follows:

$$F_{th}(V,T) = -S_0(T-T_0)-C_v\left[T\ln\left(\frac{T}{T_0}\right)-(T-T_0)\right] -C_v(T-T_0)\left[(\gamma_0-\gamma')\ln\left(\frac{V}{V_0}\right)+(\gamma-\gamma_0)\right], \quad (19)$$

$$P_{th}(V,T) = \frac{\gamma}{V}C_v(T-T_0), \quad (20)$$

$$\gamma = \gamma_0+\gamma'\left(\frac{V}{V_0}-1\right), \quad (21)$$

where the parameters of $MgSiO_3$ liquid are from Liebske and Frost (2012), except for F_0. For the liquid properties of the $CaSiO_3$ endmember, we used the result of ab initio calculations by Bajgain et al. (2015) by fitting their dataset to the above equations. The F_0 and S_0 of the $CaSiO_3$ liquid were determined by fitting the melting temperatures measured by Zerr et al. (1997), which are consistent with the results of molecular dynamics simulations by Liu et al. (2010) (Fig. 1). The thermodynamic parameters used in this study are summarized in Tables 3 and 4.

The previous study fitted F_0 for standard melting points: MgO, 3070 K at 1 bar; $MgSiO_3$, 2900 K at 25 GPa. We found that the F_0 fitted using the equations defined in the aforementioned methods is slightly different from those reported by previous studies. Nevertheless, we confirmed that we could reproduce the melting temperatures of MgO and $MgSiO_3$ and the MgO–$MgSiO_3$ melting phase relations using the parameters obtained in this study (Fig. 2, Additional file 1: Figures S3 and S4).

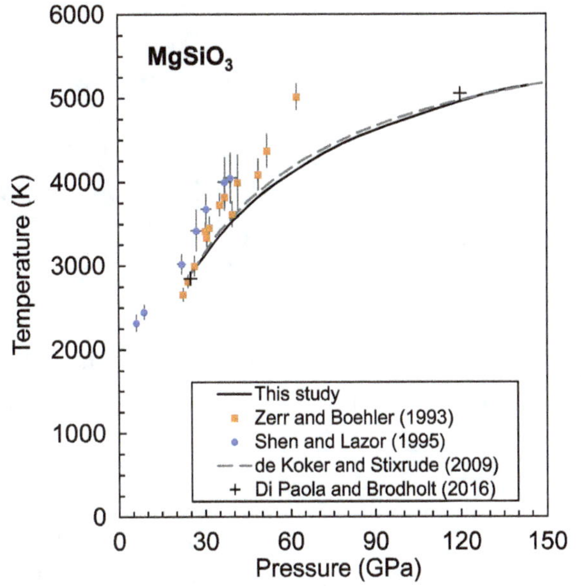

Fig. 2 Melting temperatures of $MgSiO_3$. The black line represents the melting temperatures calculated in this study, the crosses by Di Paola and Brodholt (2016), and the dashed gray line by de Koker and Stixrude (2009). The melting temperatures determined by high-pressure experiments are also plotted: orange squares, Zerr and Boehler (1993); blue circles, Shen and Lazor (1995)

Results and discussion

X-ray diffraction and phase identification

Additional file 1: Figure S5 shows the XRD patterns of the run products. Intense peaks from Re were unavoidably observed in all the patterns due to the small sample sizes in this study. Peaks from bridgmanite were evidently observed in OS3097 and OS3099, indicating that the experiments were performed outside of the garnet stability region, which is consistent with a previous study that reported the phase transformation of a $MgSiO_3$-rich composition from garnet to bridgmanite at 22.4 GPa (Gasparik 1996). Some weak peaks, presumably indexed to bridgmanite, were observed in OS3095 and OS3150 (e.g., Additional file 1: Figure S5e). The bridgmanite is interpreted as quench crystals from melt in the corresponding SEM images as shown later. $CaSiO_3$-rich perovskite was not identified in all the samples since it should be amorphized at the ambient pressure.

Electron microscopic observation and compositional analysis

Figures 3, 4, 5, and 6 show the backscattered electron images of the recovered samples. At 2600 K, the subsolidus phase assemblage of bridgmanite plus $CaSiO_3$-rich perovskite was obtained in both starting compositions (Fig. 3). Bridgmanite contains 5–6 mol% $CaSiO_3$ component, which is higher than that reported by Irifune et al. (2000) at 1773–2173 K, suggesting a positive

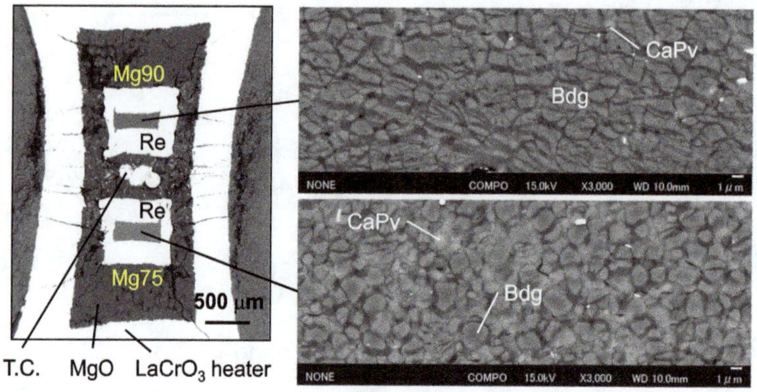

Fig. 3 Backscattered electron images for the samples recovered from high pressure and high temperature (Run No. OS3097). The subsolidus phase assemblage of bridgmanite (Bdg) plus $CaSiO_3$-rich perovskite (CaPv) was observed in both Mg90 and Mg75. T.C., thermocouple

temperature effect on solubility. We did not determine the solubility of $MgSiO_3$ component in $CaSiO_3$-rich perovskite because of its small grain size. At 2620 K, the sample OS3099-Mg90 had a quenched melt with quench textures surrounded by bridgmanite, which is interpreted as the liquidus phase, at the side part of the capsule (Fig. 4). At the central part, the subsolidus phase assemblage of bridgmanite plus $CaSiO_3$-rich perovskite was found. This observation suggests that the axial thermal gradient in the heater has been effectively reduced so that the radial thermal gradient mainly influenced the phase distribution in this sample. In the sample OS3099-Mg75, we observed a segregation of $CaSiO_3$-rich perovskite at the hot side part of the

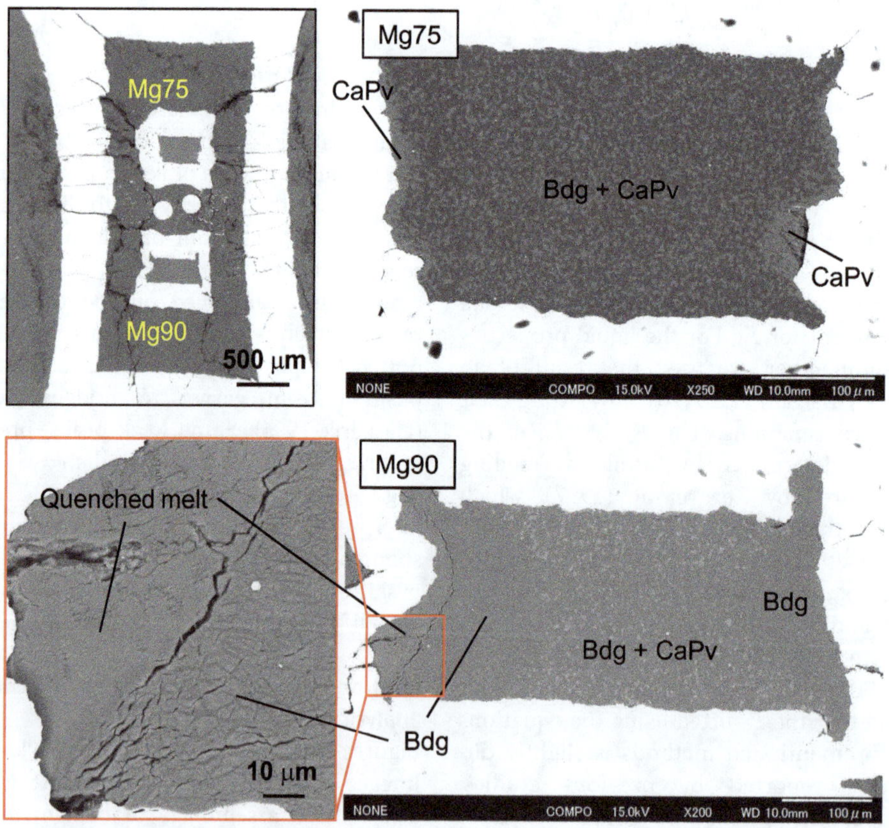

Fig. 4 Backscattered electron images for the samples recovered from high pressure and high temperature (Run No. OS3099). In Mg90, a quenched melt was found at the hottest region surrounded by bridgmanite (Bdg), which is interpreted as the liquidus phase. In contrast, $CaSiO_3$-rich perovskite (CaPv) was found to be segregated in Mg75. We failed to find the melt. The subsolidus phase assemblage was observed at the central (cold) part of both samples

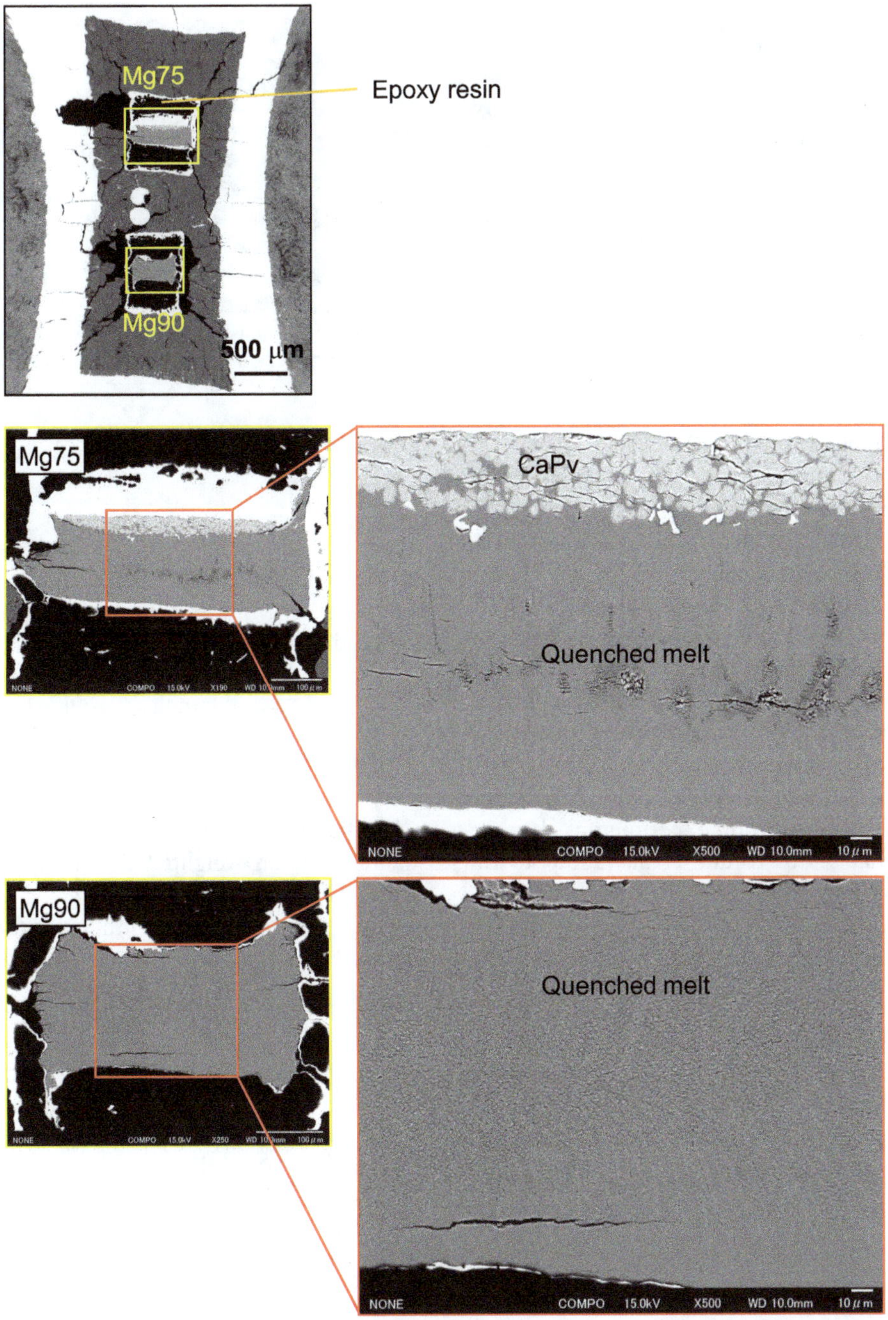

Fig. 5 Backscattered electron images for the samples recovered from high pressure and high temperature (Run No. OS3095). Quenched melts were found at most of the sample areas. In Mg75, the liquidus phase, $CaSiO_3$-rich perovskite (CaPv), was found at the coldest part

capsule (Fig. 4). This $CaSiO_3$-perovskite is interpreted as the liquidus phase, although we failed to find the quenched melt. At 2700 K, the samples were extensively molten, as clearly evidenced by the textures shown in Fig. 5. Both quenched melts exhibited textures similar to that of metal casting: fine grains attached to the mold wall; columnar grains arranged closely in parallel with the direction of cooling heat flow; and granular grains accumulated at the center. The sample OS3095-Mg90 had a bulk composition identical to that of the starting material, supporting a complete melting. In the sample OS3095-Mg75, $CaSiO_3$-rich perovskite, interpreted as the liquidus phase, was observed at the coldest part of the capsule. At 3000 K, the sample OS3150-Mg50 yielded a phase assemblage of $CaSiO_3$-rich perovskite plus melt (Fig. 6), identical to the sample OS3095-Mg75 (Fig. 5). The results of the chemical analyses are summarized in Table 5.

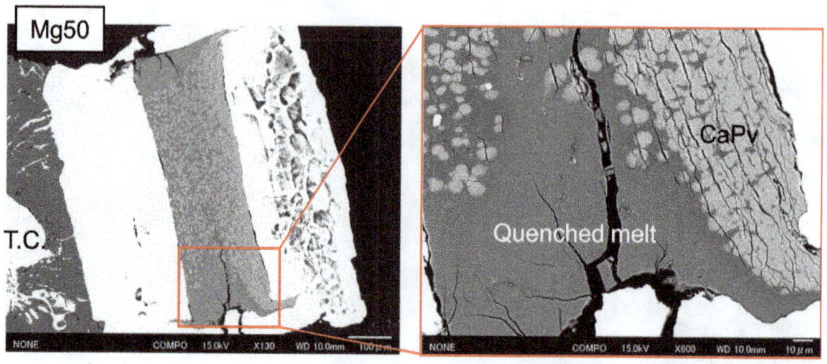

Fig. 6 Backscattered electron images for the sample recovered from high pressure and high temperature (Run No. OS3150-Mg50). A quenched melt was found at the hottest region surrounded by $CaSiO_3$-rich perovskite (CaPv), which is interpreted as the liquidus phase

Melting phase relations in the $MgSiO_3$–$CaSiO_3$ system

The phase relations are summarized in Fig. 7. The solubility of $MgSiO_3$ component into $CaSiO_3$-rich perovskite is much higher than that predicted by theoretical calculations (Jung and Schmidt 2011). This is because Jung and Schmidt (2011) considered a tetragonal structure (I4/mcm) for $CaSiO_3$ perovskite, in addition to their use of oversimplified assumptions, as commented in their paper. Indeed, Komabayashi et al. (2007) showed that $CaSiO_3$ perovskite undergoes a phase transition from tetragonal to cubic above ~ 500 K at ~ 24 GPa. We determined the interaction parameter W^{solid} by fitting our experimental data together with the dataset of Irifune et al. (2000) to the following equations:

$$W^{solid}_{MgSiO_3\text{-}CaSiO_3}\ [kJ/mol] = 108(11)–0.024(5) \times T\ [K],$$

$$W^{solid}_{CaSiO_3\text{-}MgSiO_3}\ [kJ/mol] = 52(8) + 0.010(4) \times T\ [K].$$

The eutectic point was found to be located at 81–86 mol% $MgSiO_3$ and 2600–2620 K. The interaction parameter W^{liquid} was determined by fitting our experimental data to the following equations:

$$W^{liquid}_{MgSiO_3\text{-}CaSiO_3}\ [kJ/mol] = 1839(741)–0.59(26) \times T\ [K],$$

$$W^{liquid}_{CaSiO_3\text{-}MgSiO_3}\ [kJ/mol] = –1170(167) + 0.39(6) \times T\ [K].$$

The parameters were estimated using a linear least-squares method weighted by the compositional uncertainty. Therefore, the standard error of the fitting parameter, shown in parenthesis, does not include temperature uncertainty since the dataset from the

Table 5 Chemical compositions of phases in run products

Run no.	SM[a]	Phase[b]	MgO	CaO	SiO_2	Total (wt.%)	Mg	Ca	Si	∑Cation[c]	Mg/(Mg + Ca)
OS3097	Mg90	S-Bdg (48)	37.56(101)	3.07(121)	62.12(76)	102.74(112)	0.915(19)	0.054(21)	1.016(4)	1.984(4)	0.945(22)
	Mg75	S-Bdg (32)	36.45(93)	3.35(86)	61.46(72)	101.25(130)	0.901(16)	0.060(15)	1.020(4)	1.980(4)	0.938(16)
OS3099	Mg90	Melt (9)	33.57(42)	8.49(28)	57.33(86)	99.39(86)	0.864(12)	0.157(6)	0.990(6)	2.010(6)	0.846(5)
		L-Bdg (44)	38.25(106)	2.12(79)	61.04(100)	101.42(171)	0.943(15)	0.038(14)	1.010(3)	1.990(3)	0.962(14)
		S-Bdg (34)	38.26(100)	1.99(60)	60.84(106)	101.09(181)	0.946(13)	0.035(11)	1.009(4)	1.991(4)	0.964(11)
		S-CaPv (28)	4.89(63)	40.58(67)	52.85(67)	98.32(85)	0.140(18)	0.834(17)	1.013(5)	1.987(5)	0.143(17)
	Mg75	L-CaPv (29)	5.34(75)	40.30(83)	52.72(84)	98.36(97)	0.153(22)	0.827(17)	1.010(8)	1.990(8)	0.155(21)
		S-CaPv (33)	5.68(57)	39.90(78)	52.97(64)	98.55(87)	0.161(16)	0.816(16)	1.011(7)	1.989(7)	0.165(16)
		S-Bdg (44)	39.05(66)	1.80(31)	61.89(74)	102.74(121)	0.949(8)	0.031(6)	1.010(3)	1.990(3)	0.968(6)
OS3095	Mg90	Melt (13)	35.49(38)	5.29(13)	58.47(44)	99.24(84)	0.904(4)	0.097(2)	0.999(2)	2.000(2)	0.903(2)
	Mg75	Melt (15)	31.47(46)	10.35(16)	57.63(57)	99.46(112)	0.812(4)	0.192(2)	0.998(2)	2.002(2)	0.809(2)
		L-CaPv (31)	4.39(36)	42.59(58)	53.01(65)	100.00(81)	0.124(10)	0.865(13)	1.005(4)	1.995(4)	0.125(10)
OS3150	Mg50	Melt (21)	28.27(72)	14.57(90)	56.85(40)	99.70(67)	0.737(16)	0.273(18)	0.995(2)	2.005(2)	0.730(17)
		L-CaPv (37)	3.42(43)	43.32(59)	53.52(31)	100.25(48)	0.096(12)	0.878(13)	1.013(3)	1.987(3)	0.099(12)

SM starting material, *S-* subsolidus, *L-* liquidus, *Bdg* bridgmanite, *CaPv* $CaSiO_3$-rich perovskite
[a]See Table 1 for details
[b]Number of electron microprobe analyses is shown in parentheses
[c]Number of oxygen is 3

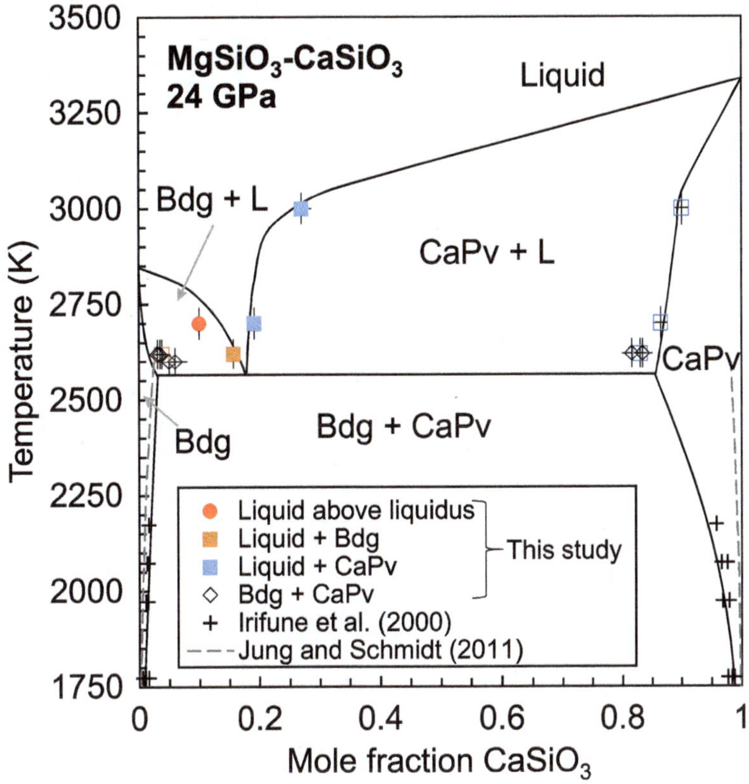

Fig. 7 Melting phase diagram in the $MgSiO_3$–$CaSiO_3$ system at 24 GPa. The filled red circle represents the composition of liquid above liquidus, the filled and open orange squares represent the liquid and coexisting bridgmanite (Bdg), the filled and open blue squares represent the liquid and coexisting $CaSiO_3$-rich perovskite (CaPv), the open diamonds represent the subsolidus phases, the cross symbols represent the subsolidus phases at 25 and 27 GPa reported in Irifune et al. (2000), and the dashed gray line represents the subsolidus phases at 25 GPa reported in Jung and Schmidt (2011)

literature did not describe how the temperature uncertainty was determined. The experimental data suggest that the interaction parameters approach to zero at high temperatures, suggesting an ideal behavior. These empirical (linear) fitting equations should not be extrapolated out from the experimental range. Figure 7 shows the resulting phase diagram from the thermodynamic calculations together with the experimental data. In prohibiting the extrapolation, the uncertainty of the thermodynamic model is described by the deviations from the experimental data (Figs. 1 and 7, Additional file 1: Figure S4).

Conclusions

Melting phase relations in the system $MgSiO_3$–$CaSiO_3$ were determined experimentally at 24 GPa. A thermodynamic model was constructed to reproduce the experimental data. The experimental dataset and the thermodynamic parameters derived in this study serve as the basis for understanding the complicated melting behaviors of the Earth's lower mantle.

The advantage of establishing a thermodynamic model is that the degree of melting can be calculated at a certain P-T in a multi-component system. The seismic observations suggested a degree of melting of < 5–30% in ultra-low velocity zones at the base of the lower mantle (Williams and Garnero 1996). The estimation of the degree of melting depends on the wetting property and the elastic properties of the melt; conversely, these properties depend on the compositions of the melt; thus, they depend on the degree of melting. The degree of melting is also an important parameter in understanding the chemical differentiation during magma ocean crystallization. The melt-solid separation by rheological transition occurs at a crystal fraction of approximately 60% (e.g., Abe 1995). Therefore, it is of great importance to understand the phase relations together with the knowledge of the degree of melting at relevant pressures and temperatures. However, there are some difficulties in obtaining the degree of melting by high-pressure melting experiments under thermal gradients because the observed degree of melting arises from the thermal gradients, not from the phase equilibrium.

Figure 8 shows the eutectic temperatures of the MgO–$MgSiO_3$ system calculated using interaction parameters constrained from the chemical analyses of the recovered diamond anvil cell (DAC) samples (Ohnishi et al. 2017) and those measured directly by

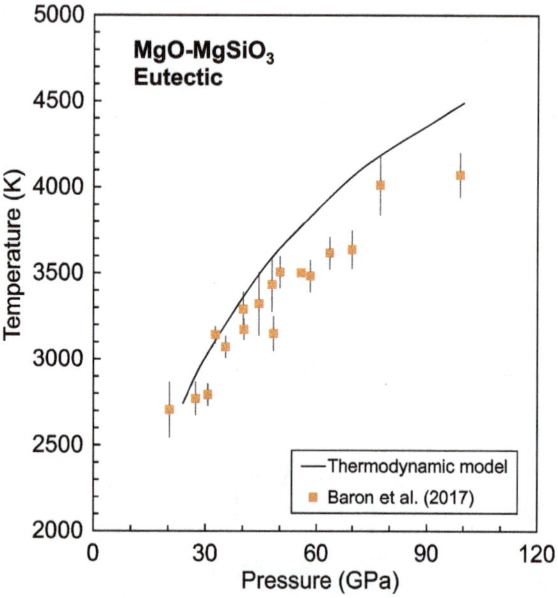

Fig. 8 Eutectic temperatures of the MgO–$MgSiO_3$ system calculated using interaction parameters constrained from chemical analyses of the recovered DAC samples (Ohnishi et al. 2017) and those measured directly by pyrometry in DAC experiments (Baron et al. 2017)

pyrometry in DAC experiments (Baron et al. 2017). As shown in Fig. 8, there is a deviation between these two studies with systematically lower temperatures in Baron et al. (2017). The deviation may have been caused by the inaccuracy of the temperature measurements by the pyrometer, the uncertain wavelength-dependent emissivity of the sample, and the effect of integrated radiation along the thermal gradients in an axial direction. The latter would be serious when transparent materials are used as a sample and pressure medium, resulting in an underestimation of the temperature. Indeed, the melting temperatures of endmembers show large discrepancies among DAC studies even if similar melting criteria were used. On the other hand, chemical analyses of the liquid and solid phases in contact (Ohnishi et al. 2017) present a much more reliable dataset because only an assumption of the local equilibrium is necessary without the information on temperature (it is noted that chemical equilibrium is assumed implicitly when melting temperatures are measured directly). Therefore, constructing the thermodynamic model using the Gibbs free energies of the endmembers by ab initio calculations and phase relations constrained by multi-anvil and DAC experiments combined with chemical analysis should be a promising approach to understand the mineralogy of deep mantle melting and comparisons with observations; however, this approach requires a step-by-step accumulation of relevant phase diagrams.

Additional file

Additional file 1: Figure S1. Pressure calibrations at room temperature (RT) and high temperatures. Fo, Mg_2SiO_4, forsterite; Wd, Mg_2SiO_4 wadsleyite; Rw, Mg_2SiO_4 ringwoodite; Ak, $MgSiO_3$ akimotoite; Bdg, $MgSiO_3$ bridgmanite; Pc, MgO periclase. Pressures of the phase transitions at high temperatures were taken from Morishima et al. (1994), Suzuki et al. (2000), and Fei et al. (2004). The red point at 790 ton, 23 GPa, and 2273 K is 0.55 GPa away from the green curve of 1873 K, which is comparable to the standard deviation of the fitting curves, 0.4 GPa. Figure S2: Heating power-temperature relationships. Figure S3: Melting temperatures of MgO. The black line represents the melting temperatures calculated in this study, and the dashed gray line represents those by de Koker and Stixrude (2009). The melting temperatures determined by high-pressure experiments and ab initio calculations are also plotted: orange squares, Zerr and Boehler (1994); green triangles, Alfe (2005); blue circles, Zhang and Fei (2008); purple squares, Du and Lee (2014); and red squares, Kimura et al. (2017). Figure S4: Melting phase diagram in the MgO–$MgSiO_3$ system at 24 GPa. The black line represents the phase boundary calculated in this study, and the gray line represents that by Liebske and Frost (2012). The experimental data by Liebske and Frost (2012) are also plotted, where the red circle represents the composition of liquid above liquidus, the green squares represent liquid coexisting with periclase (Pc), the orange squares represent liquid coexisting with bridgmanite (Bdg), and the diamonds represent subsolidus phases. Interaction parameter updated by Ohnishi et al. (2017) was used to calculate the phase relations. Figure S5: X-ray diffraction patterns of the run products. The black short bars beneath the curves denote the peak positions of bridgmanite (Bdg) and rhenium (Re).

Acknowledgements
We thank Takashi Taniuchi (Ehime Univ.) for discussions and Dr. Toru Shinmei (Ehime Univ.), Dr. Takeshi Arimoto (Ehime Univ.), Dr. Wei Sun (Ehime Univ.), and Dr. Steeve Gréaux (Ehime Univ.) for their assistance in our experiments. We thank the editor and the anonymous reviewers for their constructive comments and suggestions. We thank Dr. Hideharu Kuwahara (Ehime Univ.) and Dr. Chie Kato (Tokyo Tech.) for their assistance in the compositional analyses.

Funding
This work was supported by JSPS KAKENHI Grant Numbers JP15H05470, JP15H05830, and JP25220712.

Authors' contributions
RN and YZ designed the project. YZ and TI designed the experiments. YZ carried out the melting experiments and the chemical analysis. RN analyzed the data. RN and YZ wrote the manuscript. All authors read and approved the final manuscript.

Competing interests
The authors declare that they have no competing interests.

Author details
[1]Geodynamics Research Center, Ehime University, 2-5 Bunkyo, Matsuyama, Ehime 790-8577, Japan. [2]Earth-Life Science Institute, Tokyo Institute of Technology, 2-12-1 Ookayama, Meguro, Tokyo 152-8550, Japan.

References
Abe Y (1995) Basic equations for evolution of partially molten mantle and core. In: Yukutake T (ed) The Earth's central part: its structure and dynamics. TERRAPUB, Tokyo, pp 215–235

Alfè D (2005) Melting curve of MgO from first-principles simulations. Phys Rev Lett 94:235701. doi: 10.1103/PhysRevLett.94.235701

Bajgain SK, Ghosh DB, Karki BB (2015) First-principles simulations of CaO and $CaSiO_3$ liquids: structure, thermodynamics and diffusion. Phys Chem Mineral 42:393–404. doi: 10.1007/s00269-014-0730-9

Baron MA, Lord OT, Myhill R, Thomson AR, Wang W, Trønnes RG, Walter MJ (2017) Experimental constraints on melting temperatures in the MgO–SiO_2 system at lower mantle pressures. Earth Planet Sci Lett 472:186–196. doi: 10.1016/j.epsl.2017.05.020

de Koker N, Stixrude L (2009) Self-consistent thermodynamic description of silicate liquids, with application to shock melting of MgO periclase and $MgSiO_3$ perovskite. Geophys J Int 178:162–179. doi: 10.1111/j.1365-246X.2009.04142.x

Di Paola C, Brodholt JP (2016) Modeling the melting of multicomponent systems: the case of $MgSiO_3$ perovskite under lower mantle conditions. Sci Rep 6: 29830. doi: 10.1038/srep29830

Du Z, Lee KK (2014) High-pressure melting of MgO from (Mg, Fe) O solid solutions. Geophys Res Lett 41:8061–8066. doi: 10.1002/2014GL061954

Fei Y, Van Orman J, Li J, Van Westrenen W, Sanloup C, Minarik W, Hirose K, Komabayashi T, Walter M, Funakoshi KI (2004) Experimentally determined postspinel transformation boundary in Mg_2SiO_4 using MgO as an internal pressure standard and its geophysical implications. J Geophys Res Solid Earth 109:B02305. doi: 10.1029/2003JB002562

Gasparik T (1996) Melting experiments on the enstatite-diopside join at 70–224 kbar, including the melting of diopside. Contrib Mineral Petrol 124:139–153. doi: 10.1007/s004100050181

Gasparik T (2003) Phase diagrams for geoscientists. An atlas of the Earth's interior, Springer Ed.

Gasparik T, Wolf K, Smith CM (1994) Experimental determination of phase relations in the $CaSiO_3$ system from 8 to 15 GPa. Am Min 79:1219–1222

Hirose K, Fei Y, Ma Y, Mao HK (1999) The fate of subducted basaltic crust in the Earth's lower mantle. Nature 397:53–56. doi: 10.1038/16225

Irifune T (1994) Absence of an aluminous phase in the upper part of the Earth's lower mantle. Nature 370:131–133. doi: 10.1038/370131a0

Irifune T, Miyashita M, Inoue T, Ando J, Funakoshi K, Utsumi W (2000) High-pressure phase transformation in $CaMgSi_2O_6$ and implications for origin of ultra-deep diamond inclusions. Geophys Res Lett 27:3541–3544. doi: 10.1029/2000GL012105

Ito E, Kubo A, Katsura T, Walter MJ (2004) Melting experiments of mantle materials under lower mantle conditions with implications for magma ocean differentiation. Phys Earth Planet Inter 143:397–406. doi: 10.1016/j.pepi.2003.09.016

Jung DY, Schmidt MW (2011) Solid solution behavior of $CaSiO_3$ and $MgSiO_3$ perovskites. Phys Chem Mineral 38:311–319. doi: 10.1007/s00269-010-0405-0

Kimura T, Ohfuji H, Nishi M, Irifune T (2017) Melting temperatures of MgO under high pressure by micro-texture analysis. Nature Comm 8:15735. doi: 10.1038/ncomms15735

Komabayashi T, Hirose K, Sata N, Ohishi Y, Dubrovinsky LS (2007) Phase transition in $CaSiO_3$ perovskite. Earth Planet Sci Lett 260:564–569. doi: 10.1016/j.epsl.2007.06.015

Liebske C, Frost DJ (2012) Melting phase relations in the MgO–$MgSiO_3$ system between 16 and 26GPa: implications for melting in Earth's deep interior. Earth Planet Sci Lett 345:159–170. doi: 10.1016/j.epsl.2012.06.038

Lithgow-Bertelloni C, Stixrude L (2005) Thermodynamics of mantle minerals—I. Physical properties. Geophys J Int 162:610–632. doi: 10.1111/j.1365-246X.2005.02642.x

Liu ZJ, Yan J, Duan SQ, Sun XW, Zhang CR, Guo Y (2010) The melting curve of perovskite under lower mantle pressures. Solid State Comm 150:590–593. doi: 10.1016/j.ssc.2009.12.038

Morishima H, Kato T, Suto M, Ohtani E, Urakawa S, Utsumi W, Shimomura O, Kikegawa T (1994) The phase boundary between α- and β-Mg_2SiO_4 determined by in situ X-ray observation. Science 265:1202–1203. doi: 10.1126/science.265.5176.1202

Ohnishi S, Kuwayama Y, Inoue T (2017) Melting relations in the MgO–$MgSiO_3$ system up to 70 GPa. Phys Chem Mineral 44:1–9. doi: 10.1007/s00269-017-0871-8

Shen G, Lazor P (1995) Measurement of melting temperatures of some minerals under lower mantle pressures. J Geophys Res Solid Earth 100:17699–17713. doi: 10.1029/95JB01864

Stixrude L, Lithgow-Bertelloni C (2011) Thermodynamics of mantle minerals-II. Phase equilibria. Geophys J Int 184:1180–1213. doi: 10.1111/j.1365-246X.2010.04890.x

Suzuki A, Ohtani E, Morishima H, Kubo T, Kanbe Y, Kondo T (2000) In situ determination of the phase boundary between wadsleyite and ringwoodite in Mg_2SiO_4. Geophys Res Lett 6:803–806. doi: 10.1029/1999GL008425

Williams Q, Garnero EJ (1996) Seismic evidence for partial melt at the base of Earth's mantle. Science 273:1528. doi: 10.1126/science.273.5281.1528

Zerr A, Boehler R (1993) Melting of (Mg, Fe) SiO_3-perovskite to 625 kilobars: indication of a high melting temperature in the lower mantle. Science 262: 553–556. doi: 10.1126/science.262.5133.553

Zerr A, Boehler R (1994) Constraints on the melting temperature of the lower mantle from high-pressure experiments on MgO and magnesiowustite. Nature 371:506–508. doi: 10.1038/371506a0

Zerr A, Serghiou G, Boehler R (1997) Melting of $CaSiO_3$ perovskite to 430 kbar and first in-situ measurements of lower mantle eutectic temperatures. Geophys Res Lett 24:909–912. doi: 10.1029/97GL00829

Zhang J, Liebermann RC, Gasparik T, Herzberg CT (1993) Melting and subsolidus relations of SiO_2 at 9–14 GPa. J Geophys Res 98:19785–19783. doi: 10.1029/93JB02218

Zhang L, Fei Y (2008) Melting behavior of (Mg, Fe) O solid solutions at high pressure. Geophys Res Lett 35:L13302. doi: 10.1029/2008GL034585

Zhou Y, Irifune T, Ohfuji H, Sinmei T, Du W (2017) Stability region of $K_{0.2}Na_{0.8}AlSi_3O_8$ hollandite at 22 GPa and 2273 K. Phys Chem Mineral 44:33–42. doi: 10.1007/s00269-016-0834-5

9

Habitat evolution of a peat swamp forest and belowground carbon sequestration during the Holocene along the coastal lowland in Central Sumatra, Indonesia

Kiyoshi Fujimoto[1*], Masafumi Miura[2,5], Shigeo Kobayashi[3] and Herwint Simbolon[4]

Abstract

Habitat evolution of a peat swamp forest and belowground carbon storage was examined in a coastal lowland along the eastern coast of Central Sumatra, Indonesia. Boring surveys using a hand-operated peat sampler for 32 sites and radiocarbon dating for 29 samples revealed that the peat swamp forest had begun formation by 6500 cal BP and expanded rapidly between 6500 and 5000 cal BP and between 4000 and 2000 cal BP to nearly the present distribution area. The thickness of the peat layer formed under the peat swamp forest reached approximately 4 m, even in the margin of the peat dome, and more than 9 m in the center. Stored carbon per unit volume of the peat layer was estimated to be between 30 and 44 kg C m^{-3} in the margin of the peat dome and approximately 26 kg C m^{-3} in the center, which was less than the general carbon values of the mangrove peat. The relatively higher amount of stored carbon per unit volume in the margin was possibly caused by an inflow of allochthonous carbon from flood deposits. Belowground carbon burial rate of the peat layer was calculated as between 31 and 105 g C m^{-2} $years^{-1}$ in the peat dome margin and between 27 and 76 g C m^{-2} $years^{-1}$ in the center, which were identical to mangrove peat. These data suggest that peat swamp forests have played a significant role as places for carbon sequestration belowground as well as mangrove forest with mangrove peat.

Keywords: Peat swamp forest, Habitat evolution, Belowground carbon sequestration, Holocene, Sumatra

Introduction

Tropical coastal swamp forests, such as mangrove forests, peat swamp forests, and freshwater swamp forests, are considered to be places of significant carbon sequestration during the Holocene (Fujimoto 2004; Dommain et al. 2011; Page et al. 2011; Alongi 2012). The coastal swamp forests have generally expanded with the evolution of coastal lowland after the postglacial transgression.

The habitat evolution of mangrove forests has been revealed geomorphologically (e.g., Woodroffe 1982; Fujimoto et al. 1996, 1999a), and concrete data on belowground carbon storage has been accumulating since the late 1990s (Fujimoto et al. 1999b; Fujimoto 2004; Alongi 2012). Fujimoto (2004) compiled the data on belowground carbon storage and revealed that the *Rhizophora* community, which develops in the low sediment inflow area, can solely create mangrove peat (Mochida et al. 1999) and has the highest ability to accumulate belowground carbon in mangrove communities.

Freshwater swamp forests generally develop behind the mangrove forest on deltas with high sediment inflow, such as the Mekong and Chao Phraya deltas, and have no rich organic sediments except the delta margins away from the river channel (Nguyen et al. 2000; Tanabe et al. 2003; Fujimoto et al. 2011). On the other hand, lowland peat swamp forests develop on the coastal lowland with little sediment inflow on islands, such as Sumatra and Borneo, and they form peat domes that are over 10 m thick (Gastaldo 2010; Dommain et al. 2011).

Sediments and topography of coastal peat swamp forests were first reported by Polak (1933). Anderson (1964) was the first to discuss in literature the development of coastal

* Correspondence: kfuji@nanzan-u.ac.jp
[1]Faculty of Policy Studies, Nanzan University, 18 Yamazatocho, Showa-Ku, Nagoya 466-8673, Japan
Full list of author information is available at the end of the article

peat swamps in the tropics using radiocarbon ages. Furukawa and Supiandi (1985) revealed the topography and stratigraphy using a 60-point boring survey along a 70 km transect on the Batang Hari Lowland in Sumatra. Neuzil et al. (1993) clarified the inorganic geochemistry of domed peat in the Siak River Lowland and Bengkalis Island. Supardi et al. (1993) revealed the geology of the same area as Neuzil et al. (1993) and reported the depositional ages of peat layer formed under peat swamp forests. Staub and Esterle (1994) and Staub and Gastaldo (2003) revealed the sedimentary structure and peat accumulation processes based on a large number of cores and radiocarbon ages on the Rajang River Delta in Sarawak. Neuzil (1997) compiled the radiocarbon ages obtained from the basal peat of four peat swamps in Sumatra and Kalimantan and discussed the initiation of peat formation and the peat accumulation rate. Palynological studies were also conducted for Holocene peat in Malaysia and Indonesia (e.g., Anderson and Muller 1975; Haseldonckx 1977; Morley 1981; Cole et al. 2015).

These studies revealed the outline of evolution for the coastal lowlands covered by peat swamp forests. However, there are few studies (Staub and Esterle 1994; Staub and Gastaldo 2003) on the spatial habitat evolution of tropical peat swamp forests in a specific coastal lowland based on intensive coring and dating. Most previous studies obtained depositional ages of the peat layer from a few boring cores (e.g., Anderson 1964; Furukawa and Supiandi 1985; Cameron et al. 1989; Staub and Esterle 1993; Supiandi 1988; Neuzil 1997).

In terms of carbon storage, Dommain et al. (2011) estimated peat dome initiation and peat and carbon accumulation rates using existing data (e.g., Neuzil 1997; Shimada et al. 2001; Page et al. 2004) obtained from Peninsular Malaysia, Kalimantan, and Sumatra. Dommain et al. (2014) calculated stored and released carbon from the peat swamps of Kalimantan and Sumatra using the data compiled by Dommain et al. (2011). Dommain et al. (2015) suggested that mechanisms of peat accumulation in tropical forests were possibly different from those of northern peatlands. Cobb et al. (2017) built a numerical model of waterlogging and peat accumulation to simulate peat dome geomorphogenesis and carbon fluxes.

However, there are few data (Cameron et al. 1989; Neuzil 1997; Dommain et al. 2015) on carbon concentrations together with bulk density of peat deposits for the coastal peat swamp forest, which are essential for calculating carbon storage.

The aims of this study are to clarify the spatial habitat evolution of a peat swamp forest on a coastal lowland along the eastern coast of Central Sumatra, to estimate carbon accumulation rates related to the geomorphological situation, and to discuss the differences with the values of mangrove peat.

Regional settings

This study was carried out on the coastal lowland in Riau Province, Central Sumatra (Fig. 1). The coastal lowland develops along the Siak, Kampar, and Indragir Rivers. Most areas of the coastal lowland of Sumatra, which are distributed along the eastern coast, were covered by peat swamp forests until recent years. However, the coverage area dropped from 75 to 28% peatlands between 1990 and 2010 (Miettinen et al. 2011) and to 19% in 2015 (Miettinen et al. 2016) due to development, mainly for plantation fields. The peat swamp forests in Riau Province decreased from 30,962 km^2 (80% peatlands) in 1990 to 13,879 km^2 (36%) in 2010 (Miettinen et al. 2011) and to 9812 km^2 (24%) in 2015 (Miettinen et al. 2016) for plantations, such as oil palm, acacia pulp, and coconut, and agricultural land. Small areas of mangrove forests are distributed along the coast. Though oilfields have been developed in the area between the Siak and Kampar Rivers, the peat swamp forest in the development area has been protected, except for roads and drill sites.

Methods

Boring survey and sampling

Boring surveys using a hand-operated peat sampler (DIK-105A, Daiki, Japan), which enabled the collection of undisturbed semicircular columnar cores 50 cm long and 5 cm diameter, excluding large woody debris, was conducted for 32 sites between 2005 and 2008 (Fig. 1 and Table 1). Eight sites were located along the Siak River (SSB-1 to 8), 19 sites were along the Kampar River (SKB-1 to 19), and 3 sites were along the Indragir River (SIB-1 to 3).

SDB-1 was located around the center of the peat dome between the Siak and Kampar Rivers, approximately 17 km from the Siak River mainstream. SDB-2 was situated beside an upland. Most sites were situated along the rivers because we had to use a boat for field research, except along limited roads. However, we conducted the boring survey not only along the mainstream but also along the tributaries to collect data from as many interior sites as possible. As SDB-1 was situated on the private land of PT. Bumi Siak Pusako, an oil development company, we carried out the research with the permission of the company.

Soil samples for bulk density measurement and carbon and nitrogen contents analyses were collected at four sites, three of them located beside the Kampar River (SKB-1, 5, and 6), and another one was located around the center of the peat dome between the Siak and Kampar Rivers (SDB-1). The soil samples were sequentially collected, except for the top 50 cm horizon of SKB-5 and the woody horizons of SKB-5 and SDB-1, which were impossible to collect as undisturbed cores

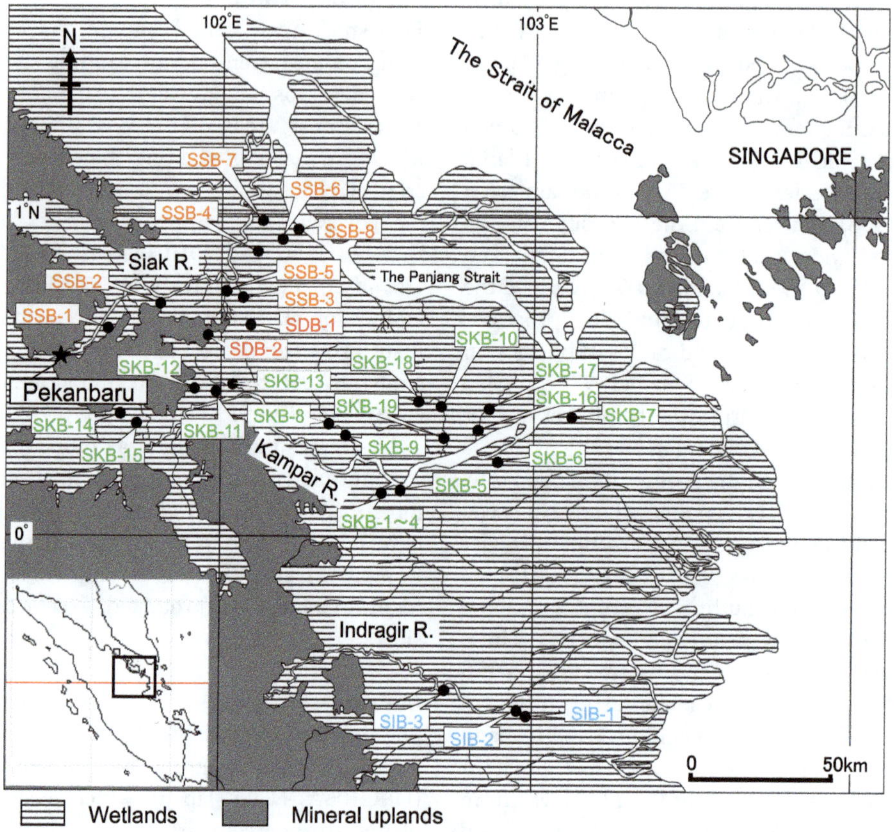

Fig. 1 Map showing the study area and boring sites

using the peat sampler. The sampled depth of cores for analysis at SKB-1, 5, and 6 are shown in Table 3. One sample for each core was collected for analysis at these sites. At SDB-1, each 50 cm long boring core was cut into 10 cm sections, and one or two cut cores were used for analysis for each layer, up to 188 cm deep, 12 cut cores were used for the analysis of the homogenous peat layer between 190 and 920 cm deep. The number of samples analyzed for each layer is shown in Table 4.

Soil analyses and radiocarbon dating

Dry weight measurements for bulk density calculations and carbon and nitrogen contents analyses were conducted at the Indonesian Soil Research Institute in Bogor. Dry weight was measured after 24 h drying at 80 °C, which is an optimal procedure for peat and highly organic soils (O'Kelly 2005). Carbon and nitrogen contents were analyzed by the Walklay-Black and Kjeldahl methods, respectively.

To estimate the sedimentary environment around the boundary between the peat layer and the underlying clay layer, electrical conductivity (EC), and total sulfur (TS) content analyses were conducted for soil samples obtained from 4 sites along the Kampar River (SKB-1, 3, 5 and 6). The samples were collected at 20 to 30 cm intervals. EC was measured using a portable EC meter (WM-22EP, TOA DKK, Japan) according to the method by Yokoyama (1993), and TS analysis was conducted using an elemental analyzer (Flash EA1112, Thermo Electron Corporation, USA) at the University of Tokyo. Generally, freshwater, brackish-water, and marine sediments were indicated by EC values less than 0.4 mS/cm, 0.4 to 1.2 mS/cm, and 1.3 to 3.0 mS/cm, respectively (Yokoyama 1993). TS content was also useful in identifying marine sediments. Generally, freshwater sediments were indicated by a value of less than 0.3%, while marine sediments were indicated by values over 0.3% (Keith and Digens 1959). We applied these values of combination to estimate sedimentary environments.

Freshwater peat deposits formed under the peat swamp forest are occasionally underlain by mangrove peat (Cameron et al. 1989; Dommain et al. 2015). These two peats can be easily distinguished visually because the freshwater peat generally contains a large amount of woody debris (Cameron et al. 1989, Dommain et al. 2015), though the upper central portion of the peat dome (Cameron et al. 1989) and tip-up pool deposits formed after tree falls (Dommain et al. 2015) are more fibrous and less woody, whereas the mangrove peat mainly consists of roots (Fujimoto et al. 1999a) with

Table 1 Description of boring sites

Loc. no.	Latitude (N)/longitude (E)	Distance from river mouth (km)	Land use
SKB-1~4	N 0° 08′ 39.3″/E 102° 31′ 01.3″	101	Peat swamp forest
SKB-5	N 0° 09′ 31.3″/E 102° 34′ 07.5″	95	Private house
SKB-6	N 0° 15′ 28.7″/E 102° 52′ 45.4″	60	Coconut and pineapple plantation
SKB-7	N 0° 23′ 00.5″/E 103° 06′ 32.6″	30	Mangrove forest
SKB-8	N 0° 21′ 13.5″/E 102° 20′ 58.6″	112 (48)	Peat swamp forest
SKB-9	N 0° 19′ 10.7″/E 102° 24′ 15.0″	107 (49)	Peat swamp forest
SKB-10	N 0° 24′ 44.6″/E 102° 42′ 06.8″	73 (39)	Peat swamp forest
SKB-11	N 0° 27′ 10.3″/E 102° 01′ 30.7″	147 (59)	Coconut plantation
SKB-12	N 0° 27′ 35.1″/E 101° 56′ 01.7″	158 (66)	Private house
SKB-13	N 0° 27′ 45.6″/E 102° 02′ 09.9″	145 (57)	Private house
SKB-14	N 0° 23′ 15.8″/E 101°41′ 49.1″	183	Private house
SKB-15	N 0° 21′ 21.5″/E 101° 43′ 58.2″	180	Oil palm plantation
SKB-16	N 0° 19′ 36.5″/E 102° 49′ 34.5″	62 (35)	Peat swamp forest
SKB-17	N 0° 23′ 20.5″/E 102° 50′ 39.7″	57 (28)	Peat swamp forest
SKB-18	N 0° 25′36.2″/E 102° 38′ 26.3″	79 (44)	Peat swamp forest
SKB-19	N 0° 18′ 40.7″/E 102° 42′ 49.6″	74 (44)	Peat swamp forest
SDB-1	N 0° 40′ 23.7″/E 102° 06′ 14.3″	(36)	Peat swamp forest
SDB-2	N 0° 38′ 17.1″/E 101° 57′ 44.2″	(49)	Private house
SIB-1	S 0° 33′ 28.9″/E 102° 59′ 11.3″	72	Private house
SIB-2	S 0° 32′ 27.9″/E 102° 56′ 47.5″	75	Private house
SIB-3	S 0° 28′ 06.0″/E 102° 43′ 32.3″	93	Private house
SSB-1	N 0° 39′ 18.8″/E 101° 38′ 19.4″	86	Secondary forest
SSB-2	N 0° 44′ 21.4″/E 101° 49′ 04.6″	67 (54)	Private house
SSB-3	N 0° 45′ 31.9″/E 102° 03′ 55.8″	53 (32)	Secondary forest
SSB-4	N 0° 53′ 28.1″/E 102° 07′ 32.0″	38 (16)	Pineapple farm
SSB-5	N 0° 46′ 06.3″/E 102° 02′ 27.1″	52 (31)	Grass land
SSB-6	N 0° 56′ 34.3″/E 102° 12′ 56.3″	(5)	Oil palm plantation
SSB-7	N 0° 59′ 15.0″/E 102° 08′ 09.0″	27 (10)	Oil palm plantation
SSB-8	N 0° 57′ 22.1″/E 102° 14′ 22.5″	(2)	Oil palm plantation

Number in parentheses indicates distance from the nearest coast of the Panjang Strait

black bark fragments. To reinforce the visual identification, EC and TS content analyses were applied for the basal peat at SKB-6.

Radiocarbon dating was conducted for 28 wood fragment samples and a peat sample at Institute of Accelerator Analysis Ltd. in Japan (Table 2) using the accelerator mass spectrometry (AMS) method; however, four samples (IAA-687, 689, 690, and 691) were analyzed by the radiometric method. The radiocarbon ages were calibrated to calendar year using CALIB 5.0. The calibrated ages in Table 2 show the minimum and maximum values of 1 σ standard deviation with median possibility in parentheses. Radiocarbon ages obtained from the basal horizon of peat swamp forest peat reported by Supardi et al. (1993) were also calibrated to the calendar year because the ages were indicated by ^{14}C age only. The calibrated ages were shown in Fig. 4.

Results

Distribution of peat layers and their formative ages

Figures 2 and 3 show the geologic columns of boring sites along the Kampar River and the Siak and Indragir Rivers, including the center of peat dome between the Siak and Kampar Rivers, and the results of radiocarbon dating using calibrated age. Figure 4 shows the spatial distribution of the calibrated ages obtained around the basal horizon of the peat layer, indicating the initiation age of peat swamp forest or mangrove forest consisting of *Rhizophora* sp., including the cores that were short of the base of the peat layer.

On the inland valley bottom lowlands, an age of approximately 7030 cal BP (IAAA-72998) was obtained from the clay layer with peat (3.20 to 3.23 m depth), which was just below the peaty clay layer overlain by the

Table 2 Results of radiocarbon dating and calibrated ages

Lab. code	Site no.	Depth (cm)	Sample	$\delta^{13}C$	Conventional age (^{14}C BP)	1σ calibrated age (cal BP)	Moving average of calibration curve (years)
IAAA-42316	SKB-1	345–350	Wood frag	− 39.90 ± 0.74	2850 ± 40	2880-(2960)-3010	20
IAA-687	SKB-2	320–330	Wood frag	− 28.9	2300 ± 80	2150-(2310)-2430	50
IAAA-42317	SKB-3	370–380	Wood frag	− 31.53 ± 0.88	2830 ± 40	2870-(2940)-2980	20
IAAA-42427	SKB-3	430–440	Wood frag	− 26.91 ± 0.61	2750 ± 40	2790-(2840)-2870	20
IAA-689	SKB-5	367–397	Wood frag	− 26.3	1670 ± 80	1440-(1580)-1700	20
IAA-690	SKB-5	443–470	Wood frag	− 26.0	2160 ± 80	2060-(2160)-2310	50
IAA-691	SKB-6	330–337	Peat	− 27.6	3690 ± 80	3920-(4030)-4140	50
IAAA-62241	SKB-8	125	Wood frag	− 26.25 ± 0.69	1290 ± 30	1180-(1230)-1280	20
IAAA-62242	SKB-8	362	Wood frag	− 25.56 ± 0.70	2160 ± 30	2120-(2170)-2300	20
IAAA-62243	SKB-9	165	Wood frag	− 25.25 ± 0.82	1360 ± 30	1280-(1290)-1300	20
IAAA-62244	SKB-9	309	Wood frag	− 26.67 ± 0.71	2440 ± 30	2360-(2480)-2680	20
IAAA-62245	SKB-10	130	Wood frag	− 29.16 ± 0.72	2330 ± 30	2340-(2350)-2350	20
IAAA-62246	SKB-10	195	Wood frag	− 20.11 ± 0.72	3520 ± 30	3720-(3790)-3840	20
IAAA-72997	SKB-11	260	Wood frag	− 38.78 ± 0.60	5690 ± 40	6410-(6470)-6500	20
IAAA-73002	SKB-15	387	Wood frag	− 34.00 ± 0.94	5560 ± 40	6310-(6350)-6390	20
IAAA-73003	SKB-17	315–320	Wood frag	− 24.02 ± 0.87	2750 ± 30	2790-(2840)-2870	20
IAAA-73004	SKB-17	340–342	Wood frag	− 29.41 ± 0.81	2860 ± 30	2930-(2980)-3050	20
IAAA-73005	SKB-18	314–315	Wood frag	− 30.16 ± 0.70	3560 ± 30	3830-(3860)-3900	20
IAAA-73006	SKB-18	380–385	Wood frag	− 25.59 ± 0.55	3870 ± 40	4240-(4300)-4400	20
IAAA-52500	SDB-1	180–190	Wood frag	− 30.33 ± 0.93	560 ± 40	530-(590)-630	20
IAAA-52501	SDB-1	580–590	Wood frag	− 27.61 ± 0.78	3990 ± 40	4420-(4470)-4520	20
IAAA-52502	SDB-1	915–920	Wood frag	− 24.86 ± 0.74	5210 ± 40	5930-(5970)-5990	20
IAAA-52503	SIB-2	180–185	Wood frag	− 29.21 ± 0.82	1500 ± 30	1350-(1380)-1410	20
IAAA-52504	SIB-2	197–205	Wood frag	− 25.76 ± 0.88	1400 ± 40	1290-(1310)-1340	20
IAAA-52505	SIB-3	61–64	Wood frag	− 20.64 ± 0.74	1180 ± 30	1060-(1110)-1160	20
IAAA-52506	SIB-3	146–150	Wood frag	− 27.68 ± 0.96	1200 ± 40	1070-(1130)-1170	20
IAAA-72998	SSB-1	320–323	Wood frag	− 29.07 ± 0.61	6130 ± 40	6950-(7030)-7150	20
IAAA-73000	SSB-4	395–400	Wood frag	− 27.24 ± 0.64	3540 ± 30	3730-(3830)-3880	20
IAAA-73001	SSB-6	262–263	Wood frag	− 21.46 ± 0.62	5720 ± 30	6460-(6510)-6550	20

peat layer at SSB-1 in the Siak River basin, and an age of approximately 6350 cal BP (IAAA-73002) was obtained from the middle horizon of the peat layer (3.87 m depth) at SKB-15 in the Kampar River basin. At SKB-11, which is located near the mouth of the valley bottom lowland of the Kampar River, an age of approximately 6470 cal BP (IAAA-72997) was obtained from the basal horizon of peat layer (2.60 m depth).

Near the center of landward peat dome on the coastal lowland between the Siak and Kampar Rivers, peat layer over 9.2 m thick was confirmed at SDB-1, and an age of approximately 5970 cal BP (IAAA-52502) was obtained from the deepest horizon of the core (9.15 to 9.20 m depth).

On the coastal lowland of the Siak River basin, a peat layer over 4 m thick was formed 38 km inland from the river mouth (SSB-4), and an age of approximately 3830 cal BP (IAAA-73000) was obtained from the deepest horizon (3.95 to 4.00 m depth). At the site located approximately 11 km seaward from SSB-4, an age of approximately 6510 cal BP (SSB-6 IAAA-73001) was obtained from the clay layer overlain by the peat layer (2.62 to 2.63 m depth).

On the coastal lowland of the Kampar River basin, a peat layer less than 4.5 m thick was distributed along the river, and ages between approximately 2170 and 2980 cal BP were obtained from the basal horizon of the peat layer (SKB-1: IAAA-42316, SKB-3: IAAA-42427, SKB-5: IAA-690, SKB-9: IAAA-62244, SKB-10: IAAA-62245, and SKB-17: IAAA-73004), though distinctly older ages, approximately 4030 (IAA-691) and 4300 cal BP (IAAA-73006), were obtained from the basal horizon of the peat layer at SKB-6 and clayey peat layer at SKB-18,

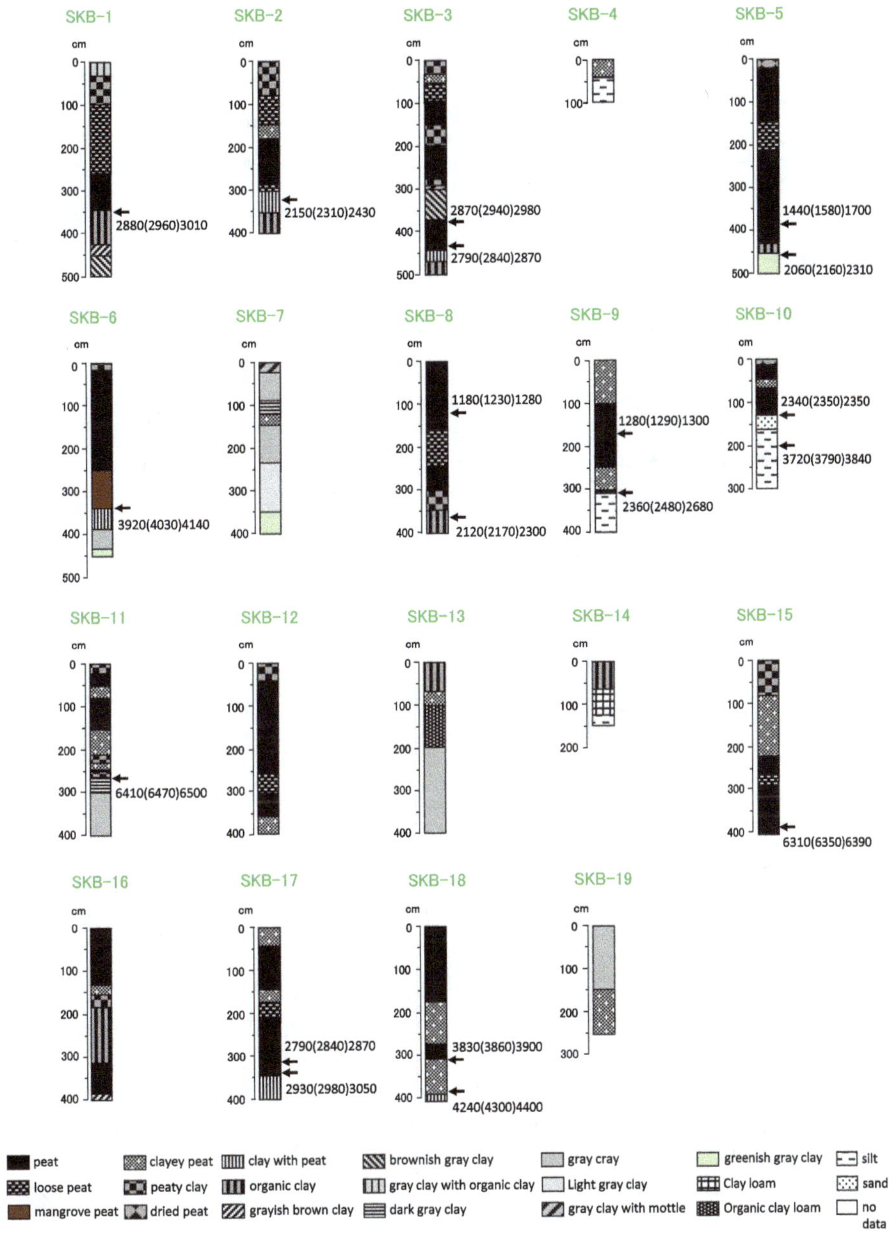

Fig. 2 Geologic columns of boring sites and the results of radiocarbon dating using calibrated age along the Kampar River

respectively. Sampling depths for the coring sites are shown in Table 2.

The basal peat between 2.5 and 3.4 m depth at SKB-6 was identified as mangrove peat from the main materials, i.e., fibric fine roots with distinctive black bark fragments. The wood sample for IAA-690, which was a black bark fragment, obtained from SKB-5 is presumed to be of mangrove origin because of this feature.

Along the Indragir River, relatively younger ages, between approximately 1130 and 1380 cal BP (SIB-2: IAAA-52504 and IAAA-52503, and SIB-3: IAAA-52505), were obtained from the basal horizon of the peat layers, their thicknesses were less than 2 m.

EC and TS analyses

Figure 5 shows the results of EC and TS content analyses. High values of EC and TS content, indicating brackish to marine environments, were measured for the clay layers overlain by peat layers at SKB-1, 5, and 6. At SKB-1, over 0.5 mS/cm of EC and over 1% of TS content were obtained from the organic clay layer below 4.0 m depth. At SKB-5, over 0.4 mS/cm of EC was obtained

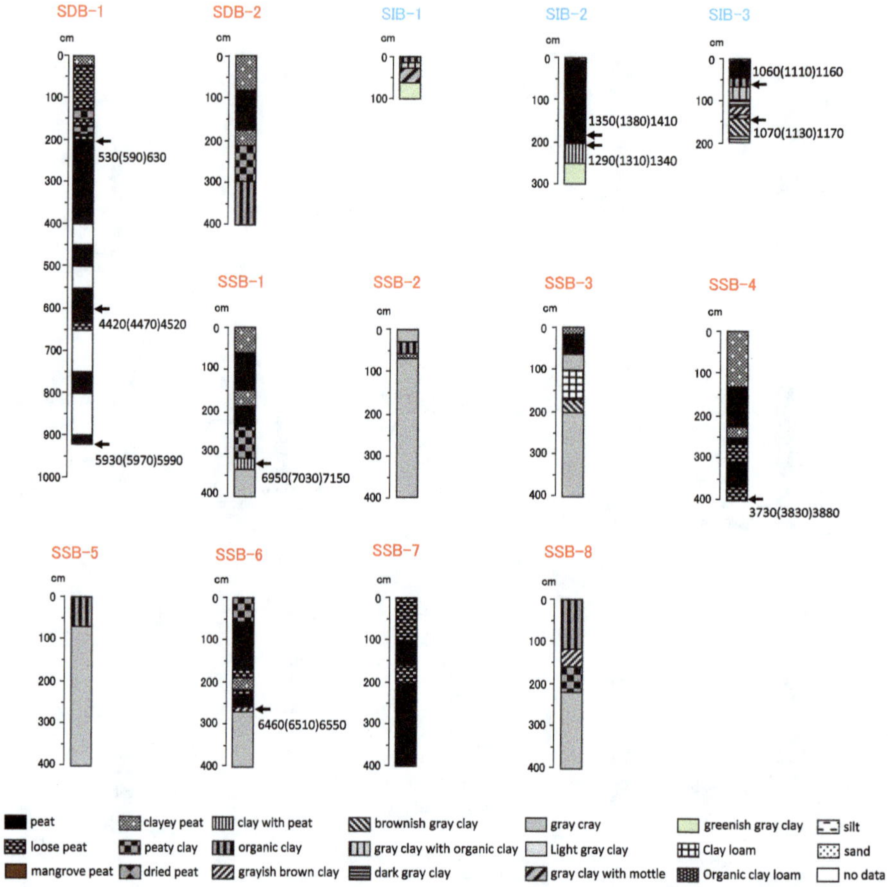

Fig. 3 Geologic columns of boring sites and the results of radiocarbon dating using calibrated age along the Siak and Indragir Rivers and the center of peat dome between the Siak and Kampar Rivers

from the organic clay layer between 4.3 and 4.4 m depth, and over 1% of TS content was obtained from the greenish gray clay layer between 4.5 and 4.8 m depth. At SKB-6, over 0.5 mS/cm of EC was obtained from the clay layer below 3.5 m depth, and over 0.8% of TS content was obtained from the peat layer below 2.8 m depth, which was identified as mangrove peat from the main materials, with the clay layers overlain by the peat layer.

Low values of both EC and TS content, indicating freshwater environments, were obtained from the peat layer and upper horizon of the organic clay layer at SKB-1 and 3, less than 0.4 mS/cm and less than 0.3%, respectively, with the peat layer overlaying the clay layer at SKB-5 and the peat layer overlaying mangrove peat at SKB-6.

Belowground carbon storage and its burial rate

Tables 3 and 4 show the bulk density, carbon, and nitrogen concentrations and their masses per square meter at the boring sites in the margin and center of the peat dome, respectively.

Carbon concentrations of peaty layers at SKB-1, 5, and 6 located in the margin of peat dome were between 18 and 43%, which was lower than those at SDB-1 located in the center, except for the top 40 cm, those were approximately 55%.

At SKB-1 situated in the margin of peat dome, a 3.45-m-thick peat layer underlain by clay layer was deposited. The amount of stored carbon in the 3.45 m deep peat layer was 102.9 kg C m^{-2}, whereas the entire 5-m-thick deposit, which included a basal clay layer contained 158 kg C m^{-2}.

SKB-5, situated in the margin of peat dome, was located in a private garden, with a 4.29-m-thick peat layer, whose top 0.2 m was dried peat and was underlain by a clay layer. Samples for analyses were collected at up to 5.0 m depth, except for the top 0.5 m because non-disturbed samples for bulk density measurement could not be collected. Stored carbon was calculated at 165.2 kg C m^{-2} in the peat layer between 0.5 and 4.29 m depth and 184.1 kg C m^{-2} at up to 5.0 m depth.

SKB-6 that was situated in the margin of the peat dome was located in a coconut and pineapple field with

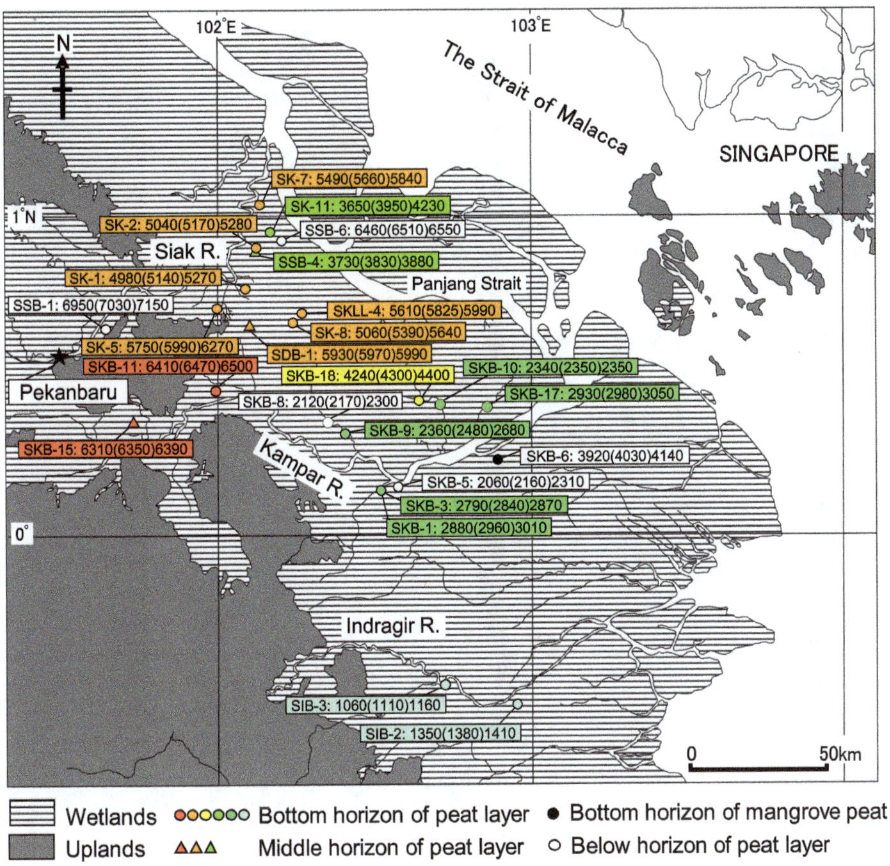

Fig. 4 Spatial distribution of the calibrated ages obtained around the basal horizon of peat layer. Red, orange, yellow, greenish yellow, green, and aqua colors are ages obtained from the deposit of peat swamp forest, indicating over 6000 cal BP, between 5999 and 5000 cal BP, between 4999 and 4000 cal BP, between 3999 and 3000 cal BP, between 2999 and 2000 cal BP, and between 1999 and 1000 cal BP, respectively

a peaty layer deposit of up to 3.37 m depth, though the top 0.17 m was peaty clay, and between 2.5 and 3.37 m was mangrove peat. Stored carbon was calculated at 97.0 kg C m^{-2} at up to 2.5 m depth, 127.8 kg C m^{-2} at up to 3.37 m depth, including the mangrove peat layer, and 175.9 kg C m^{-2} at up to 4.33 m depth, including the underlying clay layer.

At SDB-1 located in a peat swamp forest near the center of peat dome, a peat layer over 9.2 m deep accumulated. Stored carbon was calculated at 239.5 kg C m^{-2} at up to 9.2 m depth.

Belowground carbon burial rates of the peat layers were calculated to be between 31 and 105 g m^{-2} $years^{-1}$ for the sites located in the peat dome margin and between 27 and 76 g m^{-2} $years^{-1}$ for the site near the center of peat dome (Table 5).

Discussion

Habitat evolution of peat swamp forest

The calibrated ages indicating approximately 6350 cal BP and approximately 6470 cal BP obtained from the middle and bottom horizons of peat layers, which were inferred to be peat swamp forest peat from the features, were in the valley bottom lowland of the Kampar River basin and its mouth (Figs. 2 and 4, SKB-15 and 11 located 180 km and 147 km inland from the mouth of the Kampar River, respectively); they show that the peat swamp forest in the valley bottom lowland began to be formed by 6350 cal BP.

The calibrated age indicating approximately 7030 cal BP obtained from the clay layer with peat just below the peaty clay layer overlain by peat swamp forest peat at SSB-1 located approximately 86 km inland from the mouth of the Siak River (Figs. 2 and 4) has two possible interpretations, i.e., there was either a period of mangrove forest or freshwater swamp forest because the clay layer overlain by the peat swamp forest peat was deposited under blackish or freshwater environments. The calibrated age indicating approximately 5970 cal BP obtained from the middle horizon of the peat swamp forest peat located approximately 36 km inland from the coastline of the Panjang Strait (Figs. 2 and 4, SDB-1) shows that the peat swamp forest expanded to near the center of present peat dome between the Siak and Kampar Rivers by 6000 cal BP.

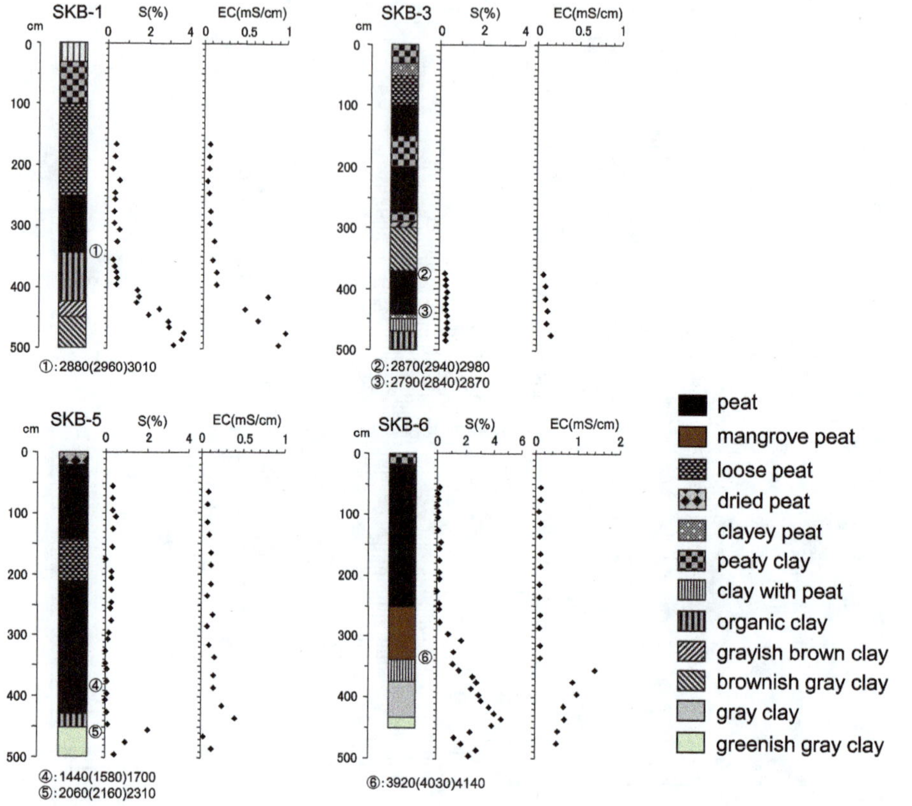

Fig. 5 Results of EC and TS content analyses

The age, approximately 6510 cal BP, obtained from SSB-6 located near the present coastline of the Panjang Strait possibly does not indicate the initiation age of the peat swamp forest but it does show the distribution of mangrove forest because the age was obtained from the clay layer overlain by the peat swamp forest peat and the site is situated in a geomorphological environment with low flooding impact because due to its distance from the rivers.

Supardi et al. (1993) reported 26 radiocarbon ages obtained from the peat swamp forest peat along the Siak River including 7 samples obtained from the basal horizon. SK-1, 2, 5, 7, 8, 11, and SKLL-4 in Fig. 4 are sites obtained the basal ages, which were indicated the ages between 3950 and 5990 cal BP.

The calibrated ages obtained from the Siak River Lowland in this study and Supardi et al. (1993) indicated that the peat swamp forest along the Siak River expanded between 6000 and 5000 cal BP.

On the other hand, the calibrated ages obtained from the bottom horizon of peat layers in the Kampar River basin suggest that the peat swamp forest on the coastal lowland expanded to the present area between 4300 and 2000 cal BP (Fig. 4, SKB-1, 3, 8, 9, 10, 17, and 18).

Approximately 4000 cal BP, a *Rhizophora* dominant mangrove forest was distributed around SKB-6 and was located approximately 60 km inland from the present river mouth of the Kampar River. The calibrated age of IAA-690 obtained from SKB-5 and the EC and TS content values of the sampling horizon (Fig. 5) suggest that mangrove forests were distributed along the river up to 100 km inland approximately 2160 cal BP.

The calibrated ages obtained from the coastal lowland of the Indragir River (Fig. 4, SIB-2 and 3) strongly indicate that the formation of this coastal lowland occurred later than those in the Siak and Kampar River basins. The ages obtained from the basal horizons of peat layers in this study area suggested that the peat swamp forests expanded from the north to south, i.e., from the Siak River Lowland to the Indragir River lowland.

In the Asia-Pacific region, relatively high sea levels approximately 6500 cal BP and 4000 cal BP and relatively low sea levels approximately 5000 cal BP and 2000 cal BP were reported (Tjia 1996; Sinsakul et al. 1985; Sinsakul 1992; Fujimoto et al. 1996, 1999a). In the Sunda Shelf, a relatively high sea level approximately 4200 cal BP was recognized (Geyh et al. 1979; Hanebuth et al. 2011), though the high level approximately 6500 cal BP has not been determined.

The peat swamp forest initiated habitat formation by 6500 cal BP in the inland valley bottom lowlands. They developed first in the Siak River Lowland between 6000

Table 3 Bulk density, carbon, and nitrogen concentrations and their masses per square meter at the boring sites in the margin of peat dome

Loc. no.	Depth (cm)	Sedimentary facis	Sampled depth of cores for analyzing (cm)	Bulk density (kg m^{-3})	Organic *C* (%)	Organic *N* (%)	*C/N*	Thickness for calculation (cm)	Stored C (kg C m^{-2})	Stored *N* (kg N m^{-2})
SKB-1	0–32	BOC	0–32	168.0	28.2	1.6	18.2	32	15.1	0.8
	32–150	PC	32–50	215.8	39.6	1.1	35.4	18	15.4	0.4
			50–100	121.4	18.2	0.4	44.3	100	22.1	0.5
	150–250	LP	150–200	65.7	25.7	0.5	47.5	50	8.4	0.2
			200–250	188.0	18.8	0.4	50.8	50	17.6	0.3
	250–345	P	250–300	86.2	25.1	0.5	50.2	50	10.8	0.2
			300–345	85.4	35.0	0.6	54.7	45	13.5	0.2
	345–424	OC	345–350	236.3	24.7	0.5	48.4	5	2.9	0.1
			350–400	124.9	16.8	0.4	47.9	50	10.5	0.2
			400–424	376.6	15.3	0.3	49.2	24	13.8	0.3
	424–450	GBC	424–450	204.3	11.6	0.3	41.4	26	6.2	0.1
	450–500	BGC	450–500	360.8	12.1	0.3	41.6	50	21.7	0.5
								0–500	158.0	4.0
SKB-5	0–20	DP	0–20	–	–	–	–	–		
	20–50	P	20–50	–	–	–	–	–		
	50–75	DBP	50–75	160.8	34.1	0.6	58.8	25	13.7	0.2
	75–140	P	75–100	123.6	36.7	0.8	44.2	25	11.3	0.3
			100–140	160.0	35.2	0.6	55.8	40	22.5	0.4
	140–208	LP	140–150	123.4	37.3	0.6	59.2	10	4.6	0.1
			150–200	88.8	42.4	0.7	57.3	58	21.8	0.4
	208–429	P	208–250	93.7	35.8	0.7	49.7	42	14.1	0.3
			250–300	135.4	36.7	0.7	51.0	50	24.9	0.5
			300–350	131.6	23.4	0.4	58.4	50	15.4	0.3
			350–367	145.3	32.8	0.6	56.6	50	23.8	0.4
			400–429	137.9	32.5	0.6	59.1	29	13.0	0.2
	429–450	DBOC	429–450	162.6	38.3	0.8	46.1	21	13.1	0.3
	450–500	GGC	450–500	607.1	1.9	0.1	14.8	50	5.8	0.4
								50–500	184.1	3.7
SKB-6	0–17	PC	0–17	207.5	36.9	1.0	37.2	17	13.0	0.3
	17–100	P	17–50	129.2	33.9	0.8	42.8	33	14.4	0.3
			50–100	74.7	32.5	0.6	55.1	50	12.1	0.2
	100–180	FP	110–150	137.2	40.0	0.7	58.8	50	27.4	0.5
			150–180	109.2	42.7	0.8	52.7	30	14.0	0.3
	180–250	P	180–200	83.3	34.7	0.7	53.4	20	5.8	0.1
			200–250	98.7	20.9	0.4	49.7	50	10.3	0.2
	250–337	MP	250–300	88.7	43.9	0.8	57.8	50	19.5	0.3
			300–330	84.1	36.3	0.8	44.3	37	11.3	0.3
	337-375	GCwP	337–350	185.1	26.2	0.5	51.3	13	6.3	0.1
			350–375	243.8	28.4	0.5	52.6	25	17.3	0.3
	375–433	GC	375–400	347.4	12.6	0.3	48.4	25	10.9	0.2
			400–433	370.8	11.1	0.2	46.1	33	13.5	0.3
	433–500	GGC	433–450	647.3	–	–	–	–		
			450–500	449.4	–	–	–	–		
								0–433	175.9	3.5

P peat, *FP* fibric peat, *LP* loose peat, *MP* mangrove peat, *DP* dried peat, *DBP* dark brown peat, *PC* peaty clay, *BOC* brown organic clay, *DBOC* dark brown organic clay, *OC* organic clay, *GBC* grayish brown clay, *BGC* brownish gray clay, *GC* gray clay, *GGC* greenish gray clay, *wP* with peat

Table 4 Bulk density, carbon, and nitrogen concentrations and masses per square meter at SDB-1 near the center of peat dome

Depth (cm)	Sedimentary facis	Number of samples	Bulk density (kg m^{-3})	Organic C (%)	Organic N (%)	C/N	Thickness for calculation (cm)	Stored *C* (kg C m^{-2})	Stored *N* (kg N m^{-2})
0–40	DP	2	45.9 (3.0)	35.9 (0.6)	1.4 (0.1)	26.6 (1.7)	40	3.3	0.1
40–130	LP	2	33.0 (9.6)	53.2 (1.7)	1.0 (0.2)	56.3 (9.1)	90	19.3	0.4
130–150	DeP	2	112.8 (47.0)	56.8 (2.9)	1.0 (0.1)	56.8 (7.2)	20	12.8	0.2
150–170	LP	1	42.8	56.5	0.8	71.5	20	4.8	0.1
170–188	DeP	1	105.8	54.7	1.9	28.2	18	10.4	0.4
188–920	P	12	46.1 (22.7)	55.9 (6.2)	0.8 (0.1)	68.7 (5.4)	732	188.8	2.8
							0–920	239.5	3.9

P peat, *DP* dried peat, *LP* loose peat, *DeP* decomposed peat. Values for the layers with plural samples are mean (standard deviation)

and 5000 cal BP and expanded after 4000 cal BP in the Kampar River and the Indragir River Lowlands (Fig. 4). Namely, the peat swamp forests in the study area seem to have formed their habitat with the sea-level changes; the sea-level falls between 6500 and 5000 cal BP and between 4000 and 2000 cal BP seems to have especially affected the habitat expansion on the coastal lowlands because of decrease of tidally affected area, though the sea-level fall between 6500 and 5000 cal BP has not been recognized in this area.

Carbon sequestration ability of coastal peat swamp forest

Stored carbon per unit volume of the peat layer at SKB-1, 5, and 6 situated in the margin of the peat dome were calculated at 29.8, 43.6, and 38.8 kg C m^{-3}, respectively, while at SDB-1 situated in the center, it was 26.0 kg C m^{-3}. These results indicated that stored carbon per unit volume of the peat layer in the center was possibly less than that of the margin.

Carbon concentrations at SKB-1, 5, and 6 were distinctly lower than those of SDB-1; the bulk densities of the former sites were higher than those of later site (Tables 3 and 4), suggesting that the deposits in the peat dome margin contained a clay fraction deposited by flooding. Some parts of the accumulated carbon in the margin were possibly originated by allochthonous organic matter brought with flood deposits.

Belowground carbon burial rates of the peat layers of peat swamp forests were calculated to be between 31 and 105 g m^{-2} years^{-1} in the peat dome margin and between 27 and 76 g m^{-2} years^{-1} near the center of peat dome. These values were equivalent to the data from previous studies conducted for coastal peat swamps in the tropics (e.g., Neuzil 1997; Page et al. 2004).

Stored carbon per unit volume and long-term carbon burial rate of the mangrove peat were estimated to be between 50 and 65 kg C m^{-3} and between 24 and 92 g C m^{-2} years^{-1}, respectively (Fujimoto 2004); though the short-term burial rate was occasionally calculated to be higher, for example 300 g C m^{-2} years^{-1} by Dommain et al. (2015). Though the belowground stored carbon per unit volume in peat swamp forest was possibly lower than that of *Rhizophora* forest, which is the sole mangrove genus that can create mangrove peat, the long-term carbon burial rate was almost identical. The results suggested that peat swamp forests have played an equally significant role as places for belowground carbon sequestration as *Rhizophora* forests during the late Holocene.

Conclusions

The peat swamp forest initiated its formation by 6500 cal BP in the inland valley bottom lowlands and expanded rapidly between 6000 and 5000 cal BP and

Table 5 Belowground carbon burial rates of peat layers

Loc. no.	Depth (cm)	Span of accumulation: From–to	Span of accumulation: Years	Stored *C* (kg C m^{-2})	C burial rate (g m^{-2} years^{-1})
SKB-1	0–345	BC1010–AD2005	3015	103	34
SKB-5	50–382	AD370[1)]–AD1740	1370	144	105
	382–456.5	BC210–AD370	580	35	61
SKB-6	0–333.5	BC2080–AD2005	4085	127	31
SDB-1	0–185	AD1360–AD2006	646	49	76
	185–585	BC2520–AD1360	3880	104	27
	585–917.5	BC4020–BC2520	1500	86	57

[1]The age of the horizon at 50 cm deep at SKB-5 was estimated under assumption that the peat accumulation rate was constant since AD1740

between 4000 and 2000 cal BP to nearly the present distribution area. The thickness of the peat layer formed under the peat swamp forest reached approximately 4 m, even in the margin of the peat dome, and more than 9 m at the center. Stored carbon per unit volume of the peat layer was estimated to be between 30 and 44 kg C m^{-3} in the margin of the peat dome and approximately 26 kg C m^{-3} in the center, which was less than the general amount for mangrove peat. Belowground carbon burial rate of the peat layer was calculated to be between 31 and 105 g m^{-2} $years^{-1}$ in the peat dome margin and between 27 and 76 g m^{-2} $years^{-1}$ in the center, which were identical to rates of the mangrove peat; however, the accumulated carbon in the margin possibly contained some allochthonous carbon brought in by flooding. These data suggest that peat swamp forests have played a significant role as places for belowground carbon sequestration similar to mangrove forests with mangrove peat.

Abbreviations

AMS: Accelerator mass spectrometry; EC: Electrical conductivity; TS: Total sulfur

Acknowledgements

We thank Mr. Wardi and Mr. Oscar, researchers at the Research Center for Biology, the Indonesian Institute of Sciences, for their assistance in our field research. We also thank PT. Bumi Siak Pusako for giving us permission to perform on the company's land.

Funding

This work was supported by Comprehensive promotion expenses for Global Environmental Research, Ministry of Environment, Japanese Government (FY2003–2007, S2-2b).

Authors' contributions

KF designed the methods of field research and analyses and wrote the manuscript. MM carried out EC and TS content analyses and drew the drafts of figures. SK planed the outline of the project and proposed it to the research fund. HS arranged the research permission and soil analyses in Indonesia. All authors joined in the field research. MM and SK read and approved the final manuscript.

Authors' information

Dr. Herwint Simbolon, who was a core member of our research team, suddenly passed away in 2011. We offer him our sincerest condolences and this paper.

Competing interests

The authors declare that they have no competing interests.

Author details

[1]Faculty of Policy Studies, Nanzan University, 18 Yamazatocho, Showa-Ku, Nagoya 466-8673, Japan. [2]Graduate School of Frontier Sciences, The University of Tokyo, Kashiwa, Japan. [3]Graduate School of Asian and African Area Studies, Kyoto University, Kyoto, Japan. [4]Research Center for Biology, The Indonesian Institute of Sciences, Cibinong, Indonesia. [5]Present Address: Kasukabe City Office, Kasukabe, Japan.

References

Alongi DM (2012) Carbon sequestration in mangrove forests. Carbon Manage 3: 313–322

Anderson JAR (1964) The structure and development of the peat swamps of Sarawaku and Brunei. J Trop Geogr 18:7–16

Anderson JAR, Muller J (1975) Palynological study of a Holocene peat and a Miocene coal deposit from NW Borneo. Rev Palaeobot Palynol 19:291–351

Cameron CC, Esterle JS, Curtis AP (1989) The geology, botany and chemistry of selected peat-forming environments from temperate and tropical latitudes. Int J Coal Geol 12:105–156

Cobb AR, Hoyt AM, Gandois L, Eri J, Dommain R, Salim KA, Kai FM, Su'ut NSH, Harvey CF (2017) How temporal patterns in rainfall determine the geomorphology and carbon fluxes of tropical peatlands. Proc Natl Acad Sci 114:E5187–E5196

Cole LE, Bhagwat SA, Willis KJ (2015) Long-term disturbance dynamics and resilience of tropical peat swamp forests. J Ecol 103:16–30

Dommain R, Cobb AR, Joosten H, Glaser PH, Chua AFL, Gandois L, Kai FM, Noren A, Salim KA, Su'ut NSH, Harvey CF (2015) Forest dynamics and tip-up pools drive pulses of high carbon accumulation rates in a tropical peat dome in Borneo (Southeast Asia). J Geophys Res Biogeosci 120:617–640

Dommain R, Couwenberg J, Glaser PH, Joosten H, Suryadiputra INN (2014) Carbon storage and release in Indonesian peatlands since the last deglaciation. Quat Sci Rev 97:1–32

Dommain R, Couwenberg J, Joosten H (2011) Development and carbon sequestration of tropical peat domes in south-east Asia: links to post-glacial sea-level changes and Holocene climate variability. Quat Sci Rev 30:999–101

Fujimoto K (2004) Below-ground carbon sequestration of mangrove forests in the Asia-Pacific region. In: Vannucci M (ed) Mangrove management & conservation: present & future, United Nations. University Press, Tokyo, pp 138–146

Fujimoto K, Imaya A, Tabuchi R, Kuramoto S, Utsugi H, Murofushi T (1999b) Belowground carbon storage of Micronesian mangrove forests. Ecol Res 14: 409–413

Fujimoto K, Miyagi T, Kikuchi T, Kawana T (1996) Mangrove habitat formation and response to Holocene sea-level changes on Kosrae Island. Micronesia Mangrov Salt Marshes 1:47–57

Fujimoto K, Miyagi T, Murofushi T, Mochida Y, Umitsu M, Adachi H, Pramojanee P (1999a) Mangrove habitat dynamics and Holocene Sea-level changes in the southwestern coast of Thailand. TROPICS 8:239–255

Fujimoto K, Umitsu M, Nguyen VL, Ta TKO, Kawase K, Huynh DH, Nakamura T (2011) Geomorphological evolution and mangrove habitat dynamics related to Holocene sea-level changes in the northern Mekong River Delta and the Dong Nai River Delta, southern Vietnam. In: Schmidt PE (ed) river deltas: types, structures and ecology. Nova Science Publishers, New York, pp. 125–141. https://www.novapublishers.org/catalog/product_info.php?products_id=21842. Accessed 20 May 2019

Furukawa H, Supiandi S (1985) Agricultural landscape in the lower Batang Hari, Sumatra part one: stratigraphy and geomorphology of coastal swampy lands. Southeast Asian Stud 23:3–37 (in Japanese)

Gastaldo R (2010) Peat or no peat: why do the Rajang and Mahakam deltas differ? Int J Coal Geol 83:162–172

Geyh MA, Kudrass HR, Streif H (1979) Sea-level changes during the late Pleistocene and Holocene in the strait of Malacca. Nature 278:441–443

Hanebuth TJJ, Voris HK, Yokoyama Y, Saito Y, Okuno J (2011) Formation and fate of sedimentary depocentres on Southeast Asia's Sunda shelf over the past sea-level cycle and biogeographic implications. Earth Sci Rev 104:92–110

Haseldonckx P (1977) The palynology of a Holocene marginal peat swamp environment in Johore, Malaysia. Rev Palaeobot Palynol 24:227–238

Keith ML, Digens ET (1959) Geochemical indicators of marine and fresh-water sediments. In: Abelson PH (ed) Researches in Geochemistry. John Wiley and Sons, New York, pp 38–61

Miettinen J, Liew SC, Kwoh LK (2011) Decline of Sumatran peat swamp forests since 1990. In: Proceedings of 34th international symposium on remote sensing of environment. http://www.isprs.org/proceedings/2011/ISRSE-34/211104015Final00486.pdf. Accessed 23 August 2018

Miettinen J, Shi C, Liew SC (2016) Land cover distribution in the peatlands of peninsular Malaysia, Sumatra and Borneo in 2015 with changes since 1990. Glob Ecol Conserv 6:67–78

Mochida Y, Fujimoto K, Miyagi T, Ishihara S, Murofushi T, Kikuchi T, Pramojanee P (1999) A phytosociological study of the mangrove vegetation in the Malay peninsula: special reference to the micro-topography and mangrove deposit. TROPICS 8:207–220

Morley RJ (1981) Development and vegetation dynamics of a lowland ombrogenous peat swamp in Kalimantan Tengah, Indonesia. J Biogeogr 8: 383–404

Neuzil SG (1997) Onset and rate of peat and carbon accumulation in four domed ombrogenous peat deposits, Indonesia. In: Rieley JO, Page SE (eds)

Biodiversity and sustainability of tropical peatlands. Samara Publishing Ltd., Cardigan, pp 55–72. https://www.cambridge.org/core/journals/new-phytologist/article/biodiversity-and-sustainability-of-tropical-peatlands-ed-by-j-o-rieley-and-s-e-page-24817-cm-pp-370-with-numerous-textfigures-cardigan-uk-samara-publishing-ltd-1997-price-hb-4500-isbn-1-873692-10-2/D426A8601E4C260CDCD560FDFCAB68B2
Neuzil SG, Supardi, Cecil CB, Kane JS, Soedjono K (1993) Inorganic geochemistry of domed peat in Indonesia and its implication for the origin of mineral matter in coal. In: Cobb JC, Cecil CB (eds) Modern and ancient coal-forming environments. Geological Society of America Special Paper 286, Boulder, pp 23–44
Nguyen VL, Ta TKO, Tateishi M (2000) Late Holocene depositional environments and coastal evolution of the Mekong River Delta, southern Vietnam. J Asian Earth Sci 18:427–439
O'Kelly BC (2005) Oven-drying characteristics of soils of different origins. Dry Technol 23:1141–1149
Page SE, Rieley JO, Banks CJ (2011) Global and regional importance of the tropical peatland. Glob Chang Biol 17:798–818
Page SE, Wüst RAJ, Weiss D, Rieley JO, Shotyk W, Limin SH (2004) A record of Late Pleistocene and Holocene carbon accumulation and climate change from an equatorial peat bog (Kalimantan, Indonesia): implications for past, present and future carbon dynamics. J Quat Sci 19:625–635
Polak E (1933) Ueber Torf und Moor in Niederländisch Indien. Verhandelingen der Koninklijke Academie van Wetenschappen te Amsterdam, vol 3. Afdeeling Natuurkunde (tweede sectie). North-Holland Publishing Company, Amsterdam, pp 1–85
Shimada S, Takahashi H, Haraguchi A, Kaneko M (2001) The carbon content characteristics of tropical peats in Central Kalimantan, Indonesia: estimating their spatial variability in density. Biogeochemistry 53:249–267
Sinsakul S (1992) Evidence of Quaternary Sea level changes in the coastal areas of Thailand: a review. J SE Asian Earth Sci 7:23–37
Sinsakul S, Sonsuk M, Hasting PJ (1985) Holocene Sea levels in Thailand: evidence and basis for interpretation. J Geol Soc Thai 8:1–12
Staub JR, Esterle JS (1993) Provenance and sediment dispersal in the Rajang River delta/coastal plain system, Sarawak, East Malaysia. Sediment Geol 85:191–201
Staub JR, Esterle JS (1994) Peat-accumulating depositional systems of Sarawak, East Malaysia. Sediment Geol 89:91–106
Staub JR, Gastaldo RA (2003) Late Quaternary incised-valley fill and deltaic sediments in the Rajang river Delta. In: Sidi HF, Nummedal D, Imbert P, Darman H, Posamentier HW (eds) Tropical deltas of southeast Asia: sedimentology, stratigraphy, and petroleum geology. Special Publication No. 76, SEPM, Tulsa, pp 71–87
Supiandi S (1988) Studies on peat in the coastal plains of Sumatra and Borneo. Part I: physiography and geomorphology of the coastal plains. Southeast Asian Stud 26:308–335
Supardi, Subekty AD, Neuzil SG (1993) General geology and peat resources of the Siak Kanan and Bengkalis Island peat deposits, Sumatra, Indonesia. In: Cobb JC, Cecil CB (eds) Modern and ancient coal-forming environments. Geological Society of America special paper 286, Boulder, pp 45–61
Tanabe S, Saito Y, Sato Y, Suzuki Y, Sinsakul S, Tiyapairach S, Chaimanee N (2003) Stratigraphy and Holocene evolution of the mud-dominated Chao Phraya delta, Thailand. Quat Sci Rev 22:789–807
Tjia HD (1996) Sea-level changes in the tectonically stable Malay-Thai peninsula. Quat Int 31:95–101
Woodroffe CD (1982) Geomorphology and development of mangrove swamps, Grand Cayman Island, West Indies. Bull Mar Sci 32:381–398
Yokoyama T (1993) Method for electric conductivity. In: Japan Association for Quaternary Research (ed) a handbook of quaternary research, vol 2. Tokyo University Press, Tokyo, pp 109–118 (in Japanese)

10

Melting carbonated epidote eclogites: carbonatites from subducting slabs

Stefano Poli

Abstract

Current knowledge on the solidus temperature for carbonated eclogites suggests that carbonatitic liquids should not form from a subducted oceanic lithosphere at sub-arc depth. However, the oceanic crust includes a range of gabbroic rocks, altered on rifts and transforms, with large amounts of anorthite-rich plagioclase forming epidote on metamorphism. Epidote disappearance with pressure depends on the normative anorthite content of the bulk composition; we therefore expect that altered gabbros might display a much wider pressure range where epidote persists, potentially affecting the solidus relationships. A set of experimental data up to 4.6 GPa, and 1000 °C, including new syntheses on mafic eclogites with 36.8 % normative anorthite, is discussed to unravel the effect of variable bulk and volatile compositions in model eclogites, enriched in the normative anorthite component (An_{37} and An_{45}). Experiments are performed in piston cylinder and multianvil machines. Garnet, clinopyroxene, and coesite form in all syntheses. Lawsonite was found to persist at 3.7 GPa, 750 °C, with both dolomite and magnesite; at 3.8 GPa, 775–800 °C, fluid-saturated conditions, epidote coexists with kyanite, dolomite, and magnesite. The anhydrous assemblage garnet, omphacite, aragonite, and kyanite is found at 4.2 GPa, 850 °C. At 900 °C, a silicate glass of granitoid composition, a carbonatitic precipitate, and Na-carbonate are observed. Precipitates are interpreted as evidence of hydrous carbonatitic liquids at run conditions; these liquids produced are richer in Ca compared to experimental carbonatites from anhydrous experiments, consistently with the dramatic role of H_2O in depressing the solidus temperature for $CaCO_3$. The fluid-absent melting of the assemblage epidote + dolomite, enlarged in its pressure stability for *An*-rich gabbros, is expected to promote the generation of carbonatitic liquids. The subsolidus breakdown of epidote in the presence of carbonates at depths exceeding 120 km provides a major source of C-O-H volatiles at sub-arc depth. In warm subduction zones, the possibility of extracting carbonatitic liquids from a variety of gabbroic rocks and epidosites offers new scenarios on the metasomatic processes in the lithospheric wedge of subduction zones and a new mechanism for recycling carbon.

Keywords: Eclogite, Carbonatite, Carbon cycle, Kyanite, Epidote, Calcite, Subduction

Introduction

Epidote is a hydroxylated Ca-Al-silicate commonly observed in a variety of high-pressure rocks, from basaltic eclogites, to metagabbros and rodingites of the metamorphosed oceanic crust, then to intermediate rock compositions representative of slices of the continental crust involved in high-pressure processes at subduction zones. The stability field of the epidote zoisite has been thoroughly explored experimentally in the model system CaO-Al_2O_3-SiO_2-H_2O (CASH, Boettcher 1970; Poli and Schmidt, 1998 and references therein). High-pressure phase relationships at the boundaries of zoisite stability field are characterized by reactions involving lawsonite at low temperature, by the subsolidus dehydration reaction:

$$\text{zoisite} = \text{grossular} + \text{kyanite} + \text{coesite} + \text{vapor} \quad (1)$$

located at approximately 6–7 GPa, above 900 °C, and by a fluid-absent melting reaction at approximately 1200 °C, although experimental data at high temperature are missing above 2.6 GPa. As a result of such an extended stability in pressure-temperature-composition space, epidote is found both in metamorphic and magmatic rocks (Poli and Schmidt 2004; Schmidt and Poli, 2004).

Correspondence: stefano.poli@unimi.it
Dipartimento di Scienze della Terra, Università degli Studi di Milano, Via Botticelli 23, Milan 20133, Italy

The melting of zoisite at high pressure (Fig. 1), i.e., above the stability of anorthite + H_2O, can be described by three model reactions of relevance in a variety of relatively Ca-rich bulk compositions:

$$\text{zoisite} + \text{kyanite} + \text{quartz/coesite} + \text{vapor} = \text{liquid} \quad (2)$$

$$\text{zoisite} + \text{quartz/coesite} = \text{grossular} + \text{kyanite} + \text{liquid} \quad (3)$$

$$\text{zoisite} = \text{grossular} + \text{kyanite} + \text{corundum} + \text{liquid} \quad (4)$$

or, eventually,

$$\text{zoisite} = \text{grossular} + \text{kyanite} + \text{liquid} \quad (5)$$

if the liquid composition degenerates on the plane kyanite-grossular-vapor (see Figure 2b in Schmidt and Poli, 2004). Eutectic occurs on reaction (2), potentially accounting for wet melting of an eclogite beyond the stability field of amphibole.

The extent of pressure stability of epidote in eclogites has been proposed to be a function of the normative anorthite content (*An*) of the bulk composition of the system of interest. Zoisite was found in eclogite assemblages in a MOR basalt composition with normative anorthite content $An = 31$ % to 3 GPa by Poli et al. (2009) and up to at least 4 GPa, 900 °C in a plagioclase-rich gabbro, $An = 71$ % by Wittenberg et al. (2003). As the wet solidus for MOR basalts approaches such conditions (Fig. 1), epidote is expected to affect melting of mafic eclogites at pressures largely exceeding the amphibole stability field, therefore down to sub-arc depths. The role and pressure effect of Fe^{3+} is unexplored and may further expand epidote stability if Fe^{3+} preferentially partitions in epidote with respect to garnet.

The abundance of epidote in the subsolidus is again governed by the abundance of anorthite component in plagioclase hydrated at low pressure following the reaction:

$$4\,\text{anorthite} + 1\,H_2O = 2\,\text{zoisite} + 1\,\text{kyanite} + 1\,\text{quartz} \quad (6)$$

The large majority of experimental studies dealt with phase transformations in eclogites derived from the so-called "altered oceanic crust"; these studies have focused on a very limited range of bulk compositions (Fig. 2), based on minor chemical variations of a mid-ocean ridge basalt (other than for volatile components added by hydrothermal processes). Mg numbers ($100 \times Mg/(Mg + Fe^{2+})$) range from 47 in Forneris and Holloway (2003) to a maximum of 64 in starting material SLEC1 used by Dasgupta et al. (2004); normative *An* range from 23 in SLEC1 to 31 in bulk OTB used by Poli et al. (2009). However, the mafic oceanic crust includes sequences of gabbros, olivine gabbros, and troctolites several kilometers in thickness. Plutonic rocks of the oceanic crust have been

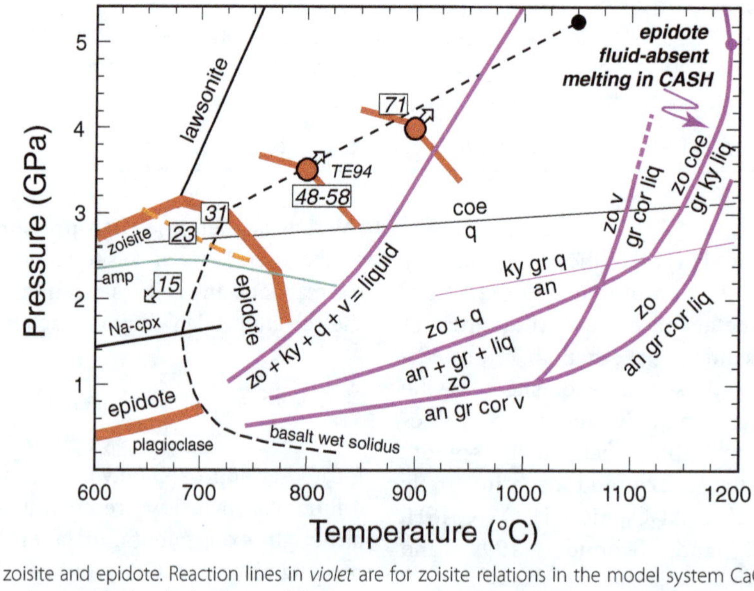

Fig. 1 The stability field of zoisite and epidote. Reaction lines in *violet* are for zoisite relations in the model system CaO-Al_2O_3-SiO_2-H_2O (CASH). Boundaries in *red* represent schematically the displacement of the maximum extent of epidote in mafic and intermediate bulk compositions as a function of CIPW normative anorthite content, *An*. *Bold numbers* indicate anorthite contents in weight percent from studies that synthesized epidote (*circles*; *An* = 48–58: Thompson and Ellis 1994; *An* =71: Wittenberg et al. 2003) or defined an epidote-out reaction (*An* = 31: Poli and Schmidt 1995; Schmidt and Poli 1998; *An* = 23: Forneris and Holloway 2003, *broken bold line*). Experiments performed on a Mg-rich andesite, with *An* =15 (Poli and Schmidt 1995), did not result in the presence of epidote, pointing to a low-pressure stability of epidote in such composition. *Dashed line* represents the wet solidus after Kessel et al. (2005), *black dot* stands for the second critical endpoint

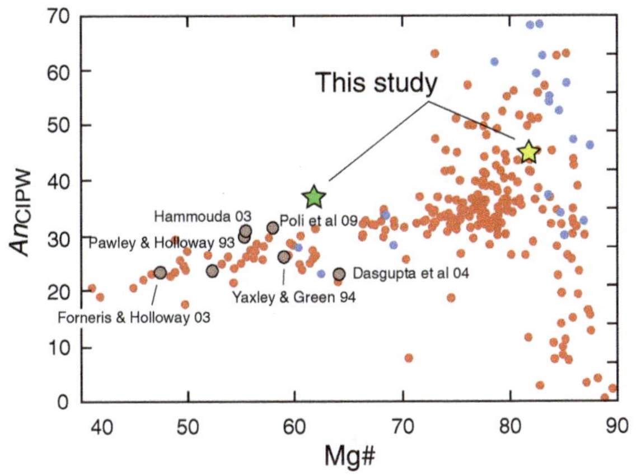

Fig. 2 The diversity of plutonic rocks in the oceanic crust. Normative CIPW anorthite content (*An*) vs. Mg# values, 100 × Mg/(Mg + Fe), of starting materials used in previous experimental studies (*brown circles*) compared to variability of compositions for gabbroic rocks sampled in modern oceanic crusts. *Red dots* are from Godard et al. (2009) and *blue dots* from Perk et al. (2007). *Stars* are the two bulk compositions adopted in this study

sampled along the East Pacific Rise, at the Pito Deep (Perk et al., 2007), on the Mid-Atlantic Ridge, Atlantis Massif (Godard et al., 2009), and at the Southwest Indian Ridge (Meyer et al., 1989). Remnants of such oceanic cumulate rocks are then recognizable in high-pressure terrains, e.g., in the Alps (Bucher and Grapes, 2009, Lagabrielle et al., 2015) testifying to their relevance in orogenic processes. Cumulates are characterized by variable amount of olivine and plagioclase, the latter usually exceeding 50 % in volume, often up to 70–80 % in volume. As such cumulates are derived from primitive liquids, the fraction of anorthite component is high, e.g., approximately 0.7–0.8 for more than 700 m in the section of the East Pacific Rise (Perk et al., 2007). The mineralogical features of gabbros and troctolites result in a strong compositional variability (Fig. 2), ranging from "MORB" to cumulate compositions at Mg# 80, where normative *An* reaches values as high as 70. Hydrothermal alteration and prograde metamorphism of these compositions will therefore produce phase assemblages enriched in epidote.

The limited set of plutonic rocks sampled in oceans hampers a correct assessment of the alteration pattern of the lower crustal sequences. Bach et al. (2001) presented a detailed reconstruction of chemical exchanges between seawater and oceanic lithosphere in the ODP Hole 735B which sampled gabbros down to 1500 m below seafloor at the SW Indian Ridge. A complex and multistage low-temperature (<250 °C) alteration history, typically confined to fractured regions, is responsible for the precipitation of secondary smectite–chlorite mixed-layer minerals, chlorite, calcite, zeolite, sulfide, and Fe-oxyhydroxide, resulting in percent level of both H_2O and CO_2 added. It is worth noting that carbonation appears to be a slow process, active on timescales in the order of tens of million years (Alt and Teagle, 1999). Furthermore, fluid-rock interaction may act at much greater depths, e.g., along active faults at the outer rise outboard of trenches (Faccenda, 2014).

The devolatilization history of altered gabbros on subduction is characterized by the progressive disappearance of the hydrous phases lawsonite, chlorite, paragonite, amphibole, and eventually talc. Melting within the pressure stability field of amphibole is unlikely at subduction zones, unless unusually "hot" regimes occur (e.g., early subduction of young oceanic crust). At pressures higher than 2.5–2.8 GPa, i.e., above amphibole breakdown in mafic eclogites (Poli et al., 2009), epidote is the only major H_2O reservoir of the subducted oceanic crust, for temperatures exceeding lawsonite stability (Fig. 1). When CO_2 is added by low-pressure fluid-rock interactions, the carbonates potentially stable at P-T conditions beyond amphibole stability field are magnesite-siderite solid solution, dolomite-ankerite-Mg-calcite solution, and the high-pressure polymorph of $CaCO_3$, aragonite.

The wet solidus for MOR basalts (Kessel et al., 2005) is located approximately 100° lower than eutectic reaction (2) in the model system CASH, at 2.5 GPa, and, because of a lower dP/dT slope, it converges and intersects reaction (2) at approximately 4.5 GPa. We therefore expect that epidote and ternary carbonates will play a fundamental role on reactions on the solidus at variable volatile component abundances, for pressure between 3 and 5 GPa, i.e., for depths of the slab typical for the subarc at subduction zones. The goal of this exploratory survey is to constrain the nature of liquids formed upon melting of zoisite-bearing carbonated eclogites. This work also builds on a summary of experimental results presented in Poli (2015).

Methods/Experimental

Bulk compositions and starting materials

The composition of an olivine gabbro from the Southwest Indian Ridge (sample 59-3, Table 1, Meyer et al., 1989) was used as a representative composition for most experiments in this work. The modal proportions of major constituents in the original gabbro 59-3 are 63 % plagioclase and 33 % clinopyroxene. This bulk composition has a normative $An_{CIPW} = 44.9$, Mg# = 81.8, and Ca/(Ca + Mg + Fe) = 0.497. Additionally, a first set of experiments was performed on an altered basalt compositions (sample 80-3, Alt and Emmermann, 1985), with $An_{CIPW} = 36.8$, Mg# = 61.8, and Ca/(Ca + Mg + Fe) = 0.449. These two representative compositions of the variability of the oceanic crust compared to an average MORB or altered MORB are displaced

Table 1 Bulk composition and CIPW norm of representative oceanic gabbros and of starting material used in this study and in previous experimental studies

	59-3	80-3	A1-2	593-ccb	59-3-ccb-CA	59-3-ccb-CB	OTB	SLEC1	SLEC3	OTBC	GA1cc	EC1	Witt
SiO_2	49.26	48.10	38.16	43.84	35.13	33.83	52.38	41.21	41.69	47.23	45.32	30.11	45.60
TiO_2	0.06	0.84	0.20	–	0.04	0.04	0.00	2.16	2.18	0.00	1.34	0.00	0.09
Al_2O_3	19.42	17.10	20.57	17.28	18.62	13.59	16.93	10.89	11.02	15.35	14.88	11.74	28.60
FeO	3.68	7.80	4.91	3.28	3.21	3.84	10.29	12.83	11.03	8.93	8.85	10.05	2.90
MnO	0.07	0.27	0.08	–	0.09	0.11	0.00	0.12	0.12	0.00	0.15	0.00	0.04
MgO	9.25	7.08	3.55	9.09	6.95	7.77	7.13	12.87	11.07	6.24	7.15	12.44	3.30
CaO	15.56	13.00	20.26	20.01	20.70	25.08	10.05	13.09	16.89	14.77	14.24	19.41	16.00
Na_2O	1.70	2.42	2.46	1.51	1.46	1.74	3.21	1.63	1.40	2.91	3.14	0.87	1.60
K_2O	0.01	0.02	0.09	0.01	0.01	0.01	0.00	0.11	0.10	0.02	0.40	0.00	0.12
CO_2	0.00	0.00	8.10	4.84	10.05	13.51	0.00	5.00	4.42	4.43	4.40	15.38	0.00
H_2O	0.00	0.00	0.96	0.38	3.74	0.48	0.00	0.00	0.00	0.06	0.50	0.00	0.00
Total	99.01	96.63	99.34	100.24	100.00	99.78	99.99	99.91	99.92	99.94	100.37	100.00	98.27
Mg/(Mg + Fe)	0.818	0.618	0.563	0.832	0.794	0.783	0.553	0.641	0.641	0.555	0.59	0.688	0.670
Ca/(Ca + Mg + Fe)	0.497	0.449	0.698	0.568	0.630	0.644	0.359	0.319	0.413	0.485	0.466	0.436	0.700
or	0.06	0.12	0.00	0.00	0.00	0.00	0.00	0.68	0.00	0.12	2.44	0.00	0.71
ab	15.11	22.38	0.00	0.00	0.00	0.00	28.67	5.04	0.00	17.20	10.61	0.00	13.98
an	44.87	36.81	49.00	41.51	50.14	33.23	31.62	23.14	24.61	29.99	26.21	32.53	71.04
lc	0.00	0.00	0.46	0.05	0.05	0.05	0.00	0.00	0.50	0.00	0.00	0.00	0.00
ne	0.00	0.00	14.49	8.37	8.92	10.65	0.00	6.18	7.90	6.02	11.10	5.42	0.30
di	25.23	23.70	17.12	37.42	4.96	4.05	14.38	35.95	46.74	37.12	39.40	0.00	7.05
hy	6.86	8.31	0.00	0.00	0.00	0.00	19.72	0.00	0.00	0.00	0.00	0.00	0.00
ol	6.18	4.96	4.81	6.77	16.02	19.19	4.05	21.84	11.62	7.91	5.09	28.63	5.08
mt	1.61	2.52	1.94	1.67	1.83	1.83	1.56	4.01	4.03	1.63	3.20	1.81	1.67
il	0.08	1.21	0.30	0.07	0.06	0.06	0.00	3.15	3.18	0.00	1.93	0.00	0.13
ap	0.00	0.00	0.46	0.00	0.00	0.00	0.00	0.00	0.00	0.00	0.00	0.00	0.04
wus	0.00	0.00	0.00	0.00	0.00	0.00	0.00	0.00	0.00	0.00	0.00	7.96	0.00

59-3 olivine gabbro from Southwest Indian Ridge (Meyer et al. 1989), *A1-2* altered gabbro from Southwest Indian Ridge (Bach et al. 2001), *59-3-ccb* bulk composition obtained from composition 59-3 by adding 11 wt.% calcite and 1.24 wt.% brucite, *59-3-ccb-CA* bulk composition used in sandwich experiments adding $Al(OH)_3$ and silica to composition 59-3-ccb, *59-3-ccb-CB* this composition differs from 59-3-ccb-CA because of it volatile composition and therefore Al, and Si addition, *OTB* starting material used in Poli et al (2009), *SLEC 1* and *SLEC3* carbonated eclogite compositions (Dasgupta et al. 2005), *OTBC* hydrated and carbonated material used in Hammouda (2003), *GA1cc* composition used in Kiseeva et al (2012), *EC1* composition used in Yaxley and Brey (2004), *Witt* composition as used by Wittenberg (1997)

toward zoisite/anorthite compositions in a ACF projection (Fig. 3). Upper bound of chemical variability is represented by the plagioclase cumulate investigated by Wittenberg et al. (2003) and the lower bound by composition SLEC1 used by Dasgupta et al. (2004). The dispersion of the bulk compositions experimentally explored well reproduces varying proportions of pyroxene-olivine versus plagioclase.

In order to promote the synthesis at subsolidus conditions, a gel of chemical composition corresponding to samples 59-3 and 80-3 (Table 1) has been prepared using tetraethylorthosilicate (TEOS) as silica source, pure Na-, Ca-, Mg-, and Al- nitric solutions, and ferric benzoate for Fe_2O_3. Gels have the major advantage of a large active surface, and perfect homogeneity, related to the co-precipitation of a hydrogel from clear solutions. Gels were fired in a gas mixing furnace at fO_2 conditions approaching FMQ buffer at 1 atm, then seeded with 3 wt.% of a mixture of garnets and clinopyroxene, that are expected to be major phases stable in the pressure range investigated. Individual seed composition (synthetic pyrope, almandine, grossular, diopside) largely differs from the stable composition expected at P-T conditions investigated to favour unquestionable identification of relicts, but the composition of seeds mixture closely reproduces the bulk composition chosen to avoid fractionation. Seeding is required to prevent from the nucleation of metastable phases, which is a potential problem notably in volatile-rich compositions. Seeded gels are of superior homogeneity and reactivity compared to fully crystalline mixtures, grinded glasses, or oxide mixtures at subsolidus or near-solidus pressure-temperature

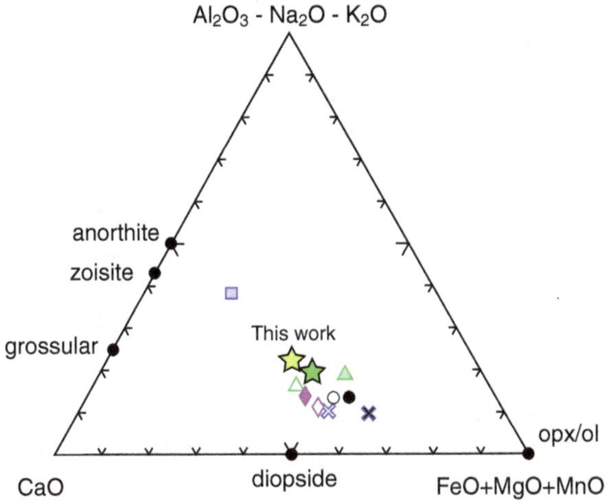

Fig. 3 ACF diagram of bulk compositions used in this study (*stars*) compared with compositions adopted in previous studies. *Cross*, Dasgupta et al. (2005); *filled cross*, Dasgupta et al. (2004); *circle*, Pawley and Holloway (1993); *solid circle*, Forneris and Holloway (2003); *diamond*, Yaxley and Brey (2004); *solid diamond*, Kiseeva et al. (2012); *triangle*, Hammouda (2003); *solid triangle*, Poli et al. 2009; *solid square*, Wittenberg et al. (2003)

conditions. Furthermore, as a conventional "bracketing" of mineral compositions (e.g., re-running experimental charges at different pressure and temperature) is hampered by the low reactivity of fully crystalline run products, growth rims on seeds of variable compositions (e.g., almandine vs. pyrope vs. grossular) are profitably used to evaluate the approach to equilibrium.

H_2O and CO_2 were added following three distinct experimental strategies:

(i) H_2O and CO_2 were added by oxalic acid dihydrate, which provides a source of volatiles decoupled from chemical variations of non-volatile components;

(ii) volatile addition has been related to the precipitation of secondary minerals chemical, in order to reproduce the chemical and mineralogical composition of altered gabbros, for example, as recorded in ODP Hole 735B (sample A1-2, Bach et al. 2001). Addition of 11 wt.% calcite (cc) and 1.24 wt.% brucite (br) therefore resulted to starting material 59-3-ccb (Table 1). Differences in the composition of starting material 59-3ccb as compared to 59-3 are expected to produce a higher amount of carbonate in the run products without altering phase relationships of silicates, in the range of pressure and temperature investigated here.

(iii) because high-pressure experiments in relatively *An*-rich bulk compositions are expected to produce kyanite, after reaction (6), a precise amounts of H_2O in the run charge can be obtained by adding a bottom layer of $Al(OH)_3$ and SiO_2, which produce kyanite + H_2O under run conditions. Then, CO_2 can be added as carbonates (Na_2CO_3, $CaCO_3$, natural dolomite) as required to model variable CO_2/H_2O ratio and carbonate proportions (see below).

Experimental setup

A double-capsule technique was adopted in experiments with bulk composition 59-3 and 80-3. Hydrogen fugacity was buffered by hematite-magnetite (HM)-H_2O mixture, using inner $Ag_{50}Pd_{50}$ capsules and outer Au capsules. Welded inner capsules (2 mm long) contain the starting materials plus 10 wt.% oxalic acid dihydrate (OAD). It is worth introducing here that HM buffer in the outer capsule fix the chemical potential of hydrogen in a pure H_2O vapor, but the inner capsule contains a mixed vapor; as a consequence the oxygen fugacity in the inner capsule is intrinsically lower than in the outer capsule (see details in Poli et al., 2009). The actual value of the oxygen fugacity in the inner capsule is then fixed either by redox equilibria, or by a fixed C/O ratio, or by the occurrence of graphite/diamond, however not observed here.

A second set of experiments was designed to investigate bulk composition 59-3 ccb at vapor-poor or vapor-absent conditions assuming that carbonate, epidote and, eventually, liquid fractionate most volatile components in this Ca-rich starting material. A single Au capsule was used for these runs.

A "sandwich" technique was used for a third set of experiments; a carbonate layer was inserted in 1:5 proportion between two layers of the same gel as used in the previous experimental setup, hereafter called "forward" setup (Fig. 4). The composition of the carbonate layer is based on the composition of carbonate quench products in the experiment at 900 °C, 4.2 GPa. It is therefore expected that phase proportions vary, to facilitate analytical characterization, but phase compositions remain unaltered (see discussion below).

High-pressure experiments were performed at pressures ranging from 3.7 to 4.6 GPa and temperatures from 750 to 1000 °C at the Dipartimento di Scienze della Terra in Milano, Italy. Pressures lower than 4.0 GPa were investigated in an end-loaded piston cylinder, using MgO–salt assemblies. Temperature was measured by K-type thermocouple and was considered accurate to ±5 °C. A Walker-type multianvil apparatus was used for higher pressures using tungsten carbide cubes of 32 mm edge length and 17 mm truncation edge length (TEL), pressure cells made of prefabricated MgO octahedra (containing 5 wt.% of Cr_2O_3) with a 25 mm edge length (25 M). Pressure uncertainties, which largely depend on the accuracy of the calibrant reaction, were assumed to be ±3 %. For further details, see Fumagalli and Poli (2005). Temperature was measured by S-type thermocouples and was considered accurate to ±20 °C

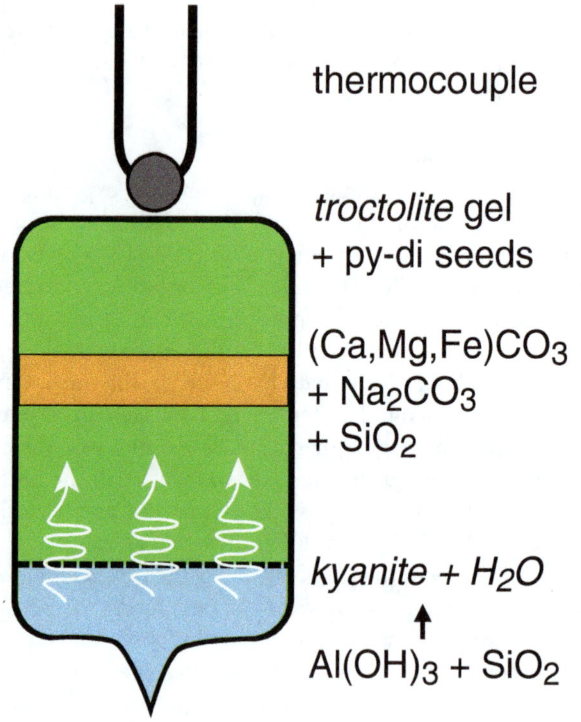

Fig. 4 Capsule geometry in "sandwich" experiments

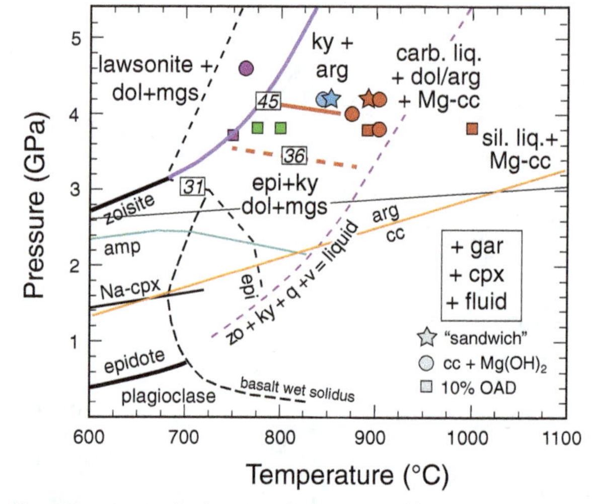

Fig. 5 Experimentally determined phase assemblages in model gabbros. *Circles*, experiments with oxalix acid dihydrate as volatile source; *squares*, experiments with brucite and carbonate; *stars*, sandwich experiments. *Red symbols* represent liquid-bearing experiments. *Lines* and *numbers* as in Fig. 1

without taking into account any effect of pressure on the e.m.f. of the thermocouple.

Run products were characterized by WDS electron microprobe analysis using a JEOL8200 Superprobe.

Results

Phase assemblages and their textural features

Garnet, clinopyroxene, and quartz/coesite were observed in all runs performed in this study (Fig. 5, Table 2). Texturally equilibrated garnet is present in most experiments, but among the minerals encountered in this study, garnet exhibits the most sluggish crystallization, with growth rims on seeds, mostly between 5 and 10 μm thick, depending on temperature. Clinopyroxene nucleates from the starting material and constitutes mostly a matrix of small euhedral crystals. Kyanite was recognized in all run products other than those where lawsonite occurs. Hydrous phases are absent in all runs on starting material 80-3, with normative $An = 36.8$, the stability field of epidote therefore expected to be restricted to approximately 3.5 GPa in this bulk composition.

Lawsonite was recovered at 3.7 GPa, 750 °C, and 4.6 GPa, 700 °C in bulk composition 59-3 and 59-3ccb, as relatively large prismatic crystals, 20–40 μm long (Fig. 6a). Epidote was found at 3.8 GPa, from 775 to 900 °C, at the latter temperature as a minor phase coexisting with a quenched carbonate phase, interpreted here as precipitated from liquid (see below).

All syntheses contain one or more carbonate phases. Magnesite and dolomite form when coexisting with Ca-Al-silicates, i.e., lawsonite or epidote (Fig. 6b). Nonetheless, magnesite was never observed in run products of experiments beyond the stability field of lawsonite or epidote, where, on the contrary a (Mg,Fe)-calcite and/or aragonite were found. The term Mg-calcite is used here for carbonate compositions where Ca > 1.1 atoms per formula unit. Subsolidus synthesis at 4.2 GPa, 850 °C, resulted in the assemblage garnet + clinopyroxene + aragonite + kyanite, showing that the transformation from calcite, to dolomite, then to magnesite bearing eclogite with increasing pressure (e.g., Kerrick and Connolly, 2001; Poli et al. 2009) depends largely on the bulk Ca/(Ca + Mg + Fe) ratio. In this experiment, globules and droplets of both a silicate phase and a carbonate were recovered in the open porosity at the top of the capsule (see also Figure S3 in Poli 2015), therefore suggesting conspicuous dissolution of both silicates and carbonate in the high-pressure fluid.

All experiments performed at 900 °C show the occurrence of partial melting, with a variety of textural features as a function of volatile abundances. Experiments arranged in a double-capsule technique and oxalic acid dihydrate as H_2O and CO_2 source are expected to have a large amount of fluid present (≈10 wt.%) and are characterized by the presence of large pools (in the order of a hundred micrometers large) of quench precipitates where silicate droplets are intermingled with carbonate aggregates and large skeletal crystals (Fig. 6c, d). The chemical composition of starting material 80-3 is only slightly enriched in normative anorthite content

Table 2 Summary of the starting material used, of experimental conditions, and of run products in order of abundance

Run	Geometry	Bulk	P (GPa)	T (°C)	Duration (h)	H_2O (wt.%)	CO_2 (wt.%)	CO_2/H_2O (wt.%)	Carbonate st. mat. (wt.%)	Volatile source	Run products
SPT1	Double capsule	59-3	3.7	750	159	2.85	6.98	2.4	–	OAD	cpx + gar + law + mgs + dol + cs ± epi
SPT1-8	Double capsule	80-3	3.7	750	159	2.85	6.98	2.4	–	OAD	cpx + gar + mgs + dol + ky + cs
SPT2	Double capsule	59-3	3.8	775	146	2.85	6.98	2.4	–	OAD	cpx + gar + epi + dol + mgs + ky + cs
SPT2-8	Double capsule	80-3	3.8	775	146	2.85	6.98	2.4	–	OAD	cpx + gar + mgs + dol + ky + cs
EFSP2	Double capsule	59-3	3.8	800	110	2.85	6.98	2.4	–	OAD	cpx + gar + epi + dol + mgs + ky + cs
EFSP1	Double capsule	80-3	3.8	800	110	2.85	6.98	2.4	–	OAD	cpx + gar + Mg-cc + ky + cs
EFSP3	Double capsule	59-3	3.8	900	92	2.85	6.98	2.4	–	OAD	cpx + gar + Mg-cc + glass + quench-cc + ky + cs
EFSP4	Double capsule	80-3	3.8	900	92	2.85	6.98	2.4	–	OAD	cpx + gar + Mg-cc + ky + cs + quench-cc + quench-sil
SPT3	Double capsule	59-3	3.8	1000	121	2.85	6.98	2.4	–	OAD	cpx + gar + glass + Mg-cc + ky + cs
SPT3-8	Double capsule	80-3	3.8	1000	121	2.85	6.98	2.4	–	OAD	cpx + gar + Mg-cc + ky + glass + cs
SPT7	Single capsule	59-3-ccb	3.8	900	175	0.38	4.84	12.7	11	cc + br	cpx + gar + Mg-cc + epi + cs + quench-cc ± ky
FB6	Single capsule	59-3-ccb	4.0	870	144	0.38	4.84	12.7	11	cc + br	cpx + gar + arg + quench-cc + ky + cs
SPT5	Single capsule	59-3-ccb	4.2	900	105	0.38	4.84	12.7	11	cc + br	cpx + gar + arg + dol + quench-cc + glass + ky + cs
SPT6	Single capsule	59-3-ccb	4.2	850	316	0.38	4.84	12.7	11	cc + br	cpx + gar + arg + ky + cs
SPT4	Single capsule	59-3-ccb	4.6	700	264	0.38	4.84	12.7	11	cc + br	cpx + gar + law + dol + cs
SPT9	Sandwich	59-3-ccb-CA	4.2	850	86	3.7	10	2.7	26.7	$Al(OH)_3$ + cc + br	cpx + gar + arg + dol + ky + cs + quench drops
SPT8	Sandwich	59-3-ccb-CA	4.2	900	234	3.7	10	2.7	26.7	$Al(OH)_3$ + cc + br	cpx + gar + quench-cc + glass + arg + Mg-cc + ky + cs
SPT10	Sandwich	59-3-ccb-CB	4.2	900	185	0.4	13.5	33.8	31	$Al(OH)_3$ + cc + br	cpx + gar + quench-cc + glass + arg + Mg-cc + ky + cs

Run indicates experiments also discussed in Poli (2015). Geometry: double capsule indicates synthesis experiments performed at vapor saturation, buffered at HM in the outer capsule; single capsule stands for experiments in single Au capsule, with an amount of volatiles adjusted to stabilize hydrous phases and carbonates only (see the "Methods" section); sandwich is for experiments in a single Au capsule, where a layer of carbonates is placed in between two layers of gels. H_2O, CO_2, and CO_2/H_2O are the total weight proportions and ratios of volatiles in the bulk composition. Carbonate st. mat. is the total amount of carbonate in the starting material, in weight proportions. Bulk: bulk composition of starting materials; 59-3 and 59-3-ccb: see Table 1; 59-3-ccb-CA: composition 59-3 with addition of a layer of Ca, Mg, Fe, and Na carbonates, simulating the liquid produced at 4.2 GPa, 900 °C (see Tables 3, 4, 5, 6, 7, and 8), and a layer of $Al(OH)_3$ + SiO_2, producing kyanite + H_2O at run conditions. Volatile sources - *OAD* oxalic acid dihydrate, *cc* calcite, *br* brucite. Run products - *arg* aragonite, *cs* coesite, *cpx* omphacite, *dol* dolomite, *epi* epidote, *gar* garnet, *glass* silicate glass, *ky* kyanite, *law* lawsonite, *Mg-cc* (Mg,Fe)-calcite, *mgs* magnesite, *quench-cc* Ca-rich carbonate in dendritic, feathery, or lobed/fingered morphology; it includes individual grains of Na-carbonate

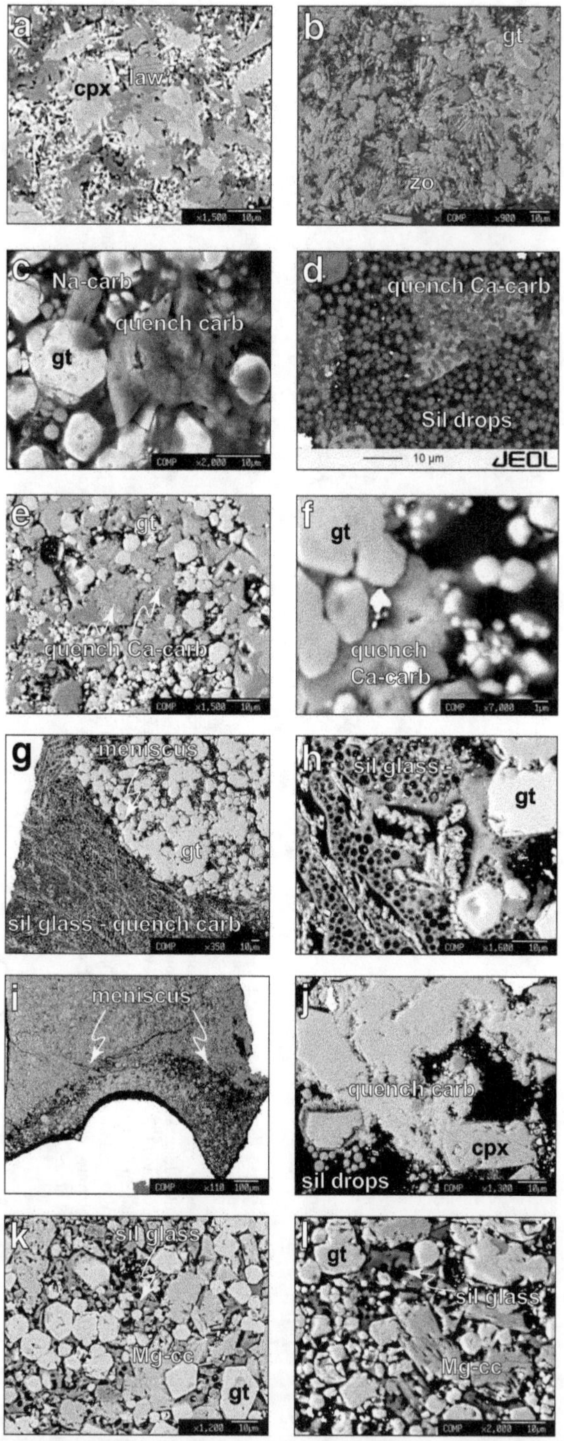

Fig. 6 Back-scattered electron images of representative run products. **a** 3.7 GPa at 750 °C, run SPT1; **b** 3.8 GPa at 800 °C, run EFSP1; **c** 3.8 GPa at 900 °C, run EFSP3; **d** 3.8 GPa at 900 °C, run EFSP4; **e** 4.0 GPa at 870 °C, run FB6; **f** 4.2 GPa at 900 °C, run SPT5; **g, h** 4.2 GPa at 900 °C, sandwich experiment SPT8, $CO_2/H_2O = 2.7$; **i, j** 4.2 GPa at 900 °C, sandwich experiment SPT10, $CO_2/H_2O = 33.8$; **k** 3.8 GPa at 1000 °C, run SPT3, bulk 593; **l** 3.8 GPa at 1000 °C, run SPT3-8, bulk 803

(*An* = 36.8 in 80-3, Table 1, compared to *An* = 31.6 in OTB, Poli et al. 2009) and epidote is absent a 3.8 GPa; nonetheless, its vapor-saturated solidus shows the presence of a carbonatitic component in the liquids formed (Fig. 6d). Therefore, although this contribution will mainly focus on bulk compositions accounting for the stability of epidote up to 4 GPa, and on its consequences on the features of the solidus reactions at such pressure, it is worth pointing out that vapor saturation, e.g., achieved by fluid-flushing, on even mildly *An*-enriched lithologies may result in extended carbonate dissolution in liquids formed.

The composition of the carbonate phase precipitated is in most cases in the ternary Ca-Mg-Fe field, but Na-carbonate is also observed (Fig. 6c). Whether quench textures observed in these volatile-rich experiments are the result of the precipitation of a single volatile-rich carbonate-silicate liquid, of a liquid coexisting with a solute-rich fluid, of two liquids (one silicate, the second carbonatitic) or, as an extreme case, of three "liquids," one silicate, the second carbonatitic, and the third, the solute-rich fluid, it is impossible to discern.

A set of experiments (Table 2) was conducted to reduce the amount of volatiles in the run charge; nonetheless, they show the appearance of dendritic aggregates of a carbonate phase at 4.0 GPa, 870 °C (Fig. 6e), and, most notably, small pools with lobe and cusp structures (Fig. 6f) at 3.8–4.2 GPa, 900 °C. Lobe and cusp structures are known to develop during viscous fingering moved by Saffman–Taylor instabilities in liquids of contrasting viscosities (e.g., Perugini and Poli, 2005), and their coexistence with whiskery crystals strongly suggests the occurrence of more than a single liquid on quenching of these experiments. Again, quench textures do not unequivocally indicate how many liquids occurred at run conditions and the determination of the composition of such liquids is hampered by the size of quench products. Sandwich experiments (Fig. 4) at variable CO_2/H_2O ratio (Table 2) were therefore introduced in order to highlight the relationships between precipitates and to preserve large areas of molten material to retrieve compositional features. Run products at 4.2 GPa, 900 °C, are characterized by a segregated fraction at the hot end of the capsule, close to the thermocouple, delimited by a meniscus (Fig. 6g, i). Precipitated intergrowths from the run charge with 3.7 wt.% H_2O added show a peculiar texture characterized by a silicate glass foamed by micrometer-sized vesicles, where long dendrites of carbonate (Fig. 6h) are often arranged to form a rhombohedral grid. It is worth noting that rims of silicate glass are not fragmented and vesicularity is less pronounced. Minor silica globules are found at some interstices. Na-carbonate dendrites are recovered (see also Figure S4 in Poli 2015). Euhedral (Mg,Fe)-calcite coexist with garnet and clinopyroxene in the residue.

The absence of a continuous porosity and/or large vesicles (Fig. 6g) supports the absence of a free fluid at run conditions. Such textural features can be therefore interpreted as a precipitation sequence upon quenching of a single volatile-rich "liquid" phase. Because (Ca,Mg,Fe)-carbonate liquids cannot be vitrified, carbonate dendrites precipitate first; a free fluid phase, solute-rich, separates from the silicate "residual liquid" afterwards; progression of quenching is finally responsible for the precipitation of the solute from the fluid in the form of silicate droplets and Na-carbonate. The carbonate layer is entirely liquefied in this experiment, so the amount of volatiles dissolved in the liquid formed can be estimated at ≈20 wt.% H_2O and ≈25 wt.% CO_2.

On the contrary, run charge with 0.4 wt.% H_2O and 13.4 wt.% CO_2, produced a continuous porosity at the top of the capsule (Fig. 6i), and dendritic aggregates of carbonate are intermingled with silicate globules (Fig. 6j). The carbonate layer is partly preserved, and partial melting in the presence of a fluid is therefore likely in this case.

Experiments at 1000 °C were produced with OAD as a volatile source and generated large pools of silicate glass coexisting with large euhedral Mg-calcite crystals (Fig. 6k, l). Again, vesicularity suggests the exsolution of the fluid phase on quenching of the silicate liquid.

Phase compositions and the representativeness of sandwich experiments

Representative analyses of run products are reported in Tables 3, 4, 5, 6, 7, and 8. All experiments at run conditions include a "fluid" or a "liquid" of variable chemical composition. Quench microstructural features unequivocally indicate the presence of large amounts of non-volatile solutes in "fluids" and of volatile components in "liquids." Heterogeneous spatial distribution of quench products hinders an analytical estimate of the amount of solutes in fluids and of volatile components in liquids. As a consequence, phase proportions by a mass balance approach are not given here.

Garnet and clinopyroxene display major variations in cation distribution with temperature (Fig. 7) and, therefore, phase assemblages. As expected, Mg/(Mg + Fe) ratio (and Mg concentration) in garnet increases with temperature from 0.5, at 700 °C, to approximately 0.7, at 800 °C, in the subsolidus, up to 0.76 at 1000 °C in the presence of the silicate liquid. Similarly, Ca decreases from 1.5 a.p.f.u. down to 0.9 a.p.f.u. (not shown). The group of experiments at 870–900 °C, where a carbonatitic liquid is observed, is shifted systematically to lower Mg/(Mg + Fe) and higher Ca compared to this trend. Unexpectedly, data retrieved at 850 °C cluster with this group, possibly suggesting that a liquid might have occurred, though overlooked. These experiments were performed on different bulk compositions related to varying proportions of carbonates and silicates, as discussed in the "Methods" section (Table 1). Similarly, clinopyroxene shows an increase of the Mg content with temperature from 700 to 800 °C. Again, all experiments at 850–900 °C are clustered at Mg/(Mg + Fe) 0.8–0.9 in the variety of experimental charge setup. Notably, the composition of both garnets and clinopyroxene produced at 4.2 GPa, 900 °C, in forward experiments and in the two different sandwich experiments, at variable CO_2/H_2O ratio, are identical within error bars.

Sandwich experiments have been proposed to investigate the composition of small degree partial melts. On principle, at equilibrium, the chemical potential of components in the liquid and in the coexisting solids is identical independently on their relative abundances; therefore, provided the phase assemblage obtained is identical, the chemical composition of liquids in "forward" and sandwich experiments should not differ, whereas liquid proportions vary, thus facilitating analytical characterization. However, on practical application, this method presents a potential pitfall (Hirschmann and Dasgupta 2007) given by the uncertainty in estimating the starting "equilibrium" composition; a deviation from equilibrium composition may lead to shifts in bulk compositions affecting final phase compositions, and ultimately phase assemblage. Iterations are then commonly required to obtain convergence to expected element concentrations at true equilibrium.

Sandwich experiment SPT8 was constructed placing a layer of a carbonate mixture (15 % in weight, on the total amount of starting material) based on the composition of the liquid as measured in lobes for the run charge SPT5 (Fig. 6f), between two layers of the model gabbro constituting the starting material of experiment SPT5. Silica and $Al(OH)_3$ were added to provide 3.7 wt.% H_2O in the starting material and kyanite saturation. Similarly, experiment SPT10 was run with 22 wt.% carbonate layer and $Al(OH)_3$ to obtain 0.48 wt.% H_2O (Table 1).

Despite the differences in the bulk composition of the experimental charge, due to the different proportions in carbonates, silicates, and volatile sources, we obtained identical Mg/(Mg + Fe) ratio in clinopyroxenes of the three experimental setups at 4.2 GPa, 900 °C, in "forward" and sandwich experiments. Although the composition of garnet in SPT10 is slightly shifted to lower Mg/(Mg + Fe), again the compositions in the three syntheses are comparable.

Carbonate phases show high Mg/(Mg + Fe) (X_{Mg}), 0.88–0.91 in magnesite, 0.88 in dolomite, and 0.80–0.85 in (Mg,Fe)-calcite at 850–900 °C, but up to 0.9 at 1000 °C. Carbonate dendrites from quenched liquids at 900 °C have much lower X_{Mg}, 0.55 in "forward" experiment SPT5 where small pools are analyzed, and 0.72 in sandwich experiment SPT8 (Table 8). Although such

Table 3 Chemical composition of clinopyroxene. Units are weight percent of oxides and atoms per formula unit. $X_{Mg} = Mg/(Mg + Fe)$

Run	SPT1	EFSP2	SPT2	EFSP3	SPT3	SPT4	SPT5
P (GPa)	3.7	3.8	3.8	3.8	3.8	4.6	4.2
T (°C)	750	800	775	900	1000	700	900
SiO_2	55.80 (1.67)	54.11 (1.15)	56.52 (0.14)	54.22 (0.22)	55.20 (1.17)	54.67 (0.32)	54.97 (0.42)
Al_2O_3	12.83 (2.70)	6.54 (1.02)	8.44 (1.01)	13.14 (1.54)	12.00 (1.15)	11.34 (0.42)	16.25 (0.59)
FeO	2.75 (0.33)	2.08 (0.54)	2.53 (0.17)	3.65 (1.00)	1.89 (0.38)	3.05 (0.21)	1.81 (0.16)
MgO	8.58 (0.95)	13.56 (0.55)	11.38 (0.69)	8.71 (1.16)	11.07 (0.40)	10.23 (0.33)	8.29 (0.26)
CaO	14.73 (1.08)	21.14 (0.61)	18.00 (0.86)	13.39 (1.24)	17.21 (0.55)	18.04 (0.45)	13.92 (0.30)
Na_2O	5.71 (0.80)	1.99 (0.11)	4.03 (0.37)	5.80 (0.77)	3.79 (0.30)	3.66 (0.28)	5.50 (0.22)
Total	100.39 (1.07)	99.43 (1.45)	100.89 (0.15)	98.90 (0.86)	101.17 (1.36)	100.99 (0.21)	100.74 (0.47)
Si	1.961 (0.051)	1.952 (0.024)	1.991 (0.004)	1.948 (0.014)	1.928 (0.016)	1.928 (0.008)	1.909 (0.009)
Al	0.531 (0.110)	0.279 (0.047)	0.350 (0.041)	0.557 (0.067)	0.493 (0.041)	0.471 (0.017)	0.665 (0.024)
Fe^{2+}	0.081 (0.010)	0.063 (0.017)	0.075 (0.005)	0.109 (0.030)	0.055 (0.011)	0.090 (0.006)	0.052 (0.004)
Mg	0.450 (0.053)	0.729 (0.021)	0.598 (0.038)	0.466 (0.061)	0.577 (0.029)	0.538 (0.018)	0.429 (0.014)
Ca	0.555 (0.045)	0.817 (0.016)	0.680 (0.034)	0.515 (0.046)	0.644 (0.032)	0.682 (0.018)	0.518 (0.012)
Na	0.389 (0.050)	0.139 (0.008)	0.275 (0.025)	0.404 (0.055)	0.257 (0.019)	0.250 (0.018)	0.370 (0.015)
X_{Mg}	0.848 (0.012)	0.921 (0.019)	0.888 (0.011)	0.812 (0.022)	0.913 (0.015)	0.857 (0.010)	0.891 (0.010)
Run	SPT6	FB6	SPT7	SPT8	SPT9	SPT10	
P (GPa)	4.2	4.0	3.8	4.2	4.2	4.2	
T (°C)	850	870	900	900	850	900	
SiO_2	55.17 (0.52)	54.52 (0.99)	54.88 (0.80)	56.35 (0.44)	56.58 (0.61)	56.49 (0.44)	
Al_2O_3	16.31 (1.22)	15.04 (2.26)	12.48 (2.07)	14.13 (0.45)	13.65 (1.25)	13.96 (1.50)	
FeO	1.66 (0.16)	2.38 (0.69)	2.74 (0.70)	1.78 (0.12)	2.08 (0.62)	1.76 (0.16)	
MgO	7.89 (0.29)	8.43 (0.63)	9.70 (1.04)	8.10 (0.28)	7.99 (0.25)	8.35 (0.98)	
CaO	13.25 (0.44)	14.51 (0.83)	16.20 (1.00)	13.71 (0.41)	13.55 (0.46)	14.00 (1.00)	
Na_2O	5.71 (0.39)	4.94 (0.42)	4.61 (0.52)	6.20 (0.38)	6.27 (0.31)	6.05 (0.45)	
Total	99.99 (0.21)	99.82 (0.58)	100.61 (0.42)	100.27 (0.37)	100.12 (0.53)	100.61 (0.33)	
Si	1.924 (0.020)	1.918 (0.037)	1.932 (0.026)	1.966 (0.009)	1.979 (0.020)	1.965 (0.007)	
Al	0.670 (0.049)	0.623 (0.087)	0.518 (0.084)	0.581 (0.018)	0.562 (0.047)	0.572 (0.058)	
Fe^{2+}	0.048 (0.005)	0.070 (0.020)	0.081 (0.021)	0.052 (0.004)	0.061 (0.019)	0.051 (0.005)	
Mg	0.410 (0.015)	0.442 (0.035)	0.509 (0.056)	0.421 (0.016)	0.417 (0.014)	0.433 (0.054)	
Ca	0.495 (0.017)	0.547 (0.035)	0.611 (0.040)	0.512 (0.016)	0.508 (0.020)	0.522 (0.041)	
Na	0.386 (0.027)	0.337 (0.029)	0.315 (0.034)	0.419 (0.025)	0.425 (0.022)	0.408 (0.028)	
X_{Mg}	0.895 (0.008)	0.864 (0.035)	0.863 (0.034)	0.890 (0.007)	0.873 (0.034)	0.894 (0.010)	

difference is apparently large, the two compositions are very enriched in calcium and within analytical uncertainty, at $Ca/(Ca + Mg + Fe) = 0.75–0.80$. Silicate glass intermingled with carbonate dendrites is variable in composition, though grossly tonalitic. Ca, Fe, and Mg contents usually do not sum up to more than 5 wt.% total. Glass is most enriched in silica in the rims of large globules and in the smallest droplets, suggesting that composition varies along cooling, and therefore along precipitation sequence.

Iterations to further bracket liquid-solid element partitioning in forward and sandwich experiments was not performed yet, in this first study. Nonetheless, the consistency between garnet, clinopyroxene, and carbonate precipitated from the carbonatitic liquid support the representativeness of sandwich setup in reproducing phase relationships on the solidus of gabbroic eclogites.

Epidote compositions reveal a maximum Fe^{3+} content in the order of 0.10 a.p.f.u. based on 13 oxygens.

Discussion

Phase diagrams for the system $CaCO_3$-$MgCO_3$-$FeCO_3$

Whenever CO_2 is available, ternary (Ca,Mg,Fe)-carbonates are common constituents of high-pressure

Table 4 Chemical composition of garnet. Units are weight percent of oxides and atoms per formula unit. $X_{Mg} = Mg/(Mg + Fe)$

Run	EFSP2	SPT2	EFSP3	SPT3	SPT4	SPT5
P (GPa)	3.8	3.8	3.8	3.8	4.6	4.2
T (°C)	800	775	900	1000	700	900
SiO_2	41.02 (1.35)	41.26 (0.30)	40.41 (0.78)	41.96 (0.29)	40.37 (0.31)	40.07 (0.09)
Al_2O_3	22.84 (0.45)	22.42 (0.23)	22.38 (0.73)	23.13 (0.38)	22.91 (0.38)	23.66 (0.38)
Fe_2O_3	0.00 (0.00)	0.00 (0.00)	0.80 (0.96)	0.33 (0.46)	0.32 (0.31)	0.41 (0.22)
FeO	10.69 (2.28)	13.22 (1.00)	16.82 (3.15)	7.95 (0.81)	11.39 (1.68)	10.35 (0.77)
MgO	12.46 (1.31)	9.82 (0.79)	9.67 (1.59)	14.43 (0.18)	6.71 (1.18)	8.47 (0.41)
CaO	12.17 (0.60)	14.04 (0.32)	10.48 (0.84)	12.57 (0.44)	19.28 (1.51)	17.27 (0.24)
Na_2O	0.02 (0.02)	0.01 (0.01)	0.09 (0.08)	0.06 (0.04)	0.02 (0.02)	0.05 (0.06)
Total	99.21 (3.02)	100.78 (0.57)	100.65 (2.53)	100.43 (0.80)	101.00 (0.60)	100.27 (0.33)
Si	3.023 (0.046)	3.042 (0.009)	3.014 (0.050)	3.018 (0.018)	2.992 (0.011)	2.962 (0.013)
Al	1.984 (0.020)	1.949 (0.012)	1.967 (0.037)	1.961 (0.021)	2.001 (0.025)	2.061 (0.024)
Fe^{3+}	0.000 (0.000)	0.000 (0.000)	0.044 (0.054)	0.018 (0.024)	0.018 (0.017)	0.023 (0.012)
Fe^{2+}	0.658 (0.138)	0.816 (0.064)	1.051 (0.211)	0.478 (0.047)	0.706 (0.105)	0.640 (0.050)
Mg	1.370 (0.159)	1.079 (0.083)	1.074 (0.175)	1.548 (0.021)	0.740 (0.125)	0.934 (0.041)
Ca	0.961 (0.046)	1.110 (0.030)	0.837 (0.067)	0.969 (0.037)	1.532 (0.128)	1.368 (0.019)
Na	0.003 (0.003)	0.002 (0.001)	0.012 (0.011)	0.008 (0.005)	0.003 (0.003)	0.008 (0.009)
X_{Mg}	0.675 (0.071)	0.569 (0.038)	0.496 (0.079)	0.758 (0.019)	0.504 (0.062)	0.585 (0.027)

Run	SPT6	FB6	SPT7	SPT8	SPT9	SPT10
P (GPa)	4.2	4.0	3.8	4.2	4.2	4.2
T (°C)	850	870	900	900	850	900
SiO_2	40.84 (0.60)	40.80 (0.53)	40.19 (0.42)	40.58 (0.42)	40.34 (0.41)	40.14 (0.23)
Al_2O_3	23.19 (0.90)	22.81 (0.44)	22.92 (0.42)	23.50 (0.18)	22.09 (0.56)	22.93 (0.25)
Fe_2O_3	0.27 (0.58)	0.00 (0.00)	0.42 (0.54)	0.38 (0.40)	0.03 (0.10)	0.26 (0.22)
FeO	10.85 (0.87)	10.79 (1.92)	14.20 (1.31)	10.99 (0.61)	15.28 (1.58)	13.06 (1.53)
MgO	8.11 (0.40)	8.68 (2.18)	7.63 (0.72)	8.42 (0.34)	5.54 (0.71)	7.50 (1.12)
CaO	17.45 (0.62)	16.58 (1.35)	15.38 (0.75)	16.92 (0.52)	17.00 (1.08)	16.11 (0.29)
Na_2O	0.08 (0.03)	0.06 (0.04)	0.07 (0.06)	0.04 (0.04)	0.11 (0.10)	0.06 (0.04)
Total	100.79 (0.61)	99.73 (0.67)	100.81 (0.68)	100.83 (0.37)	100.39 (0.38)	100.06 (0.18)
Si	3.012 (0.038)	3.032 (0.031)	2.991 (0.024)	2.984 (0.026)	3.040 (0.024)	2.989 (0.007)
Al	2.016 (0.075)	1.997 (0.017)	2.010 (0.030)	2.036 (0.016)	1.962 (0.042)	2.013 (0.009)
Fe^{3+}	0.015 (0.032)	0.000 (0.000)	0.024 (0.030)	0.021 (0.022)	0.002 (0.006)	0.015 (0.012)
Fe^{2+}	0.669 (0.052)	0.672 (0.124)	0.884 (0.084)	0.676 (0.037)	0.963 (0.106)	0.814 (0.099)
Mg	0.891 (0.044)	0.960 (0.224)	0.845 (0.076)	0.922 (0.038)	0.621 (0.075)	0.832 (0.117)
Ca	1.379 (0.054)	1.321 (0.118)	1.226 (0.059)	1.333 (0.040)	1.372 (0.084)	1.286 (0.026)
Na	0.012 (0.005)	0.009 (0.006)	0.010 (0.008)	0.005 (0.006)	0.016 (0.015)	0.008 (0.005)
X_{Mg}	0.566 (0.027)	0.583 (0.096)	0.483 (0.042)	0.570 (0.021)	0.392 (0.049)	0.501 (0.067)

rocks, because of the relatively low dP/dT slope of decarbonation reactions compared to dehydration reactions (Figure 4 in Poli and Schmidt 2002). Although experimental data reveal major complexities in carbonate phase diagrams, and natural occurrences highlight extreme compositional variability, these series have been mostly regarded as simple binary mangesite-siderite, and dolomite-ankerite solid solutions, the latter eventually coexisting with calcite/aragonite. Thermodynamic modeling of solids suffers from a very limited set of experimental data, especially in Fe-bearing systems; both the topology and the location of the liquidus surface in the ternary system is entirely unknown. The role of additional components (silica, alkali, H_2O) on the liquidus

Table 5 Chemical composition of dolomite. Units are weight percent of oxides and atoms per formula unit

Run	SPT1	EFSP2	SPT2	SPT5	ST9
P (GPa)	3.7	3.8	3.8	4.2	4.2
T (°C)	750	800	775	900	850
FeO	4.44 (0.67)	4.04 (0.76)	4.51 (0.14)	3.66 (0.39)	0.43 (0.15)
MgO	18.14 (0.16)	17.82 (3.18)	18.58 (0.47)	17.78 (1.67)	20.59 (0.21)
CaO	28.45 (1.39)	30.37 (1.85)	29.57 (0.76)	31.25 (0.69)	30.93 (0.84)
Na_2O	0.12 (0.11)	0.10 (0.05)	0.17 (0.09)	0.04 (0.02)	0.01 (0.01)
CO_2	48.24 (0.01)	47.52 (3.30)	47.37 (0.94)	46.36 (1.71)	47.08 (0.68)
Total	99.39 (0.37)	99.84 (6.55)	100.19 (2.29)	99.10 (3.20)	99.05 (1.55)
Fe^{2+}	0.113 (0.017)	0.104 (0.017)	0.117 (0.004)	0.097 (0.013)	0.011 (0.004)
Mg	0.821 (0.007)	0.812 (0.113)	0.857 (0.031)	0.836 (0.048)	0.955 (0.015)
Ca	0.926 (0.044)	1.005 (0.044)	0.980 (0.011)	1.059 (0.036)	1.031 (0.017)
Na	0.007 (0.007)	0.006 (0.004)	0.010 (0.005)	0.002 (0.001)	0.000 (0.000)
C	2.000 (0.000)	2.000 (0.000)	2.000 (0.000)	2.000 (0.000)	2.000 (0.000)

surface introduces further complexities, far from being quantitatively addressable at present. Experimental results obtained here nonetheless introduce some major constraint on the topology of phase relationships at the solidus for hydrous carbonatitic melts.

Subsolidus relations in the system $CaCO_3$-$MgCO_3$-$FeCO_3$ have been computed using the Perple_X software (Connolly 1990, 2005) and an internally consistent thermodynamic data set (Holland and Powell 1998; and update). The recently developed solid solution model for ternary Ca-Fe-Mg carbonates (Franzolin et al. 2011) was used to determine the compositional variations, the transition of carbonate minerals, and the effect of Fe on the topology of phase diagrams. Figure 8 was calculated at a pressure of 3.5 GPa because at such conditions, the possible topologies arising from the complexity in the shape of the solvus for ternary carbonates are better displayed.

Table 6 Chemical composition of magnesite Units are weight percent of oxides and atoms per formula unit

Run	SPT1	EFSP2	SPT2
P (GPa)	3.7	3.8	3.8
T (°C)	750	800	775
FeO	9.10 (0.77)	7.04 (3.15)	7.32 (0.67)
MgO	38.27 (1.67)	40.87 (2.97)	39.38 (0.79)
CaO	2.40 (0.59)	2.09 (0.79)	2.73 (0.52)
Na_2O	0.12 (0.14)	0.06 (0.06)	0.05 (0.06)
CO_2	50.75 (1.31)	51.04 (0.74)	51.33 (1.11)
Total	100.68 (2.59)	101.12 (1.07)	100.85 (2.88)
Fe^{2+}	0.110 (0.008)	0.085 (0.039)	0.087 (0.006)
Mg	0.823 (0.024)	0.874 (0.053)	0.838 (0.032)
Ca	0.037 (0.008)	0.032 (0.013)	0.042 (0.007)
Na	0.003 (0.004)	0.002 (0.002)	0.002 (0.002)
C	1.000 (0.000)	1.000 (0.000)	1.000 (0.000)

At low temperature (below 700 °C, Fig. 8a), the extent of solution between magnesite-siderite and dolomite-ankerite series is very limited. At low X_{Mg}, the phase diagram is characterized by the three-carbonate assemblage siderite-dolomite-aragonite. The phase diagram therefore accounts for the occurrence of dolomite and dolomite + magnesite in experiments on mafic compositions (this study, and Poli et al., 2009) in lawsonite-bearing eclogites. Between 700 and 900 °C, the one phase field for dolomite-ankerite-Mg-calcite solid solution largely expands.

The key observation in all experiments at 900 °C is the occurrence of two carbonate solids, aragonite, and dolomite in forward experiment SPT5, and aragonite and (Mg,Fe)-calcite in sandwich experiment SPT8, coexisting with the hydrous carbonatitic liquid. In this temperature region, phase relationships in the ternary region dolomite-ankerite-aragonite are complex. The solvus of carbonates with R-3C structure resembles that of feldspar, i.e., with large immiscibility field along only one of the three binaries, namely $CaCO_3$-$CaMg(CO_3)_2$; such immiscibility coupled to the stability of aragonite over pure calcite at high pressure leads to the possible occurrence of three carbonates at 870 °C (Fig. 8b) and high X_{Mg}, i.e., aragonite + Mg-calcite + dolomite. The composition of the hydrous carbonatitic liquids obtained at 900 °C, both for forward and sandwich geometries, projects in the upper portion of this compositional space, close to the Mg-calcite composition expected to be in equilibrium with aragonite and dolomite (compare Fig. 8c and 8b). Again, the phase diagram recalls that of ternary feldspar. As long as two carbonates coexist with the melt,

Table 7 Chemical composition of (Mg,Fe)-calcite. Units are weight percent of oxides and atoms per formula unit

Run	SPT3	FB6	SPT7	SPT8	SPT10
P (GPa)	3.8	4.0	3.8	4.2	4.2
T (°C)	1000	870	900	900	900
FeO	2.40 (0.41)	2.77 (0.87)	2.34 (0.17)	2.69 (0.17)	3.42 (0.37)
MgO	13.92 (0.36)	6.61 (1.29)	5.64 (0.53)	8.39 (0.45)	10.40 (0.50)
CaO	37.07 (0.74)	45.47 (2.49)	47.18 (0.85)	42.90 (1.45)	40.86 (1.26)
Na_2O	0.07 (0.05)	0.06 (0.02)	0.06 (0.02)	0.07 (0.06)	0.03 (0.02)
CO_2	45.98 (0.70)	44.92 (2.83)	44.97 (0.64)	44.95 (0.37)	45.69 (0.83)
Total	99.44 (1.51)	99.82 (2.32)	100.18 (1.39)	99.01 (0.93)	100.39 (1.84)
Fe^{2+}	0.064 (0.010)	0.075 (0.021)	0.064 (0.004)	0.073 (0.004)	0.092 (0.010)
Mg	0.661 (0.011)	0.321 (0.054)	0.274 (0.023)	0.408 (0.022)	0.497 (0.022)
Ca	1.266 (0.023)	1.590 (0.052)	1.647 (0.026)	1.498 (0.040)	1.403 (0.025)
Na	0.004 (0.003)	0.004 (0.001)	0.004 (0.001)	0.004 (0.004)	0.002 (0.001)
C	2.000 (0.000)	2.000 (0.000)	2.000 (0.000)	2.000 (0.000)	2.000 (0.000)

the liquidus surface has to form a univariant line descending toward Fe-rich compositions. However, when the solidus assemblage in equilibrium with the liquid line of descent leaves the solvus, this univariant reaction degenerates to a minimum valley in the liquidus surface (schematically when the line becomes dashed in Fig. 8c), as in feldspars close to the albite-K-feldspar binary (see Figure 10.28 in Philpotts and Ague, 2009).

Table 8 Chemical composition of solids precipitated on quenching experiments. Units are weight percent of oxides and atoms per formula unit for carbonate. CIPW norm is given for silicate glass

	Quench carbonate			Silicate glass
Run	SPT5	SPT8	Run	SPT8
P (GPa)	4.2	4.2	*P* (GPa)	4.2
T (°C)	900	900	*T* (°C)	900
SiO_2	1.79 (0.70)	1.79 (1.81)	SiO_2	65.89 (3.32)
Al_2O_3	1.41 (0.18)	0.38 (0.21)	Al_2O_3	12.22 (1.68)
FeO	5.38 (0.39)	4.22 (0.43)	FeO	0.60 (0.77)
MgO	3.76 (0.26)	6.53 (0.35)	MgO	0.41 (0.04)
CaO	40.16 (0.85)	40.22 (1.43)	CaO	2.85 (0.65)
Na_2O	2.26 (0.07)	2.27 (1.23)	Na_2O	2.31 (1.90)
CO_2	44.68 (0.43)	46.25 (1.08)	-	
Total	99.42	101.66	Total	84.28 (3.87)
Si	0.06 (0.02)	0.06 (0.06)	Q	51.38
Al	0.05 (0.01)	0.01 (0.01)	ab	25.19
Fe^{2+}	0.15 (0.01)	0.11 (0.01)	an	17.17
Mg	0.18 (0.01)	0.31 (0.02)	cor	4.28
Ca	1.41 (0.04)	1.36 (0.02)	hy	1.38
Na	0.14 (0.01)	0.14 (0.07)	mt	0.08
C	2.00 (0.00)	2.00 (0.00)	hem	0.52

Whether this univariant melting reaction has the topology of co-precipitation (cotectic), or a reversal to resorptional (peritectic) occurs with decreasing X_{Mg}, it is at present impossible to ascertain.

The composition of hydrous carbonatitic liquids found here closely approach those measured by Hammouda (2003) in a study on high-pressure melting of a carbonated eclogite, at 6.5–10 GPa, 1000 °C. At lower pressure, 3.5–5 GPa, similar results were obtained by Kiseeva et al. (2012) at 1100 °C. Starting materials of both experimental studies were characterized by the presence of minor H_2O, 0.12 wt.% in Hammouda (2003), 0.5 wt.% in Kiseeva et al. (2012). Again, a common feature of their syntheses and the results presented here is the crystallization of aragonite or Mg-calcite (83–92 mol% $CaCO_3$ in Kiseeva et al. 2012) with the carbonatitic liquid. Because of the differences in bulk compositions (Table 1), and pressure and temperature conditions, H_2O is expected to play a fundamental role in modifying the liquids surface of these carbonatitic liquids, as observed in silicate systems.

Anhydrous experiments on carbonated eclogites (Dasgupta et al., 2004; Dasgupta et al. 2005, Yaxley and Brey 2004) produced carbonatitic liquids with a much lower Ca/(Ca + Mg + Fe) in equilibrium with dolomite-Mg-calcite solid solution approximately at 60–70 mol% $CaCO_3$ (Yaxley and Brey 2004) or magnesite (Dasgupta et al. 2004, above 4.1 GPa). These liquid compositions therefore schematically approach the minimum on the anhydrous liquidus surface as constrained by the end-member binary system $MgCO_3$ – $CaCO_3$ (Irving and Wyllie 1975; Byrnes and Wyllie 1981), neglecting for simplicity the influence of iron and alkali in lowering the absolute temperature of melting (Dasgupta et al., 2005). Above 1100 °C (Fig. 8d), complete

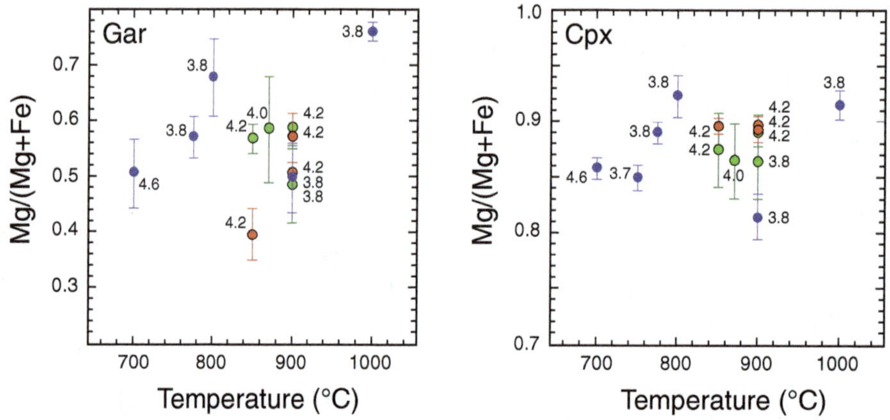

Fig. 7 Mg/(Mg + Fe) variations in garnet and clinopyroxene with temperature. *Blue dots*, forward experiments, bulk 593; *green*, forward experiments, bulk 593ccb; *red*, sandwich experiments

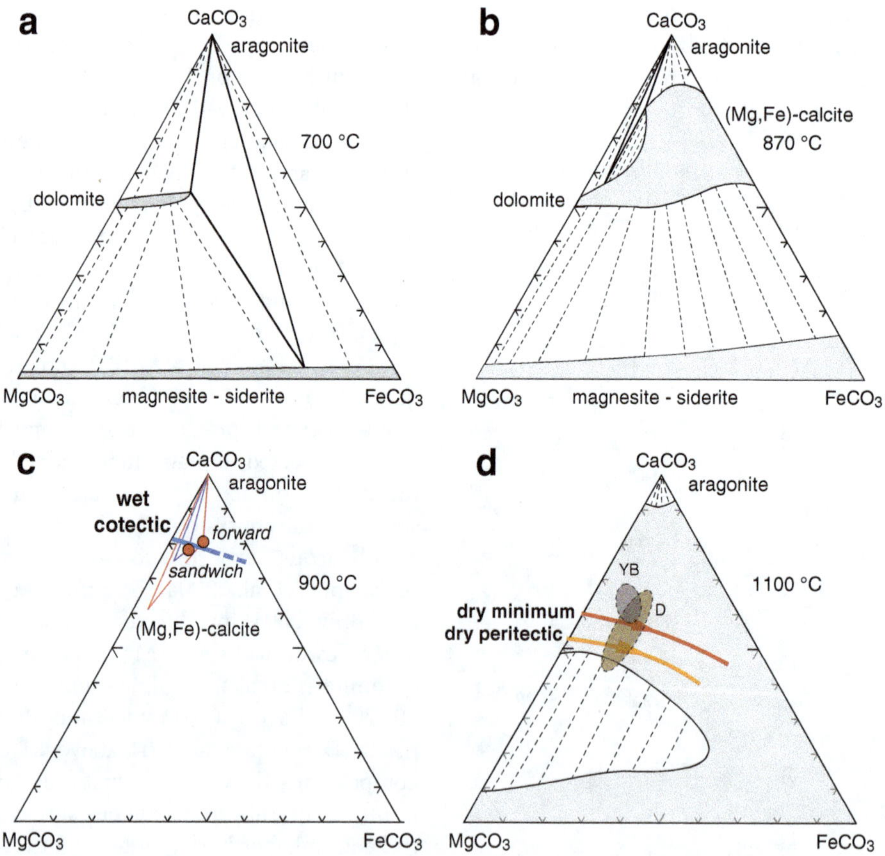

Fig. 8 Phase diagrams for ternary carbonates. **a** and **b** calculated phase diagram at 700 and 870 °C, respectively; **c** experimentally derived phase assemblages at 900 °C and schematic arrangement of a wet cotectic line based on the composition of liquids (*red circles*) retrieved from forward and sandwich experiments; **d** calculated high-temperature phase diagram compared with the composition of liquids from anhydrous experiments (D, Dasgupta et al. 2004, 2005; YB, Yaxley and Brey 2004)

miscibility occurs in the field dolomite-ankerite-calcite; therefore, a single Ca-carbonate (eventually with addition of minor magnesite) is expected with the liquid. At higher pressures, $P > 9$ GPa, Kiseeva et al. (2013) found that the anhydrous carbonatitic solidus is located at approximately 1300 °C. Liquids produced coexist with fully decarbonated residues, and their composition therefore reflect the relatively high Ca/(Ca + Mg + Fe) of the starting material, compared to previous studies (Dasgupta et al., 2004; Dasgupta et al. 2005, Yaxley and Brey 2004).

H_2O was found to cause a dramatic lowering of the liquidus surface for $CaCO_3$ (Wyllie and Boettcher, 1969), and for $CaCO_3$ + SiO_2 (Boettcher and Wyllie 1969; Huang et al., 1980). At 3 GPa, calcite melting occurs at 1600 °C (Irving and Wyllie, 1975), whereas addition of excess H_2O leads to the formation of a eutectic at 530–540 °C, with calcite, portlandite, and an aqueous fluid on the solidus. Similarly, vapor-absent melting of calcite + quartz at 3 GPa was found at 1480 °C (Huang et al., 1980) and wet melting at 530 °C (Boettcher and Wyllie 1969). Notably, the influence of H_2O on the melting of magnesite is not as pronounced, eutectic melting at 3 GPa given by magnesite + periclase + vapor = liquid at approximately 1400 °C, and above 5 GPa, by magnesite + brucite + vapor = liquid at 1150 °C (Ellis and Wyllie, 1979). This asymmetry in the response of $CaCO_3$-rich liquids to the addition of H_2O compared to $MgCO_3$-rich liquids should therefore cause a progressive compositional shift of the minimum of the liquidus surface toward $CaCO_3$.

The solidus for carbonated epidote eclogites

The composition of the putative single liquid generated in experiments at 4.2 GPa, 900 °C, precipitating (Mg,Fe)-calcite dendrites intermingled with tonalitic glass, reveal the importance of both carbonates and Al-rich silicates on the solidus for carbonated eclogites. The stoichiometry of reactions accounting for melting can be described in the model system CaO-MgO-Al_2O_3-SiO_2-H_2O-CO_2 (CMASHC) (Fig. 9a), assuming, for simplicity, that the CO_2 content of the liquid balances Ca and Mg, and that H_2O content is arbitrarily set to 1/3 of that of CO_2. The latter assumption simply affects the stoichiometric coefficient of "vapor" (H_2O in the present exercise) in the reaction set. The liquid composition is given by 0.30 CaO, 0.10 MgO, 0.05 SiO_2, 0.01 Al_2O_3, 0.14 H_2O, and 0.40 CO_2.

The subsolidus breakdown of epidote is given by the reaction (in parentheses, the phase is absent, following Schreinemakers' convention).

$$\begin{aligned} 6\text{ zoisite} + 6\text{ dolomite} = {} & 12\text{ aragonite} \\ & + 4\text{ garnet (gr50)} \\ & + 2\text{ kyanite} + 4\text{ coesite} \\ & + 3\text{ vapor (liq)}. \end{aligned}$$

Eutectic melting is defined by the degenerate reaction:

$$\begin{aligned} 1\text{ kyanite} + 4\text{ coesite} + 10\text{ dolomite} + 20\text{ aragonite} \\ + 14\text{ vapor} = 100\text{ liquid (zo, gr)}. \end{aligned}$$

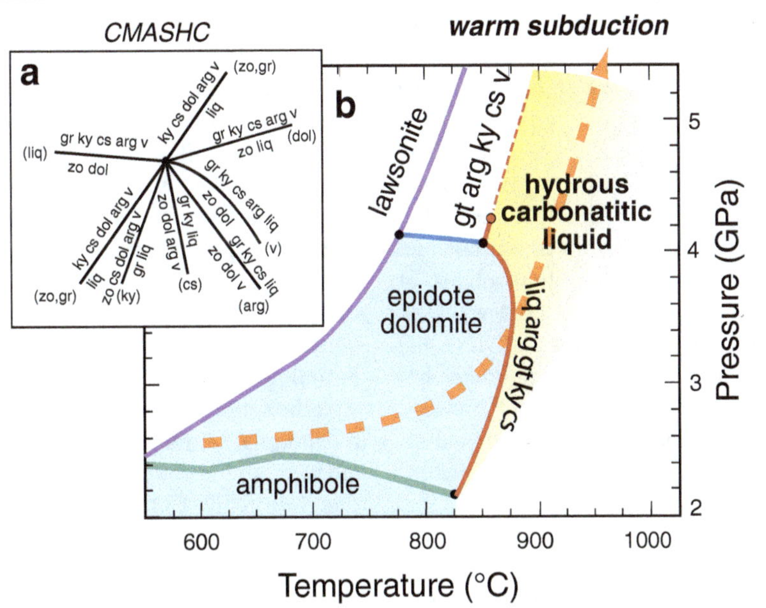

Fig. 9 The arrangement of melting reactions for carbonated epidote eclogites. **a** Schreinemakers' analysis for the system CaO-MgO-Al_2O_3-SiO_2-H_2O-CO_2 (CMASHC); **b** Experimentally derived phase diagram build on the reaction set illustrated in **a**; P-T path for a warm subduction zone is after Syracuse et al. (2010)

The wet melting for carbonated zoisite-bearing eclogites can be represented by the following reaction:

$$6 \text{ zoisite} + 4 \text{ coesite} + 26 \text{ dolomite} + 28 \text{ aragonite} + 25 \text{ vapor} = 4 \text{ garnet} + 200 \text{ liquid } (\text{ky}).$$

The most relevant reaction accounting for the generation of melt is the "vapor-absent" reaction:

$$84 \text{ zoisite} + 114 \text{ dolomite} = 56 \text{ garnet} + 25 \text{ kyanite} + 44 \text{ coesite} + 108 \text{ aragonite} + 300 \text{ liquid } (\nu),$$

accounting for melting observed in experiment SPT 8 at 4.2 GPa, 900 °C.

The Schreinemakers' analysis of the four reactions above, and of the additional four reactions defined in the six-component system CMASHC, can be build on the model reactions in the system CASH, controlling the phase relationships of zoisite at high pressure and high temperature (Schmidt and Poli, 2004). Figure 9a shows the schematic arrangement of these reactions in pressure and temperature space. The pseudo-invariant point for the compositional pseudo-section of gabbroic eclogite investigated here is located at approximately 4 GPa, 870–900 °C, and the temperature range between eutectic reaction (zo, gar) and vapor-absent reaction (ν) is expected to be very limited, given the location of eutectic melting in CASH given by reaction (2) (Fig. 5).

The minimum temperatures found for carbonatitic liquids in a variety of mafic bulk compositions range from 1020–1080 °C, 3–5 GPa, in anhydrous carbonated MORB eclogites (Dasgupta et al. 2004), 1180–300 °C at 3–4 GPa in a silica-poor carbonated eclogite (Yaxley and Brey 2004), 1100 °C at 3–5 GPa in Kiseeva et al. (2012), and 1000 °C at 6.5–10 GPa in Hammouda (2003). The latter two studies contain H_2O in the starting material and its effect on the composition of the liquids has been discussed above. In carbonated pelites (i.e., H_2O-bearing), carbonatitic melts have been found to exsolve from the silicate liquid above 1100 °C at 3–5 GPa (Thomsen and Schmidt 2008; Skora et al., 2015). Hydrated, carbonated peridotites again show carbonatites on the solidus at 1070 °C at 2–3 GPa (Tumiati et al. 2013). All of these bulk compositions are characterized by a Ca/(Ca + Mg + Fe) ratio lower than 0.5; therefore, the >150 °C difference between minimum temperatures found here and previous studies should be ascribed primarily to the combined effect of H_2O and high Ca/(Ca + Mg + Fe) values, promoting discrete amounts of eutectic carbonatitic liquids.

Conclusions

Altered cumulate gabbros represent kilometer-thick portions of the subducted oceanic crust. Hydration and carbonation occur not only in ocean floor metamorphism, but they are also expected to further progress at depth at the trench–rise system in response to bending-related fracturing and faulting of the subducting lithosphere (Faccenda 2014). Plagioclase-rich rocks should be therefore regarded as volatile reservoirs as important as the altered oceanic basalts or the overlying sedimentary cover, which is often limited to a few hundreds meters in thickness (Whittaker et al. 2013).

CO_2 and H_2O are recycled back into the mantle mostly by carbonates (the role of graphite is neglected here) and by hydrous minerals, of which epidote is the most important in mafic bulk compositions for thermal regimes exceeding lawsonite stability. The stability field of epidote is maximized in mafic bulk compositions originally enriched in anorthitic plagioclase, and reaches 3.8–4 GPa in phase relations presented here. Epidote, dolomite, and kyanite occur on the solidus of mafic eclogites in a large pressure range. Calculated pressure-temperature paths for the subducted slab in relatively warm subduction zones (Fig. 9b; see also Figure 6 in Syracuse et al., 2010) suggest that the altered gabbroic rocks of the oceanic crust may undergo partial melting at sub-arc depth. Liquids generated at 870–900 °C are hydrous, with both carbonatitic and silicate components, whose proportions are a function of the H_2O available. H_2O component can be provided internally by epidote breakdown similarly to mica fluid-absent melting reactions, or by an external source, even beyond epidote stability field, assuming that the oceanic crust is flushed by aqueous fluids derived from the deserpentinization of the underlying harzburgitic rocks. Such liquids should be more common if actual temperatures in subduction zones are higher than previously expected from numerical modeling (Penniston-Dorland et al. 2015).

Thermodynamic reconstruction of the phase diagram for ternary, Ca-Mg-Fe carbonates, along the hydrous carbonatitic solidus discussed here, reveals that two carbonates, dolomite and aragonite/Mg-calcite, can coexist on the liquidus surface. As a result, at such relatively low temperatures a Ca-rich carbonate is expected to be a common product of eclogite melting, for a wide range of starting bulk compositions. The stability of a Ca-rich carbonate over the first melting interval for carbonatitic liquids should therefore affect the fractionation paths of both liquids and residues on segregation processes.

Carbonatitic liquids, enriched in Ca, Si, volatiles, and secondarily Al, migrate in the mantle wedge, react with olivine, and produce clinopyroxene and magnesite, therefore fixing CO_2 back again temporarily in the mineral assemblage. If excess H_2O does not fractionate from the precipitated carbonate, then the downgoing flow in the hanging wall of the slab drags these carbonated pyroxenites above 950–1050 °C isotherm where the hydrous carbonated ultramafics melt (Wallace and Green 1988, Tumiati et al 2013), therefore generating a derivate hydrous carbonatitic liquid of dolomitic composition. It is highly unlikely that such liquids reach shallow levels to give carbonatitic rocks; they are rather expected to progressively dilute in the primary arc melts and to provide a "pyroxenitic" geochemical signature for the source region of arc magmas.

Abbreviations

AMP: Amphibole; AN: Anorthite; ARG: Aragonite; CARB: Carbonate/carbonatitic; CC: Calcite; COE: Coesite; COR: Corundum; CPX: Clinopyroxene; GR: Grossular; GT: Garnet; KY: Kyanite; LIQ: Liquid; MG-CC: Mg-calcite; MGS: Magnesite; Q: Quartz; Sil: Silicate; V: Vapor; ZO: Zoisite

Acknowledgements

Reviews by two anonymous referees significantly improved the manuscript.

Funding

This research was supported by the Italian Ministry of Education, University and Research (MIUR) program PRIN2012R33ECR. S.P. acknowledges support from the Deep Carbon Observatory (DCO).

Competing interest

The author declares that he has no competing interests.

References

Alt JC, Emmermann R (1985) Geochemistry of hydrothermally altered basalts: Deep Sea Drilling Project hole 504B, Leg 83. In Rep DSDP 83:249–262

Alt JC, Teagle DAH (1999) The uptake of carbon during alteration of ocean crust. Geochim Cosmochim Acta 63:1527–1535

Bach W, Alt JC, Niu YL, Humphris SE, Erzinger J, Dick HJB (2001) The geochemical consequences of late-stage low-grade alteration of lower ocean crust at the SW Indian Ridge: results from ODP hole 735B (Leg 176). Geochim Cosmochim Acta 65:3267–3287

Boettcher AL (1970) The system $CaO\text{-}Al_2O_3\text{-}SiO_2\text{-}H_2O$ at high pressures and temperatures. J Petrol 11:337–379

Boettcher AL, Wyllie PJ (1969) The system $CaO\text{-}SiO_2\text{-}CO_2\text{-}H_2O$ - III. Second critical end-point on the melting curve. Geochim Cosmochim Acta 33:611–632

Bucher K, Grapes R (2009) The eclogite-facies Allalin gabbro of the Zermatt-Saas Ophiolite, Western Alps: a record of subduction zone hydration. J Petrol 50: 1405–1442

Byrnes AP, Wyllie PJ (1981) Subsolidus and melting relations for the join $CaCO_3$-$MgCO_3$ at 10 kbars. Geochim Cosmochim Acta 45:321–328

Connolly JAD (1990) Multivariable phase diagrams: an algorithm based on generalized thermodynamics. Am J Sci 290:666–718

Connolly JAD (2005) Computation of phase equilibria by linear programming: a tool for geodynamic modelling and its application to subduction zone decarbonation. Earth Planet Sci Lett 236:524–541

Dasgupta R, Hirschmann MM, Withers AC (2004) Deep global cycling of carbon constrained by the solidus of anhydrous, carbonated eclogite under upper mantle conditions. Earth Planet Sci Lett 227:73–85

Dasgupta R, Hirschmann MM, Dellas N (2005) The effect of bulk composition on the solidus of carbonated eclogite from partial melting experiments at 3 GPa. Contrib Mineral Petrol 149:288–305

Ellis DE, Wyllie PJ (1979) Carbonation, hydration, and melting relations in the system $MgO\text{-}H_2O\text{-}CO_2$ at pressures up to 100 kbar. Am Mineral 64:32–40

Faccenda M (2014) Water in the slab: a trilogy. Tectonophysics 614:1–30

Forneris JF, Holloway JR (2003) Phase equilibria in subducting basaltic crust: implications for H_2O release from the slab. Earth Planet Sci Lett 214:187–201

Franzolin E, Schmidt MW, Poli S (2011) Ternary Ca-Fe-Mg carbonates: subsolidus phase relations at 3.5 GPa and a thermodynamic solid solution model including order/disorder. Contrib Mineral Petrol 161:213–227

Fumagalli P, Poli S (2005). Experimentally determined phase relations in hydrous peridotites to 6.5 GPa and their consequences on the dynamics of subduction zones. Journal of Petrology 46:555–578.

Godard M, Awaji S, Hansen H, Hellebrand E, Brunelli D, Johnson K, Yamasaki T, Maeda J, Abratis M, Christie D, Kato Y, Mariet C, Rosner M (2009) Geochemistry of a long in situ section of intrusive slow-spread oceanic lithosphere: results from IODP site U1309 (Atlantis Massif, 30°N Mid-Atlantic-Ridge). Earth Planet Sci Lett 279:110–122

Hammouda T (2003) High-pressure melting of carbonated eclogite and experimental constraints on carbon recycling and storage in the mantle. Earth Planet Sci Lett 214:357–368

Hirschmann MM, Dasgupta R (2007) A modified iterative sandwich method for determination of near-solidus partial melt compositions. I. Theoretical considerations. Contrib Mineral Petrol 154:635–645

Holland TJB, Powell R (1998) An internally consistent thermodynamic data set for phases of petrological interest. J Met Geol 16:309–343

Huang W-L, Wyllie PJ, Nehru CE (1980) Subsolidus and liquidus phase relationships in the system $CaO\text{-}SiO_2\text{-}CO_2$ to 30 kbar with geological applications. Am Miner 65:285–301

Irving AJ, Wyllie PJ (1975) Subsolidus and melting relationships for calcite, magnesite and the join $CaCO_3$-$MgCO_3$ to 36kbar. Geochim Cosmochim Acta 39:35–53

Kerrick DM, Connolly JAD (2001) Metamorphic devolatilization of subducted oceanic metabasalts: implications for seismicity, arc magmatism and volatile recycling. Earth Planet Sci Lett 189:19–29

Kessel R, Ulmer P, Pettke T, Schmidt MW, Thompson AB (2005) The water-basalt system at 4 to 6 GPa: phase relations and the second critical endpoint in a K-free eclogite at 700–1400 °C. Earth Planet Sci Lett 237:873–892

Kiseeva ES, Yaxley GM, Hermann J, Litasov KD, Rosenthal A, Kamenetsky VS (2012) An experimental study of carbonated eclogite at 3.5–5. 5 GPa—implications for silicate and carbonate metasomatism in the Cratonic mantle. J Pet 53:727–759

Kiseeva ES, Litasov KD, Yaxley GM, Ohtani E, Kamenetsky VS (2013) Melting and phase relations of carbonated eclogite at 9–21 GPa and the petrogenesis of alkali-rich melts in the deep mantle. J Pet 54:1555–1583

Lagabrielle Y, Brovarone AV, Ildefonse B (2015) Fossil oceanic core complexes recognized in the blueschist metaophiolites of Western Alps and Corsica. Earth Sci Rev 141:1–26

Meyer PS, Dick HJB, Thompson G (1989) Cumulate gabbros from the southwest Indian Ridge, 54° S–7°16′ E: implications for magmatic processes at slow spreading ridge. Contrib Mineral Petrol 103:44–63

Pawley AR, Holloway JR (1993) Water sources for subduction zone volcanism—new experimental constraints. Science 260:664–667

Penniston-Dorland SC, Kohn MJ, Manning CE (2015) The global range of subduction zone thermal structures from exhumed blueschists and eclogites: rocks are hotter than models. Earth Planet Sci Lett 428:243–254

Perk NW, Coogan LA, Karson JA, Klein EM, Hanna HD (2007) Petrology and geochemistry of primitive lower oceanic crust from Pito Deep: implications for the accretion of the lower crust at the Southern East Pacific Rise. Contrib Mineral Petrol 154:575–590

Perugini D, Poli G (2005) Viscous fingering during replenishment of felsic magma chambers by continuous inputs of mafic magmas: field evidence and fluid-mechanics experiments. Geology 33:5–8

Philpotts AR, Ague JJ (2009) Principles of igneous and metamorphic petrology. Cambridge University Press, Cambridge

Poli S (2015) Carbon mobilized at shallow depths in subduction zones by carbonatitic liquids. Nat Geo 8:633–636

Poli S, Schmidt MW (1995) H_2O transport and release in subduction zones—experimental constraints on basaltic and andesitic systems. J Geophys Res 100:22299–22314

Poli S, Schmidt MW (1998) The high-pressure stability of zoisite and phase relationships of zoisite-bearing assemblages. Contrib Mineral Petrol 130:162–175

Poli S, Schmidt MW (2002) Petrology of subducted slabs. Annu Rev Earth Planet Sci 30:207–235

Poli S, Schmidt MW (2004) Experimental subsolidus studies on epidote minerals. Rev Mineral Geochem 56:171–195

Poli S, Franzolin E, Fumagalli P, Crottini A (2009) The transport of carbon and hydrogen in subducted oceanic crust: an experimental study to 5 GPa. Earth Planet Sci Lett 278:350–360

Schmidt MW, Poli S (1998) Experimentally based water budgets for dehydrating slabs and consequences for arc magma generation. Earth Planet Sci Lett 163:361–379

Schmidt MW, Poli S (2004) Magmatic epidote. Rev Mineral Geochem 56:399–430

Skora S, Blundy JD, Brooker RA, Green ECR, de Hoog JCM, Connolly JAD (2015) Hydrous phase relations and trace element partitioning behaviour in calcareous sediments at subduction-zone conditions. J Petrol 56:953–980

Syracuse EM, van Keken PE, Abers GA (2010) The global range of subduction zone thermal models. Phys Earth Planetary Int 183:73–90

Thompson AB, Ellis DJ (1994) CaO + MgO + Al_2O_3 + SiO_2 + H_2O to 35 kbar: amphibole, talc and zoisite dehydration and melting reactions in the silica-excess part of the system and their possible significance in subduction zones, amphibolite melting and magma fractionation. Am J Sci 294:1229–1289

Thomsen TB, Schmidt MW (2008) Melting of carbonated pelites at 2.5–5.0 GPa, silicate-carbonatite liquid immiscibility, and potassium-carbon metasomatism of the mantle. Earth Planet Sci Lett 267:17–31

Tumiati S, Fumagalli P, Tiraboschi C, Poli S (2013) An experimental study on COH-bearing peridotite up to 3.2 GPa and implications for crust-mantle recycling. J Petrol 54:453–479

Yaxley GM, Brey GP (2004) Phase relations of carbonate-bearing eclogite assemblages from 2.5 to 5.5 GPa: implications for petrogenesis of carbonatites. Contrib Mineral Petrol 146:606–619

Wallace ME, Green DH (1988) An experimental determination of primary carbonatite magma composition. Nature 335:343–345

Whittaker J, Goncharov A, Williams S, Dietmar Müller R, Leitchenkov G (2013) Global sediment thickness dataset updated for the Australian-Antarctic Southern Ocean. Geochem Geophys Geosys. doi:10.1002/ggge.20181

Wittenberg A (1997) Geochemische und isotopengeochemische Untersuchungen an mafischen Gesteinen der kontinentalen Unterkruste und deren Hochdruck-äquivalenten (Grospydite). Universität Hannover, Dissertation

Wittenberg A, Rohnert D, Gesing TM, Holtz F (2003) Zoisite—an effective trace element vehicle at HP metamorphic processes. Geochim Comsochim Acta 67(18):A535

Wyllie PJ, Boettcher AL (1969) Liquidus phase relationships in the system CaO-CO_2-H_2O to 40 kilobars pressure with petrological applications. Am J Sci 267-A:489–508

11

Contrasting partition behavior of F and Cl during hydrous mantle melting: implications for Cl/F signature in arc magmas

Célia Dalou[1,2,3,4*], Kenneth T Koga[2,3,4], Marion Le Voyer[5,6] and Nobumichi Shimizu[7]

Abstract

We present the results of five experiments on F and Cl partitioning during hydrous mantle melting under conditions relevant to subduction zone magmatism (1.2–2.5 GPa, 1,180°C–1,430°C). For each experiment, we determined the F and Cl partition coefficients between lherzolitic mineral phases (olivine, orthopyroxene (opx), clinopyroxene (cpx), and garnet), amphibole, and hydrous basaltic melts (0.2–5.9 wt.% dissolved H_2O). At constant pressure, $D_F^{opx/melt}$ and $D_{Cl}^{opx/melt}$ show contrasting response to the combined effects of decreasing temperature from 1,310°C to 1,180°C and increasing H_2O content in the melt from 0.2 to 5.9 wt.%: $D_F^{opx/melt}$. decreases from 0.123 ± 0.004 to 0.021 ± 0.014 while $D_{Cl}^{opx/melt}$ increases from 0.0021 ± 0.0031 to 0.07 ± 0.01. Similar results are observed for clinopyroxene: $D_F^{cpx/melt}$ decreases from 0.153 ± 0.004 to 0.083 ± 0.004 while $D_{Cl}^{cpx/melt}$ increases from 0.009 ± 0.0005 to 0.015 ± 0.0008. Experimentally determined F and Cl partition coefficients were used in a hydrous melting model of a lherzolitic mantle metasomatized by slab fluid. In this model, we vary the amount of metasomatic slab fluid added into the mantle while its composition is kept constant. Increasing the amount of fluid results in an increase of both the degree of melting (due to the effect of H_2O addition) and the F and Cl input in the mantle wedge. Because of the change of F and Cl partition coefficients with the increase of H_2O, the observed variation in the F and Cl contents of the modeled melts is produced not only by F and Cl input from the fluid, but also by the changes in F and Cl fractionation during hydrous melting. Overall, the model predicts that the Cl/F ratio of modeled melts increases with increasing fluid fraction. Therefore, a variation in the amount of fluid added to the mantle wedge can contribute to the variability in Cl/F ratios observed in arc melt inclusions.

Keywords: Fluorine; Chlorine; Partition coefficients; Lherzolite minerals; Hydrous silicate melt; Fluid induced melting; Subduction

Background

Volatile species (H_2O, CO_2, S, Cl, and F) recycling in subduction zones is the main driver of mantle wedge melting and production of arc magmas. Previous studies have shown that primitive arc magmas recorded in olivine-hosted melt inclusions are enriched in volatiles, especially H_2O, compared to mid-ocean ridge basalts and/or their melt inclusions (e.g. Hilton et al. 2002, Straub and Layne 2003; Wallace 2005). The addition of volatiles in the mantle wedge,via H_2O-rich fluids and/or melts derived from the downgoing slab facilitates partial melting in the mantle wedge by lowering the solidus temperature of mantle peridotites (e.g. Kushiro et al. 1968; Kushiro 1969; 1972; 1990; Mysen and Boettcher 1975; Gaetani et al. 1994) and affects the physical and chemical properties of arc magmas (e.g. Shaw 1963; 1972; Green 1973; Mysen 1977; Sisson and Grove 1993; Ochs and Lange 1999; Bercovici and Karato 2003; Hirschmann 2006). The composition of the metasomatized mantle wedge can vary greatly according to the nature (aqueous fluids formed from by the dehydration of the slab, or silicate melts, formed by the melting of the slab) and the amounts of slab-derived components added to the mantle (e.g. Ayers 1998; Grove et al. 2002;

* Correspondence: cdalou@jsg.utexas.edu
[1]Department of Geological Sciences, Jackson School of Geosciences, The University of Texas at Austin, 1 University Station C1100, Austin, TX 78712, USA
[2]Université Blaise Pascal, Laboratoire Magmas et Volcans, Clermont Université, Clermont-Ferrand, BP 10448, F-63000, France
Full list of author information is available at the end of the article

Rüpke et al. 2002; Straub et al. 2004). As the slab-derived components are still largely unconstrained, quantifying the volatile input from the slab remains a challenge.

Along with H_2O, halogen elements such as F and Cl are also released from the slab and are enriched in the mantle wedge (Métrich et al. 1999; Straub and Layne 2003). In contrast to H_2O and other volatile species, F and Cl are not significantly degassed from mafic magmas at the depth of melt inclusion formation, if at all (Carroll and Webster 1994; Métrich and Wallace 2008), because of their comparatively high solubility and low concentration in mafic magmatic liquid (Webster 2004; Dalou et al. 2014). Moreover, olivine-hosted melt inclusion studies have shown that F and Cl, unlike H_2O, are less affected by post-entrapment diffusion through the host mineral (e.g. Gaetani et al. 2012; Bucholz et al. 2013; Lloyd et al. 2013; Le Voyer et al. 2014). Therefore, the F and Cl contents recorded by arc melt inclusions represent the product of fractional crystallization from primary melts, and thus they retain valuable information on the source composition of arc magmas.

Existing F and Cl data from basaltic melt inclusions recovered from subduction zone settings (e.g. Straub and Layne 2003; Wallace 2005; Le Voyer et al. 2008; 2010; Bouvier et al. 2008; 2010; Sadofsky et al. 2008; Rose-Koga et al. 2012) has revealed a high variability of F and Cl signatures (presented as Cl/F ratio in this paper) among volcanic arcs (Cl/F from approximately 0.5 to 10), but also from a single volcano (Cl/F from 1.1 to 4.1 for Shasta melt inclusions, Le Voyer et al. 2010). One possible explanation for the Cl/F variability in arc melt inclusions is that the addition of an H_2O-rich component derived from slab dehydration into the mantle wedge generates arc basalts with high Cl/F, while the addition of slab melts generates basalts with low Cl/F (Straub and Layne 2003; Le Voyer et al. 2010). This hypothesis is consistent with experimental data on F and Cl partitioning between minerals and aqueous fluids, which have shown that addition of an aqueous fluid to the mantle wedge efficiently enriches its Cl content but not its F content (Bernini et al. 2012; Wu and Koga 2013). Instead, it is suggested that F enrichment is most likely achieved by the percolation of silicate melts (Wu and Koga 2013). Experimentally determined F partition coefficients between mantle minerals (olivine, pyroxenes, and garnet) and anhydrous silicate melts (Beyer et al. 2012; Dalou et al. 2012) allow the calculation of F content in melts produced by simple batch melting of an anhydrous lherzolite (with F =16 ppm, Saal et al. 2002). The addition of those calculated anhydrous silicate melts (maximum F content approximately 400 ppm) into the mantle wedge does not generate high F content arc magmas (e.g. basaltic andesite melt inclusions from Cascades contain up to 1,220 ppm of F). In order to better assess the link between Cl/F signature and slab input, more experimental data on F and Cl partition coefficients under hydrous conditions, i.e. relevant to subduction zone magmatism, are needed. This is the objective of the present study. We address the effect of H_2O on F and Cl partition coefficients between anhydrous minerals (olivine, orthopyroxene (opx), clinopyroxene (cpx), and garnet), amphibole, and hydrous basaltic melts (0.2–5.9 wt.% dissolved H_2O), at mantle wedge conditions (1.2 and 2.5 GPa and 1,180°C and 1,430°C). It is shown that the strong and contrasting effect of H_2O on F and Cl partitioning is key to the production of F-rich melts and to explain the variability of Cl/F ratios in mantle wedge products.

Methods

Starting materials

Three starting compositions (Table 1) were prepared as gel mixtures, following the procedure of Hamilton and Henderson (1968). The two basalts and the peridotite gels were made to be as close as possible in composition to the basalt-peridotite pair of Gaetani and Grove (1998) (basalt 82-72f and peridotite PUM), because a mixed 70:30 proportion of this basalt-peridotite pair allows to obtain liquidus saturation of mantle wedge phases (olivine, orthopyroxene, clinopyroxene, garnet/spinel), while maintaining large melt fractions in both anhydrous and hydrous systems (Gaetani and Grove 1998). The hydrous basalt composition was prepared without Al_2O_3 to

Table 1 Composition of starting materials

	Basalt	Peridotite	Hydrous basalt[a]
SiO_2	47.70(47)	45.97(47)	46.10(45)
TiO_2	0.62(2)	0.18(1)	0.43(1)
Al_2O_3	19.05(21)	14.76(16)	approximately 14.8[a]
Cr_2O_3	0.12(0.3)	0.47(1)	0.10(0.2)
FeO*	7.82(8)	7.54(7)	7.38(7)
MnO	0.15(0.3)	0.13(0.3)	0.14(0.3)
MgO	10.49(3)	37.79(11)	7.32(2)[b]
CaO	11.75(35)	3.21(10)	13.09(39)
Na_2O	2.35(5)	0.33(1)	2.27(7)
K_2O	0.08(0.2)	0.03(0.1)	0.60(2)
P_2O_5	0.01(0.02)	0.03(0.1)	ND
H_2O	N.D.	N.D.	approximately 7.8[a]
Total	100.13	99.99	100.00

Analyses were done by ICP-AES on basalt and peridotite powders. Units in parentheses represent one standard deviation, determined from the ICP-AES standard BHVO-1.

[a]Hydrous basalt composition was analyzed before the addition of $Al(OH)_3$. $Al(OH)_3$ was then added to complete Al_2O_3 contents, so Al_2O_3 and H_2O are calculated compositions likely associated with a larger error (since reported contents are calculated from weighted $Al(OH)_3$).

[b]MgO of basalt is low to allow the addition of MgF_2 and $MgCl_2$.

*FeO** total Fe as FeO.

N.D. not determined.

allow the introduction of H_2O as gibbsite ($Al(OH)_3$, Table 1). This protocol allows a better constraint on H_2O content than the addition of a drop of liquid water for low H_2O content samples. The gibbsite absorbs less atmospheric CO_2 than brucite [$Mg(OH)_2$, more commonly used for hydrous experiments, Béarat et al. 2002]. The gels were conditioned for 12 h at 800°C in a 1 atm gas-mixing furnace, with fO_2 about 1 log unit above the quartz-fayalite-magnetite buffer (QFM). This fO_2 was selected to mimic the oxidation state of sub-arc upper mantle (fO_2 from 0.5 to 1.7 log units above QFM; Parkinson and Arculus 1999).

Starting composition powders were dried for at least 12 h at 110°C to remove any H_2O. Starting compositions were weighed to obtain a 70:30 proportion of basalt-peridotite and the desired H_2O content by adjusting the proportion of H_2O-bearing basalt versus H_2O-free basalt. A drop of liquid water (approximately 1 μL) was also added in the H_2O-rich samples (Table 2) to recover water lost by possible gibbsite destabilization occurring during arc welding. The F and Cl were weighted and added to the mixtures as MgF_2 and $MgCl_2$ to be in trace amounts (approximately 2,000 ppm each). Nevertheless, prior contaminations in F (from the capsules' preparation, discussed in the 'experimental procedure' section) or Cl (from the powder storage in the oven containing NaCl cylinders) make the assessment of the initial F and Cl content difficult. Each mixture was then ground in an agate mortar with ethanol for 45 min. Finally, clinopyroxene seeds were added at the bottom of the sample capsules in experiments *CD1*H0 and CD2H3 (Table 2) to catalyze cpx growth (Dalou et al. 2012).

Experimental procedure

Experiments were conducted at 1.2 and 2.5 GPa and from 1,180°C to 1,430°C on ¾-inch and ½-inch piston cylinders at the Laboratoire Magmas et Volcans (LMV). Four- to six-day durations were chosen for experiments to optimize crystal size and to reach equilibrium (see results for more details on equilibrium considerations, Table 2). To reduce H_2O and Fe losses from the starting material, $Au_{80}Pd_{20}$ double capsules, pre-saturated in Fe, were used. Moreover, because of the high H_2O incompatibility in olivine (e.g. Hauri et al. 2006), San Carlos olivines ground to approximately 10 μm were packed in the outer capsules (Figure 1). The pre-saturation in Fe consists of melting the starting basalt in those capsules, at the hotspot of a vertical gas-mixing furnace at 1,250°C, with fO_2 controlled at 1 log unit below the FMQ buffer, for 72 h. The pre-saturation fO_2 was chosen to be lower than the starting material fO_2, since the basalt used for Fe pre-saturation was the unreduced starting gel and a pre-saturation at higher fO_2, being more efficient (Ratajeski and Sisson 1999), may have caused a Fe increase in samples. After the pre-saturation, the capsules were placed in an HF bath for 48 h at 90°C, followed by three ultrasound baths in distilled water of 1 h each. The capsules were then cleaned in a HNO_3^- at 60°C to dissolve SiF precipitates and finally rinsed with distilled water. This procedure allowed the perfect dissolution of basaltic melt, but revealed that F contaminated the metallic capsules (up to several percent for samples CD2H4, CD2H5, and CD2H6, Table 3). However, this contamination does not significantly affect the determination of F and Cl partition coefficients (see 'Results' and 'Discussion' sections for more details). The pre-saturation procedure minimizes Fe loss from samples (<1.5 wt.%) but sometimes leads to an apparent Fe contamination from the capsule to the sample (Fe contamination up to 1.8 wt.%, Table 2). Overall, this Fe enrichment remains minimal compared to the Fe loss for experiments without pre-saturation (e.g. up to 4 wt.% and 2.9 wt.% average of Fe loss for similar conditions, Kawamoto and Hirose 1994).

A negligible water loss from the inner capsule to the surrounding olivine was measured by ion probe in sample CD2H3 (238 ± 1 ppm H_2O loss, compared with the 2.6 wt.% dissolved in the melt). We also checked potential CO_2 contamination by Fourier transform infrared analysis (LMV) of sample CD2H5. Less than 100 ppm of CO_2 was measured in the melt.

Analytical procedure

Major element concentrations (given in Table 3) were analyzed by a Cameca SX100, LMV electron microprobe. Minerals were analyzed with a focused beam and a 15 nA current, whereas melts were analyzed by using a 8 nA current and a beam defocused to 5–20-μm diameter. The calibration standards, counting times, and choices of spectrometers are described in Hammouda (2003).

The abundance of F, Cl, and H_2O (given in Table 3) was determined at the Woods Hole Oceanographic Institution (WHOI) with a Cameca IMF 1,280 ion probe following the procedure described in Rose-Koga et al. (2008), Shimizu et al. (2009), and Dalou et al. (2012). A Cs + primary ion beam rastered over an area of 30 × 30 μm^2 was used. The primary beam intensity was approximately 1.43 nA during crystal analyses and decreased to 520 pA for melts. The beam size was between 20 × 20 to 10 × 10 μm^2, depending on the size of the phases. At those conditions, the detection limit is <5 ppm for F, <2 ppm for Cl, and <0.01 wt.% for H_2O. To achieve reliable F and Cl concentration measurements: 1) three to ten spots were measured per phase, and 2) only data with stable signals (less than 20% signal variation during analysis) were selected and used to determine element abundances.

The total uncertainty, taking into account both the standard deviation over ten cycles of analyses (i.e. internal error) and the errors on the regression of the calibration line, was approximately 10% for all volatiles.

Table 2 Experimental conditions and phase assemblages in runs from Dalou et al. (2012) and this study

Run	New names[a]	Starting materials[b]	P (GPa)	T (°C)	Duration (days)	H_2O[c]	Run products (%)[d]						ΣR (%)[e]	Fe Exchange[f]
							Olivine	Opx	Cpx	Grt	Amp	Glass		
CC4#7b	*CD1H0*	$\beta + \pi + cs$	1.2	1,310	7	0.2		16(2)	16(19)			68(32)	0.82	−0.4
H3V04	CD2H3	$H\beta + \pi + cs$	1.2	1,240	6	2.6	34(2)	16(3)	13(2)			38(2)	0.15	+1.8
H4V4p	CD2H4	$H\beta + \pi + H_2O^g$	1.2	1,180	5	4.3		6(2)			36(5)	58(4)	0.35	+0.3
H5V4p	CD2H5	$H\beta + \pi + H_2O$	1.2	1,190	5.5	4.9		11(2)	34(3)			55(2)	0.76	+1.5
H6V04	CD2H6	$H\beta + \pi + H_2O^g$	1.2	1,200	4	5.9		23(2)			11(5)	66(5)	0.17	+0.1
C4p#25kb	*CD1*-2.5GPa	$\beta + \pi$	2.5	1,430	5.5	0.8		4(1)	40(1)	28(2)		28(2)	0.64	+0.1
Grt3V4p	CD2-2.5GPa	$H\beta + \pi$	2.5	1,370	4.5	1.6		24(4)	9(9)	37(2)		30(1)	1.58	−1.5

[a]New names are used to identify samples from Dalou et al. (2012) (*CD1*) from samples from this study (CD2).
[b]*β* basalt, *Hβ* hydrous basalt, *π* peridotite, *cs* cpx seeds, *H_2O* drop of liquid H_2O.
[c]Measured concentration of H_2O dissolved in the glass.
[d]*Opx* orthopyroxene, *Cpx* clinopyroxene, *Amp* amphibole, *Glass* quenched basaltic glass.
[e]Square root of an average of phase proportion calculations, modified from Albarède and Provost (1977), to account for the addition of F, H_2O, and Cl in the starting materials (that diluted the initial element concentrations, see section 'Equilibrium considerations' for details). There is a significant uncertainty regarding volatile element initial contents.
FeO was excluded from mass-balance calculations (FeO loss or gain is assessed after the phase proportions are determined).
[f]Difference between the FeO content of the bulk composition used as starting material in the mass balance (diluted with F and H_2O content) and the FeO in the sample bulk composition as determined by mass balance calculation, in relative weight percent. Positive value indicates apparent Fe gain, while negative value indicates apparent Fe loss.
[g]Capsules F contaminated during HF cleaning.

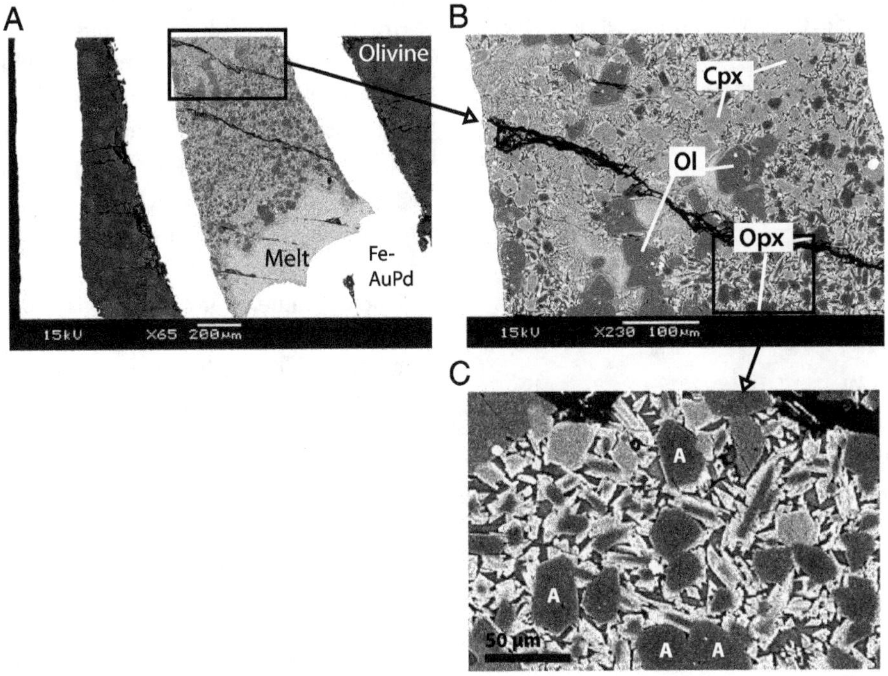

Figure 1 Back-scattered electron images of sample CD2H3 (A), (B). The rectangle zone in A enlarged to show coexisting minerals and interstitial melt in light gray, and **(C)** the rectangle zone in B enlarged to show cpx rims around small opx. Abbreviations used: *Ol* olivine, *Opx* orthopyroxene, *Cpx* clinopyroxene, and *Melt* quenched melt. The white part is the metal capsule noted as Fe-AuPd, i.e. AuPd pre-saturated in Fe. In the outer capsule, recrystallized olivine is visible. Mineral grains have >30-μm diameter, which is appropriate for SIMS measurements. The smallest opx has quench growth cpx around them **(C)**. For analysis, we only used the largest opx (>50 μm) showing no cpx rims around them. Dendritic textures are observed as a boundary layer between crystallized areas and the melt pool. Except for this dendritic boundary, the melt is perfectly quenched.

Results

The results of hydrous experiments are compared with anhydrous analog samples CC4#7b and C4p#25 kb from Dalou et al. (2012), synthesized at 1.2 and 2.5 GPa, respectively. To facilitate the comparison, we use the following notation: *CD1*, refers to renamed samples from Dalou et al. (2012); and CD2, refers to samples from this study (Table 2).

Textural description of experimental samples

Figure 1 shows back-scattered electron images (BSE) of sample CD2H3. Sample textures are appropriate for microanalysis since pools of homogeneous quenched melt are segregated from crystal grains. No bubbles were observed in basaltic glasses, suggesting that the maximum of H_2O content added to the starting materials (approximately 6 wt.%) was less than that needed for water saturation in these melts at pressure and temperature conditions of this study's experiments. Iron contents measured in capsules (between 0.2 and 0.5 wt. %) suggest a fO_2 during the experiment varying from approximately QFM +0.4 to QFM +1, depending on the experiment [following the solution models of Barr and Grove (2010) and Balta et al. (2011)].

The degree of crystallization in runs varies from 32% to 72% (Table 2). The grain sizes are sufficiently large (>30 μm) to determine concentration of volatiles except for cpx in sample CD2-2.5GPa (≤30 μm). The mineral assemblage (Table 2) varies between samples due to variable pressure and temperature *(P-T)* conditions and to the abundance of volatiles (H_2O, F, and Cl). Olivine is present in sample CD2H3 only, as large euhedral crystals (>50-μm diameter). Orthopyroxene is present in all experiments, forming polyhedral grains of 10- to 50-μm diameter, whereas cpx forms polyhedral grains of 5- to 50-μm diameter. Clinopyroxene rims (approximately 5-μm wide), overgrown during quenching, are also observed around some small opx crystals (<20 μm, Figure 1C), which were not analyzed for F and Cl. Dendritic clinopyroxene is also observed at the boundary between crystallized and quenched melt (Figure 1b). Garnet (<30-μm diameter) forms at 2.5 GPa, and is evenly distributed within the capsules. Amphibole is crystallized only in cpx-free samples at temperatures (1,180°C and 1,200°C) higher than typical stability conditions (stable up to 950°C at 1 GPa, Eggler 1972), forming large euhedral crystals (approximately 300 μm). This higher temperature is likely because F increases amphibole stability (Ferguson 1978).

Table 3 Comparison of volatile elements' composition of experimental run products between Dalou et al. (2012) and this study

Samples	Phase	H_2O (wt.%)	F (ppm)	Cl (ppm)
Dalou et al. (2012)				
*CD1*H0	Glass [4]	0.17(0.3)	101(1)	8,515(57)
	Opx [3]	0.10(1)	14(0.3)	19(3)
	Cpx [3]	0.05(0.3)	16(0.3)	77(4)
CD1-2.5GPa	Glass [5]	0.79(2)	6,406(122)	4,772(93)
	Grt [12]	0.47(1)	73(1)	13(0.3)
	Opx [4]	0.49(1)	121(4)	13(1)
	Cpx [4]	0.39(1)	319(4)	10(0.4)
This study				
CD2H3	Glass [3]	2.59(1)	18,241(64)	3,491(13)
	Opx [5]	0.20(1)	1,203(42)	83(2)
	Cpx [3]	0.21(0.2)	2,319(22)	44(3)
	Ol [6]	0.10(0.2)	33(1)	9(1)
CD2H4	Glass [3]	4.33(29)	146,867(17,657)	405(68)
	Amph [4]	1.27(0.2)	62,264(2,214)	57(8)
CD2H5	Glass [3]	4.89(3)	14,765(385)	2,491(10)
	Opx [3]	0.49(1)	668(35)	99(3)
	Cpx [4]	0.55(1)	1,185(23)	35(2)
CD2H6	Glass [2]	5.88(29)	119,267(10,293)	640(72)
	Opx [3]	1.12(17)	2,001(24)	35(3)
	Amp [3]	4.43(1)	62,959(5,033)	201(29)
CD2-2.5GPa	Glass [2]	1.58(1)	7,647(42)	3,379(13)
	Grt [5]	0.52(2)	986(12)	229(14)
	Opx [3]	0.45(1)	352(5)	22(2)

Numbers in brackets indicate number of individual ion probe analyses included in average.
Numbers in parentheses represent one standard deviation in terms of least units cited.
b.d.l. below detection limits.

Equilibrium considerations

Several observations indicate that the experiments approached equilibrium after 4–5 days. These observations include 1) polygonal crystal shape suggesting equilibrium growth (Figure 1); 2) lack of detectable compositional zoning in crystals (>20 μm); and 3) lack of compositional heterogeneity of melt (based on small standard deviations of replicate analyses, Table 4). Major element compositions of minerals and melt, presented in Table 4, agree with previous data (samples B333, B305, B304, B330, and B329 of Gaetani and Grove 1998, synthetized at same pressure and similar temperatures and H_2O contents) in which the attainment of equilibrium was demonstrated via a reversal experiment (B277, Gaetani and Grove 1998). Only sample CD2H5 shows some compositional deviations from previous experimental results. Specifically, the melt is richer in SiO_2 by 5 wt.%, compared with data from Gaetani and Grove (1998) at similar conditions.

The attainment of chemical equilibrium is further inferred by comparing the Fe-Mg exchange coefficient between olivine and melt, ${}^{Ol/Melt}Kd_{Fe\text{-}Mg} = ({}^{Ol}Fe^{2+}.{}^{Melt}Mg) / ({}^{Melt}Fe^{2+}.{}^{Ol}Mg)$, with that of previously reported values from anhydrous experiment data (Roeder and Emslie, 1970; Toplis, 2005). In the olivine-bearing sample (H3V04), the ${}^{Ol/Melt}Kd_{Fe\text{-}Mg}$ value is 0.30 and the predicted value, from the Toplis (2005) model, is 0.29 (±0.03). In olivine-free experiments, the Fe-Mg exchange coefficients between opx (or cpx) and melt measured in some of the volatile-poor experiments (*CD1*H0, CD2H3, and CD2-2.5GPa: 0.25–0.32 ± 0.01 for opx, and 0.29–0.39 ± 0.02 for cpx) are similar to those of Gaetani and Grove (1998; 0.30–0.34 for opx and 0.32–0.35 for cpx). In F-contaminated samples (approximately 12–15 wt.% F), Kd values are significantly higher than the Gaetani and Grove (1998) values (0.52 ± 0.004 for CD2H6 opx and 0.59 ± 0.01 for CD2H5 cpx). Filiberto et al. (2012) showed that the Kd values increase with increasing F content in the melt. They show that at 2.9 wt.% F in anhydrous basalt glass ${}^{Ol/Melt}Kd_{Fe\text{-}Mg}$ ${}^{Opx/Melt}Kd_{Fe\text{-}Mg}$ is 0.40. They suggest that F complexes with Mg in the melt and thus increases the silica activity of the melt, depresses the liquidus, and changes the composition of the crystallizing minerals. Finally, one sample (CD2H5) has a ${}^{Opx/Melt}Kd$ (0.18 ± 0.03) that is lower than that of Gaetani and Grove (1998). This may indicate that a significant portion of iron in the melt is Fe^{3+}, most likely due to a loss of H during this experiment.

Finally, equilibrium is assessed by error-weighted mass-balance calculations (Albarède and Provost 1977) of the major element abundances. It verifies the closure of the chemical system in our experiments. Within uncertainty, results of mass-balance calculations account for the mixing proportions of basalt and peridotite and volatile elements (F, Cl, and H_2O). The square root of the sum of residual square calculations ($\sqrt{\Sigma R^2}$) ranges between 0.1% and 1.58% (Table 2). The main portions of these residuals likely relate to unknown uncertainties in F, Cl, and H_2O abundance in the starting materials, which explains why $\sqrt{\Sigma R^2}$ is significant (>0.5%).

Minerals have balanced stoichiometry (examined with total cations), except for orthopyroxene in sample CD2H4 (total cations in M1 site ≤1.9, instead of 2) and clinopyroxene in sample CD2-2.5GPa (total cations in M1 site >1.5, generally <1.2). For this reason, F and Cl data in orthopyroxene in sample CD2H4 and clinopyroxene in sample CD2-2.5GPa will not be discussed any further.

Fluorine and chlorine contents are homogeneous in minerals and melt (Table 3), with standard deviations

Table 4 Major elements compositions of experimental run products

Samples	Phase	SiO_2	TiO_2	Al_2O_3	Cr_2O_3	FeO*	MnO	MgO	CaO	Na_2O	K_2O	Total
CD2H3	Glass [4]	45.88(42)	0.71(7)	15.56(20)	0.04(5)	10.13(12)	0.14(3)	10.28(12)	10.94(4)	2.69(10)	0.13(2)	96.51(57)
	Opx [7]	53.74(11)	0.11(3)	5.25(27)	0.86(5)	7.50(19)	0.16(4)	30.78(22)	1.74(10)	0.07(1)	0.01(1)	100.26(21)
	Cpx [5]	51.04(16)	0.22(3)	6.36(36)	0.94(2)	5.07(27)	0.14(3)	17.30(6)	18.57(24)	0.62(3)	0.001(2)	100.30(39)
	Ol [5]	40.36(16)	0.00(0)	0.09(2)	0.04(3)	13.58(9)	0.16(1)	45.78(42)	0.24(2)	b.d.l.	b.d.l.	100.46(49)
CD2H4	Glass [6]	44.95(23)	0.58(7)	13.35(10)	0.04(2)	7.87(16)	0.18(4)	12.75(13)	13.35(25)	1.46(8)	0.09(3)	94.62(51)
	Opx[a] [6]	53.52(83)	0.08(5)	3.70(66)	0.49(5)	12.16(36)	0.24(3)	27.65(20)	1.87(19)	0.11(2)	0.01(1)	99.85(35)
	Amp [5]	46.78(22)	0.20(7)	10.38(47)	0.63(11)	6.84(11)	0.15(3)	21.10(14)	8.52(33)	2.79(8)	0.10(2)	97.48(59)
CD2H5	Glass [5]	50.92(16)	0.60(4)	14.63(22)	0.04(4)	6.73(17)	0.14(7)	9.30(12)	9.56(11)	2.79(10)	0.17(3)	94.89(53)
	Opx [7]	54.39(34)	0.06(2)	4.93(34)	0.23(8)	4.53(11)	0.18(4)	34.41(19)	1.29(23)	0.08(2)	0.005(9)	100.10(33)
	Cpx [8]	49.43(34)	0.20(4)	6.01(32)	0.30(6)	6.80(36)	0.14(3)	15.72(17)	20.16(19)	0.88(4)	0.01(2)	99.66(41)
CD2H6	Glass [7]	43.69(19)	0.59(6)	12.44(12)	0.11(3)	7.71(14)	0.13(4)	14.84(13)	12.65(16)	1.81(9)	0.09(3)	94.09(42)
	Opx [10]	54.58(61)	0.05(3)	3.83(62)	0.89(17)	8.42(11)	0.21(2)	31.03(45)	1.25(8)	0.05(1)	0.007(6)	100.24(42)
	Amp [8]	45.79(20)	0.21(4)	11.12(22)	0.98(12)	5.04(13)	0.10(3)	21.61(23)	9.40(18)	3.09(6)	0.12(2)	97.46(31)
CD2-2.5GPa	Glass [5]	53.04(44)	1.28(6)	15.01(24)	0.04(3)	6.51(11)	0.11(5)	6.77(4)	7.46(34)	3.51(15)	0.51(5)	94.22(41)
	Grt [12]	41.24(35)	0.39(18)	23.04(70)	0.37(9)	9.92(25)	0.23(3)	19.45(48)	4.80(57)	N.D.	N.D.	99.52(38)
	Opx [4]	53.87(45)	0.10(5)	6.32(42)	0.36(6)	8.64(19)	0.14(3)	28.25(11)	1.71(7)	0.40(1)	0.01(1)	99.79(31)
	Cpx[a] [4]	52.91(26)	0.24(2)	9.23(52)	0.31(8)	7.31(4)	0.15(5)	19.35(40)	8.90(22)	1.66(6)	0.002(4)	100.11(22)

Numbers in brackets indicate number of individual electron microprobe analyses included in average.
Numbers in parentheses represent one standard deviation in terms of least units cited.
*FeO** total, assuming all Fe as FeO.
[a]Analyses with low stoichiometry, not discussed in the text.
b.d.l. below detection limit.
N.D. not determined.

less than 5% relative for F and less than 12% for Cl in olivine, opx, cpx, and garnet. These variations are smaller than the internal precision of a typical analysis.

F and Cl partition coefficients for hydrous experiments

Table 5 reports F and Cl mineral/melt partition coefficients ($D_F^{mineral/melt}$ and $D_{Cl}^{mineral/melt}$) In samples synthetized at 1.2 GPa and at about the same temperature (±60°C), $D_F^{opx/melt}$ decreases from 0.056 ± 0.003 to 0.021 ± 0.031 and $D_F^{cpx/melt}$ decreases from 0.114 ± 0.002 to 0.08 3 ± 0.004 with increasing H_2O content (from 2.6 to 5.9 wt.%), while $D_{Cl}^{opx/melt}$ increases from 0.020 ± 0.0006 to 0.069 ± 0.010 and $D_{Cl}^{cpx/melt}$ increases from 0.011 ± 0.001 to 0.015 ± 0.008 (Figure 2). The $D^{mineral/melt}$ can be compared between olivine, opx, and cpx in sample CD2H3, where all three phases are coexisting: $D^{cpx/melt} > D^{opx/melt} > D^{ol/melt}$ for both F and Cl, consistently with Hauri et al. (2006), Dalou et al. (2012), Beyer et al. (2012). The amphibole/melt partition coefficients for F and Cl are higher than for nominally anhydrous minerals (NAMs) ($D_F^{amp/melt} = 0.360 \pm 0.066–0.635 \pm 0.087$ and $D_{Cl}^{amp/melt} = 0.199 \pm 0.044–0.378 \pm 0.081$). The F partition coefficients between amphiboles (fluor-pargasitic hornblende) are lower than experimentally determined $D_F^{amp/melt}$ in basaltic system (0.9–1.2, Hauri et al. 2006) at higher H_2O content, than *DF* between K-richterite and melt (0.9–1.8, Edgar and Pizzolato 1995) and than $D_F^{amp/melt}$ modeled by Giesting and Filiberto (2014) ranging from 4 to 13 (undersaturated conditions). The amphibole/melt partition coefficients for Cl are consistent with the range of $D_{Cl}^{amp/melt}$ determined by the Giesting and Filiberto (2014) model (approximately 0.08–3), undersaturated conditions). At 2.5 GPa and 1,340°C, $D_F^{opx/melt}$ is 0.046 ± 0.001 and $D_{Cl}^{opx/melt}$ is 0.006 ± 0.0005, smaller than $D_F^{garnet/melt}$ (0.166 ± 0.002) and $D_{Cl}^{garnet/melt}$ (0.087 ± 0.004). Note that $D_F^{mineral/melt}$ are greater than $D_{Cl}^{mineral/melt}$ for all NAMs, which is consistent with the results of Dalou et al. (2012) and Hauri et al. (2006).

The effect of H_2O content on the partitioning of F and Cl can be observed by comparing hydrous data from this study with the experiments from Dalou et al. (2012) performed at similar but anhydrous conditions. For example, at 1.2 GPa, $D^{opx/melt}$ and $D^{cpx/melt}$ can be compared with *CD1*H0 of Dalou et al. (2012), also performed at 1.2 GPa (0.123 ± 0.004 versus 0.153 ± 0.004 for F, 0.002 ± 0.001 versus 0.009 ± 0.0005 for Cl, respectively, Table 5). Moreover, $D_F^{opx/melt}$ values in anhydrous

Table 5 Fluorine and Chlorine partition coefficients from Dalou et al. (2012) and this study

Run	D_F	D_{Cl}
	Olivine	
CD2H3	0.002(0.1)	0.003(4)
	Orthopyroxene	
*CD1*H0	0.123(4)	0.002(1)
CD2H3	0.056(0.3)	0.020(0.6)
CD2H5	0.042(4)	0.037(2)
CD2H6	0.021(31)	0.069(10)
CD1-2.5GPa	0.016(1)	0.002(0.2)
CD2-2.5GPa	0.046(1)	0.006(0.5)
	Clinopyroxene	
*CD1*H0	0.153(4)	0.009(0.5)
CD2H3	0.114(2)	0.011(1)
CD2H5	0.083(4)	0.015(0.8)
	Garnet	
CD1-2.5GPa	0.012(0.4)	0.003(0.1)
CD2-2.5GPa	0.166(2)	0.087(4)
	Amphibole	
CD2H4	0.360(66)	0.119(44)
CD2H6	0.635(87)	0.378(81)

Numbers in parentheses represent one standard deviation in terms of least units cited.
CD1 data presented to compare anhydrous and hydrous data.

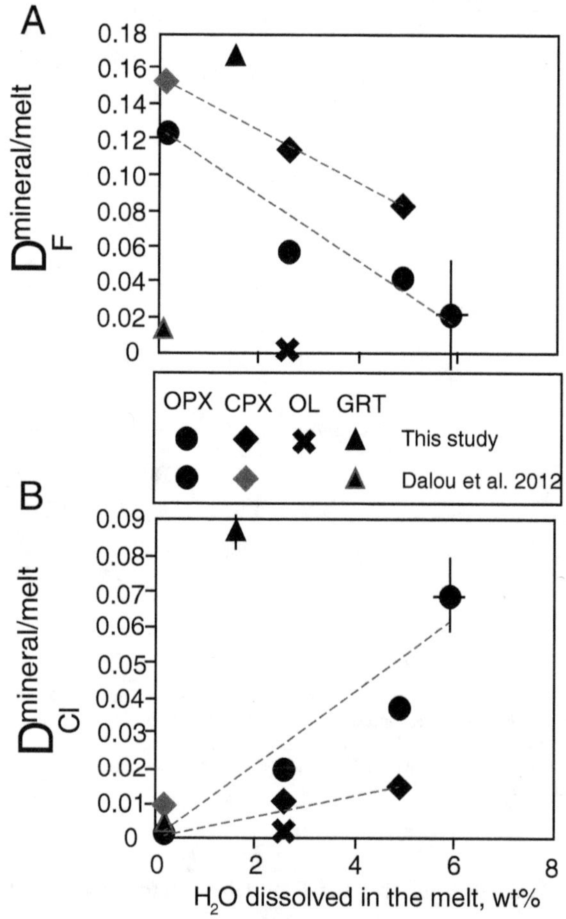

Figure 2 Comparison of D_F and D_{Cl} variations. Comparison of *DF* **(A)** and *DCl* **(B)** variations with increasing H2O content dissolved in the melt, at 1.2 GPa, for nominally anhydrous minerals (NAMS): orthopyroxene (diamonds), clinopyroxene (circles), olivine (cross) and garnet (triangle). Samples *CD1* from Dalou et al. (2012) are shown in grey, while samples of this study are shown in black. The two trends are linear regression on opx and cpx data.

and hydrous samples (0.123 ± 0.004–0.021 ± 0.031 at 1.2 GPa; 0.016 ± 0.001–0.046 ± 0.001 at 2.5 GPa) are in overall agreement with Beyer et al. (2012; 0.031 ±0.005 at 1 GPa, 0.037 ±0.001 at 2.5 GPa) who conducted anhydrous experiments in the CaO-MgO-Al_2O_3-SiO_2 system. For F, the *D*opx/melt (Beyer et al. 2012; Dalou et al. 2012) and *D*cpx/melt (O'Leary et al. 2010: 0.087 ± 0.004; Dalou et al. 2012) of dry experiments are lower that the ones in hydrous experiments (O'Leary et al. 2010: 0.051 ± 0.001–0.069 ± 0.001; this study); the $D_{Cl}^{opx/melt}$ is higher in the case of Cl (Hauri et al. 2006; Dalou et al., 2012; this study). This difference is consistent with the contrasting effect of increasing water content on the mineral/melt partitioning of F and Cl described above. The $D_F^{ol/melt}$ value (0.116) of Dalou et al. (2012) was dismissed because it is two orders of magnitude higher than that reported by Beyer et al. (2012) at similar *P-T* conditions in the CMASF (CaO, MgO, Al_2O_3, SiO_2, and F) and NCMASF (Na_2O, CaO, MgO, Al_2O_3, SiO_2, and F) systems and by Hauri et al. (2006) in basaltic and basanitic systems. The discrepancy between these data is perhaps due to the presence of micro-inclusions, clinohumite lamellae, or clumped OH/F defects (Crépisson et al. 2014) in the high F olivine (Dalou et al. 2012).

Thus, in the rest of the discussion, we used the average $D_F^{ol/melt}$ and $D_{Cl}^{ol/melt}$ from Hauri et al. (2006), of 0.0026 ± 0.0001 and 0.0003 ± 0.0001, respectively. These averages are in a very good agreement with our data in the hydrous sample CD2H3: $D_F^{ol/melt}$ 0.002 ± 0.0001 and $D_{Cl}^{ol/melt}$ 0.003 ± 0.004. Although the error on $D_{Cl}^{ol/melt}$ is significant, we decide to use this value because to date, no other values are available, except for maximum values given by Hauri et al. (2006; ranging from 0.002 to 0.003). For garnet, the $D_F^{garnet/melt}$ (0.166 ± 0.002) and $D_{Cl}^{garnet/melt}$ (0.087 ± 0.004) of hydrous sample CD2-2.5GPa is higher than $D_F^{garnet/melt}$ (0.012 ± 0.0004) and $D_{Cl}^{garnet/melt}$ (0.003 ± 0.0001) of anhydrous sample *CD1*-2.5GPa (2.5 GPa, Dalou et al. 2012, which is the only dry sample for which F and Cl partition coefficients were determined for garnet).

Discussion

Bulk F and Cl partitioning during hydrous melting

We calculated the bulk $D_{\mathrm{F}}^{\mathrm{bulk}}$ and $D_{\mathrm{Cl}}^{\mathrm{bulk}}$ (D_0^{bulk} see Table 6) between an anhydrous spinel lherzolite containing 57% olivine, 28% opx, 13% cpx, and 2% spinel; (Workman and Hart 2005), and 1) anhydrous or 2) hydrous basaltic melt. These D values are 0.056 ± 0.002 for F and 0.0019 ± 0.0004 for Cl in the case of anhydrous melting and 0.018 ± 0.009 for F and 0.023 ± 0.005 for Cl in for hydrous melting, with 5.9 wt.% of H_2O dissolved in the melt. Thus, from anhydrous to hydrous lherzolite melting, F becomes more incompatible (from 0.056 to 0.018) and Cl less incompatible (from 0.002 to 0.023). These results indicate that behavior of F and Cl become decoupled during hydrous melting. In the case of amphibole-bearing lherzolite (60% olivine, 20% opx, 15% cpx, and 5% amphibole), modeled bulk $D_{\mathrm{F}}^{\mathrm{bulk}}$ and $D_{\mathrm{Cl}}^{\mathrm{bulk}}$ (D_0^{bulk}) between amphibole-bearing lherzolite and hydrous basaltic melt are 0.048 ± 0.004–0.050 ± 0.011 and 0.013 ± 0.005–0.037 ± 0.009, respectively with increasing H_2O content (from 4.3 to 5.9 wt.%). Nevertheless, in the melting model presented below, we used an amphibole-bearing wehrlite (Médard et al. 2006), because to date no melting modes are available for amphibole-bearing lherzolite. In the case of amphibole-bearing wehrlite (20% olivine, 28% cpx, and 52% amphibole; Médard et al. 2006), the high proportion of amphibole enhances its role during melting; modeled bulk $D_{\mathrm{F}}^{\mathrm{bulk}}$ and $D_{\mathrm{Cl}}^{\mathrm{bulk}}$ (D_0^{bulk}) used in the hydrous melting model are 0.219 ± 0.0002–0.353 ± 0.0003 and 0.066 ± 0.023–0.201 ± 0.043, respectively with increasing H_2O content (from 4.3 to 5.9 wt.%).

Effect of melt structure on F and Cl partition coefficients

It is not possible to completely separate the effects of temperature and H_2O on the F and Cl partition coefficients (Figure 2) because in order to obtain multi-phase assemblages in the experiments, it is necessary to decrease the temperature as the H_2O content of the system is increased, to account for the effect of water on the solidus temperature of the peridotite phase assemblage (e.g. Kushiro 1969). Moreover, we found no clear relationship between $D_{\mathrm{F}}^{\mathrm{mineral/melt}}$ or $D_{\mathrm{Cl}}^{\mathrm{mineral/melt}}$ and the F content dissolved in melts (Table 3), which supports the conclusion that the fluorine contamination from HF capsule cleaning (up to several weight percent, Table 3) cannot explain the F and Cl partition coefficient variations among the samples. The contrasting changes of F and Cl partition coefficients between pyroxenes and melt with increasing water content (i.e. decreasing temperature) is, therefore, the result of either 1) the effect of H_2O and temperature on melting phase relations and phase composition, 2) their effect on the melt structure, or 3) a combination of both.

Figure 3 presents $D_{\mathrm{F}}^{\mathrm{mineral/melt}}$ (A) and $D_{\mathrm{Cl}}^{\mathrm{mineral/melt}}$ (B) versus the melt structure parameter NBO/T, which describes the number of non-bridging oxygen (NBO) per tetrahedrally coordinated cations (T). Here, a variation of NBO/T can result from the depolymerizing effect of

Table 6 Bulk partition coefficients and data used in fluid induced melting model

	α	$D_{o\mathrm{Cl}}^{\mathrm{bulk}\ a}$	$D_{o\mathrm{F}}^{\mathrm{bulk}\ a}$	$P_{\mathrm{Cl}}^{\mathrm{bulk}}$	$P_{\mathrm{F}}^{\mathrm{bulk}}$	Samples used to calculate D_o^{bulk}	For phases:
Spinel lherzolite							
Dry melting	0.002	0.002(0.6)	0.056(2)	0.006(0.7)	0.157(4)	*CD1*H0	Ol,[b]opx, cpx
Low degree of metasomatism	0.026	0.009(0.5)	0.032(0.4)	0.016(0.8)	0.099(1)	CD2H3	Ol, opx, cpx
Intermediate degree of metasomatism	0.049	0.014(0.8)	0.024(2)	0.027(1)	0.072(4)	CD2H3	Ol
						CD2H5	Opx, cpx
High to very high degree of metasomatism	0.059	0.023(3)	0.018(9)	0.044(5)	0.061(18)	CD2H3	Ol
						CD2H6	Opx
						CD2H5	Cpx
Amphibole-bearing wehrlite							
Intermediate degree of metasomatism	0.043	0.066(23)	0.219(0.2)	0.130(48)	0.398(72)	CD2H3	Ol
						CD2H4	Amp
						CD2H5	Cpx
High to very high degree of metasomatism	0.059	0.201(43)	0.353(0.3)	0.411(87)	0.695(95)	CD2H3	Ol
						CD2H6	Amp
						CD2H5	Cpx

[a]From Johnson et al. (1990).
[b]From Hauri et al. (2006), we used average value of D_{F} and D_{Cl} between olivine and basaltic melt.
$D_o^{\mathrm{bulk}} = \Sigma\, D_i^{\mathrm{mineral/melt}} \cdot X_i^{\mathrm{mineral}}$, where X is the proportion of mineral in the anhydrous lherzolite and $D_i^{\mathrm{mineral/melt}}$ are presented Table 5.
$P_i^{\mathrm{bulk}} = \Sigma\, D_i^{\mathrm{mineral/melt}} \cdot p_i^{\mathrm{mineral}}$, where p are the stoichiometric coefficients of melting (see text).
$D_i^{\mathrm{bulk}} = [D_o^{\mathrm{bulk}} - (P_i^{\mathrm{bulk}} \cdot \chi)]/(1-\chi)$, where χ is the degree of melting.

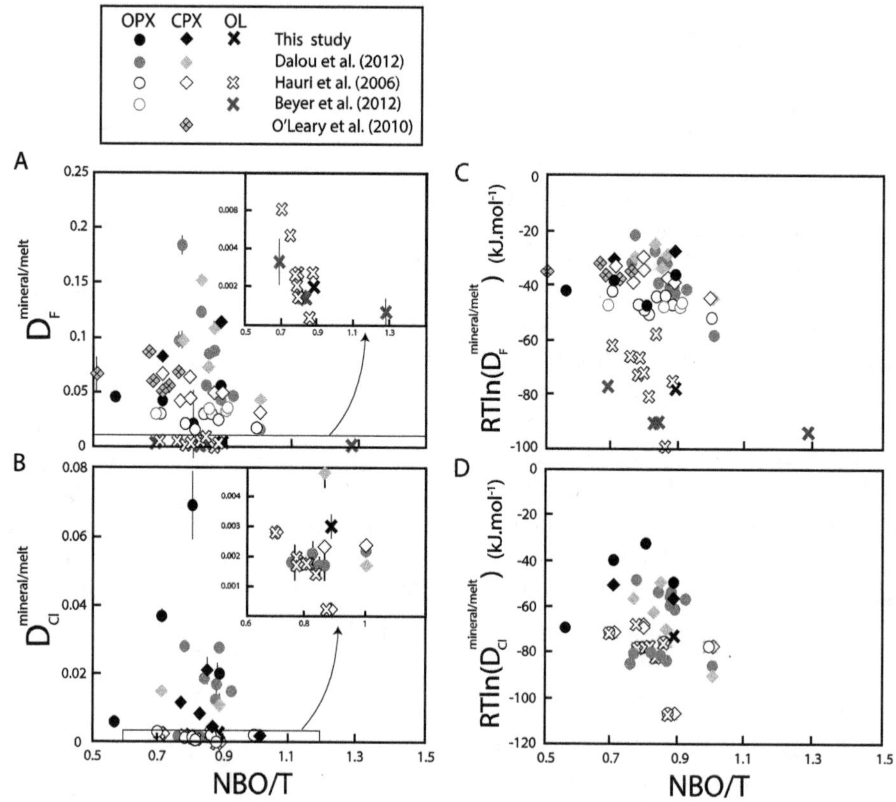

Figure 3 Mineral/melt partition coefficients of fluorine (A) and chlorine (B) versus NBO/T, (C) RTln$\left(D_F^{mineral/melt}\right)$ versus NBO/T and (D). $RTln\left(D_{Cl}^{mineral/melt}\right)$ versus NBO/T, where *R* is the gas constant and *T* the temperature in K. Here RTln$\left(D_{volatile}^{mineral/melt}\right)$ is equal to Δ*G*, the free Gibbs energy; NBO/T were calculated following the procedure in Mysen and Richet (2005, Chapter 4) as explained in the text. Data from this study are compared with literature data for olivine (Hauri et al. 2006; Beyer et al. 2012), orthopyroxene (Hauri et al. 2006; Beyer et al. 2012; Dalou et al. 2012), and clinopyroxene (Hauri et al. 2006; O'Leary et al. 2010; Dalou et al. 2012).

H_2O content (e.g. Mysen, 2007) and F content (e.g. Mysen et al. 2004), especially in F-rich samples. Because F depolymerizes melts equally or more efficiently than H_2O (Dingwell and Virgo 1987; Mysen et al. 2004) the NBO/T data presented in Figure 3 were corrected for F content, assuming that F has the same depolymerizing effect as H_2O (Mysen 2007). The NBO/T values, before correction, were calculated following Mysen and Richet (2005; Chapter 4): NBO/T = $(4{\bullet}X_T - 2{\bullet}X_O)/X_T$, where X_T and X_O are the atomic proportions of tetrahedrally coordinated cations (Si and Al) and oxygen, respectively. A simple way to correct for the F effect is to assume that F has the same depolymerizing effect as H_2O in aluminosilicate melts:

$$NBO/T_{corr} = NBO/T - (NBO/T{\bullet}m{\bullet}[F]), \quad (1)$$

where *m* is the slope value of NBO/T versus H_2O (mol fraction) from Mysen (2007; for NS4NA7.5 composition, similar NBO/T and similar Al_2O_3 content as our basaltic melts), and [F] is the F mol fraction of each sample. This correction allows deciphering the effect of H_2O from the effect of F on NBO/T in Figure 3.

Figure 3A,C shows that, for each mineral (from the literature data and this study), F partition coefficients tend to decrease with increasing H_2O-induced NBO/T variations, i.e. with increasing melt depolymerization. In contrast, there is no obvious correlation between Cl partition coefficients for pyroxenes and NBO/T (Figure 3B, D). We interpret that the scattering of D_{Cl} versus NBO/T reflects the fact that mineral chemistry has a significant role in D_{Cl} variation, as suggested by Bernini et al. (2012) and Dalou et al. (2012). For example, a mild correlation between Cl substitution and octahedral site elasticity has been demonstrated by Dalou et al. (2012). In contrast, Dalou et al. (2012) found no direct correlation between D_F and pyroxene chemistry or elastic parameters. Fluorine, being smaller than Cl (133 and 181 pm, respectively), is more sensitive to melt structure than Cl (Dalou et al. 2012). The same relationship between ion size and melt depolymerization was previously found for cations (e.g. Gaetani et al. 2003).

Previous work on F and Cl solution behavior in aluminosilicate melts (Mysen and Virgo 1985; Kohn et al. 1991; Schaller et al. 1992; Zeng and Stebbins 2000; Mysen et al. 2004; Zimova and Webb 2006; Baasner et al. 2013) have shown that F preferentially bonds with tetrahedrally coordinated cations such as Al and Si. Although in more complex melts such as basalt, F can also preferentially bond with Mg (Kiczenski et al. 2004; Filiberto et al. 2012) and form more complex bonds as Al-F-Ca or Al-F-Na (Zeng and Stebbins 2000; Mysen et al., 2004); Si-F and Al-F are always found. Moreover, as NBO/T increases, F bonding and dissolution in silicate melts is favored (Mysen et al. 2004). In contrast, Cl preferentially bonds with network-modifier cations (Ca, Na, K; Stebbins and Du 2002; Sandland et al. 2004; Zimova and Webb 2006; Dalou and Mysen 2012; Baasner et al. 2013) or with Fe^{2+} in F-rich basalt (Filiberto and Treiman 2009; Filiberto et al. 2014), and its solubility in aluminosilicate melts decreases with increasing NBO/T (Dalou and Mysen 2012). Assuming that F and Cl solubility response to NBO/T in compositionally simple aluminosilicate melts can be extended to basaltic composition, F and Cl solubility response to NBO/T can explain changes of F incompatibility and partly Cl incompatibility. As shown above, the variation of melt structure in our experiments is mainly due to the variation in dissolved H_2O content. Therefore, the variation in H_2O content dissolved in melts is the main cause of our F partition coefficient variation and also could have a role of the Cl partition coefficient variation. This is consistent with recent F and Cl solubility results in aluminosilicate melts with increasing H_2O content (Dalou and Mysen 2013) where increasing H_2O content in the melt results in increased F. This means that mineral/melt partition coefficients decrease with increasing H_2O and results, therefore, in decreased mineral/melt partition coefficients for chlorine.

F and Cl during hydrous melting

Fluid-induced melting model

The $D_F^{\text{lherzolite/melt}}$ and $D_{Cl}^{\text{lherzolite/melt}}$ were used to calculate the F and Cl composition of melts derived from a spinel lherzolite (57% olivine, 28% opx, 13% cpx, and 2% spinel; Workman and Hart 2005) metasomatized by various proportions of aqueous fluids. In this model, we assume that the partial melts are undersaturated in aqueous fluids because our experiments are undersaturated in H_2O. Therefore, the model follows the flux-induced melting proposed by Stolper and Newman (1994) and Eiler et al. (2000). The equation chosen is derived from Eiler et al. (2000) modified to represent non-modal melting following the equation by Johnson et al. (1990) (Table 6).

$$C_i^{\text{arc magmas}} = \frac{C_i^{\text{DMM}}(1-X_f) + X_f * C_i^{\text{slab flux}}}{\bar{D}_i^{\text{lherzolite/arc magma}} + \chi\left(1-\bar{P}_i^{\text{lherzolite/arc magma}}\right)} \quad (2)$$

In this equation, i stands for F or Cl. The partition coefficients of element i at χ degree of melting, $\bar{D}_i^{\text{lherzolite/arc magma}}$ are the mineral proportion weighted partition coefficients of i. $\bar{P}_i^{\text{lherzolite/arc magma}}$ are the weighted partition coefficients of liquid. The X_f represents the amount of metasomatic fluid added to the lherzolite. From the fluid-induced melting model (Stolper and Newman 1994; Eiler et al. 2000), it is assumed that the fluid is the main factor controlling the degree of melting at constant pressure and temperature. For example, every addition of 1 wt.% H_2O results in 10 wt.% of partial melting (Eiler et al. 2000). This notion is introduced by the factor alpha ($\boldsymbol{\alpha}$), the rate of fluid-induced melting, expressed as $\boldsymbol{\alpha}$ = amount of H_2O added to the lherzolite divided by the degree of melting. In equation (2), X_f is replaced by $\alpha\chi\left(C_{H_2O}^{\text{slab flux}}\right)^{-1}$ to obtain equation (3). The $\alpha\chi\left(C_{H_2O}^{\text{slab flux}}\right)^{-1}$ term takes into account that 1) the degree of melting x depends on the amount of H_2O added by the slab flux, and 2) the slab fluid composition is not pure H_2O. Thus, we normalized X_f to the amount of H_2O in slab fluxes $C_{H_2O}^{\text{slab flux}}$:

$$C_i^{\text{arc magmas}} = \frac{C_i^{\text{DMM}} + \alpha\chi\left(C_{H_2O}^{\text{slab flux}}\right)^{-1}\left(C_i^{\text{slab flux}} - C_i^{\text{DMM}}\right)}{\bar{D}_i^{\text{lherzolite/arc magma}} + \chi\left(1-\bar{P}_i^{\text{lherzolite/arc magma}}\right)} \quad (3)$$

In this study, alpha ($\boldsymbol{\alpha}$) values were chosen to be 0.026, 0.049, 0.059, and 0.10, because $\boldsymbol{\alpha}$ corresponds to the H_2O content dissolved in melts during the experiments (2.6, 4.9, and 5.9 wt.%, Table 6). Although no experiment was performed with 10 wt.% H_2O dissolved in the melt, $\boldsymbol{\alpha}$ = 0.10 was chosen to test the model reaction to a high rate of fluid-induced melting. The $\bar{P}_i^{\text{lherzolite/arc magma}}$ is calculated from stoichiometric coefficients for spinel lherzolite melting reactions under anhydrous and hydrous conditions using Kinzler (1997) and Gaetani and Grove (1998), respectively. For amphibole-bearing lithology, the stoichiometric coefficients of an amphibole-bearing wehrlite from Médard et al. (2006) were chosen. In the melting calculation, the degree of melting, χ, of the metasomatized lherzolite was varied from 1% to 20%. The composition of the anhydrous lherzolite is the depleted upper mantle composition (F = 16 ppm and Cl = 0.83 ppm) from Saal et al. (2002). Three different compositions of slab components, $C_i^{\text{slab flux}}$, were tested. Two different aqueous fluids: F = 990 ± 270 ppm, Cl =

9,390 ± 2,500 ppm, and H_2O = 25 ± 7 wt.% from Straub and Layne (2003) and F = 2,200 ppm, Cl = 8,900 ppm, and H_2O = 53 wt.% from Le Voyer et al. (2010) were used; together with one slab derived melt: F = 13,700 ppm, Cl = 14,800 ppm, and H_2O = 44.5 wt.% from Le Voyer et al. (2010) (Figure 4).

This simple model illustrates how an increase in the proportion of metasomatic fluid affects magma composition in two ways: (1) the degree of melting increases; and (2) more metasomatic component (mostly H_2O and Cl) resides in the final magma. In other words, the model illustrates the competition between fractionation and dilution of F and Cl. Note that this model calculation is aimed to assess the role of variable F and Cl partition coefficients in wet and dry mantle, in order to extract robust systematics. It emphasizes the basic framework of the fluid-induced melting, but ignores potential complication of arc magma genesis such as reactive fluid and melt transport, super-adiabatic and adiabatic melting.

With increasing degree of melting, χ, the Cl/F ratio in the final melt increases independently of the composition of the slab fluid (Figure 4). This ratio also increases with increasingly Cl-rich metasomatic fluids. Therefore, both the fluid composition and degree of melting are the major variables controlling the melt Cl/F values. The increasing degree of melting has more effect on the Cl/F ratio than α, the rate of fluid-induced melting (Figure 4). For example, at constant degree of melting (20%), the Cl/F ratio of melts increases by a factor of 1.3 (from 5.2 to 6.7) as the melt H_2O content varies from 2.6 to 5.9 wt.%. In comparison, for α = 0.059, Cl/F increases by factor of 6.4 (from 1.0 to 6.7) from 1% to 20% degree of melting.

Comparisons to F and Cl in arc melt inclusions

Figure 5A reports F and Cl concentrations in melt inclusions from various volcanic arcs (grouped in arcs: Trans-Mexican: Sadofsky et al. 2008; Vigouroux et al. 2008; Kamchatka: Churikova et al. 2007; Portnyagin et al. 2007; Kermadec: Wysoczanski et al. 2006; Izu-Mariana: Straub and Layne 2003; Shaw et al. 2012; Sunda: Elburg et al. 2006; Aeolian: Rose-Koga et al. 2012; Cascade: Le Voyer et al. 2010; S. America: Le Voyer et al. 2008; Vanuatu: Sorbadere et al. 2011; Lesser Antilles: Bouvier et al. 2008; 2010) and illustrates the variety of Cl/F ratios among arc magmas (between 0.25 and 10). Although the variations of degree of melting and crystallization are likely responsible for some of F and Cl variation within each group of arc melt inclusions, it is unlikely that crystallization is solely responsible for these variations. For instance, Le Voyer et al. (2010) investigated the variation of volatile elements signature among Mount Shasta melt inclusions, by minimizing the effect of melting-crystallization normalizing F and Cl content to Y content (incompatible, fluid-immobile element). It appears that Mount Shasta melt inclusions display two trends of Cl/Y versus F/Y that corresponds to two distinct Cl/F ratios of approximately 1 and 4. Therefore, although some of the variation observed on Figure 5A might be produced by variable degree of crystallization of different

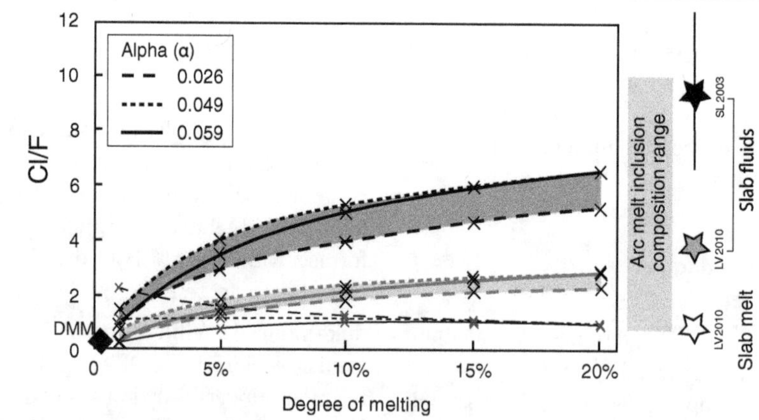

Figure 4 Modeled melt Cl/F ratios are shown as a function of degrees of fluid induced melting. Model tests three slab components (*stars*): slab fluid composition from Straub and Layne (2003) (*black star*), slab fluid composition from Le Voyer et al. (2010) (*grey star*), and slab melt composition from Le Voyer et al. (2010) (*white star*). Lines exhibit the variation of the rate of fluid-induced melting, by changing α values: 0.026 (*dashed lines*), 0.049 (*dotted lines*), and 0.059 (*solid lines*). Crosses are calculated points from the model joined by the lines for more clarity. The *black diamond* shows the depleted mantle composition. The *grey rectangle* on the side of the diagram shows the range of Cl/F in arc melt inclusions (Straub and Layne 2003; Elburg et al. 2006; Wysoczanski et al. 2006; Churikova et al. 2007; Portnyagin et al. 2007; Sadofsky et al. 2008; Vigouroux et al. 2008; Bouvier et al. 2008; 2010; Le Voyer et al. 2008; 2010; Sorbadere et al. 2011; Shaw et al. 2012; Rose-Koga et al. 2012). The Cl/F ratio is determined by the slab component, the degree of melting, but also the rate of fluid-induced melting. In other words, it means that the amount of fluid component has an influence of the Cl/F ratio in melts. Interestingly, at low degrees of slab fluid-induced melting (1% to 10%), Cl/F ratios in melts are higher for α = 0.049 than for α = 0.059. This is because Cl content in the melt is controlled by increasing incompatibility in opx and the dissolution of opx with increasing α.

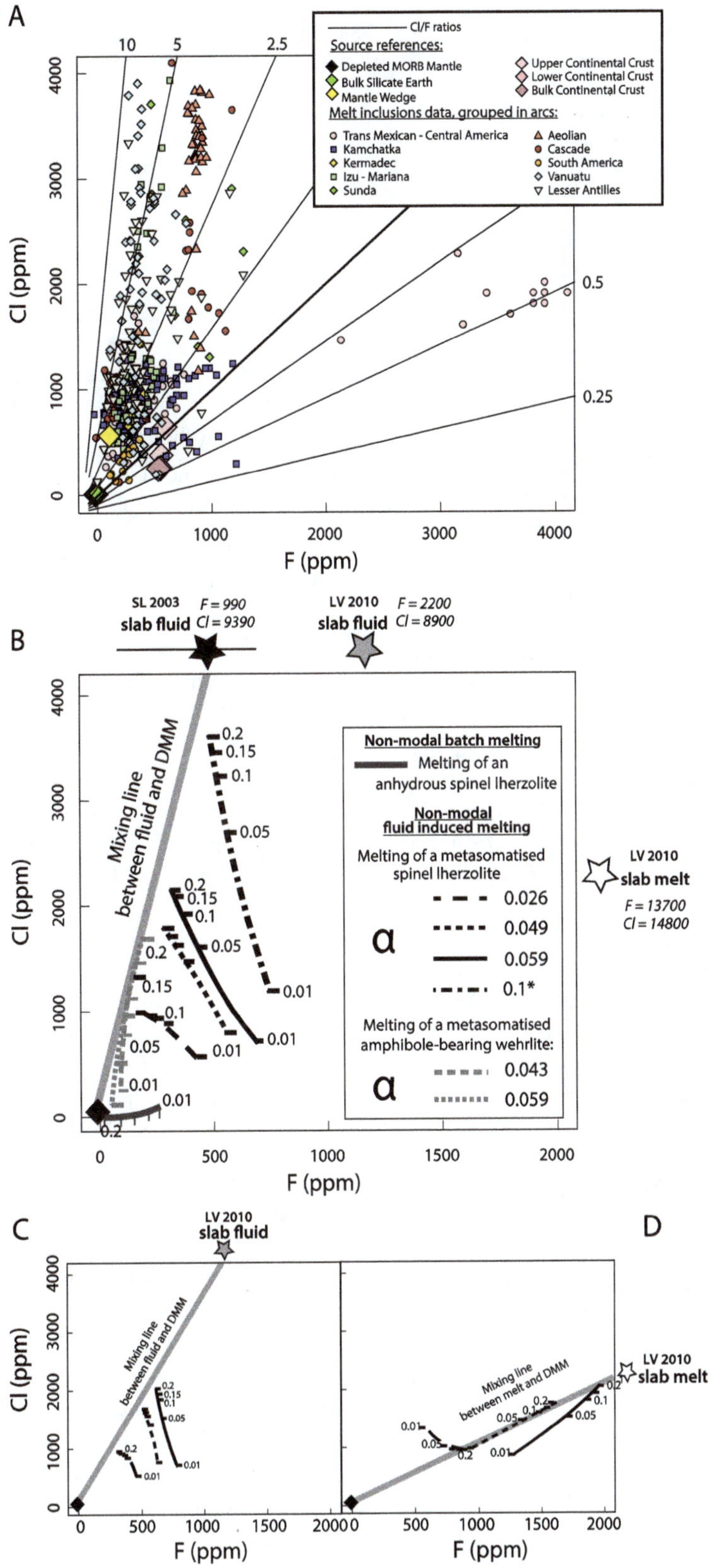

Figure 5 (See legend on next page.)

(See figure on previous page.)
Figure 5 Cl versus F in, (A) arc melt inclusions, slab and mantle wedge, (B, C, and D) predicted melts. (A) Melt inclusions data are compiled from recent studies (see text). Each *thin line* represents a constant Cl/F ratio. **(B)** Melt compositions derived from anhydrous mantle (DMM, *blue diamond*) are predicted by the *dark gray curve*. Predicted melts' compositions calculated for fluid-induced melting of a spinel lherzolite are indicated by the *dotted curve* for low degree (α = 0.026), *dashed* for intermediate (α = 0.049), and *solid curves* for high (α = 0.059) and very high degree of melt production rate for a unit flux (α = 0.1). Fluid composition is from Straub and Layne (2003) (*black star*). Fluid-induced melting of an amphibole-bearing wehrlite is indicated by *gray curves*: *dashed* for α = 0.043 and *dotted* for α = 0.059. Each curve is graduated as a function of degree of melting (from 1% to 20%). See text for details on the model parameters. The *gray line* represents the mixing line between the fluid composition (Straub and Layne, 2003) and the DMM. Compositions of aqueous fluid from Le Voyer et al. (2010) and slab melt from Le Voyer et al. (2010) are shown for comparison (*gray star* and *white star*, respectively). **(C)** F and Cl compositions of melts produced by fluid-induced melting of a lherzolite metasomatized by an aqueous fluid from Le Voyer et al. (2010). **(D)** F and Cl compositions of melts produced by fluid-induced melting of a lherzolite metasomatized by slab-derived melt from Le Voyer et al. (2010).

samples, it is likely that a large part of the Cl/F variations is related to the magma genesis in the mantle wedge.

The variations in volatile and trace elements composition are commonly explained by a compositional variation of the metasomatic agent (e.g. Straub et al. 2004; Le Voyer et al. 2008; Sadofsky et al. 2008; Bouvier et al. 2010; Sorbadere et al. 2011). The composition of the metasomatic agent is usually calculated from mass balance assuming a mixing between the depleted mantle (DMM) and this agent, either a slab aqueous fluid or a slab melt (e.g. Grove et al. 2002; Straub and Layne 2003; Straub et al. 2004; Le Voyer et al. 2010). Considering that seawater is about 20,000 times richer in Cl than F (e.g. Straub and Layne 2003) and that Cl is highly soluble in slab aqueous fluids (e.g. Brenan 1993), a Cl/F >1 in arc melt inclusions is expected. The Cl/F ratio in aqueous fluid from slab dehydration was estimated to be approximately 4 (e.g. Mount Shasta, Le Voyer et al. 2010) or up to 9.5 (e.g. Izu arc front, Straub and Layne 2003). In contrast, the low Cl/F (<1) found in some arc melt inclusions (e.g. Central and South American arcs, Kermadec, Wysoczanski et al. 2006; Sadofsky et al. 2008; Vigouroux et al. 2008; Mount Shasta, Le Voyer et al. 2010) cannot result from the mixing between a depleted source and an aqueous fluid. Instead, it was suggested that it was produced by mixing with a H_2O-rich slab melt (e.g. Le Voyer et al. 2010), with a Cl/F approximately 1 (Figure 5B).

Part of the Cl/F variability in arc melt inclusions could also result from the presence of hydrous minerals as amphiboles or micas, such as pargasite or phlogopite. On Figure 5B, melting trends of an amphibole-bearing wehrlite are reported. This trend is almost parallel to the mixing trend between DMM and fluid and shows that Cl is more fractionated than F during amphibole-bearing wehrlite melting. Melt inclusions from the western Trans-Mexican arc display very low Cl/F (approximately 0.5), which has been interpreted as the melting of a phlogopite-bearing lithology (Vigouroux et al. 2008). Although more experimentally determined partition coefficients of F and Cl are needed for amphibole and phlogopite, the melting of a mantle with hydrous phases likely would fractionate F and Cl differently than in an anhydrous mantle. The hydroxyl site of hydrous phases such as amphiboles and micas provide ideal substitution sites for F and Cl, therefore $D_F^{\text{hydrous mineral/melt}}$ and $D_{Cl}^{\text{hydrous mineral/melt}}$ should be higher than the ones of NAMs. Thus, we suggest that F and Cl fractionation during the melting of amphibole or mica-bearing lithologies may explain some of Cl/F variability in melt inclusions.

The effect of H_2O on $D_F^{\text{mineral/melt}}$ and $D_{Cl}^{\text{mineral/melt}}$ implies that the F and Cl signature in melts is somehow related to the amount of aqueous fluid involved during arc magma genesis. When model results shown in Figure 4, tested for fluid composition from Straub and Layne (2003), are plotted in the Cl versus F diagram (Figure 5B), the direction of melting trends is determined by the value of α, whereas the extent of the trends are determined by the degree of melting. There is a rotation in the F versus Cl melting trends with the increasing proportion of metasomatic fluids: from a negative slope of Cl versus F at low α to a near-vertical slope at α = 0.1. At low α, the fluid input, and in particular the Cl input from the fluid, is low. Therefore, the melting trend is determined by $D_F^{\text{lherzolite/melt}}$ and $D_{Cl}^{\text{lherzolite/melt}}$. As α increases, there is competition between the F and Cl fractionation (controlled by $D_F^{\text{lherzolite/melt}}$ and $D_{Cl}^{\text{lherzolite/melt}}$) and the dilution of Cl carried by the fluid. This is why Cl begins to fractionate more than F as α increases and why at very high α, only Cl seems to fractionate. This change of melting trend directions suggest that various Cl/F ratios in melts can be produced by variable rate of fluid-induced melting without changing the composition of the fluid. Interestingly, subvertical alignments of melt inclusion compositions are seen among the Lesser Antilles, Aeolian, and Vanuatu arcs and some of the Cascade arc (Figure 5A). While high α melting trend and data trends may not be an exact match, it is possible that these subvertical alignments result from melting at high fluid content, with a strong fractionation of Cl compared to F. This suggests that (1) the natural highly variable Cl trends we observe probably represent melting trends and (2) those magmas were produced by high degree of metasomatism.

In modeled melts, consequently to increasing α, not only Cl increases because of increasing fluid input, but F increases as well because $D_F^{mineral/melt}$ decreases with increasing H_2O content. Note that the model described here does not challenge the important role of the metasomatic agent in the composition of arc magmas (Figure 5C,D). The nature of this agent, fluid or melt, and its composition also control the slope of melting trends (Figure 5B–D). At high degree of melting (>20%), the melt compositions are near that of melt in the mixing lines between the DMM and slab components. In summary, while the main variability of Cl/F ratio among melt inclusions is produced by the various composition of slab component, within the same arc region or the same volcano at a given time, some Cl/F variability can be explained by variable amount of the same slab component (which has also a strong influence on the degree of melting of the mantle wedge).

Conclusions

We report F and Cl partition coefficients between olivine, orthopyroxene, clinopyroxene, garnet and amphibole, and hydrous basaltic melt for *P-T* conditions ranging from 1.2 to 2.5 GPa and 1180 to 1430°C, with a H_2O content dissolved in the melt ranging from 0.2 to 5.9 wt%. We show that increasing H_2O content has a contrasting effect on F and Cl partition coefficients between anhydrous minerals and melt. The $D_F^{mineral/melt}$ decrease with the increasing melt depolymerization resulting from increasing H_2O content. In contrast, $D_{Cl}^{mineral/melt}$ increase with melt H_2O content. This relationship is complex since Cl behavior depends on both melt and crystal chemistry.

Because our experimental conditions are relevant to that of the mantle wedge, our $D_F^{mineral/melt}$ and $D_{Cl}^{mineral/melt}$ data were used to model Cl/F ratio in arc magmas produced by fluid induced melting. We establish that fluid induced melting can produce high Cl/F and high F content melts, even when the metasomatic agent is an aqueous fluid. Our model shows that the Cl/F variability in arc melt inclusions results from 3 related variables: the composition but also the amount of metasomatic agent released from the slab into the mantle wedge and the degree of melting induced by the amount of metasomatic agent.

Competing interests

The authors declare that they have no competing interests.

Authors' contributions

KTK conceived and designed the study. CD carried out the experimental study, obtained analytical data with NS, and wrote the manuscript. KTK, ML, and NS collaborated with the corresponding author in the construction of manuscript. All authors read and approved the final manuscript.

Acknowledgements

This research was financed by the French Government Laboratory of Excellence initiative n°ANR-10-LABX-0006, the Région Auvergne and the European Regional Development Fund. This is Laboratory of Excellence ClerVolc contribution number 127. Critical reviews from AE Bjorn Mysen and three anonymous reviewers were greatly appreciated. CD thanks Robert Dennen for his help with English revisions.

This research article belongs to no. 4 section: Solid Earth Sciences.

Author details

[1]Department of Geological Sciences, Jackson School of Geosciences, The University of Texas at Austin, 1 University Station C1100, Austin, TX 78712, USA. [2]Université Blaise Pascal, Laboratoire Magmas et Volcans, Clermont Université, Clermont-Ferrand, BP 10448, F-63000, France. [3]CNRS, UMR 6524, LMV, Clermont-Ferrand F-63038, France. [4]IRD, R 163, LMV, Clermont-Ferrand F-63038, France. [5]Department of Terrestrial Magnetism, Carnegie Institution of Washington, Washington, DC 20015, USA. [6]Department of Mineral Sciences, National Museum of Natural History, Smithsonian Institution, Washington, DC 20013, USA. [7]Woods Hole Oceanographic Institution, Woods Hole, MA 02453, USA.

References

Albarède F, Provost A (1977) Petrological and geochemical mass-balance equations: an algorithm for least-square fitting and general error analysis. Comput Geosci 3:309–326

Ayers J (1998) Trace element modeling of aqueous fluid–peridotite interaction in the mantle wedge of subduction zones. Contribut Mineral Petrol 132(4):390–404

Baasner A, Schmidt BC, Webb SL (2013) Compositional dependence of the rheology of halogen (F, Cl) bearing-aluminosilicate melts. Chem Geol 346(C):172_183

Balta JB, Beckett JR, Asimow PD (2011) Thermodynamic properties of alloys of gold-74/palladium-26 with variable amounts of iron and the use of Au-Pd-Fe alloys as containers for experimental petrology. Am Mineral 96(10):1467–1474

Barr JA, Grove TL (2010) AuPdFe ternary solution model and applications to understanding the fO_2 of hydrous, high-pressure experiments. Contrib Mineral Petrol 160(5):631–643

Béarat H, McKelvy MJ, Chizmeshya AV, Sharma R, Carpenter RW (2002) Magnesium hydroxide dehydroxylation/carbonation reaction processes: implications for carbon dioxide mineral sequestration. J Am Ceramic Soc 85(4):742–748

Bercovici D, Karato S (2003) Whole mantle convection and the transition-zone water filter. Nature 425:39–44

Bernini D, Wiedenbeck M, Dolejš D, Keppler H (2012) Partitioning of halogens between mantle minerals and aqueous fluids: implications for the fluid flow regime in subduction zones. Contrib Mineral Petrol 165:117–128

Beyer C, Klemme S, Wiedenbeck M, Stracke A, Vollmer C (2012) Fluorine in nominally fluorine-free mantle minerals: experimental partitioning of F between olivine, orthopyroxene and silicate melts with implications for magmatic processes. Earth Planet Sci Lett 337–338:1–9

Bouvier A-S, Metrich N, Deloule E (2008) Slab-derived fluids in the Magma Sources of St. Vincent (Lesser Antilles Arc): volatile and light element imprints. J Petrol 49:1427–1448

Bouvier A-S, Metrich N, Deloule E (2010) Light elements, volatiles, and stable isotopes in basaltic melt inclusions from Grenada, Lesser Antilles: inferences for magma genesis. Geochem Geophys Geosyst 11, Q09004

Brenan J (1993) Partitioning of fluorine and chlorine between apatite and aqueous fluids at high pressure and temperature: implications for the F and Cl content of high P-T fluids. Earth PlanetSci Lett 117:251–263

Bucholz CE, Gaetani GA, Behn MD, Shimizu N (2013) Post-entrapment modification of volatiles and oxygen fugacity in olivine-hosted melt inclusions. Earth PlanetSci Lett 374:145–155

Carroll MR, Webster JD (1994) Solubilities of sulfur, noble gases, nitrogen, chlorine, and fluorine in magmas. Rev Mineral Geochem 30(1):231–279

Churikova T, Wörner G, Mironov N, Kronz A (2007) Volatile (S, Cl and F) and fluid mobile trace element compositions in melt inclusions: implications for variable fluid sources across the Kamchatka arc. Contribut Mineral Petrol 154:217–239

Crépisson C, Blanchard M, Bureau H (2014) Clumped fluoride-hydroxyl defects in forsterite: implications for the upper-mantle. Earth Planet Sci Lett 390:287–295

Dalou C, Mysen BO (2012) F and Cl solubility and solution mechanisms in aluminosilicate melts at high pressure and high temperature. Eos Trans AGU Fall Meet, Abstract MR11C –2509

Dalou C, Mysen BO (2013) The effect of water on fluorine and chlorine solubility and solution mechanisms in aluminosilicate melts at high pressure and high temperature. IAVCEI, abstract 1A1_1A-O18

Dalou C, Koga KT, Shimizu N, Boulon J, Devidal J-L (2012) Experimental determination of F and Cl partitioning between lherzolite and basaltic melt. Contribut Mineral Petrol 163:591–609

Dalou C, Mysen BO, Foustoukos DI (2014) In-situ measurements of fluorine and chlorine speciation and partitioning between melts and aqueous fluids in the Na_2O-Al_2O_3-SiO_2-H_2O system. Am Mineral, *in press*

Dingwell DB, Virgo D (1987) The effect of oxidation state on the viscosity of melts in the system Na_2O-FeO-Fe_2O_3-SiO_2. Geochimica Cosmochimica Acta 51:195–205

Edgar AD, Pizzolato LA (1995) An experimental study of partitioning of fluorine between K-richterite, apatite, phlogopite, and melt at 20 kbar. Contribut Mineral Petrol 121(3):247–257

Eggler DH (1972) Amphibole stability in H_2O-undersaturated calc-alkaline melts. Earth Planet Sci Lett 15(1):28–34

Eiler JM, Schiano P, Kitchen N, Stolper EM (2000) Oxygen-isotope evidence for recycled crust in the sources of mid-ocean-ridge basalts. Nature 403(6769):530–534

Elburg M, Kamenetsky V, Nikogosian IK, Foden J, Sobolev AV (2006) Coexisting high- and low-calcium melts identified by mineral and melt inclusion studies of a subduction-influenced syn-collisional magma from South Sulawesi, Indonesia. J Petrol 47:2433

Ferguson AK (1978) The crystallization of pyroxenes and amphiboles in some alkaline rocks and the presence of a pyroxene compositional gap. Contribut Mineral Petrol 67(1):11–15

Filiberto J, Treiman AH (2009) The effect of chlorine on the liquidus of basalt: first results and implications for basalt genesis on Mars and Earth. Chem Geol 263(1):60–68

Filiberto J, Wood J, Dasgupta R, Shimizu N, Le L, Treiman AH (2012) Effect of fluorine on near-liquidus phase equilibria of a Fe–Mg rich basalt. Chem Geol 312–313:118–126

Filiberto J, Dasgupta R, Gross J, Treiman AH (2014) Effect of chlorine on near-liquidus phase equilibria of an Fe–Mg-rich tholeiitic basalt. Contribut Mineral Petrol 168(1):1–13

Gaetani GA, Grove TL (1998) The influence of water on melting of mantle peridotite. Contribut Mineral Petrol 131:323–346

Gaetani GA, Grove TL, Bryan WB (1994) Experimental phase relations of basaltic andesite from Hole 839b under hydrous and anhydrous conditions. In: Hawkins J, Parsons L, Allan J (eds) Proc ODP Sci Results, 135th edn. Ocean Drilling Program, College Station TX, pp 557–564

Gaetani GA, Kent AJ, Grove TL, Hutcheon ID, Stolper EM (2003) Mineral/melt partitioning of trace elements during hydrous peridotite partial melting. Contribut Mineral Petrol 145(4):391–405

Gaetani GA, O'Leary JA, Shimizu N, Bucholz CE, Newville M (2012) Rapid reequilibration of H_2O and oxygen fugacity in olivine-hosted melt inclusions. Geology 40(10):915–918

Giesting PA, Filiberto J (2014) Quantitative models linking igneous amphibole composition with magma Cl and OH content. Am Mineral 99(4):852–865

Green DH (1973) Experimental melting studies on a model upper mantle composition at high pressure under water-saturated and water-undersaturated conditions. Earth Planet Sci Lett 19(1):37–53

Grove T, Parman S, Bowring S, Price R, Baker M (2002) The role of an H_2O-rich fluid component in the generation of primitive basaltic andesites and andesites from the Mt. Shasta region. N California Contribut Mineral Petrol 142(4):375–396

Hamilton DL, Henderson CMB (1968) The preparation of silicate compositions by gelling method. Mineral Mag 36:832–838

Hammouda T (2003) High-pressure melting of carbonated eclogite and experimental constraints on carbon recycling and storage in the mantle. Earth Planet Sci Lett 214(1):357–368

Hauri E, Gaetani G, Green T (2006) Partitioning of water during melting of the Earth's upper mantle at H_2O-undersaturated conditions. Earth Planet Sci Lett 248:715–734

Hilton DR, Fischer TP, Marty B (2002) Noble gases and volatile recycling at subduction zones. Rev Mineral Geochem 47(1):319–370

Hirschmann MM (2006) Water, melting, and the deep Earth H_2O cycle. Annu Rev Earth Planet Sci 34:629–653

Johnson K, Dick HJB, Shimizu N (1990) Melting in the oceanic upper mantle: an ion microprobe study of diopsides in abyssal peridotites. J Geophys Res 95(B3):2661–2678

Kawamoto T, Hirose K (1994) Au-Pd sample containers for melting experiments on iron and water bearing systems. Eur J Mineral Ohne Beihefte 6(3):381–386

Kiczenski TJ, Du LS, Stebbins JF (2004) F-19 NMR study of the ordering of high field strength cations at fluoride sites in silicate and aluminosilicate glasses. J Non-Crystalline Solids 337(2):142–149

Kinzler RJ (1997) Melting of mantle peridotite at pressures approaching the spinel to garnet transition: Application to mid-ocean ridge basalt petrogenesis. J Geophys Res 102(B1):853–874

Kohn SC, Dupree R, Mortuza MG, Henderson CMB (1991) NMR evidence for five-and six-coordinated aluminum fluoride complexes in F-bearing aluminosilicate glasses. Am Mineral 76(1–2):309–312

Kushiro I (1969) The system forsterite–diopside–silica with and without water at high pressures. Am J Sci 267A:269–294

Kushiro I (1972) Effect of water on the composition of magmas formed at high pressures. J Petrol 13(2):311–334

Kushiro I (1990) Partial melting of mantle wedge and evolution of island arc crust. J Geophys Res 95:15929–15939

Kushiro I, Yoder HS Jr, Nishikawa M (1968) Effect of water on the melting of enstatite. Geologic Soc Am Bull 79:1685–1692

Le Voyer M, Rose-Koga EF, Laubier M, Schiano P (2008) Petrogenesis of arc lavas from the Rucu Pichincha and Pan de Azucar volcanoes (Ecuadorian arc): major, trace element, and boron isotope evidences from olivine-hosted melt inclusions. Geochem Geophys Geosyst 9:1–27

Le Voyer M, Rose-Koga EF, Shimizu N, Grove TL, Schiano P (2010) Two contrasting H_2O-rich components in primary melt inclusions from Mount Shasta. J Petrol 51:1571–1595

Le Voyer M, Asimow PD, Mosenfelder JM, Guan Y, Wallace PJ, Schiano P, Stolper EM, Eiler JM (2014) NanoSIMS determination of H_2O and F concentrations in olivines around melt inclusions. J Petrol, doi: 10.1093/petrology/egu003

Lloyd AS, Plank T, Ruprecht P, Hauri EH, Rose W (2013) Volatile loss from melt inclusions in pyroclasts of differing sizes. Contribut Mineral Petrol 165(1):129–153

Médard E, Schmidt MW, Schiano P, Ottolini L (2006) Melting of amphibole-bearing wehrlites: an experimental study on the origin of ultra-calcic nepheline-normative melts. J Petrol 47(3):481–504

Métrich N, Wallace PJ (2008) Volatile abundances in basaltic magmas and their degassing paths tracked by melt inclusions. Rev Mineral Geochem 69(1):363–402

Métrich N, Schiano P, Clocchiatti R, Maury RC (1999) Transfer of sulfur in subduction settings: an example from Batan Island (Luzon volcanic arc, Philippines). Earth Planet Sci Lett 167(1):1–14

Mysen BO (1977) The solubility of H_2O and CO_2 under predicted magma genesis conditions and some petrological and geophysical implications. Rev Geophys 15(3):351–361

Mysen BO (2007) The solution behavior of H_2O in peralkaline aluminosilicate melts at high pressure with implications for properties of hydrous melts. Geochimica Cosmochimica Acta 71(7):1820–1834

Mysen BO, Boettcher AL (1975) Melting of a hydrous mantle: I. Phase relations of natural peridotite at high pressures and temperatures with controlled activities of water, carbon dioxide, and hydrogen. J Petrol 16(1):520–548

Mysen BO, Richet P (2005) Silicate Glasses and Melts - Properties and Structure. Elsevier, New York

Mysen BO, Virgo D (1985) Structure and properties of fluorine-bearing aluminosilicate melts: the system Na_2O-Al_2O_3-SiO_2-F at 1 atm. Contribut Mineral Petrol 91(3):205–220

Mysen BO, Cody GD, Smith A (2004) Solubility mechanisms of fluorine in peralkaline and meta-aluminous silicate glasses and in melts to magmatic temperatures. Geochimica Cosmochimica Acta 68(12):2745–2769

O'Leary JA, Gaetani GA, Hauri EH (2010) The effect of tetrahedral Al^{3+} on the partitioning of water between clinopyroxene and silicate melt. Earth Planet Sci Lett 297:111–120

Ochs FA, Lange RA (1999) The density of hydrous magmatic liquids. Science 283:1314–1317
Parkinson IJ, Arculus RJ (1999) The redox state of subduction zones: insights from arc-peridotites. Chemical Geology 160:409–423
Portnyagin M, Hoernle K, Plechov P, Mironov N, Khubunaya S (2007) Constraints on mantle melting and composition and nature of slab components in volcanic arcs from volatiles (H_2O, S, Cl, F) and trace elements in melt inclusions from the Kamchatka Arc. Earth Planet Sci Lett 255:53–69
Ratajeski K, Sisson TW (1999) Loss of iron to gold capsules in rock-melting experiments. Am Mineral 84:1521–1527
Roeder PL, Emslie RF (1970) Olivine-liquid equilibrium. Contribut Mineral Petrol 19:275–289
Rose-Koga EF, Shimizu N, Devidal J-L, Koga KT, Le Voyer M, Döbeli M (2008) Investigation of F, S and Cl standards by ion probe and electron probe. In: *Presented at the Eos Trans. AGU, Fall meeting supplement*
Rose-Koga EF, Koga KT, Schiano P, Le Voyer M, Shimizu N, Whitehouse MJ, Clocchiatti R (2012) Mantle source heterogeneity for South Tyrrhenian magmas revealed by Pb isotopes and halogen contents of olivine-hosted melt inclusions. Chem Geol 334:266–279
Rüpke LH, Morgan JP, Hort M, Connolly JA (2002) Are the regional variations in Central American arc lavas due to differing basaltic versus peridotitic slab sources of fluids? Geology 30(11):1035–1038
Saal AE, Hauri EH, Langmuir CH, Perfit MR (2002) Vapour undersaturation in primitive mid-ocean-ridge basalt and the volatile content of Earth's upper mantle. Nature 419(6906):451–455
Sadofsky SJ, Portnyagin M, Hoernle K, van den Bogaard P (2008) Subduction cycling of volatiles and trace elements through the Central American volcanic arc: evidence from melt inclusions. Contribut Mineral Petrol 155:433–456
Sandland TO, Du LS, Stebbins JF, Webster JD (2004) Structure of Cl-containing silicate and aluminosilicate glasses: A ^{35}Cl MAS-NMR study. Geochimica Cosmochimica Acta 68(24):5059–5069
Schaller T, Dingwell DB, Keppler H, Knöller W, Merwin L, Sebald A (1992) Fluorine in silicate glasses: a multinuclear nuclear magnetic resonance study. Geochimica Cosmochimica Acta 56(2):701–707
Shaw HR (1963) Obsidian-H_2O viscosities at 1000 and 2000 bars in the temperature range 700°C–900°C. J Geophys Res 68:6337–6343
Shaw HR (1972) Viscosities of magmatic liquids: an empirical method of prediction. Am J Sci 272:870–893
Shaw AM, Hauri EH, Behn MD, Hilton DR, Macpherson CG, Sinton JM (2012) Long-term preservation of slab signatures in the mantle inferred from hydrogen isotopes. Nat Geosci 5:224–228
Shimizu K, Shimizu N, Komiya K, Suzuki K, Maruyama S, Tatsumi Y (2009) CO_2-rich komatiitic melt inclusions in Cr-spinels within beach sand from Gorgona Island, Colombia. Earth Planet Sci Lett 299:33–43
Sisson TW, Grove TL (1993) Temperatures and H_2O contents of low-MgO high-alumina basalts. Contribut Mineral Petrol 113(2):167–184
Sorbadere F, Schiano P, Metrich N, Garaebiti E (2011) Insights into the origin of primitive silica-undersaturated arc magmas of Aoba volcano (Vanuatu arc). Contribut Mineral Petrol 162:995–1009
Stebbins JF, Du LS (2002) Chloride ion sites in silicate and aluminosilicate glasses: a preliminary study by ^{35}Cl solid-state NMR. Am Mineral 87(2–3):359–363
Stolper EM, Newman S (1994) The role of water in the petrogenesis of Mariana trough magmas. Earth Planet Sci Lett 121:293–325
Straub SM, Layne GD (2003) The systematics of chlorine, fluorine, and water in Izu arc front volcanic rocks: Implications for volatile recycling in subduction zones. Geochimica Cosmochimica Acta 67:4179–4203
Straub SM, Layne GD, Schmidt A, Langmuir CH (2004) Volcanic glasses at the Izu arc volcanic front: New perspectives on fluid and sediment melt recycling in subduction zones. Geochem Geophys Geosyst 5(1): doi: 10.1029/2002GC000408
Toplis MJ (2005) The thermodynamics of iron and magnesium partitioning between olivine and liquid: criteria for assessing and predicting equilibrium in natural and experimental systems. Contribut Mineral Petrol 149:22–39
Vigouroux N, Wallace P, Kent AJR (2008) Volatiles in High-K Magmas from the Western Trans-Mexican volcanic belt: evidence for fluid fluxing and extreme enrichment of the mantle wedge by subduction processes. J Petrol 49:1589
Wallace PJ (2005) Volatiles in subduction zone magmas: concentrations and fluxes based on melt inclusion and volcanic gas data. J Volc Geother Res 140(1):217–240
Webster JD (2004) The exsolution of magmatic hydrosaline chloride liquids. Chem Geol 210(1):33–48
Workman RK, Hart SR (2005) Major and trace element composition of the depleted MORB mantle (DMM). Earth Planet Sci Lett 231(1):53–72
Wu J, Koga KT (2013) Fluorine partitioning between hydrous minerals and aqueous fluid at 1 GPa and 770°C–947°C: A new constraint on slab flux. Geochimica Cosmochimica Acta 119:77–92
Wysoczanski R, Wright I, Gamble JA, Hauri EH, Luhr J, Eggins SM, Handler M (2006) Volatile contents of Kermadec Arc–Havre trough pillow glasses: fingerprinting slab-derived aqueous fluids in the mantle sources of arc and back-arc lavas. J Volc Geother Res 152:51–73
Zeng Q, Stebbins JF (2000) Fluoride sites in aluminosilicate glasses: high-resolution ^{19}F NMR results. Am Mineral 85(5–6):863–867
Zimova M, Webb S (2006) The effect of chlorine on the viscosity of Na_2O-Fe_2O_3-Al_2O_3-SiO_2 melts. Am Mineral 91(2–3):344–352

12

The effects of meteoric diagenesis on the geochemical composition and microstructure of Pliocene fossil *Terebratalia coreanica* and *Laqueus rubellus* brachiopod shells from northeastern Japan

Hiroshi Fujioka[1], Hideko Takayanagi[1], Koshi Yamamoto[2] and Yasufumi Iryu[1*]

Abstract

Stable carbon ($\delta^{13}C$) and oxygen isotope ($\delta^{18}O$) compositions of fossil brachiopod shells can be used to interpret paleoclimatic and paleoceanographic conditions. However, the initial isotopic composition of the living shells might be modified by diagenetic alteration. To assess the degree of this modification, we analyzed $\delta^{13}C$ and $\delta^{18}O$ and three common indicators of alteration in shells: minor element (manganese (Mn), iron (Fe), and strontium (Sr)) concentrations; cathodoluminescence (CL)/non-luminescence; and the shell microstructure of fossil *Terebratalia coreanica* and *Laqueus rubellus* (rhynchonellate brachiopod) shells from an exposure of conglomerates of the Tentokuji Formation (Pliocene, 3.85–2.75 Ma) in northeastern Japan. Two indices were used to quantify the preservation state of shell microstructure: the altered fiber ratio (AFR) and the altered fiber and puncta-filling cement ratio (AF-PCR). We qualified the degree of luminescence by measuring the mean cathodoluminescence index (MCLI), defined as the mean *R* values in a particular area of a CL image with RGB colors. The $\delta^{13}C$ and $\delta^{18}O$ values were negatively correlated with AFR or AF-PCR and MCLI. The sampling spots with high Mn and Fe concentrations also exhibited relatively low $\delta^{13}C$ and $\delta^{18}O$. The $\delta^{13}C$ and $\delta^{18}O$ were more strongly correlated with Mn concentration than with Fe concentration. Therefore, of the three minor elements examined in this study, Mn concentration is the most reliable indicator of meteoric diagenesis in the studied shells. Unlike Mn and Fe concentrations, Sr concentrations in the modern and fossil brachiopod shells were similar, with overlapping ranges. This indicates that Sr concentration is not likely to be a good indicator of diagenetic alteration in the studied shells. Our study provides further evidence that multiple criteria should be applied and cross-checked when assessing diagenetic alteration of brachiopod shells.

Keywords: Brachiopod, Carbon isotopes, Oxygen isotopes, Minor element concentration, Cathodoluminescence, Meteoric diagenesis, *Terebratalia coreanica*, *Laqueus rubellus*, Pliocene, Japan

* Correspondence: yasufumi.iryu.d8@tohoku.ac.jp
[1]Department of Earth Science, Graduate School of Science, Tohoku University, Sendai 980-8578, Japan
Full list of author information is available at the end of the article

Introduction

Stable carbon isotope ($\delta^{13}C$) and oxygen isotope ($\delta^{18}O$) compositions of biogenic carbonates are useful for paleoenvironmental analysis. The $\delta^{13}C$ and $\delta^{18}O$ values of shells from brachiopods belonging to the subphylum Rhynchonelliformea (Williams et al. 1996; hereafter, simply called brachiopods) have been widely used for reconstructing paleoclimatic and paleoceanographic conditions during the Phanerozoic, and especially during the Paleozoic (e.g., Veizer et al. 1999), because of (1) their widespread occurrence in marine sediment and their continuous geologic record; (2) their low-magnesium calcite mineralogy, which is less susceptible to diagenetic alteration than high-magnesium calcite and aragonite; and (3) shell secretion in isotopic equilibrium with ambient seawater (Lowenstam 1961). The disequilibrium effects of shell fractionation have been very well investigated since Lowenstam (1961), and it is understood in detail which taxa and shell portions reliably record past ocean environments with the highest fidelity (Auclair et al. 2003; Yamamoto et al. 2010a, 2010b, 2011, 2013; Cusack et al. 2012; Takayanagi et al. 2012, 2013, 2015).

To evaluate the preservation of brachiopod shells, the following three screening criteria have been commonly applied (Ullmann and Korte 2015): (1) minor element (manganese (Mn), iron (Fe), and strontium (Sr)) concentrations, which vary with diagenetic alteration; (2) luminescence/non-luminescence with "cold cathode"; and (3) shell microstructures observed via scanning electron microscope (SEM). Well-preserved shells or shell portions are non-luminescent, having high Sr and low Mn and Fe concentrations and well-preserved microstructure.

However, these techniques do not work effectively in all cases. It has been shown that there might be no significant difference in the isotopic composition between well-preserved and diagenetically altered shells determined using the minor element concentrations (e.g., Bruckschen et al. 1999; Korte et al. 2005). Some unaltered modern brachiopod shells show bright cathodoluminescence (CL) images (e.g., Barbin and Gaspard 1995; Griesshaber et al. 2007), which is usually taken to be an indication of diagenetic alteration. The $\delta^{13}C$ and $\delta^{18}O$ values of luminescent and non-luminescent shells might overlap or be similar in range (Rush and Chafetz 1990; Mii et al. 1997, 2013; Yamamoto et al. 2017). Therefore, applying multiple criteria to assess the diagenetic alteration and cross-checking them are required to distinguish between diagenetically altered and unaltered brachiopod shells (Yamamoto et al. 2017).

The isotopic and minor element signatures in carbonate minerals vary with the isotopic and chemical composition of the ambient pore water (Wenzel 2000; Ullmann et al. 2014), which can be marine, meteoric, or hydrothermal water, or various mixtures of these. This variation implies that diagenetic modification of the chemical and isotopic composition in brachiopod shells embedded in carbonate deposits is likely to be different from that in siliciclastic and volcaniclastic deposits. Cathodoluminescence images of brachiopod carbonates might not be useful for detecting diagenetic alteration where sources of Mn (the activator for CL) are very limited. There are some instances showing that the $\delta^{13}C$ and $\delta^{18}O$ values or the preservation of shell microstructure are unrelated with CL intensity (Yamamoto et al. 2017).

To better understand the impact of diagenetic modification on the isotopic and chemical composition of brachiopod shells in siliciclastic deposits, we examined the $\delta^{13}C$ and $\delta^{18}O$ values, minor element concentrations, CL images, and shell microstructure of fossil brachiopod shells of two species, *Terebratalia coreanica* (Adams and Reeve 1850) (Family Terebrataliidae Richardson 1975) and *Laqueus rubellus* (Sowerby 1846) (Family Laqueidae Thomson 1927) collected from the conglomerates of the Pliocene Tentokuji Formation in Minehama, northeastern Japan. We compared these shells with modern shells from Otsuchi Bay (northeastern Japan) and Sagami Bay (central Japan). *Terebratalia coreanica* is generally found at water depths of 13–287 m in the northwestern Pacific Ocean (Hokkaido, Tsugaru Strait to Boso Peninsula on the Pacific side, Korean Strait north to Tsugaru Strait on the Japan Sea side, Japan; Tsingtao, China; Korea; and the Gulf of Tartary, Russia) (Hatai 1936a, 1936b, 1938; Logan 2007). *Laqueus rubellus* inhabits water depths of 44–907 m in the Pacific Ocean (the seas around Japan; the Japan Sea; Pailolo channel, Hawaii) (Dall 1920; Hatai 1936a; Logan 2007). The wide spatiotemporal distribution of these species means that their $\delta^{13}C$ and $\delta^{18}O$ values are potentially useful for reconstructing the paleoceanographic environments of lower sublittoral to upper bathyal zones during the late Mesozoic to Cenozoic.

Materials and sample sites

Modern specimens

Terebratalia coreanica from Otsuchi Bay

Modern *T. coreanica* (OTTCo; Fig. 1a) were collected by dredging at 70 m water depth in Otsuchi Bay (39° 21.86′ N, 141° 59.97′ E) along the Sanriku Ria Coast in northeastern Japan, using the research vessel *Yayoi* of the International Coastal Research Center, Atmosphere and Ocean Research Institute (ICRS/AORI), The University of Tokyo, on 12 September 2005. High-resolution, three-dimensional $\delta^{13}C$ and $\delta^{18}O$ values of this species have been published by Yamamoto et al. (2011, 2013). The $\delta^{13}C$ and $\delta^{18}O$ values of calcite precipitated in isotopic equilibrium with ambient seawater (equilibrium calcite) at the brachiopod collection site ranged from 1.4 to 2.6‰ ($\delta^{13}C_{EC}$) and from − 1.4 to 2.4‰ ($\delta^{18}O_{EC}$) (Yamamoto et al. 2011, 2013). The

Fig. 1 Photographs of the *Terebratalia coreanica* and *Laqueus rubellus* shells examined in this study. The modern *T. coreanica* shells were collected from Otsuchi Bay, northeastern Japan, and *L. rubellus* shells were collected from Sagami Bay, central Japan. The fossil shells are from the Pliocene Tentokuji Formation in Minehama, northeastern Japan. **a** Modern *T. coreanica*. **b–f** Fossil *T. coreanica*. **g** Modern *L. rubellus*. **h–l** Fossil *L. rubellus*. Only the ventral valves were examined in this study

$\delta^{13}C_{EC}$ values were calculated by using the $\delta^{13}C$ values of inorganic carbon (DIC) at the brachiopod collection site (Otsuchi Bay), the difference between $\delta^{13}C_{HCO3-}$ and $^{13}C_{DIC}$ values (0.2‰, Grossman 1984, Romanek et al. 1992, Zhang et al. 1995), and the calcite-HCO_3^- enrichment factor (1.0 ± 0.20‰, Romanek et al. 1992) (Additional file 7: Doc. S1). The $\delta^{18}O_{EC}$ values were estimated based on the equation given by Friedman and O'Neil (1977) (Additional file 7: Doc. S1).

Laqueus rubellus from Sagami Bay

Modern *L. rubellus* (SGLR; Fig. 1g) were collected by dredging at 80 m water depth at the eastern margin of Sagami Bay, central Japan (35° 08.17′ N, 139° 34.86′ E), during a cruise of the fishing vessel *Rinkai-maru*, based at the Misaki Marine Biological Station, The University of Tokyo, on 16 June 2005. Yamamoto et al. (2010a) have presented high-resolution, three-dimensional $\delta^{13}C$ and $\delta^{18}O$ values of this species. The $\delta^{13}C_{EC}$ values ranged from 1.6 to 2.6‰, and $\delta^{18}O_{EC}$ ranged from − 0.6 to 0.8‰ (Yamamoto et al. 2010a). These values were calculated by the same method as described above.

Fossil shells

Fossil *T. coreanica* (MNTCo1–MNTCo5; Fig. 1b–f) and *L. rubellus* (MNLR1–MNLR5; Fig. 1h–l) were collected from sandy conglomerates of the upper part of the Pliocene Tentokuji Formation exposed at Minehama (MH1, Amano et al. 2011), Happo-cho, Akita Prefecture, northeastern Japan (Fig. 2; 40° 17.18′ N, 140° 4.52′ E). The

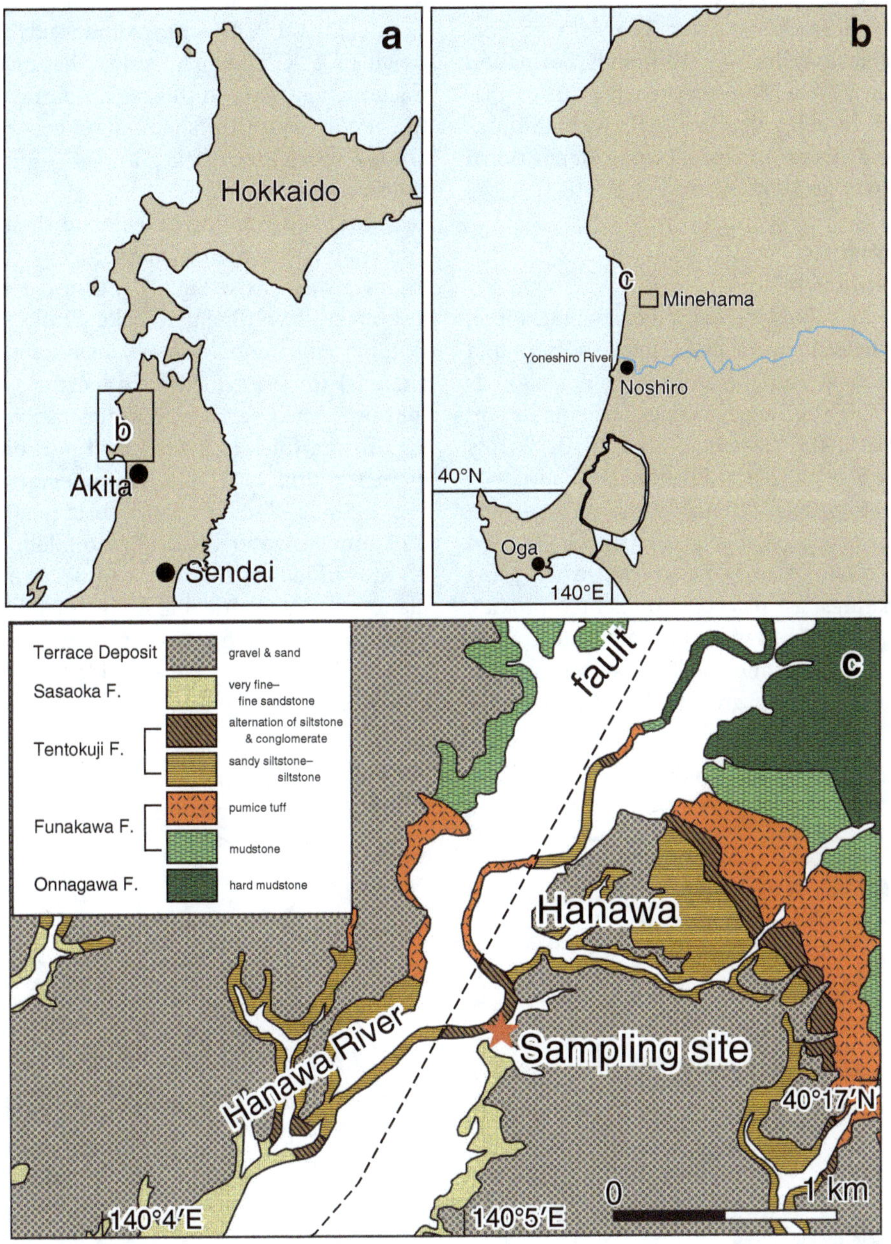

Fig. 2 Location and geologic map of the sample site. **a** Akita Prefecture, northeastern Japan. **b** Enlarged view of the square in **a**. **c** Geologic map of the Minehama area (modified from Sato et al. 2003). The red star indicates the site at which the fossil *Terebratalia coreanica* and *Laqueus rubellus* shells were collected

horizon with the fossil brachiopod shells is dated 3.85–2.75 Ma, owing to its location between the Upper Pliocene datum planes (datum planes 21 and A) (Sato et al. 2003), which represent ages of 3.85 Ma and 2.75 Ma (Sato and Kameo 1996).

The Upper Miocene to Pleistocene succession around Minehama consists of the Onnagawa, Funakawa, Tentokuji, and Sasaoka formations in ascending order (Fig. 2c; Sato et al. 2003). The Onnagawa Formation is composed mainly of hard mudstone (generally termed "hard shale") conformably overlain by the Funakawa Formation, which comprises mostly mudstone and pumice tuff. The Tentokuji Formation, conformably overlying the Funakawa Formation, consists mainly of dark bluish-gray siltstone alternating with conglomerates in the middle-upper intervals of the formation. The alternating beds are regarded as a turbidite sequence (Sato et al. 2003). The conglomerates yielding the brachiopod shells analyzed in the present study are 5–10 cm thick and consist of rounded to sub-rounded, granule to pebble-sized gravel of andesite and, to a lesser extent, pumice, with a matrix of coarse-grained volcaniclastic sand. The Sasaoka Formation conformably overlies the Tentokuji Formation and consists of poorly sorted, very coarse to coarse-grained sandstone. Bivalves that co-occur with the brachiopods indicate a lower sublittoral zone depositional environment (Higo et al. 1999; Amano et al. 2011).

Methods/Experimental

High-resolution microsampling

The $\delta^{13}C$ and $\delta^{18}O$ values of the modern brachiopod shells display significant intra-shell, intra-species, and inter-species variations, and the isotope composition is not in all cases within the range of values of equilibrium calcite (Auclair et al. 2003; Yamamoto et al. 2010a, 2010b, 2011, 2013; Cusack et al. 2012; Takayanagi et al. 2012, 2013, 2015). Therefore, specific shell portions, for which the $\delta^{13}C$ and $\delta^{18}O$ values are identical to the $\delta^{13}C_{EC}$ and $\delta^{18}O_{EC}$ values, respectively, must be selected when using the isotopic signatures as paleoenvironmental proxies. Therefore, in recent studies, samples have been extracted from spot locations on the inner shell surfaces in vertical sections along the maximum growth (longitudinal) axis (e.g., Angiolini et al. 2007; Garbelli et al. 2014, 2016; Takizawa et al. 2017). This procedure yields a relatively narrow range of $\delta^{13}C$ and $\delta^{18}O$ values because of the minimal fractionation effect in this region of the shell, arising from the very low growth (shell thickening) rate, except in and around specialized shell portions (e.g., muscle scars and the foramen).

Ventral valves of the brachiopod shells were embedded in a resin epoxy matrix and vertically cut along the maximum growth axis (Fig. 3) with a 0.4 mm thick IsoMet low-speed saw (Buehler, Lake Bluff). The samples for isotopic and minor element analyses (~ 0.1 mg for each analysis) were obtained from locations within 0.5 mm of the inner shell surface at 1 mm intervals (red arrows in Fig. 3) in a thick (> 1 mm) shell portion using a 0.2 mm diameter drill and at 2 mm intervals (blue arrows in Fig. 3) in a thin (< 1 mm) shell portion with a 0.10–0.12 mm diameter drill. The sampled spots were ~ 0.6–2.3 mm in length and less than 0.5 mm wide. We used a diamond-coated drill bit for extracting samples to prevent contamination by iron present in the drill bits.

Individual shells were divided into three intervals based on the shell morphology: interval 1 corresponding

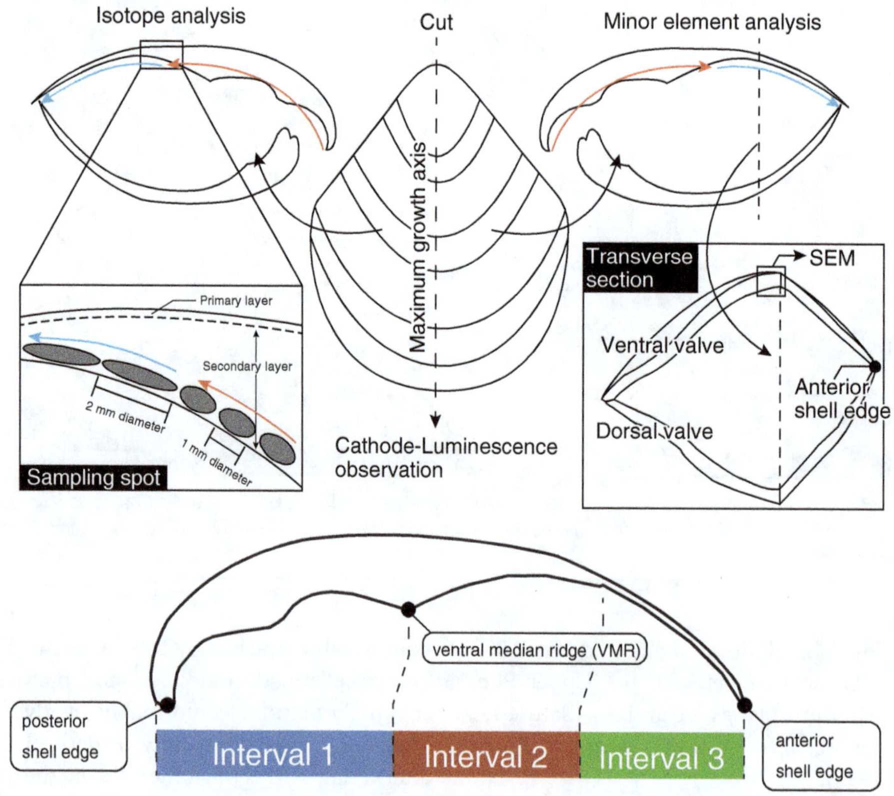

Fig. 3 Schematic illustration of the sampling procedure for geochemical analyses of the modern and fossil shells. The red and blue arrows indicate sampling intervals of 1 and 2 mm, respectively. Shells were cut along the maximum growth line and then along a line perpendicular to the maximum growth axis. Thin sections along the maximum growth axis were examined in CL images. SEM observations were carried out on the cross-section along the line perpendicular to the maximum growth axis. Individual shells were divided into three intervals (intervals 1–3) based on the shell morphology

to a thick posterior shell region, interval 2 located between intervals 1 and 3, and interval 3 denoting the thin anterior shell region (Fig. 3). This division is identical to that proposed for *Kikaithyris hanzawai* shells (Takizawa et al. 2017). The boundary between intervals 1 and 2 in both species corresponded to a convex portion located in the center of ventral valves, which was identical to the "ventral median ridge (VMR)" reported from the ventral valves of *K. hanzawai* (Endo 1989), and hereafter, we use this term. The VMR was located 12.5–16.3 mm from the posterior shell edge in *T. coreanica* and 13.2–14.6 mm in *L. rubellus*. The boundary between intervals 2 and 3 was a region where the shell thickness became less than 1 mm. This region was located 4.6–12.8 and 3.8–10.2 mm from the VMR toward the anterior shell edge in *T. coreanica* shells and *L. rubellus* shells, respectively.

Powder samples were extracted from calcite cement in an internal cavity (7 mm × 4 mm in diameter) of a fossil brachiopod shell using the same method as described above. This brachiopod shell was small (a distance from the posterior shell edge to the anterior shell edge of < 1 cm) and was not used for geochemical analyses in this study. The inner shell surface of the fossil brachiopod was covered thinly with blocky equant calcites with undulated protuberances (~ 2 mm long and ~ 1 mm in diameter) showing a helictite-like appearance. This is the largest cement found in this study. Based on its location and morphology, this calcite cement is considered to have been formed by meteoric diagenesis. The intergranular space of the sandstone matrix was filled with submillimeter-sized, blocky cements. No sign of hydrothermal activity was found around the brachiopod sampling site (Minehama) or within the surrounding region.

SEM, energy dispersive spectrometer (EDS), and CL observations

After the shells had been micro-sampled for isotopic and minor element analyses, they were cut parallel to the maximum growth axis and made into thin sections (Fig. 3). The spatial offset between the geochemical sampling and the CL sections were less than 0.5 mm. CL images of the thin sections were examined using a Relion Industries Reliotron VII cold cathodoluminescence microscope mounted on a polarizing microscope at the Department of Earth Sciences, Environmental and Resources Engineering, Waseda University, Japan. The beam current was ~ 0.06–0.2 mA, and the voltage was 10–13 kV. The beam diameters were ~ 7 mm for the observation of modern shells and ~ 1.5 cm for the fossil shells. We measured the mean CL intensity (MCLI) in the corresponding areas (~ 400 μm × 1000 μm and 200 μm × 2000 μm in thick and thin shell portions, respectively) of the isotopic and minor element sampling locations to quantify the degree of luminescence. The CL images represent color information for each dot as a mixture of R (red), G (green), and B (blue), each represented in 256 gradations from 0 to 255. The MCLI was defined as the mean *R* values of a CL image. The *R* values were measured using ImageJ software developed at the National Institutes of Health, USA (ver. 1.59 m9 with Java 1.8.0_101).

After making the thin sections for CL observations, the shells were cut transversely (perpendicular to the maximum growth axis; Fig. 3) along a transect located ~ 13–35 mm from the posterior shell edge. The cut surface was etched with 0.2 mol/L hydrochloric acid (HCl) for 30 s and then coated with carbon. The surface was then observed using a Keyence VE-8800 SEM at the Institute of Geology and Paleontology, Graduate School of Science, Tohoku University (IGPS), with an accelerating voltage of 10 kV. The secondary shell layer of brachiopods is composed of calcite fibers that extend along the maximum growth axis. During diagenesis, these calcite fibers are dissolved and replaced with blocky calcite. The altered fibers become amalgamated, and their outlines become indistinct (Samtleben et al. 2001; Yamamoto et al. 2017). Traditionally, longitudinal sections are used to examine shell microstructure (e.g., Bates and Brand 1991; Qing and Veizer 1994; Azmy et al. 1998; Korte et al. 2005), but this diagenetic alteration of fibers can be observed more clearly in transverse sections, which enables a better determination to be made of whether or not the fibers retained their initial forms (e.g., Samtleben et al. 2001; Garbelli et al. 2014; Takizawa et al. 2017; Yamamoto et al. 2017).

Chemical microanalyses were performed using an Oxford Instruments energy dispersive spectrometer (EDS) installed in a JEOL JSM-7001F field emission SEM (FE-SEM) at the IGPS. The constant accelerating voltage for the EDS analysis was 15 kV and the probe current 1.47 A.

We compared the microstructures of the modern and fossil brachiopod shells and defined two divisions of preservation state: well-preserved (Fig. 4e, f) and diagenetically altered (Fig. 4g, h). In the well-preserved shell portions, the outlines of the calcite fibers were well defined, and each fiber was plainly discernible. In the altered portions, the individual fibers were unobservable owing to their amalgamation. We used two indices to quantify the preservation state of shell microstructure: altered fiber ratio (AFR) and altered fiber and puncta-filling cement ratio (AF-PCR). The former is defined as a ratio (%) of the area of altered fibers to the total area of fibers observed in a particular area (200 μm × 250 μm (magnification = × 400) or 212 μm × 266 μm (magnification = × 600)). The AF-PCR, which corresponds to the AFR of Yamamoto et al. (2017), represents a ratio (%) of the area occupied by altered fibers plus puncta-filling cement to the total area of the fibers where we measured

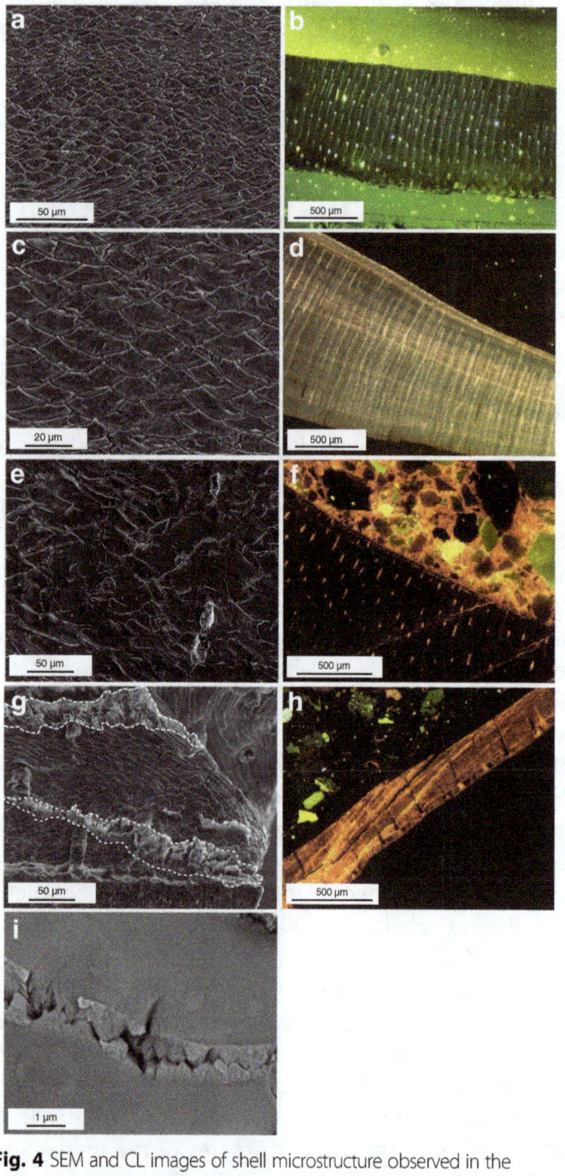

Fig. 4 SEM and CL images of shell microstructure observed in the studied brachiopod shells. **a** SEM image of modern *Terebratalia coreanica* (OTTCo). **b** CL image of modern *T. coreanica* (OTTCo)*. **c** SEM image of modern *Laqueus rubellus* shell (SGLR). **d** CL image of modern *L. rubellus* shell (SGLR). **e** SEM image of well-preserved shell microstructure of fossil *T. coreanica* (MNTCo3). **f** CL image in the shell portion adjacent to **e**, with luminescent punctae. **g** SEM image of altered shell microstructure (surrounded by dotted lines) in a fossil *L. rubellus* (MNLR2) shell. **h** CL image of the shell portion adjacent to **g**, showing bright luminescence. **i** SEM image of cement precipitated in the inter-fiber spaces of fossil *T. coreanica* (MNTCo1). Image brightness and contrast were adjusted with Adobe Photoshop software (Adobe Systems Incorporated, San Jose, USA) so that the luminescent bands become clearly visible

the AFR. We used the two indices because the impact of puncta-filling cement on the diagenetic modification of the isotopic compositions and minor element concentrations of the fossil shells is not negligible and needs to be properly evaluated. We did not measure these indices for interval 1, as the isotopic compositions for this interval cannot be used for paleoenvironmental analysis, as discussed below.

Isotopic analysis

Stable carbon and oxygen isotope analyses of carbonate powder samples were performed using a Thermo Fisher Delta V Advantage isotope ratio mass spectrometer coupled to a ThermoQuest Kiel-III automated carbonate device, at the IGPS. The samples (~ 0.1 mg) were reacted with 100% phosphoric acid at ~ 72 °C. The isotope ratios were expressed in conventional (δ‰) notation and calibrated to the NBS-19 international standard relative to the Vienna Pee Dee Belemnite (VPDB). The external precision (1σ) based on replicate measurements ($n = 91$) of the laboratory reference material (JCt-1; Okai et al. 2004) was 0.02‰ for the carbon isotope analysis and 0.05‰ for oxygen isotope analysis.

It is known that the $\delta^{18}O$ values for magnesian calcite (Mg-calcite) are higher than those for pure calcite formed under the same conditions (e.g., Tarutani et al. 1969). However, if a fractionation correction for Mg-calcite relative to pure calcite (0.06‰ per mol% $MgCO_3$, according to Tarutani et al. 1969, 0.17‰ per mol% $MgCO_3$, according to Jiménez-López et al. 2004) is applied to $\delta^{18}O$ values of the studied shells, this effect is negligible or limited due to their low Mg contents (< 1.18 mol% $MgCO_3$). Therefore, the Mg effect on the $\delta^{18}O$ values of brachiopod shells (Brand et al. 2013) does not need to be accounted for in this study.

Minor element analysis

Powdered brachiopod shell samples (20.5–189.8 μg) were dissolved in ~ 5.0–10.0 mL of volume-specific 2% (v/v) nitric acid (HNO_3). The volume of the acid was varied to adjust the Ca concentration in the solutions to ~ 3–4 ppm for most analyses. Concentrations of the three minor elements (Mn, Fe, and Sr) were analyzed using an Agilent 7700x inductively coupled plasma mass spectrometer at the Department of Earth and Environmental Sciences, Nagoya University, Japan. Minor elemental concentrations are expressed as ppm (= μg/g). The precision of the analytical method, expressed as the relative standard deviation (σ) of repeated analyses of the laboratory reference material (JCt-1; Okai et al. 2004), was 1.8–4.6% for Mn, 10.8–11.9% for Fe, and 2.2–2.4% for Sr.

Results

All analytical data are given in Additional file 1: Table S1, Additional file 2: Table S2, Additional file 3: Table S3, and Additional file 4: Table S4.

SEM, EDS, and CL observations

Modern shells

The shells of modern *T. coreanica* (OTTCo) and *L. rubellus* (SGLR) consisted of two calcite layers: the thinner outer primary shell layer and a thicker inner secondary shell layer (Fig. 4a, c). The primary shell layers of both brachiopod species displayed a fine granular structure measuring up to 30 μm thick in *T. coreanica* and up to 25 μm in *L. rubellus*. The secondary shell layer accounted for > 90% of the shell thickness and was composed of thin elongated calcite fibers oriented parallel or slightly oblique to the maximum shell growth axis. The two shell layers were penetrated with punctae perpendicular to the outer and inner shell surfaces. The calcite fibers within the secondary shell layer were clearly identifiable. We regarded this state as the initial fiber structure (Fig. 4a, c).

The modern shells of the two species exhibited intrinsic blue luminescence (Fig. 4b, d; Additional file 5: Figure S1-1b; Additional file 6: Figure S2–1b,) caused by a lack of Mn-activated visible luminescence and by crystal imperfections (Sippel and Glover 1965; Machel et al. 1991; Habermann et al. 1998). Here, we term this blue luminescence as "non-luminescence."

Fossil shells

The primary shell layer, which is commonly lost in fossil shells (Angiolini et al. 2007; Garbelli et al. 2014, 2017), was preserved in all fossil specimens (MNTCo1–MNTCo5 and MNLR1–MNLR5). The secondary shell layer of the fossil brachiopods consisted of calcite fibers (Fig. 4e) similar to those of the modern shells. However, traces of the dissolution and amalgamation of calcite fibers (Fig. 4g) were identified in many shell portions. The punctae (< 10 μm in diameter), perforations penetrating brachiopod shells, were scattered, commonly infilled with calcite cements. The abundance of altered structures was highly variable among the shell portions, with AFR values ranging from 0 to 92% and AF-PCR values from 1 to 95%.

Some parts of the fossil shells exhibited bright luminescence in CL images (Fig. 4f, h), with MCLI values ranging from 18 to 213 in *T. coreanica* and from 12 to 216 in *L. rubellus*. The strongly luminescent shell portions were also diagenetically altered, as indicated by the traces of dissolution and amalgamation of calcite fibers observed via SEM and the high AFR and AF-PCR values. The fossil shells had luminescent dots and lines (< 15 μm in diameter), each corresponding to a calcite cement filling a puncta (Fig. 5).

Submicron-sized, subhedral, blocky, isopachous fringing calcite cements were found in the inter-fiber spaces of some fossil brachiopod shells with SEM (Fig. 4i). These cements formed in a narrow space between the calcite fibers. The remnants of organic sheaths surrounding rarely occurred. EDS analysis indicated that the Mn concentrations of these cements were as high as 22,000 ppm and their Fe concentrations up to 11,000 ppm, although the concentrations were highly variable. In contrast, Sr was not detected in such cements.

Isotopic composition

Modern shells

The $\delta^{13}C$ values of samples from the inner shell surfaces of modern *T. coreanica* from Otsuchi Bay (OTTCo) ranged from − 0.15 to 1.38‰ and from − 0.12 to 1.75‰ for $\delta^{18}O$ (Table 1, Fig. 6a, and Additional file 5: Figure S1-1). Interval 1 exhibited wider ranges of isotopic values than did intervals 2 and 3, containing two local $\delta^{13}C$ minima and four local $\delta^{18}O$ minima (Fig. 6a, and Additional file 5: Figure S1-1). The $\delta^{13}C$ profile in this interval consisted of lower values near the posterior shell edge and higher values near the VMR, which were separated by a local minimum $\delta^{13}C$ value between the two sections. The $\delta^{13}C$ values in the lower-value section varied within a narrow range of 0.16 to 0.51‰. In contrast, the $\delta^{13}C$ values in the higher-value section exhibited large-amplitude fluctuations within a range of − 0.15 to 1.38‰. The $\delta^{18}O$ profile exhibited large-amplitude fluctuations throughout interval 1. The $\delta^{13}C$ and $\delta^{18}O$ profiles in intervals 2 and 3 showed small-amplitude fluctuations with values from 0.65 to 1.36‰ for the $\delta^{13}C$ values and from 0.93 to 1.75‰ for the $\delta^{18}O$ values. Although the $\delta^{18}O$ values in intervals 1–3 were within the range of equilibrium calcite, the $\delta^{13}C$ values were not (Additional file 5: Figure S1-1).

The $\delta^{13}C$ and $\delta^{18}O$ values of samples from modern *L. rubellus* from Sagami Bay (SGLR) covaried ($r = 0.57$, $n = 27$, $p < 0.01$), ranging from 0.52 to 1.96‰ and from 0.26 to 0.94‰, respectively (Table 1, Fig. 6b, and Additional file 6: Figure S2-1). The $\delta^{13}C$ profile showed a gradual increase from 0.52 to 1.57‰ toward the VMR in interval 1, followed by a plateau ranging from 1.49 to 1.94‰ with an average value of 1.68‰ in interval 2 and a range of 1.28 to 1.96‰ (average 1.56‰) in interval 3 (Fig. 6b). The $\delta^{18}O$ values varied within a narrow range from 0.26 to 0.94‰ (average 0.62‰) throughout intervals 1–3. The $\delta^{18}O$ profile showed a gradually increasing trend in interval 1 and a decreasing trend in interval 3. The $\delta^{13}C$ values in interval 1 were lower than those of equilibrium calcite, and the $\delta^{13}C$ values in intervals 2 and 3 were slightly lower than or identical to the lowest values of equilibrium calcite (Additional file 6: Figure S2-1). The $\delta^{18}O$ values fell mostly in the upper half of the $\delta^{18}O$ range of equilibrium calcite.

Fossil shells

The $\delta^{13}C$ and $\delta^{18}O$ values of the five fossil *T. coreanica* (MNTCo1–MNTCo5) shells fell in wide ranges

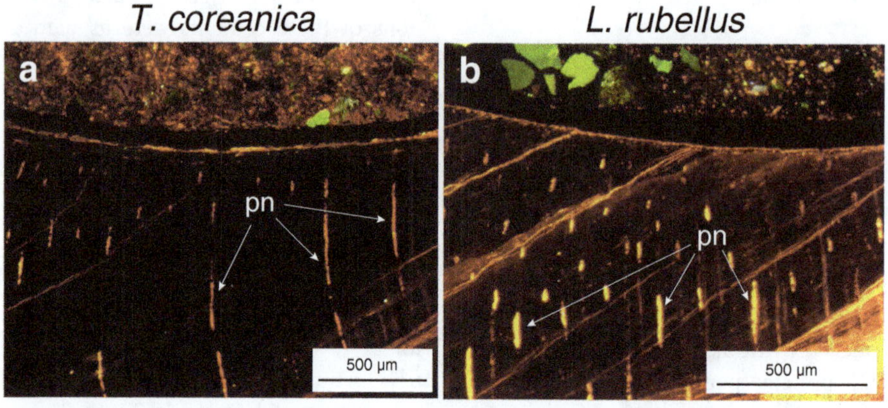

Fig. 5 CL images of puncta-filling cements. CL images of the vertical sections through the studied brachiopod shells (**a** *Terebratalia coreanica*, MNTCo2; **b** *Laqueus rubellus*, MNLR2). Note that strongly luminescent dots and lines each corresponding to a calcite cement filling a puncta

from − 3.31 to 2.43‰ for $\delta^{13}C$ and from − 0.42 to 2.18‰ for $\delta^{18}O$ (Table 1, Fig. 6a, and Additional file 5: Figures S1-2–6). The $\delta^{13}C$ profiles from interval 1 of the fossil shells had a wider range (2.41–4.37‰) than that of the modern shell (1.53‰). These $\delta^{13}C$ profiles for interval 1 generally followed the pattern of a lower-value section near the posterior shell edge and a higher-value section near the VMR. These two sections were separated by a local minimum $\delta^{13}C$ value, with an overall increasing trend toward the VMR. This was the same pattern seen in the modern shell. However, the higher-value section near the VMR of interval 1 from two shells (MNTCo2 and MNTCo5; Additional file 5: Figure S1-3, 6) was associated with a local minimum in $\delta^{13}C$ values. The $\delta^{13}C$ profiles from intervals 2 and 3 of the fossil shells exhibited a different trend from the same intervals of the modern shell. Shells MNTCo1 and MNTCo2 both showed a high-value plateau extending over interval 2. The $\delta^{13}C$ profile of MNTCo3 was similar to those of MNTCo1 and MNTCo2 but exhibited a local minimum in this interval (Additional file 5: Figures S1-2–4). Interval 2 of shells MNTCo4 and MNTCo5 exhibited a short high-value plateau near the VMR, followed by a marked decrease toward the anterior shell edge. The $\delta^{13}C$ values in interval 3 were lower than those of the high-value plateau in interval 2. Shell MNTCo1 had a local $\delta^{13}C$ minimum in this interval, and shell MNTCo3 had two local minima (Additional file 5: Figures S1-2, 4).

$\delta^{18}O$ profiles of intervals 1 and 2 of the fossil *T. coreanica* were similar to those of the modern shell. Interval 1 had larger $\delta^{18}O$ amplitudes than did intervals 2 and 3 as a result of a series of erratic, negative $\delta^{18}O$ spikes. The $\delta^{18}O$ values were more consistent in interval 2 than in intervals 1 and 3. Each profile from interval 2 fell in a narrow range (0.07–1.22‰), and the total range of the $\delta^{18}O$ values from the five shells was only 1.44‰. The $\delta^{18}O$ values in interval 3 of the three fossil shells (MNTCo2, MNTCo4, and MNTCo5) fell in a narrow range (0.28–0.57‰), as did the modern shell (0.82‰). The other two fossil shells (MNTCo1 and MNTCo3) exhibited distinct negative spikes in the $\delta^{18}O$ profiles in interval 3 (one in MNTCo1 and two in MNTCo3; Additional file 5: Figure S1-2, 4).

The $\delta^{13}C$ values of samples from fossil *L. rubellus* shells (MNLR1–MNLR5) ranged from − 3.58 to 2.42‰, and $\delta^{18}O$ ranged from 0.03 to 2.37‰ (Table 1, Fig. 6b, and Additional file 6: Figures S2-2–6). The $\delta^{13}C$ values in interval 1 were highly variable, and each profile had a wider range (3.07–5.47‰, average 3.71‰) compared with the modern shell (1.05‰, Fig. 6b). The gradually increasing trend in $\delta^{13}C$ toward the VMR in this interval was recognized in the fossil shells, but it was not as obvious as in the modern shell. The $\delta^{13}C$ profiles showed three trends in interval 2: relatively constant (MNLR1 and MNLR3), increasing toward the anterior (MNLR4), and decreasing toward the anterior (MNLR2 and MNLR5). The $\delta^{13}C$ values were again highly variable in interval 3 without any common trend among the fossil shells, and almost all values were lower than those of the modern shell except for one (MNLR1).

The $\delta^{18}O$ values of fossil *L. rubellus* shells were less varied than the $\delta^{13}C$ values throughout the three intervals. Each profile exhibited a narrow range (1.23–1.81‰), and the total range of the $\delta^{18}O$ values from the five shells was < 2.34‰.

Calcite cement

The $\delta^{13}C$ and $\delta^{18}O$ values of the calcite cement sample were much lower than those of the fossil shells. The mean values from eight repeated measurements were −17.87‰ ($\sigma = 0.12$‰) for $\delta^{13}C$ and − 6.62‰ ($\sigma = 0.09$‰) for $\delta^{18}O$ (Fig. 7).

Table 1 Modern and fossil *Terebratalia coreanica* and *Laqueus rubellus* $\delta^{13}C$ and $\delta^{18}O$ values and minor-element concentrations

	Interval 1			Interval 2			Interval 3		
Sample No.	Range	Average	Median	Range	Average	Median	Range	Average	Median
T. coreanica									
$\delta^{13}C$ (‰ VPDB)									
OTTCo	− 0.15 to 1.38	0.56	0.45	1.01 to 1.36	1.19	1.19	0.65 to 1.17	1.03	1.05
MNTCo1	−2.02 to 1.69	−0.43	−0.46	2.10 to 2.40	2.19	2.13	−2.98 to 1.70	0.31	0.84
MNTCo2	−2.30 to 1.30	−0.64	− 0.93	1.17 to 2.15	1.81	1.96	0.06 to 0.65	0.34	0.28
MNTCo3	−2.78 to 1.59	−0.05	− 0.07	0.59 to 1.98	1.57	1.73	−3.31 to 0.46	−1.16	−0.94
MNTCo4	−1.76 to 0.65	− 0.25	− 0.25	− 0.03 to 2.43	1.21	1.01	−0.10 to 0.54	0.22	0.22
MNTCo5	−1.89 to 2.08	0.16	0.00	−0.01 to 2.14	1.23	1.22	0.27 to 1.19	0.77	0.82
$\delta^{18}O$ (‰ VPDB)									
OTTCo	−0.12 to 1.75	1.04	1.12	1.17 to 1.43	1.36	1.41	0.93 to 1.75	1.39	1.47
MNTCo1	−0.42 to 2.18	1.18	1.35	2.06 to 2.13	2.08	2.07	−0.21 to 2.00	1.34	1.76
MNTCo2	0.51 to 1.42	1.00	1.07	0.98 to 1.48	1.29	1.35	0.97 to 1.54	1.20	1.18
MNTCo3	0.50 to 1.62	1.11	1.22	0.93 to 1.64	1.37	1.35	0.10 to 1.36	0.70	0.75
MNTCo4	−0.26 to 1.60	0.70	0.60	0.69 to 1.91	1.55	1.63	1.06 to 1.53	1.31	1.32
MNTCo5	0.41 to 1.83	1.33	1.38	0.99 to 1.97	1.57	1.67	1.33 to 1.61	1.51	1.55
Mn (ppm)									
OTTCo	2 to 10	5	4	3 to 7	4	3	2 to 8	5	5
MNTCo1	80 to 2000	480	240	90 to 260	160	130	380 to 3000	1200	1100
MNTCo2	270 to 1900	770	700	390 to 3300	1100	420	580 to 2300	1600	1700
MNTCo3	170 to 1800	450	240	290 to 1300	530	390	1600 to 2500	1900	1700
MNTCo4	150 to 440	250	250	180 to 1700	700	570	790 to 1400	1100	1100
MNTCo5	150 to 1000	450	400	230 to 2700	910	770	600 to 1600	1000	890
Fe (ppm)									
OTTCo	30 to 1100	250	110	50 to 800	250	70	50 to 360	150	70
MNTCo1	590 to 5800	1800	1200	1200 to 1900	1500	1300	2000 to 6900	3900	2900
MNTCo2	200 to 2000	680	530	240 to 2400	870	400	860 to 3600	1800	1500
MNTCo3	200 to 4400	1300	770	220 to 970	470	400	1200 to 3200	2000	1500
MNTCo4	310 to 1500	670	580	870 to 3300	2000	1900	3800 to 4200	4000	4000
MNTCo5	1400 to 4900	2600	2100	1100 to 4700	2300	2000	2200 to 4200	3200	3400
Sr (ppm)									
OTTCo	810 to 1100	980	960	880 to 1000	960	960	890 to 1000	970	980
MNTCo1	820 to 1000	910	920	860 to 950	910	920	810 to 960	910	910
MNTCo2	730 to 950	850	850	840 to 940	890	890	850 to 920	890	880
MNTCo3	790 to 1000	900	910	820 to 920	870	870	810 to 870	830	830
MNTCo4	890 to 1100	980	970	780 to 990	880	870	830 to 940	890	890
MNTCo5	790 to 1000	890	880	820 to 910	860	860	850 to 880	860	860
L. rubellus									
$\delta^{13}C$ (‰ VPDB)									
SGLR	0.52 to 1.57	1.06	1.09	1.49 to 1.94	1.68	1.68	1.28 to 1.96	1.56	1.59
MNLR1	−0.97 to 2.12	0.80	1.07	1.89 to 2.39	2.07	1.93	–	–	–
MNLR2	−1.41 to 1.74	0.21	0.33	−0.26 to 1.86	1.02	1.46	−1.51 to 0.57	−0.52	−0.41
MNLR3	−3.58 to 1.89	−0.57	−0.32	1.94 to 2.42	2.21	2.27	0.45 to 1.23	0.73	0.64
MNLR4	−2.65 to 0.42	−1.32	−1.50	−0.23 to 1.87	0.85	0.84	−2.82 to 0.58	−1.24	−1.49
MNLR5	−1.69 to 2.06	0.69	0.33	−1.15 to 2.23	1.10	1.60	−1.87 to 1.07	−0.21	0.18

Table 1 Modern and fossil *Terebratalia coreanica* and *Laqueus rubellus* $\delta^{13}C$ and $\delta^{18}O$ values and minor-element concentrations *(Continued)*

Sample No.	Interval 1 Range	Average	Median	Interval 2 Range	Average	Median	Interval 3 Range	Average	Median
$\delta^{18}O$ (‰ VPDB)									
SGLR	0.26 to 0.94	0.60	0.61	0.55 to 0.93	0.74	0.73	0.26 to 0.81	0.53	0.43
MNLR1	1.14 to 2.09	1.67	1.71	2.09 to 2.33	2.19	2.17	–	–	–
MNLR2	0.74 to 1.79	1.16	1.16	0.91 to 1.96	1.53	1.73	0.24 to 1.44	0.73	0.66
MNLR3	0.03 to 1.50	0.88	1.04	1.54 to 1.85	1.69	1.68	0.96 to 1.57	1.15	0.98
MNLR4	0.29 to 1.31	0.83	0.86	1.19 to 2.00	1.59	1.55	0.53 to 1.66	1.01	0.83
MNLR5	0.63 to 1.92	1.34	1.42	0.56 to 2.12	1.61	2.00	0.31 to 1.44	0.95	1.10
Mn (ppm)									
SGLR	4 to 20	9	7	6 to 10	8	7	10 to 30	20	10
MNLR1	150 to 760	370	310	260 to 500	380	370	–	–	–
MNLR2	170 to 1100	620	650	180 to 1200	610	530	450 to 2000	1300	1300
MNLR3	130 to 2700	1100	680	170 to 260	210	200	800 to 2100	1300	1200
MNLR4	490 to 1900	1100	930	260 to 700	380	280	410 to 1700	1200	1300
MNLR5	160 to 940	410	340	120 to 1900	710	240	760 to 1000	900	920
Fe (ppm)									
SGLR	70 to 10,100	1300	620	70 to 1200	310	120	100 to 3600	1100	510
MNLR1	410 to 3900	1000	640	1200 to 3200	2000	1600	–	–	–
MNLR2	180 to 5000	1100	400	230 to 750	500	500	1300 to 5800	3200	2800
MNLR3	340 to 2700	1200	1200	440 to 4900	1900	510	2000 to 4200	2800	2600
MNLR4	600 to 1500	1100	1000	400 to 940	640	610	900 to 3400	2500	2900
MNLR5	510 to 3100	1300	1000	830 to 4700	2000	1600	2700 to 5100	3500	3200
Sr (ppm)									
SGLR	860 to 1100	990	1000	800 to 860	830	820	870 to 980	920	930
MNLR1	930 to 1000	980	970	880 to 920	900	900	–	–	–
MNLR2	850 to 1000	920	930	830 to 890	860	850	820 to 970	890	880
MNLR3	730 to 1000	870	890	850 to 860	850	850	810 to 1000	890	890
MNLR4	800 to 880	840	830	830 to 870	850	850	730 to 940	820	800
MNLR5	870 to 970	920	920	800 to 890	840	830	900 to 1000	950	950

Minor elements

Modern shells

The minor element concentrations ($n = 30$) in the modern *T. coreanica* shell fell within the ranges of 2–10 ppm (mean = 5 ppm, $\sigma = 3$ ppm, median = 4 ppm) for Mn, 30–1100 ppm (mean = 220 ppm, $\sigma = 260$ ppm, median = 100 ppm) for Fe, and 810–1100 ppm (mean = 980 ppm, $\sigma =$ 80 ppm, median = 980 ppm) for Sr (Table 1, Fig. 6a, and Additional file 5: Figure S1-1). The profiles of these elements showed no systematic variations.

In the modern *L. rubellus* shell, minor element concentrations ($n = 27$) ranged from 4 to 30 ppm (mean = 10 ppm, $\sigma = 6$ ppm, median = 8 ppm) for Mn, 70 to 10,100 ppm (mean = 1000 ppm, $\sigma = 2000$ ppm, median = 510 ppm) for Fe, and 800–1100 ppm (mean = 940 ppm, $\sigma = 90$ ppm, median = 920 ppm) for Sr (Table 1, Fig. 6b, and Additional file 6: Figure S2-1) and also showed no systematic variations.

Fossil shells

The minor element concentrations ($n = 127$) in the fossil *T. coreanica* (MNTCo1–MNTCo5) shells varied between 80 and 3300 ppm (mean = 580–1100 ppm, $\sigma = 470$–800 ppm, median = 280–740 ppm) for Mn, 200 and 6900 ppm (mean = 1000–2600 ppm, $\sigma = 930$–1700 ppm, median = 600–2200 ppm) for Fe, and 730 and 1100 ppm (mean = 870–930 ppm, $\sigma = 50$–80 ppm, median = 860–920 ppm) for Sr (Table 1, Fig. 6a, and Additional file 5: Figures S1-2–6).

In the fossil *L. rubellus* shells, minor element concentrations ($n = 103$) of (MNLR1–MNLR5) fell in the ranges of

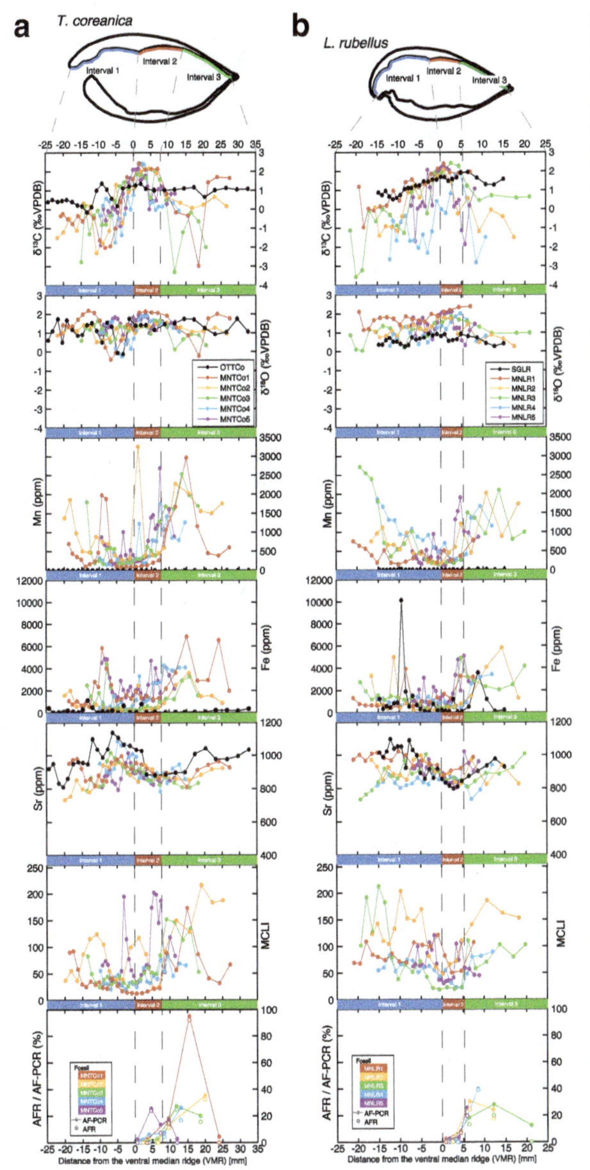

Fig. 6 Profiles of geochemical, microstructural, and optical signals in the studied brachiopod. These panels compare the $\delta^{13}C$ and $\delta^{18}O$ values, minor element concentrations, mean CL intensity (MCLI), altered fiber ratio (AFR), and altered fiber and puncta-filling cement ratio (AF-PCR) between the modern (OTTCo and SGLR) and fossil (MNTCo1–MNTCo5 and MNLR1–MNLR5) shells. **a** *Terebratalia coreanica* shells. **b** *Laqueus rubellus* shells. The geochemical profiles of each sample were fitted to the average length of interval 2 (7.74 mm for *T. coreanica* and 5.27 mm for *L. rubellus*)

120–2700 ppm (mean = 400–1000 ppm, σ = 200–830 ppm, median = 370–910 ppm) for Mn, 180–5800 ppm (mean = 1300–1900 ppm, σ = 830–1800 ppm, median = 670–1500 ppm) for Fe, and 730–1000 ppm (mean = 840–960 ppm, σ = 40–70 ppm, median = 830–970 ppm) for Sr (Table 1, Fig. 6b, and Additional file 6: Figures S2-2–6).

Roughly speaking, the Mn and Fe concentrations in the fossil shells of both species decreased from the posterior shell edge toward the VMR and increased toward the anterior shell edge. In contrast, the Sr concentrations in the fossil shells (*T. coreanica*, 730–1100 ppm; *L. rubellus*, 730–1000 ppm) were comparable with those of the modern shells (*T. coreanica*, 810–1100 ppm; *L. rubellus*, 800–1100 ppm). As with the modern shells, no discernable patterns were seen in the Sr profiles.

Calcite cement

The minor element concentrations of the calcite cement sample were 6200 ppm for Mn, 14,400 ppm for Fe, and 440 ppm for Sr. Mn concentration was 6 times higher than the median concentrations of the fossil shells, and the Fe concentration was 7 times higher. The Sr concentration of the cement was lower than that of the fossil shells.

Discussion

The diagenetic modification of isotope compositions

The $\delta^{13}C$ and $\delta^{18}O$ profiles in interval 1 of the modern *T. coreanica* shell are characterized by the occurrence of two local $\delta^{13}C$ minima and four local $\delta^{18}O$ minima. Similar depletions are found in the same interval of the modern *L. rubellus* shell, although they are not so distinct as those in the *T. coreanica* shell. Interval 1 includes the region of the shell to which the diductor and adjustor muscles are attached (Shiino and Kitazawa 2010) and a shell cavity where the digestive and reproductive organs are housed (Williams and Rowell 1965). This suggests that the observed $\delta^{18}O$ minima are potentially related to the biological activities of the brachiopods, although we cannot specify the exact mechanism. The $\delta^{13}C$ values of the muscle scar are up to 2‰ lower than those of the secondary shell layer (Parkinson et al. 2005; Yamamoto et al. 2010a, 2013) due to secondary calcification, which has been reported in the muscle scars, loop, and teeth (Williams 1968). Therefore, it is highly likely that the depletion of the $\delta^{13}C$ values in interval 1 is caused by precipitation and/or dissolution (Hughes et al. 1988) around the specialized portions.

The $\delta^{13}C$ amplitude in intervals 2 and 3 of the fossil brachiopods is ~7.0–7.5 times greater than those of modern shells, and the $\delta^{18}O$ amplitude is ~3 times greater. The $\delta^{13}C$ and $\delta^{18}O$ values of these intervals in three *T. coreanica* and four *L. rubellus* shells display significant positive correlations (*T. coreanica*: r = 0.89–0.98, $p < 0.01$, excluding MNTCo2 and MNTCo4; *L. rubellus*: r = 0.88–0.99, $p < 0.01$, excluding MNLR1) (Fig. 7). Both $\delta^{13}C$ and $\delta^{18}O$ trend negatively toward the $\delta^{13}C$ and $\delta^{18}O$ values of the calcite cement. The $\delta^{18}O$ value of calcite precipitated in oxygen isotope equilibrium with meteoric water in northeastern Japan, including the Minehama region ($\delta^{18}O$ of meteoric water = − 7.7 to − 10‰ VSMOW; Mizota and Kusakabe 1994) at the mean

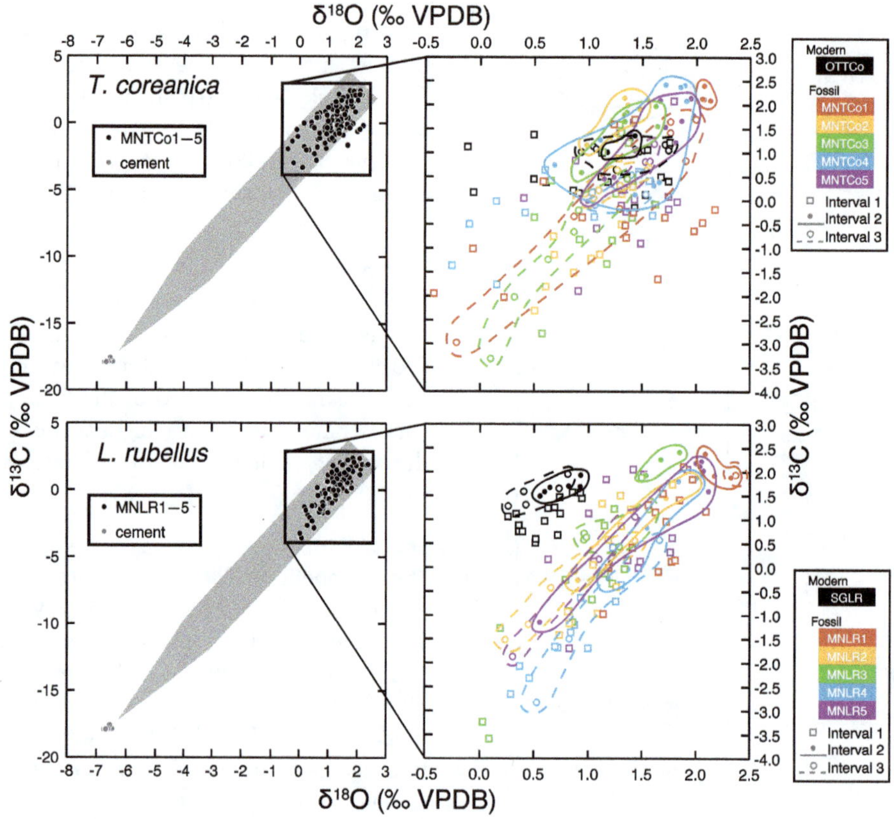

Fig. 7 Cross-plots of $\delta^{13}C$ versus $\delta^{18}O$ values of the studied brachiopod shells. The samples were extracted from the modern and fossil *T. coreanica* and *L. rubellus* shells. The cement $\delta^{13}C$ versus $\delta^{18}O$ values are also plotted

annual temperature (11.2 °C, Japan Meteorological Agency 2018) is estimated to be − 6.62 to − 8.92‰ following Friedman and O'Neil (1977). The $\delta^{18}O$ of the calcite cement (− 6.62‰) falls within this range, indicating that the $\delta^{13}C$ and $\delta^{18}O$ values of the highest end-members on the cross-plots are close to their initial values and that the isotopic values of brachiopod shells at Minehama decreased toward those of the cement during meteoric diagenesis.

Another explanation for the positive correlation of the $\delta^{13}C$ and $\delta^{18}O$ values is a kinetic isotopic fractionation effect. Kinetic isotope fractionation during CO_2 hydration and hydroxylation results in an enhanced incorporation of ^{12}C and ^{16}O during precipitation of skeletal carbonates (McConnaughey 1989a, 1989b; McConnaughey et al. 1997; Auclair et al. 2003; Yamamoto et al. 2013; Takayanagi et al. 2015). This effect is closely related to rapid growth rates and high metabolic activity and is typically characterized by a positive correlation between $\delta^{13}C$ and $\delta^{18}O$ values (or $\delta^{13}C–\delta^{13}C_{EC}$ and $\delta^{18}O–\delta^{18}O_{EC}$ values). However, it is reported that growth rates along the inner surface (i.e., shell thickening rates) of brachiopod shells, including *T. coreanica* and *L. rubellus*, are very slow and that within a single shell, the kinetic effect is negligible there (Yamamoto et al. 2010b, 2013). The $\delta^{13}C$ and $\delta^{18}O$ values of interval 1 are less strongly correlated than those of intervals 2 and 3 (Fig. 7). This is because the isotopic composition in interval 1 of the fossil shells is influenced by the biogenic effect mentioned above as well as by meteoric diagenetic modification.

The isotopic data indicate that the highest end-members on the cross-plots (Fig. 7) are likely to be closest to the initial shell $\delta^{13}C$ and $\delta^{18}O$ values. The highest end-members are from samples extracted from the thick portion of interval 2. This portion corresponds to "interval 2" of Pleistocene *K. hanzawai* shells from the Ryukyu Islands (Takizawa et al. 2017), which was regarded as a good shell portion to use for paleoenvironmental reconstructions because it exhibited relatively small intra-shell and intra-species variations. Therefore, this interval is likely to be a good region of the shells for paleoenvironmental analysis in many brachiopod taxa.

The diagenetic modification of minor element concentrations

Minor element concentrations in carbonate minerals have been used to identify diagenetically altered brachiopod shells or shell portions (e.g., Wenzel 2000; Ullmann et al. 2014). As meteoric water generally contains more Mn and Fe and less Sr compared with seawater, Fe and Mn concentrations increase and Sr concentration decreases in diagenetically altered marine carbonates (Brand and Veizer 1980; Marshall 1992).

The Mn concentrations in the studied modern shells are relatively uniform. Concentrations in *T. coreanica* fall within a narrow range of 2–12 ppm, and *L. rubellus* concentrations are between 4 and 30 ppm (Fig. 6a, b). These ranges are comparable with values reported in previous studies of modern brachiopod shells (5–500 ppm, Morrison and Brand 1986; 0–200 ppm, Brand et al. 2003; 4–30 ppm, Lee et al. 2004; 0–70 ppm, Ullmann et al. 2017) and are lower than the common Mn concentrations in modern shells (< 80 ppm) compiled by Brand et al. (2003).

In contrast, the Mn concentration in the fossil *T. coreanica* and *L. rubellus* shells is higher than that of the modern shells. The observed range is 80–3300 ppm in the *T. coreanica* shells and is 120–2700 ppm in the *L. rubellus* shells. These values exceed the common Mn concentration in modern shells (Brand et al. 2003) and indicate that there are significant intra-shell and intra-species variations in the fossil brachiopod Mn concentrations (Fig. 6). The samples with high Mn concentrations did not correspond to specific regions of the shells but exhibited low $\delta^{13}C$ and $\delta^{18}O$ values, high Fe concentrations, and strong luminescence (Fig. 6, Additional file 5: Figures S1-2–6, and Additional file 6: Figures S2-2–6). This indicates that the increased Mn concentrations are the result of meteoric diagenetic alteration.

Although some anomalously high Fe concentrations (> 600 ppm) are observed, the range of Fe concentrations in the modern *T. coreanica* shells ranged from 30 and 490 ppm and from 70 to 570 ppm in *L. rubellus* (Fig. 6a, b). These values exceed the common Fe concentrations of modern brachiopod shells (< 140 ppm) presented by Brand et al. (2003). This difference is at least partly due to the fact that the Fe concentration threshold reported in Brand et al. (2003) was derived from bulk shell samples, which represent spatially averaged values.

The Fe concentrations in the fossil *T. coreanica* (200 to 6900 ppm) and *L. rubellus* (180 to 5800 ppm) shells are much higher than those of the modern shells and the common Fe concentration range (Brand et al. 2003). The samples with high Fe concentrations also exhibited high Mn concentrations, low $\delta^{13}C$ and $\delta^{18}O$ values, and strong luminescence. As with Mn, anomalously high Fe concentrations also occur in the fossil shells. These anomalously high concentrations are not always associated with any microstructural (dissolution and amalgamation of fibers), optical (strong luminescence on CL images; see below), or geochemical (decreased $\delta^{13}C$ or $\delta^{18}O$ values or increased Mn concentration) signals of diagenetic alteration. As such, we consider that these anomalously high Fe concentrations are likely not diagenetic products but represent initial concentrations from when the brachiopods were alive. Similar anomalously high Fe concentrations without signs of diagenetic alteration in well-preserved fossil (~ 20 and ~ 70 ka) *K. hanzawai* shells have also been described by Takizawa et al. (2017).

Unlike Mn and Fe concentrations, Sr concentrations in the modern and fossil brachiopod shells were similar and their ranges overlap (Table 1). This indicates that Sr concentration in these brachiopod shells is not likely to be a good indicator of diagenetic alteration. The Mn and Sr concentrations show a moderate negative correlation (fossil *T. coreanica*: $r = -0.46$, $p < 0.01$, Fig. 8c; fossil *L. rubellus*: $r = -0.52$, $p < 0.01$, Fig. 8d). This differs from the result of Jelby et al. (2014), who found that Mn/Ca of diagenetically altered Lower Maastrichtian brachiopod shells was strongly negatively correlated with Sr/Ca ($r = -0.94$).

The Mn concentrations in the calcite cement were ~ 2–3 times higher than the maximum Mn concentrations in the fossil shells, and Fe concentrations were ~ 2 times higher. This indicates that Mn and Fe concentrations in brachiopod shells at Minehama increased during meteoric diagenesis, approaching those of the cement. The Sr concentrations in the calcite cements were ~ 50% or less of mean Sr concentrations in the fossil shells. This implies that Sr concentrations in brachiopod shells at Minehama decreased during meteoric diagenesis, approaching those of the cement. However, the decrease in the Sr concentrations was relatively more gradual than the increase in the Mn and Fe concentrations. This is the reason why Sr is not sensitive to diagenesis at Minehama and shows a moderate negative correlation with the Mn concentrations. In conclusion, our study shows that of the three minor elements analyzed in this study, Mn concentration is the most reliable indicator of diagenetically altered shells or shell portions.

Diagenetic alteration detected by CL image observations

Cathodoluminescence observations are useful for determining the paragenetic sequences of diagenetic products precipitated from the solutions containing Mn. Altered shell portions are easily detectable by luminescence activated by the presence of Mn. It has been shown that the $\delta^{13}C$ and $\delta^{18}O$ values of luminescent shell portions are commonly lower than those of non-luminescent portions (e.g., Grossman et al. 1993; Mii et al. 2012).

The modern *T. coreanica* and *L. rubellus* shells exhibit faint to weak luminescent bands that are arranged almost parallel to the inner shell surface (Fig. 4b, d; Additional file 5: Figure S1-1b, and Additional file 6: Figure S2-1b). Similar luminescent bands are observed in shells of modern inarticulate brachiopods and in those of other

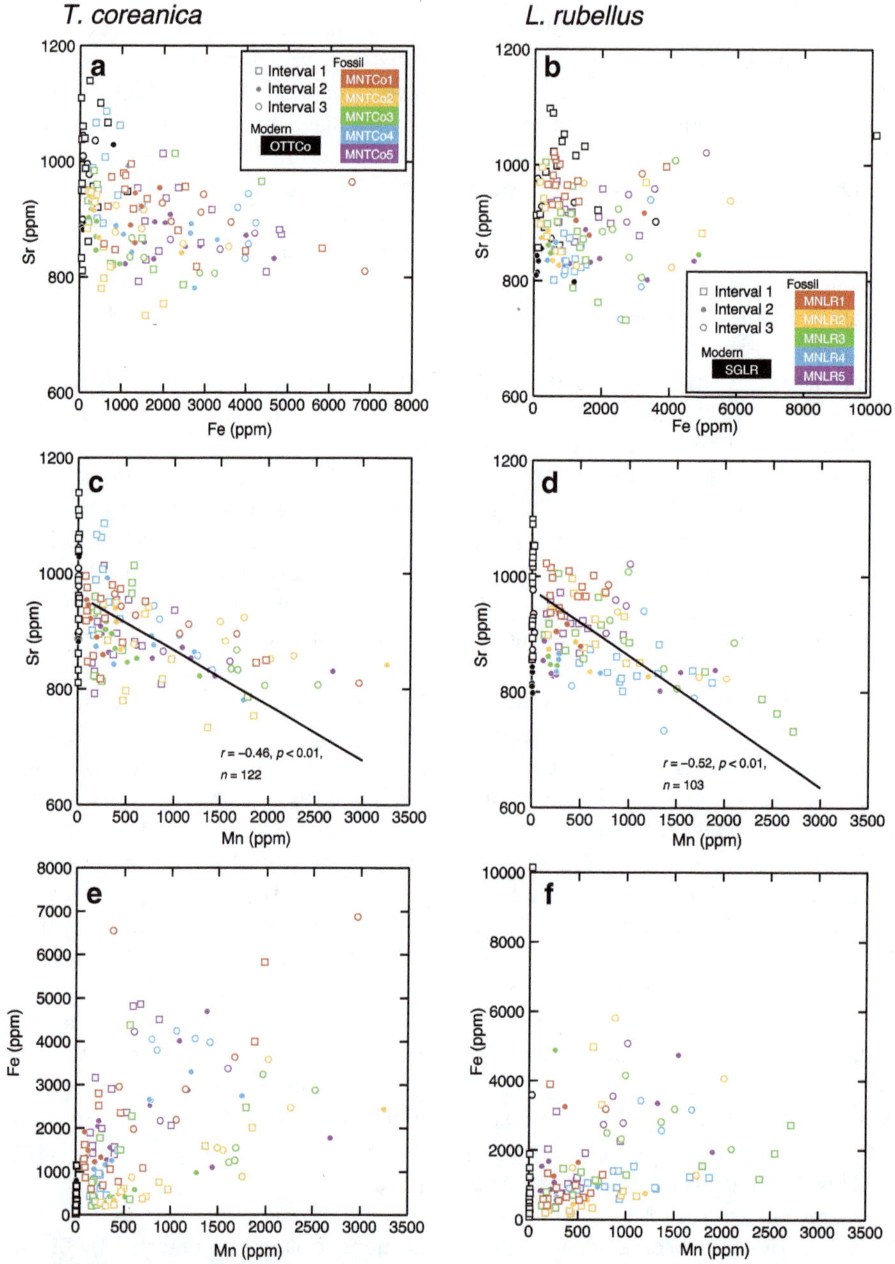

Fig. 8 Cross-plots of minor element (Mn, Fe, and Sr) concentrations in the studied brachiopod shells. The samples were extracted from the modern and fossil *T. coreanica* (**a**, **c**, **e**) and *L. rubellus* (**b**, **d**, **f**) shells

marine invertebrates (Barbin and Gaspard 1995; Barbin 2013; Riechelmann et al. 2016). These bands or lines are the result of biological activity (e.g., bleeding cycles and changes in growth rates) and/or environmental stresses (e.g., input of coastal water, tidal rhythms, and changes in seawater temperature). In the present study, it is difficult to determine the main reason for the difference in the CL intensity between the modern shells of the two species, noting that CL can also be influenced by the measurement conditions (England et al. 2006).

The fossil shells display brighter luminescence than that of the modern shells. No luminescent bands are observed in the fossil shells. The MCLI profiles of the fossil brachiopod shells show statistically significant positive correlations with Mn concentration (*T. coreanica*: $r = 0.73$, $n = 122$, $p < 0.01$; *L. rubellus*: $r = 0.51$, $n = 103$, $p < 0.01$) (Figs. 6a, b and 9a–c). The luminescent regions also correspond to the samples with higher Fe concentrations and lower $\delta^{13}C$ and $\delta^{18}O$ values than those of adjacent non-luminescent regions within an individual shell (Fig. 6a, b),

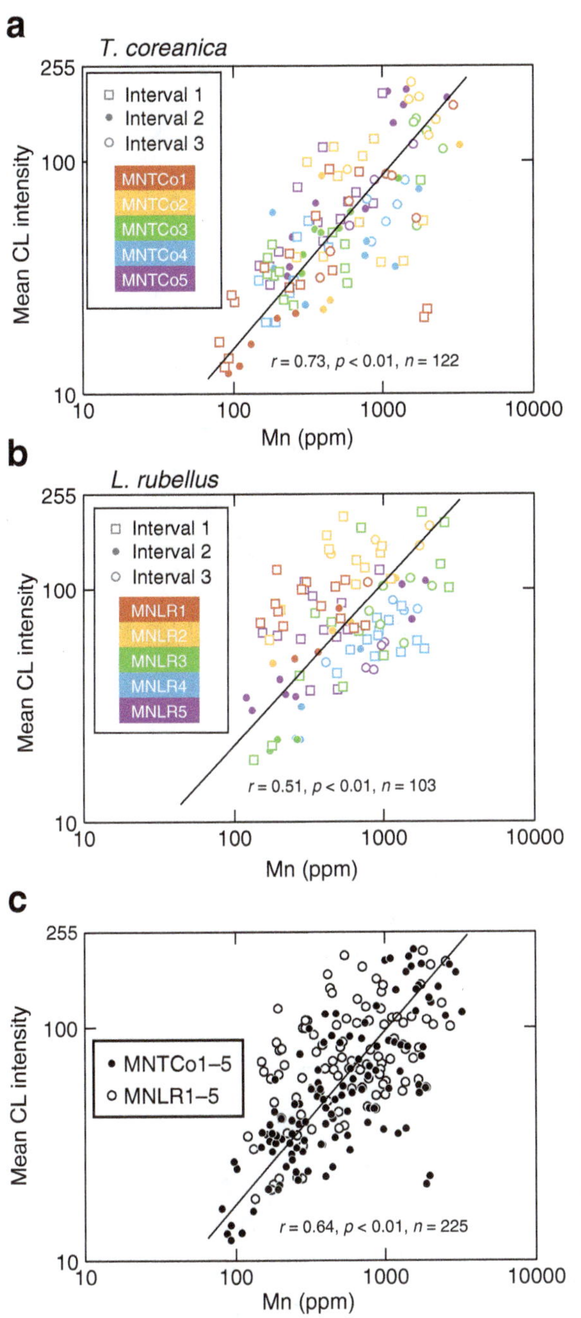

Fig. 9 Double-logarithmic cross-plots of mean CL intensity versus Mn concentration in the studied fossil shells. **a** *T. coreanica*. **b** *L. rubellus*. **c** Combined data from both species

indicating that these shell portions have been altered by meteoric diagenesis.

The punctae of the fossil brachiopods are partly to fully filled with cement and display bright luminescence (Fig. 4f, h) as has been described in other species (e.g., Takizawa et al. 2017; Casella et al. 2018). This is because the punctae are filled with organic material in live brachiopods (Williams 1997; Pérez-Huerta et al. 2009), although no such material was observed in the studied modern shells, and void spaces are created during post-mortem decomposition, with subsequent precipitation of diagenetic cements (Garbelli et al. 2016). We conclude that the presence of carbonate cements with extremely high Mn and Fe concentrations in the punctae and inter-fiber spaces is one of the major causes of high concentrations of these elements in diagenetically altered shell portions, as discussed below.

The diagenetic alteration of shell microstructure

The secondary shell layer is composed of calcite fibers, each with a concave outer surface and convex inner surface (Gaspard 1991; Samtleben et al. 2001). However, altered fibers become irregularly shaped and amalgamated with an increased diagenetic alteration. The amalgamation of two or more fibers occurs locally in modern brachiopods, but the outline of such fibers remains defined, which means that pristine amalgamated fibers can be distinguished from diagenetically altered fibers (Samtleben et al. 2001).

Altered shell portions (Additional file 5: Figures S1-2–6 and Additional file 6: Figures S2-2–6) are characterized by low $\delta^{13}C$ and $\delta^{18}O$ values, high Mn and Fe concentrations, and luminescent CL images. In contrast, the well-preserved shell portions exhibit comparatively high $\delta^{13}C$ and $\delta^{18}O$ values, low Mn and Fe concentrations, and weakly luminescent or non-luminescent CL images (Fig. 4f). AFR and AF-PCR are correlated negatively with $\delta^{13}C$ and $\delta^{18}O$ but positively with Mn and Fe concentrations and MCLI in the fossil *T. coreanica* and *L. rubellus* shells (Fig. 10).

Some studies have argued that the impact of in-filling cements in punctae upon the chemical and isotopic composition of shells is negligible because their volume accounts for only a small portion of the shell (e.g., < 10% of a Silurian brachiopod, *Resseralla elegantula*, shells; Samtleben et al. 2001) compared with that of the shell-forming calcite fibers. However, if the isotopic compositions and chemical concentrations of the puncta-filling cements are one or more orders of magnitude greater than those of the shell-forming fibers, the impact of such cements on the geochemistry of spot samples might not be negligible. Here, we assume that the isotopic composition and Mn and Fe concentrations of the puncta-filling cements are identical to those of the cement formed by meteoric diagenesis ($\delta^{13}C = -18‰$, $\delta^{18}O = -7‰$, Mn concentration = 6000 ppm, Fe concentration = 14,000 ppm). The $\delta^{13}C$ values are at least 19‰ lower than those of the fossil shells, and the $\delta^{8}O$ values are 9‰ lower. The Mn concentrations are at least 5000 ppm higher than those of the fossil shells, and Fe concentrations are 12,000 ppm higher. Under this assumption, if the punctae of the studied modern shells are filled with such

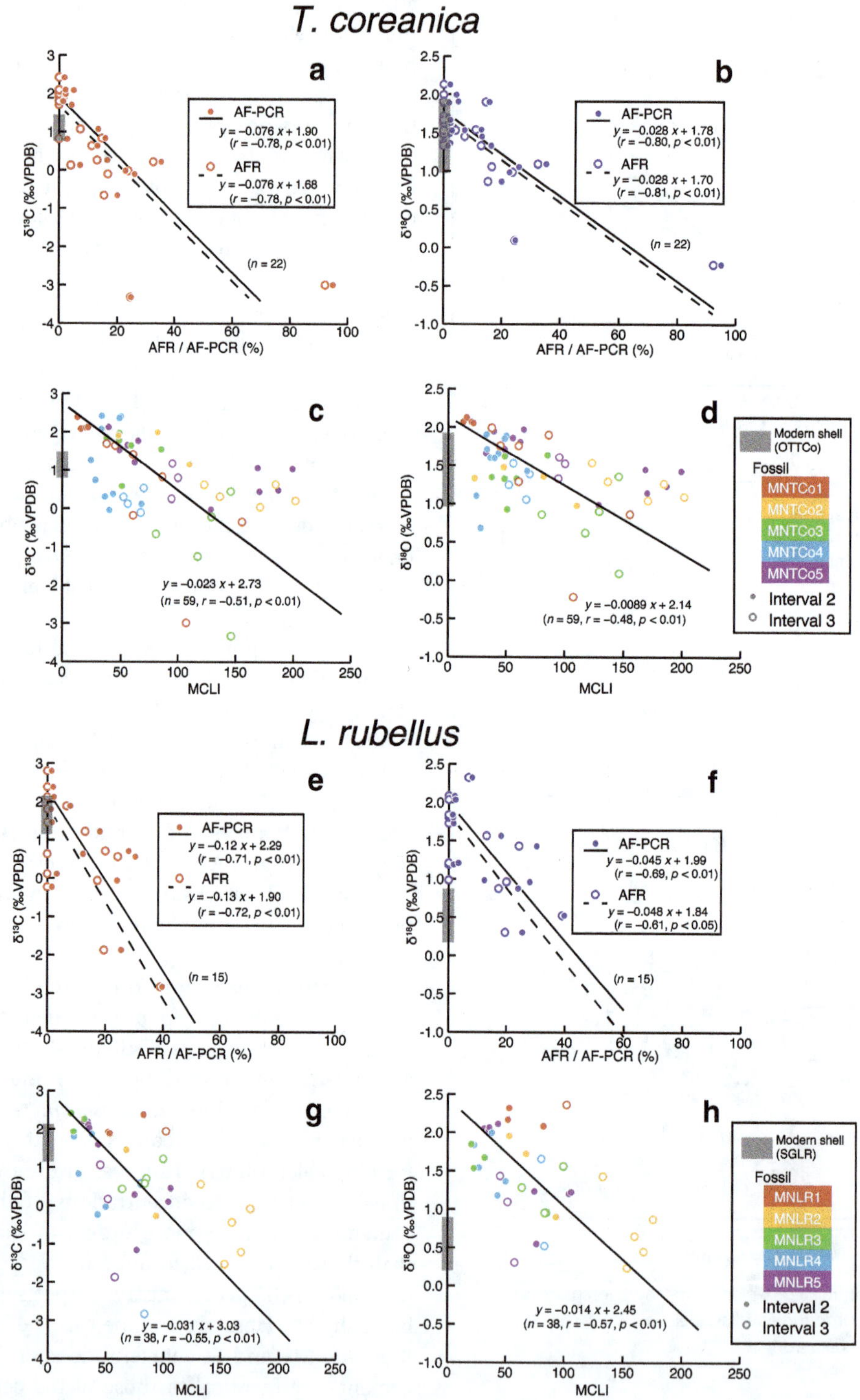

Fig. 10 Cross-plots of isotopic composition versus AFR or AF-PCR (**a**, **b**, **e**, and **f**) and of isotopic composition versus MCLI (**c**, **d**, **g**, and **h**) . The gray bar on the *y*-axis indicates a range of $\delta^{13}C$ or $\delta^{18}O$ values of the modern brachiopod shell (OTTCo or SGLR)

cements and they account for 5% of the shell volumes, then the $\delta^{13}C$ and $\delta^{18}O$ values decrease by ~ 1‰ and ~ 0.4‰, respectively (Table 2). Conversely, the Mn and Fe concentrations increase by ~ 300 and ~ 600 ppm, respectively. In fact, the punctae account for 1–9% of the fossil *T. coreanica* shell volumes and 1–12% of the *L. rubellus* shell volumes. These indicate that the isotopic composition and chemical concentration of the puncta-

Table 2 Effect of puncta-filling cements to the isotopic composition and minor element concentrations of shell samples. We assume that the isotopic composition and minor element concentrations of the puncta-filling cements are identical to those of the cement formed by meteoric diagenesis. Then, we calculate the isotopic composition and minor element concentrations of the mixtures of shell and puncta-filling cements when the punctae are filled with such cements

	Modern shell	Cement	Puncta filling cements accounting for 5% of the shell volumes		Puncta filling cements accounting for 10% of the shell volumes	
			Shell + cement	Δ*	Shell + cement	Δ*
T. coreanica						
Interval 2						
$\delta^{13}C$ (‰)	1.01 to 1.36	− 18	0.06 to 0.39	− 0.96	− 0.89 to − 0.58	− 1.92
$\delta^{18}O$ (‰)	1.17 to 1.43	− 7	0.76 to 1.01	− 0.42	0.35 to 0.59	− 0.83
Mn (ppm)	3**	6000	310	300	610	600
Fe (ppm)	70**	14,000	770	700	1500	1400
Sr (ppm)	960**	400	930	− 30	900	− 60
L. rubellus						
Interval 2						
$\delta^{13}C$ (‰)	1.49 to 1.94	− 18	0.52 to 0.94	− 0.99	− 0.46 to 0.05	− 1.97
$\delta^{18}O$ (‰)	0.55 to 0.93	− 7	0.17 to 0.53	− 0.39	− 0.21 to 0.14	− 0.78
Mn (ppm)	7**	6000	300	300	600	590
Fe (ppm)	120**	14,000	810	690	1500	1380
Sr (ppm)	820**	400	800	− 20	780	− 40

*Δ = "shell + cement" − "modern shell"
**Median

filling cements can potentially significantly bias the isotopic composition and minor element concentrations of spot or bulk shell samples.

AF-PCR was correlated more negatively with $\delta^{13}C$ and $\delta^{18}O$ and more positively with Mn and Fe concentrations and MCLI than was AFR (Table 3, Fig. 9). This suggests that the occurrence of puncta-filling cements is one of the major causes of diagenetic modification of isotopic compositions and minor element concentrations. However, it is not clear whether the puncta-filling cements have similar isotopic compositions and minor element concentrations to cements formed by meteoric diagenesis. The punctae of some modern brachiopod shells are luminescent (e.g., Griesshaber et al. 2007; Barbin 2013), indicating that they are filled with carbonate minerals that are enriched in Mn compared with shell calcite. It is deduced that such puncta-filling cements are chemical and/or microbial precipitates formed in a closed reductive space (e.g., Kaźmierczak and Iryu 1999) or biogenic products. As such, the puncta-filling cements in this study might include carbonate minerals of different origins, with likely differences in their isotopic composition and minor element concentrations.

Figure 10 shows the cross-plots of isotopic composition versus AFR/AF-PCR and isotopic composition versus MCLI. The intercepts of these cross-plots can be regarded as the isotopic composition of non-altered brachiopod shells (AFR or AF-PCR = 0, MCLI = 0), theoretically corresponding to the initial isotopic composition of the fossil brachiopod shells. The intercepts are not identical but are close to each other (the difference in $\delta^{13}C$ values is < 1.1‰ and in $\delta^{18}O$ values is < 0.6‰). The intercepts of the $\delta^{18}O$ values of the fossil *T. coreanica* shells are up to 0.5‰ greater than the $\delta^{18}O$ values of the modern shell from Otsuchi Bay. Those of *L. rubellus* are 2‰ greater than the $\delta^{18}O$ values from Sagami Bay. Assuming that the brachiopods occupied similar habitats during the Pliocene to those occupied by their modern counterparts, this indicates the presence of cooler and/or more saline seawater in the Minehama region during the Pliocene compared with that of the modern Otsuchi and Sagami bays. This interpretation is supported by the occurrence of cool-water molluscan species within the outcrop from which the fossil brachiopod shells of this study were collected (MH1; Amano et al. 2011). This suggests that AFR/AF-PCR and MCLI are good indices for quantitatively estimating the extent of diagenetic alteration within brachiopod shells.

The submicron-sized cements in a narrow space in the inter-fiber spaces (Fig. 4i) show the morphological and geochemical features that are common in cements formed during meteoric diagenesis (Tucker and Wright 1990), and the occurrence of such cements is an additional cause of high concentrations of Mn and Fe in diagenetically altered shells.

Table 3 Cross-correlations between geochemical, microstructural and optical signals in the studied fossil brachiopodCross-correlations are calculated between isotope composition ($\delta^{13}C$ and $\delta^{18}O$), minor-element (Mn, Fe, and Mn) concentrations, MCLI, and AFR or AF-PCR of the fossil *Terebratalia coreanica* and *Laqueus rubellus* shells. Coefficient values in brackets are not statistically significant ($p > 0.01$)

	$\delta^{13}C$ (‰ VPDB)	$\delta^{18}O$ (‰ VPDB)	Mn (ppm)	Fe (ppm)	Sr (ppm)	Mean CL intensity (MCLI: 0–255)	Altered fiber and puncta-filling cement ratio (AF-PCR: %)
Fossil T. coreanica and L. rubellus ($n = 36$)							
$\delta^{13}C$ (‰VPDB)	1						
$\delta^{18}O$ (‰VPDB)	0.9	1					
Mn (ppm)	−0.71	−0.71	1				
Fe (ppm)	−0.51	−0.43	0.61	1			
Sr (ppm)	[0.21]	[0.18]	[−0.42]	[−0.02]	1		
Mean CL intensity (MCLI: 0–255)	−0.53	−0.54	0.77	[0.34]	[−0.23]	1	
Altered fiber and puncta-filling cement ratio (AF-PCR: %)	−0.75	−0.76	0.75	0.6	[−0.15]	0.61	1
Fossil *T. coreanica* ($n = 22$)							
$\delta^{13}C$ (‰VPDB)	1						
$\delta^{18}O$ (‰VPDB)	0.9	1					
Mn (ppm)	−0.77	−0.72	1				
Fe (ppm)	[−0.48]	[− 0.35]	0.54	1			
Sr (ppm)	0.54	0.55	−0.68	[− 0.19]	1		
Mean CL intensity (MCLI: 0–255)	−0.65	−0.64	0.80	[0.20]	−0.47	1	
Altered fiber and puncta-filling cement ratio (AF-PCR: %)	−0.79	−0.81	0.77	0.62	[−0.44]	0.63	1
Fossil *L. rubellus* ($n = 14$)							
$\delta^{13}C$ (‰VPDB)	1						
$\delta^{18}O$ (‰VPDB)	0.91	1					
Mn (ppm)	[−0.65]	− 0.72	1				
Fe (ppm)	[−0.57]	[− 0.56]	0.76	1			
Sr (ppm)	[−0.15]	[− 0.20]	[− 0.16]	[0.17]	1		
Mean CL intensity (MCLI: 0–255)	[−0.32]	[−0.37]	0.69	[0.62]	[0.05]	1	
Altered fiber and puncta-filling cement ratio (AF-PCR: %)	−0.71	− 0.71	0.71	[0.57]	[0.31]	[0.55]	1

The cross-correlation coefficients of isotopic composition, minor element concentration, and MCLI and AF-PCR indices indicating the alteration of shell portions are presented in Table 3. The $\delta^{13}C$ and $\delta^{18}O$ values of the fossil *T. coreanica* and *L. rubellus* shells show moderate to strong negative correlations with Mn concentration, MCLI, and AF-PCR. Mn concentration, MCLI, and AF-PCR were positively correlated. The negative correlation of Fe concentration with isotopic composition is weaker than that of Mn concentration because of the occurrence of localized high Fe concentrations that are more anomalous than the high Mn concentrations. Fe concentration shows a weak (although not statistically significant ($p > 0.01$)) positive correlation with MCLI. Mn and Fe concentrations increase with increased diagenetic alteration and act as an activator (Mn) and a quencher (Fe) for CL images (Machel et al. 1991). The weak positive correlation of Fe concentration with MCLI indicates that the activator (Mn) acts more efficiently as an activator than Fe does as a quencher. Sr concentration was not correlated significantly with any other features or measures. In summary, shell portions with high Mn concentration, high MCLI, and high AF-PCR generally exhibit low $\delta^{13}C$ and $\delta^{18}O$ values, indicating that these measures are useful for detecting shell portions altered by meteoric diagenesis.

Lithologic control on the diagenetic alteration of brachiopod shells

Our study shows that the four indices commonly used to specify diagenetically altered carbonates—$\delta^{13}C$ and $\delta^{18}O$ values, minor element concentrations, the preservation state of shell microstructure, and cathodoluminescence—are applicable to brachiopod shells. However, it is known that diagenesis alters the initial paleoenvironmental signals extracted by these indices (Ullmann and Korte 2015). Yamamoto et al. (2017) studied the four indices of fossilized *Kikaithyris hanzawai* shells from the Pleistocene reef complex deposits (Iryu et al. 2006) and showed that although altered fibers are commonly observed in brachiopod shells, the initial isotopic composition can be largely preserved, and luminescent shells cannot always be regarded as having undergone degradation of shell microstructure or isotopic modification by diagenesis. The findings of those authors indicate that CL images are not always sufficient for detecting diagenetic alteration in shells. This is at least partly because of the differences in the diagenetic processes modifying the chemical and isotopic compositions between brachiopod shells embedded in siliciclastic and volcaniclastic deposits and those in carbonate deposits. The geochemistry of brachiopod shells embedded in siliciclastic and volcaniclastic deposits is affected by formation water relatively rich in Mn, which originates from meteoric water. This can explain why diagenetically altered shells are luminescent in CL images, have low $\delta^{13}C$ and $\delta^{18}O$ values, and high Mn concentrations. In contrast, carbonate deposits have limited sources of Mn, except when the deposits are associated with siliciclastics (including paleosols) or volcaniclastics. As such, CL images might not necessarily be useful for the detection of diagenetic alteration of shells in carbonate deposits.

The relationship between the degree of preservation of shell microstructures and the initial isotopic composition has been controversial. It is known that some Paleozoic brachiopods, in spite of their well-preserved shell microstructure, have low $\delta^{13}C$ and $\delta^{18}O$ values, similar to those of diagenetically altered marine carbonate fossils and sparry cements (Rush and Chafetz 1990), or exhibit large variations in isotopic composition caused by a diagenetic alteration (Banner and Kaufman 1994). To explain this conflict, it has been suggested that only large-scale dissolution and subsequent recrystallization of shells can result in isotopic resetting in the per mil range (Wadleigh and Veizer 1992; Qing and Veizer 1994). However, the significantly lower $\delta^{13}C$ and $\delta^{18}O$ values in intervals 2 and 3 of the fossil brachiopod shells compared with those of the modern shells in the present study indicate that the resetting of isotopic values in the per mil range can occur without large-scale dissolution and subsequent recrystallization.

Shell portions in which microstructures are preserved but their initial isotopic composition has been modified present a significant challenge with respect to reconstructing paleoceanographic conditions. Applying and cross-checking multiple criteria to assess diagenetic alteration are required to distinguish between diagenetically altered and unaltered brachiopod shells (Yamamoto et al. 2017).

Conclusions

To estimate the extent of modification in the isotope composition of fossil brachiopod shells with increasing diagenetic alteration, we analyzed $\delta^{13}C$ and $\delta^{18}O$ and three commonly applied indicators of meteoric diagenesis: minor element (Mn, Fe, and Sr) concentrations, cathodoluminescence, and shell microstructure. The samples for these analyses were collected from the inner surfaces of fossil *T. coreanica* and *L. rubellus* (rhynchonellate brachiopod) shells collected from conglomerates of the Tentokuji Formation (Pliocene, 3.85–2.75 Ma) exposed in northeastern Japan. The results of the study can be summarized as follows:

(1) Individual shells of *T. coreanica* and *L. rubellus* can be divided into three intervals (intervals 1–3) based on the shell morphology, each with characteristic chemical and isotopic compositions.

(2) Interval 2 is characterized by relatively small intra-shell and intra-species variations in $\delta^{13}C$ and $\delta^{18}O$ values in both modern and fossil shells. The $\delta^{13}C$ and $\delta^{18}O$ values from this interval of the fossil shells plots as the highest end-member, and these values are therefore likely to be the closest to the initial shell $\delta^{13}C$ and $\delta^{18}O$ values.

(3) Of the three minor elements examined in this study, Mn concentration is the most reliable indicator of diagenetically altered shells. Mn and Fe concentrations in the fossil brachiopod shells are higher than those of the modern shells, displaying large intra-shell and intra-species variations, with several anomalously high localized concentrations. The samples with high Mn and Fe concentrations do not correspond to specific regions of the shells but also exhibit low $\delta^{13}C$ and $\delta^{18}O$ values and strong luminescence. This indicates that the increased Mn concentrations are the result of meteoric diagenetic alteration. The anomalously high Fe concentrations, which are also observed in modern shells, are not necessarily associated with microstructural (dissolution and amalgamation of fibers), optical (strong luminescence on CL images), or isotopic (decreased $\delta^{13}C$ or $\delta^{18}O$ values). This indicates that these anomalously high Fe concentrations are likely not diagenetic products but represent initial concentrations from when the brachiopods were alive. Unlike Mn and Fe concentrations, Sr concentrations in the studied modern and fossil brachiopod shells are similar and their ranges overlap, indicating that Sr concentration in carbonate minerals is not always a good indicator for identifying diagenetic alteration.

(4) We quantified the preservation state of shell microstructure using two indices: altered fiber ratio

(AFR) and altered fiber and puncta-filling cement ratio (AF-PCR). To quantify the degree of luminescence, we measured the mean cathodoluminescence index (MCLI) defined as the mean R value in a particular area of a CL image. The $\delta^{13}C$ and $\delta^{18}O$ values of the fossil brachiopod shells are moderately or strongly negatively correlated with Mn concentration, MCLI, and AF-PCR. Mn concentration, MCLI, and AFR/AF-PCR are positively correlated with each other. Fe concentration shows a weaker negative correlation with isotopic composition than does Mn concentration on account of the occurrence of anomalously high localized Fe concentrations.

Our study provides further evidence that applying and cross-checking multiple criteria to assess diagenetic alteration is the optimal method for distinguishing between diagenetically altered and unaltered brachiopod shells.

Additional files

Additional file 1: Table S1. $\delta^{13}C$ and $\delta^{18}O$ values of the modern *Terebratalia coreanica* (OTTCo) and *Laqueus rubellus* (SGLR) shells.

Additional file 2: Table S2. $\delta^{13}C$ and $\delta^{18}O$ value, AFRs (altered fiber ratios), and AF-PCRs (altered fiber and puncta-filling cement ratios) of the fossil *Terebratalia coreanica* (MNTCo1–MNTCo5) and *Laqueus rubellus* (MNLR1–MNLR5) shells.

Additional file 3: Table S3. Minor element (Mn, Fe, and Sr) concentrations of the modern *Terebratalia coreanica* (OTTCo) and *Laqueus rubellus* (SGLR) shells.

Additional file 4: Table S4. Minor element (Mn, Fe, and Sr) concentrations and MCLIs (mean CL intensities) of the fossil *Terebratalia coreanica* (MNTCo1–MNTCo5) and *Laqueus rubellus* (MNLR1–MNLR5) shells.

Additional file 5: Figure S1. Sampling spots, $\delta^{13}C$ and $\delta^{18}O$ values, minor element (Mn, Fe, and Sr) concentrations, mean CL intensities, altered fiber ratios, and altered fiber and puncta-filling cement ratios of the modern (Figure S1-1, OTTCo) and fossil (Figures S1-2–6, MNTCo1–MNTCo5) *T. coreanica* shells. Composite CL images are presented for each shell. SEM and CL images of selected shell portions are also shown. Red and blue arrows indicate local minima in $\delta^{13}C$ and $\delta^{18}O$, respectively.

Additional file 6: Figure S2. Sampling spots, $\delta^{13}C$ and $\delta^{18}O$ values, minor element (Mn, Fe, and Sr) concentrations, mean CL intensities, altered fiber ratios, and altered fiber and puncta-filling cement ratios of the modern (Figure S2-1, SGLR) and fossil (Figures S2-2–6, MNLR1–MNLR5) *L. rubellus* shells. Composite CL images are presented for each shell. SEM and CL images of selected shell portions are also shown.

Additional file 7: Doc. S1. Estimation of $\delta^{13}C$ and $\delta^{18}O$ values of equilibrium calcite.

Abbreviations

AF-PCR: Altered fiber and puncta-filling cement ratio; AFR: Altered fiber ratio; CL: Cathodoluminescence; MCLI: Mean CL intensity

Acknowledgements

We are grateful to T. Sato for providing the information on the geologic setting of Minehama. The manuscript was significantly improved by the comments and suggestions of K. Ikehara (editor) and two anonymous reviewers.

Authors' contributions

FH, HT, and YI conceptualized and designed this study. FH and YI conducted the fieldwork and sampling. FH, HT, and KY carried out the chemical and isotope analyses. HT and YI collaborated with FH in the interpretation of the data and the construction of the manuscript. All authors read and approved the final manuscript.

Funding

This work was financially supported by the Japan Society for the Promotion of Science KAKENHI (Grant-in-Aid for Scientific Research) Grant numbers 2134052 and 26302008 to YI.

Competing interests

The authors declare that they have no competing interests.

Author details

[1]Department of Earth Science, Graduate School of Science, Tohoku University, Sendai 980-8578, Japan. [2]Department of Earth and Environmental Sciences, Nagoya University, Nagoya 464-8601, Japan.

References

Adams A, Reeve L (1850) Mollusca. Part 1. In: Adams A (ed) The zoology of the voyage of H.M.S. Samarang, Reeve and Benham, London

Amano K, Yoshida A, Sato T (2011) Influence of cooling event at 2.75 ma on the molluscan fauna of the Japan Sea borderland: molluscan fauna from central and northern parts of Akita Prefecture. J Geol Soc Jpn 117:508–522. https://doi.org/10.5575/geosoc.117.508. In Japanese with English abstract

Angiolini L, Darbyshire DPF, Stephenson MH, Leng MJ, Brewer TS, Berra F, Jadoul F (2007) Lower Permian brachiopods from Oman: their potential as climatic proxies. Ear Env Sci Trans Royl Soc Edinburgh 98:327–344. https://doi.org/10.1017/S1755691008075634

Auclair A, Joachimski MM, Lécuyer C (2003) Deciphering kinetic, metabolic and environmental controls on stable isotope fractionations between seawater and the shell of Terebratalia transversa (Brachiopoda). Chem Geol 202:59–78. https://doi.org/10.1016/S0009-2541(03)00233-X

Azmy K, Veizer J, Bassett MG, Copper P (1998) Oxygen and carbon isotopic composition of Silurian brachiopods: implications for coeval seawater and glaciations. Geol Soc Am Bull 110(11):1499–1512. https://doi.org/10.1130/0016-7606(1998)110<1499:OACICO>2.3.CO;2

Banner JL, Kaufman J (1994) The isotopic record of ocean chemistry and diagenesis preserved in non-luminescent brachiopods from Mississippian carbonate rocks, Illinois and Missouri. Geol Soc Am Bull 106:1074–1082. https://doi.org/10.1130/0016-7606(1994)106<1074:TIROOC>2.3.CO;2

Barbin V (2013) Application of cathodoluminescence microscopy to recent and past biological materials: a decade of progress. Mineral Petrol 107:353–362. https://doi.org/10.1007/s00710-013-0266-6

Barbin V, Gaspard D (1995) Cathodoluminescence of recent articulate brachiopod shells. Implication for growth stages and diagenesis evaluation. Geobios 28(Supple 1):39–45. https://doi.org/10.1016/S0016-6995(95)80151-0

Bates NR, Brand U (1991) Environmental and physiological influences on isotopic and elemental compositions of brachiopod shell calcite: implications for the isotopic evolution of Paleozoic oceans. Chem Geol 94:67–78. https://doi.org/10.1016/S0009-2541(10)80018-X

Brand U, Azmy K, Bitner MA, Logan A, Zuschin M, Came R, Ruggiero E (2013) Oxygen isotopes and $MgCO_3$ in brachiopod calcite and a new paleotemperature equation. Chem Geol 359:23–31. https://doi.org/10.1016/j.chemgeo.2013.09.014

Brand U, Logan A, Hiller N, Richardson J (2003) Geochemistry of modern brachiopods: applications and implications for oceanography and paleoceanography. Chem Geol 198:305–334. https://doi.org/10.1016/S0009-2541(03)00032-9

Brand U, Veizer J (1980) Chemical diagenesis of a multicomponent carbonate system-1: minor elements. J Sediment Res 50:1219–1236. https://doi.org/10.1306/212F7BB7-2B24-11D7-8648000102C1865D

Bruckschen P, Oesmann S, Veizer J (1999) Isotope stratigraphy of the European carboniferous: proxy signals for ocean chemistry, climate and tectonics. Chem Geol 161:127–163. https://doi.org/10.1016/S0009-2541(99)00084-4

Casella LA, Griesshaber E, Simonet Roda M, Ziegler A, Mavromatis V, Henkel D, Laudien J, Häussermann V, Neuser RD, Angiolini L, Dietzel M, Eisenhauer A, Immenhauser A, Brand U, Schmahl WW (2018) Micro- and nanostructures reflect the degree of diagenetic alteration in modern and fossil brachiopod shell calcite: a multi-analytical screening approach (CL, FE-SEM, AFM, EBSD). Palaeogeograph Palaeoclimatol Palaeoecol 502:13–30. https://doi.org/10.1016/j.palaeo.2018.03.011

Cusack M, Pérez-Huerta A, EIMF (2012) Brachiopods recording seawater temperature—a matter of class or maturation? Chem Geol 334:139–143. https://doi.org/10.1016/j.chemgeo.2012.10.021

Dall WH (1920) Annotated list of the recent Brachiopoda in the collection of the United States National Museum, with descriptions of thirty-three new forms. Proc US Nat Mus 57:261–377

Endo K (1989) Growth and life mode of a Pleistocene brachiopod, *Kikaithyris hanzawai* (Yabe). Trans Proc Palaeontol Soc Jpn, NS 156:286–290

England J, Cusack M, Paterson NW, Edwards P, Lee MR, Martin R (2006) Hyperspectral cathodoluminescence imaging of modern and fossil carbonate shells. J Geophys Res Biogeosci 111:G03001. https://doi.org/10.1029/2005JG000144

Friedman I, O'Neil JR (1977) Compilation of stable isotope fractionation factors geochemical interest. In: Fleischer M (ed) Data of geochemistry, sixth edition. US Geol Surv Prof Paper 440-KK. US Gov Print Off, Washington D. C

Garbelli C, Angiolini L, Brand U, Jadoul F (2014) Brachiopod fabric, classes and biogeochemistry: implications for the reconstruction and interpretation of seawater carbon-isotope curves and records. Chem Geol 371:60–67. https://doi.org/10.1016/j.chemgeo.2014.01.022

Garbelli C, Angiolini L, Brand U, Shen SZ, Jadoul F, Posenato R, Azmy K, Cao CQ (2016) Neotethys seawater chemistry and temperature at the dawn of the end Permian mass extinction. Gondwana Res 35:272–285. https://doi.org/10.1016/j.gr.2015.05.012

Garbelli C, Angiolini L, Shen S (2017) Biomineralization and global change: a new perspective for understanding the end-Permian extinction. Geology 45:19–22. https://doi.org/10.1130/G38430.1

Gaspard D (1991) Microstructural organisation of the exoskeleton of some articulate brachiopods (Terebratulda, Rhynchonellida). The importance of the calcite granules and the effects of diagenesis. In: Suga S, Nakahara H (eds) Biomineralization 90: mechanisms and phylogeny of mineralization in biological systems. Springer-Verlag, Tokyo

Griesshaber E, Schmahl WW, Neuser R, Pettke T, Blüm M, Mutterlose J, Brand U (2007) Crystallographic texture and microstructure of terebratulide brachiopod shell calcite: an optimized materials design with hierarchical architecture. Am Mineral 92:722–734. https://doi.org/10.2138/am.2007.2220

Grossman EL (1984) Carbon isotopic fractionation in live benthic foraminifera—comparison with inorganic precipitate studies. Geochim Cosmochim Acta 48:1505–1512. https://doi.org/10.1016/0016-7037(84)90406-X

Grossman EL, Mii HS, Yancey TE (1993) Stable isotopes in Late Pennsylvanian brachiopods from the United States: implications for Carboniferous paleoceanography. Geol Soc Am Bull 105:1284–1296. https://doi.org/10.1130/0016-7606(1993)105<1284:SIILPB>2.3.CO;2

Habermann D, Neuser RD, Richter DK (1998) Low limit of Mn^{2+}-activated cathodoluminescence of calcite: state of the art. Sediment Geol 116:13–24. https://doi.org/10.1016/S0037-0738(97)00118-8

Hatai KM (1936a) The geographic distribution of Brachiopoda. Part 1. Recent Brachiopoda of Japan. Bull. Biogeog. Soc. Jpn 6:64–70

Hatai KM (1936b) Neogene Brachiopoda from Japan. Jpn J Geol Geogr 13:283–324

Hatai KM (1938) The tertiary and recent Brachiopoda of northeast Honsyu, Japan. Saito Ho-on Kai Mus Res Bull 16:89–246

Higo S, Callomon P, Goto Y (1999) Catalogue and bibliography of the marine shell-bearing mollusca of Japan. Elle Sci Publ, Yao

Hughes WW, Rosenberg GD, Tkachuck RD (1988) Growth increments in the shell of the living brachiopod *Terebratalia transversa*. Mar Biol 98:511–518

Iryu Y, Matsuda H, Machiyama H, Piller WE, Quinn TM, Mutti M (2006) An introductory perspective on the COREF Project. Island Arc 15:393–406. https://doi.org/10.1111/j.1440-1738.2006.00537.x

Japan Meteorological Agency (2018) Climatological normals in Noshiro (Akita Prefecture) for the period 1998–2010. http://www.data.jma.go.jp/obd/stats/etrn/view/nml_amd_ym.php?prec_no=32&block_no=0183&year=&month=&day=&view=p1. Accessed 1 Dec 2018.

Jelby ME, Thibault N, Surlyk F, Ullmann CV, Harlou R, Korte C (2014) The lower Maastrichtian Hvidskud succession, Møns Klint, Denmark: calcareous nannofossil biostratigraphy, carbon isotope stratigraphy, and bulk and brachiopod oxygen isotopes. Bull Geol Soc Denmark 62:89–104

Jiménez-López C, Romanek CS, Huertas FJ, Ohmoto H, Caballero E (2004) Oxygen isotope fractionation in synthetic magnesian calcite. Geochim Cosmochim Acta 68:3367–3377. https://doi.org/10.1016/j.gca.2003.11.033

Kaźmierczak J, Iryu Y (1999) Cyanobacterial origin of microcrystalline cements from Pleistocene rhodoliths and coralline algal crusts of Okierabu-jima, Japan. Acta Palaeontol Pol 44(2):117–130

Korte C, Jasper T, Kozur HW, Veizer J (2005) $\delta^{18}O$ and $\delta^{13}C$ of Permian brachiopods: a record of seawater evolution and continental glaciation. Palaeogeogr Palaeoclimatol Palaeoecol 224:333–351. https://doi.org/10.1016/j.palaeo.2005.03.015

Lee X, Hu R, Brand U, Zhou H, Liu X, Yuan H, Yan C, Cheng H (2004) Ontogenetic trace element distribution in brachiopod shells: an indicator of original seawater chemistry. Chem Geol 209:49–65. https://doi.org/10.1016/j.chemgeo.2004.04.029

Logan A (2007) Geographic distribution of extant articulated brachiopods. In Treatise on Invertebrate Paleontology. In: Selden PA (ed) Treatise on invertebrate paleontology, Part H, Brachiopoda (revised), vol.6. Geological Society of America, Boulder, Colorado, and University of Kansas Press, Lawrence, pp 3082–3115

Lowenstam HA (1961) Mineralogy, $^{18}O/^{16}O$ ratios, and strontium and magnesium contents of recent and fossil brachiopods and their bearing on the history of the oceans. J Geol 69:241–260. https://doi.org/10.1086/626740

Machel HG, Mason RG, Mariano AN, Mucci A (1991) Cause and emission of luminescence in calcite and dolomite. In: Barker CE, Kopp OC (eds) Luminnescent microscopy and spectroscopy: qualitative and quantitative application. SEPM Short Course Note, no. 25, Society for Sedimentary Geology (SEPM), Tulsa, pp 9–25

Marshall JD (1992) Climatic and oceanographic isotopic signals from the carbonate rock record and their preservation. Geol Mag 129:143–160. https://doi.org/10.1017/S0016756800008244

McConnaughey T (1989a) ^{13}C and ^{18}O isotopic disequilibrium in biological carbonates: I. Patterns. Geochim Cosmochim Acta 53:151–162. https://doi.org/10.1016/0016-7037(89)90282-2

McConnaughey T (1989b) ^{13}C and ^{18}O isotopic disequilibrium in biological carbonates: II. In vitro simulation of kinetic isotope effects. Geochim Cosmochim Acta 53:163–171. https://doi.org/10.1016/0016-7037(89)90283-4

McConnaughey TA, Burdett J, Whelan JF, Paull CK (1997) Carbon isotopes in biological carbonates: respiration and photosynthesis. Geochim Cosmochim Acta 61:611–622. https://doi.org/10.1016/S0016-7037(96)00361-4

Mii HS, Grossman EL, Yancey TE (1997) Stable carbon and oxygen isotope shifts in Permian seas of West Spitsbergen-Global change or diagenetic artifact? Geology 25:225–230. https://doi.org/10.1130/0091-7613(1997)025<0227:SCAOIS>2.3.CO;2

Mii HS, Shi GR, Cheng CJ, Chen YY (2012) Permian Gondwanaland paleoenvironment inferred from carbon and oxygen isotope records of brachiopod fossils from Sydney Basin, Southeast Australia. Chem Geol 291:87–103. https://doi.org/10.1016/j.chemgeo.2011.10.002

Mii HS, Shi GR, Wang CA (2013) Late Paleozoic middle-latitude Gondwana environment-stable isotope records from Western Australia. Gondwana Res 24:125–138. https://doi.org/10.1016/j.gr.2012.10.013

Mizota C, Kusakabe M (1994) Spatial distribution of δD-$\delta^{18}O$ values of surface and shallow groundwaters from Japan, South Korea and East China. Geochem J 28:387–410. https://doi.org/10.2343/geochemj.28.387

Morrison JO, Brand U (1986) Geochemistry of recent marine invertebrates. Geosci Can 13:237–253

Okai T, Suzuki A, Terashima S, Inoue M, Nohara M, Kawahata H, Imai N (2004) Collaborative analysis of GSJ/AIST geochemical reference materials JCp-1 (coral) and JCt-1 (giant clam). Chikyukagaku (Geochemistry) 38:281–286. https://doi.org/10.14934/chikyukagaku.38.281. In Japanese with English abstract

Parkinson D, Curry GB, Cusack M, Fallick AE (2005) Shell structure, patterns and trends of oxygen and carbon stable isotopes in modern brachiopod shells. Chem Geol 219:193–235. https://doi.org/10.1016/j.chemgeo.2005.02.002

Pérez-Huerta A, Cusack M, McDonald S, Marone F, Stampanoni M, MacKay S (2009) Brachiopod punctae: a complexity in shell biomineralisation. J Struct Biol 167:62–67. https://doi.org/10.1016/j.jsb.2009.03.013

Qing H, Veizer J (1994) Oxygen and carbon isotopic composition of Ordovician brachiopods: implications for coeval seawater. Geochim Cosmochim Acta 58:4429–4442. https://doi.org/10.1016/0016-7037(94)90345-X

Richardson JR (1975) Loop development and the classification of terebratellacean brachiopods. Palaeontology 18:285–314

Riechelmann S, Mavromatis V, Buhl D, Dietzel M, Eisenhauer A, Immenhauser A

(2016) Impact of diagenetic alteration on brachiopod shell magnesium isotope (δ^{26}Mg) signatures: experimental versus field data. Chem Geol 440: 191–206. https://doi.org/10.1016/j.chemgeo.2016.07.020

Romanek CS, Grossman EL, Morse JW (1992) Carbon isotopic fractionation in synthetic aragonite and calcite: effects of temperature and precipitation rate. Geochim Cosmochim Acta 56:419–430. https://doi.org/10.1016/0016-7037(92)90142-6

Rush PF, Chafetz HS (1990) Fabric-retentive, non-luminescent brachiopods as indicators of initial δ^{13}C and δ^{18}O composition: a test. J Sediment Petrol 60: 968–981. https://doi.org/10.1306/D4267659-2B26-11D7-8648000102C1865D

Samtleben C, Munnecke A, Bickert T, Pätzold J (2001) Shell succession, assemblage and species dependent effects on the C/O-isotopic composition of brachiopods—examples from the Silurian of Gotland. Chem Geol 175:61–107. https://doi.org/10.1016/S0009-2541(00)00364-8

Sato T, Higuchi T, Ishii T, Yuguchi S, Amano K, Kameo K (2003) Calcareous nannofossil biostratigraphy of the Upper Pliocene to lowermost Pleistocene distributed in the northern part of Akita Prefecture, Honshu, Japan—with special reference to the late Pliocene paleoceanography event. J Geol Soc Jpn 109:280–292. https://doi.org/10.5575/geosoc.109.280 In Japanese with English abstract

Sato T, Kameo K (1996) Pliocene to Quaternary calcareous nannofossil biostratigraphy of the Arctic Ocean, with reference to late Pliocene glaciation. In: Thiede J, Myhre AM, Firth JV (eds) Proceedings of the Ocean Drilling Program, initial reports, Leg 151. Ocean Drilling Program, College Station, pp 39–59. https://doi.org/10.2973/odp.proc.sr.151.112.1996

Shiino Y, Kitazawa K (2010) Behavior of the Terebratulide brachiopod *Laqueus rubellus*, with special reference to the pedicle function. Jpn J Benthol 26:18–26. https://doi.org/10.5179/benthos.65.18 In Japanese with English abstract

Sippel RF, Glover ED (1965) Structures in carbonate rocks made visible by luminescence petrography. Science 150(3701):1283–1287. https://doi.org/10.1126/science.150.3701.1283

Sowerby GB (1846) Descriptions of thirteen new species of brachiopods. Proc Zool Soc London 14:91–95

Takayanagi H, Asami R, Abe O, Kitagawa H, Miyajima T, Iryu Y (2012) Carbon- and oxygen-isotope compositions of a deep-water modern brachiopod *Campagea japonica* collected off Aguni-jima, Central Ryukyu Islands, southwestern Japan. Geochem J 46:77–87. https://doi.org/10.2343/geochemj.1.0153

Takayanagi H, Asami R, Abe O, Miyajima T, Kitagawa H, Sasaki K, Iryu Y (2013) Intraspecific variations in carbon-isotope and oxygen-isotope compositions of a brachiopod *Basiliola lucida* collected off Okinawa-jima, southwestern Japan. Geochim Cosmochim Acta 115:115–136. https://doi.org/10.1016/j.gca.2013.03.026

Takayanagi H, Asami R, Otate T, Abe O, Miyajima T, Kitagawa H, Sasaki K, Iryu Y (2015) Quantitative analysis of intraspecific variations in the carbon and oxygen isotope compositions of the modern cool-temperate brachiopod *Terebratulina crossei*. Geochim Cosmochim Acta 170:301–320. https://doi.org/10.1016/j.gca.2015.08.006

Takizawa M, Takayanagi H, Yamamoto K, Abe O, Sasaki K, Iryu Y (2017) Paleoceanographic conditions at approximately 20 and 70 ka recorded in Kikaithyris hanzawai (Brachiopoda) shells. Geochim Cosmochim Acta 215:189–213. https://doi.org/10.1016/j.gca.2017.08.002

Tarutani T, Clayton RN, Mayeda TK (1969) The effect of polymorphism and magnesium substitution on oxygen isotope fractionation between calcium carbonate and water. Geochim Cosmochim Acta 33:987–996. https://doi.org/10.1016/0016-7037(69)90108-2

Thomson JA (1927) Brachiopod morphology and genera (Recent and Tertiary). New Zealand Board Sci Art, Manual 7:1–338

Tucker ME, Wright VP (1990) Carbonate Sedimentology. Blackwell, Oxford

Ullmann CV, Campbell HJ, Frei R, Korte C (2014) Geochemical signatures in Late Triassic brachiopods from New Caledonia. New Zealand J Geol Geophys 57: 420–431. https://doi.org/10.1080/00288306.2014.958175

Ullmann CV, Frei R, Korte C, Lüter C (2017) Element/Ca, C and O isotope ratios in modern brachiopods: species-specific signals of biomineralization. Chem Geol 460:15–24. https://doi.org/10.1016/j.chemgeo.2017.03.034

Ullmann CV, Korte C (2015) Diagenetic alteration in low-Mg calcite from macrofossils: a review. Geol Q 59:3–20. https://doi.org/10.7306/gq.1217

Veizer J, Ala D, Azmy K, Bruckschen P, Buhl D, Bruhn F, Carden GAF, Diener A, Ebneth S, Godderis Y, Jasper T, Korte C, Pawellek F, Podlaha OG, Strauss H (1999) ^{87}Sr/^{86}Sr, δ^{13}C and δ^{18}O evolution of Phanerozoic seawater. Chem Geol 161:59–88. https://doi.org/10.1016/S0009-2541(99)00081-9

Wadleigh MA, Veizer J (1992) ^{18}O/^{16}O and ^{13}C/^{12}C in lower Paleozoic articulate brachiopods: implications for the isotopic composition of seawater. Geochim Cosmochim Acta 56:431–443

Wenzel B (2000) Differential preservation of primary isotopic signatures in Silurian brachiopods from Northern Europe. J Sediment Res 70(1):194–209. https://doi.org/10.1306/2DC4090A-0E47-11D7-8643000102C1865D

Williams A (1968) Evolution of the shell structure of articulate brachiopods. Spec Pap Palaeontol 2:1–55

Williams A (1997) Shell structure. In: Kaesler RL (ed) Treatise on invertebrate paleontology part H Brachiopoda [revised 1: introduction]. The Geological Society of America and The University of Kansas Press, Boulder (CO) and Lawrence (KS), pp 267–320

Williams A, Carlson SJ, Brunton CHC, Holmer LE, Popov L (1996) A supra-ordinal classification of the Brachiopoda. Phil Trans Royl Soc London Ser B 351:1171–1193. https://doi.org/10.1098/rstb.1996.0101

Williams A, Rowell AJ (1965) Brachiopod anatomy. In: Moore RC (ed) Treatise on invertebrate paleontology, part H, Brachiopoda. Geological Society of America, Boulder, Colorado, and University of Kansas Press, Lawrence, pp 6–57

Yamamoto K, Asami R, Iryu Y (2010a) Carbon and oxygen isotopic compositions of modern brachiopod shells from a warm-temperate shelf environment, Sagami Bay, central Japan. Palaeogeogr Palaeoclimatol Palaeoecol 291:348–359. https://doi.org/10.1016/j.palaeo.2010.03.006

Yamamoto K, Asami R, Iryu Y (2010b) Within-shell variations in carbon and oxygen isotope compositions of two modern brachiopods from a subtropical shelf environment off Amami-o-Shima, southwestern Japan. Geochem Geophys Geosyst 11:Q10009. https://doi.org/10.1029/2010GC003190

Yamamoto K, Asami R, Iryu Y (2011) Brachiopod taxa and shell portions reliably recording past ocean environments: toward establishing a robust paleoceanographic proxy. Geophys Res Lett 38:L13601. https://doi.org/10.1029/2011GL047134

Yamamoto K, Asami R, Iryu Y (2013) Correlative relationships between carbon- and oxygen- isotope records in two cool-temperate brachiopod species off Otsuchi Bay, northeastern Japan. Paleontol Res 17:12–26. https://doi.org/10.2517/1342-8144-17.1.12

Yamamoto K, Takizawa M, Takayanagi H, Asami R, Iryu Y (2017) Impact of meteoric diagenesis on the geochemistry and microstructure of Pleistocene brachiopod shells from the Ryukyu Islands, southwestern Japan. Island Arc 26:e12217. https://doi.org/10.1111/iar.12217

Zhang J, Quay PD, Wilbur DO (1995) Carbon isotope fractionation during gas-water exchange and dissolution of CO_2. Geochim Cosmochim Acta 59:107–114. https://doi.org/10.1016/0016-7037(95)91550-D

13

Reactions between komatiite and CO_2-rich seawater at 250 and 350 °C, 500 bars: implications for hydrogen generation in the Hadean seafloor hydrothermal system

Hisahiro Ueda[1,2*], Takazo Shibuya[2,3,4], Yusuke Sawaki[1], Masafumi Saitoh[2,3,4], Ken Takai[2,3,4,5] and Shigenori Maruyama[6]

Abstract

To understand the chemical nature of hydrothermal fluids in the komatiite-hosted seafloor hydrothermal system in the Hadean, we conducted two hydrothermal serpentinization experiments involving synthetic komatiite and a CO_2-rich acidic NaCl fluid at 250 and 350 °C, 500 bars. During the experiments, the komatiites were strongly carbonated to yield iron-rich dolomite (3–9 wt.% FeO) at 250 °C and calcite (<0.8 wt.% FeO) at 350 °C, respectively. The carbonation of komatiites suppressed H_2 generation in the fluids. The steady-state H_2 concentrations in the fluid were approximately 0.024 and 2.9 mmol/kg at 250 and 350 °C, respectively. This correlation between the Fe content in carbonate mineral and the H_2 concentration in the fluid suggests that the incorporation of ferrous iron into the carbonate mineral probably limited magnetite formation and consequent generation of hydrogen during the serpentinization of komatiites. The H_2 concentration of the fluid at 350 °C corresponds to that of modern H_2-rich seafloor hydrothermal systems, such as the Kairei hydrothermal field, where hydrogenotrophic methanogens dominate in the prosperous microbial ecosystem. Accordingly, the high-temperature serpentinization of komatiite would provide the H_2-rich hydrothermal environments that were necessary for the emergence and early evolution of life in the Hadean ocean. In contrast, H_2-rich fluids may not have been generated by serpentinization at temperatures below 250 °C because carbonate minerals become more stable with decreasing temperature in the komatiite-H_2O-CO_2 system.

Keywords: Komatiite, CO_2-rich condition, Early Earth, Hydrothermal alteration, Serpentinization

Introduction

Deep-sea hydrothermal environments have been considered as a favorable environment for the emergence and early evolution of life on Earth (e.g., Yanagawa and Kojima 1985; Russell and Hall 1997). In particular, H_2-rich hydrothermal fluids generated through the serpentinization of ultramafic rocks would have driven prebiotic chemical evolution and the development of biotic energy metabolisms (Takai et al. 2006; Amend and McCollom 2009; Russell et al. 2014; Nakamura and Takai 2014; Shibuya et al. 2016). Molecular hydrogen (H_2) generation during serpentinization is caused by the reduction of water in conjunction with the oxidation of ferrous iron in silicates. This process is written as the following simplified reaction (e.g., Allen and Seyfried 2003; Seyfried et al. 2007; McCollom and Bach 2009):

$$2(FeO)_{rock} + H_2O \rightarrow (Fe_2O_3)_{rock} + H_2,$$

where the ferric iron-bearing solid phase generally precipitates as magnetite (Fe_3O_4). Therefore, H_2 generation is strongly related to magnetite formation during serpentinization (McCollom and Bach 2009).

Among the various prospects for the seafloor hydrothermal systems in the early Earth (e.g., Kump and Seyfried

* Correspondence: ueda.h.ai@m.titech.ac.jp
[1]Department of Earth and Planetary Sciences, Tokyo Institute of Technology, 2-12-1 Ookayama, Meguro-ku, Tokyo 152-8551, Japan
[2]Laboratory of Ocean-Earth Life Evolution Research (OELE), Japan Agency for Marine-Earth Science and Technology (JAMSTEC), 2-15 Natsushima-cho, Yokosuka 237-0061, Japan
Full list of author information is available at the end of the article

2005; Shibuya et al. 2010), two types of serpentinization are hypothesized to be dominant in the Hadean ocean: a komatiite-hosted high-temperature type at oceanic islands/plateaus (Takai et al. 2006; Yoshizaki et al. 2009; Shibuya et al. 2015) and a low-temperature alkaline type hosted by komatiites or peridotites at off-ridge ocean floors (e.g., Russell et al. 2010, 2014; Shibuya et al. 2016). Based on the geological records, Archean oceanic crust was likely much thicker than modern equivalents owing to the higher potential mantle temperature at that time (Ohta et al. 1996; Komiya et al. 2002; Komiya 2004; Moores 2002; Shibuya et al. 2012). This thick lid of oceanic crust probably limited the exposure of mantle peridotites on the seafloor, thus suggesting that komatiite volcanism would have been much more abundant than the exposed mantle peridotites that are frequently observed near modern slow-spreading ridges without a sufficient magmatic supply (Takai et al. 2006). Therefore, it has been pointed out that H_2-rich seafloor hydrothermal environments would have been predominantly driven by komatiite volcanism in the Hadean ocean (Takai et al. 2006). On the other hand, the low-temperature alkaline type could provide distinctive chemical environments such as large pH gradients in the seawater-hydrothermal fluid mixing zones, which may also have been advantageous for the possible development of proton-motive energy metabolisms (Russell et al. 2014; Shibuya et al. 2016).

Previously, hydrothermal alteration experiments have been conducted to understand the potential of H_2 generation during the serpentinization of komatiites under CO_2-free conditions (Yoshizaki et al. 2009; Shibuya et al. 2015; Suzuki et al. 2015). However, as suggested by theoretical calculations (Walker 1985; Kasting 1993; Sleep and Zahnle 2001) and geological records (Lowe and Tice 2004; Ohmoto et al. 2004; Shibuya et al. 2007, 2012, 2013a), the atmospheric CO_2 levels in the early Earth were likely much higher than the modern level. Furthermore, it has been experimentally demonstrated that carbonate formation during the serpentinization of olivine under CO_2-rich conditions suppresses H_2 generation in fluids (Jones et al. 2010; Klein and McCollom 2013; Neubeck et al. 2014). Recently, some experiments have been conducted in komatiite-CO_2-H_2O systems (Lazar et al. 2012; Hao and Li 2015), but H_2 generation in fluids was not the objective in these studies. Therefore, the potential for hydrogen generation through the serpentinization of komatiites in the Hadean has not yet been experimentally evaluated under CO_2-rich conditions.

In this study, we conducted two experiments that simulate the reactions between komatiite and CO_2-rich seawater at 250 and 350 °C, 500 bars, using a batch-type (closed system) hydrothermal reactor (Yoshizaki et al. 2009). The experiments revealed the chemical composition of high-temperature hydrothermal fluids and CO_2 absorption ability of komatiite through serpentinization under CO_2-rich conditions. The results demonstrate the significance of komatiite-hosted hydrothermal systems for seawater chemistry and provide insight into the H_2-rich hydrothermal environments in the CO_2-rich Hadean ocean.

Methods/Experimental

Synthesis of komatiite

The komatiite used in the experiments was synthesized from a mixture of 12 reagents (SiO_2, TiO_2, Al_2O_3, Fe_2O_3, MnO, MgO, $CaCO_3$, Na_2CO_3, K_2CO_3, P_2O_5, NiO, and Cr_2O_3). The chemical composition of the mixed reagents was adjusted to Al-depleted (Barberton-type) komatiite (Smith et al. 1980; Wei et al. 1990; Yoshizaki et al. 2009) because the volcanism of Al-depleted komatiite was likely more predominant than Al-undepleted komatiite in the Hadean (Shibuya et al. 2015). The mixed powder was placed in a Pt-Rh crucible at 1000 °C for 1 h in an electronic furnace to decarbonate the reagents. The mixture was melted at 1600 °C for 0.5 h at the oxygen fugacity of quartz-fayalite-magnetite (QFM) buffer regulated by a H_2-CO_2 gas mixture (Canil 1997). To create a spinifex texture of olivine, the temperature was lowered to 1450 °C over 1.5 h. Next, the crucible was quenched to room temperature, thus yielding a fresh spinifex-textured komatiite (Fig. 1). The composition and texture of the synthetic komatiite and its mineral/glass phases were analyzed using X-ray diffraction (XRD), X-ray fluorescence (XRF), and an electron probe microanalyzer (EPMA) (Table 1). In the interstitial glass, dendritic olivine crystals configurate a spinifex texture. Broadly depending on size, these crystals have Mg# values ranging from 89 to 94. Small amounts of chromian spinels are disseminated in the vicinity of the olivine crystals. Owing to such a mineral assemblage, almost all Ca in the komatiite was partitioned into the interstitial glass (Fig. 1 and Table 1). The synthesized komatiite was crushed in a tungsten mortar and sieved to obtain a <100 μm powder. To remove any contamination of organic materials during sample preparation, the powdered komatiite was washed with acetone and distilled water several times and dried for 12 h in an oven at 50 °C.

Experimental system

The Inconel alloy autoclave used for the hydrothermal alteration experiment is based on the study by Seyfried et al. (1979). It resists corrosion and withstands high-temperature and high-pressure conditions (up to 600 °C and 600 bars; Fig. 2). The reaction cell is made of a gold bag with a titanium head because of the resistance of these materials to high-temperature fluids. Furthermore,

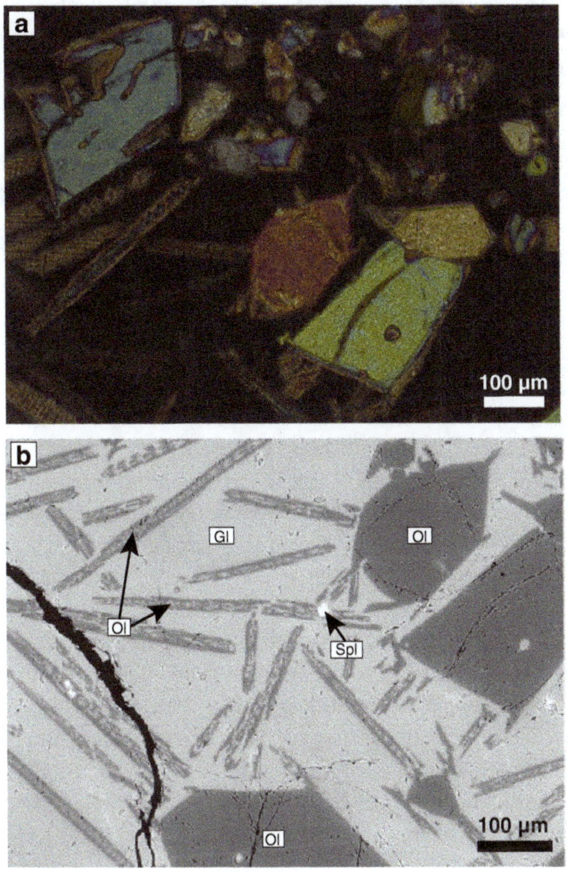

Fig. 1 **a** Photomicrograph taken in cross-polarized light and **b** a backscattered electron (BSE) image of the synthetic komatiite. *Ol* olivine, *Spl* spinel, *Gl* glass

Table 1 Compositions of the synthetic komatiite, mineral/glass phases therein, starting materials used in a previous experiment (wt.%)

Sample	Al-depleted komatiite	Olivine	Glass	Spinel	Basalt[b]
SiO_2	46.29	41.02	50.74	0.22	49.84
TiO	0.37	0.00	0.68	0.22	0.73
Al_2O_3	4.30	0.20	7.74	6.43	15.82
Cr_2O_3	0.21	0.48	0.52	52.22	–
FeO_{total}[a]	12.52	9.70	14.81	22.03	11.00
MnO	0.13	0.09	0.18	0.14	0.22
MgO	27.47	48.80	12.57	15.49	9.44
CaO	7.20	0.30	11.84	0.22	10.84
Na_2O	0.06	0.00	0.23	0.00	1.88
K_2O	0.02	0.02	0.05	0.02	0.18
NiO	0.11	0.34	0.01	0.22	–
Total	98.68	100.95	99.37	97.23	99.95
Mg#[c]	79.64	89.96	60.21	55.62	60.47

–, no mention
[a]Total iron as FeO
[b]Synthetic basalt used in Shibuya et al. (2013b)
[c]Mg# = [Mg / (Mg+Fe)] × 100

gold has sufficient flexibility to allow the fluid inside the reaction cell to be pressurized by the surrounding water. In addition, to avoid possible H_2 generation via the reaction of metallic titanium with water, the surface of the titanium was completely oxidized before the experiments. To eliminate organic contamination, all materials in contact with the reacting fluid were baked in a muffle furnace at 450 °C for 5 h.

The initial solution used in the experiments was controlled at pH 4.9 and 400 mmol/kg of ΣCO_2 ($=CO_{2\,(aq)} + HCO_3^- + CO_3^{2-}$) by the addition of $NaHCO_3$, NaCl, and HCl to ultrapure water, according to the previous studies suggesting that Hadean seawater was acidic due to the high partial pressure of CO_2 in the atmosphere (e.g., Kasting 1993; Macleod et al. 1994). To avoid the degassing of CO_2 by a reaction between $NaHCO_3$ and HCl prior to the sealing of the reaction cell, these materials were separately placed in the gold tube and then mixed after sealing (Shibuya et al. 2013b). The Cl concentration of the solution was regulated to be approximately 1000 mmol/kg because salinity of seawater in the early Earth was potentially 1.5–2 times higher than the modern value (approximately 550 mmol/kg) because of the absence of continents and the associated salt deposits/saline water (Knauth 2005). The komatiite powder was reacted with this hypothetical seawater in the reaction cell at 250 and 350 °C, 500 bars for 1530–2760 h (hereafter abbreviated as Exp-250 and Exp-350, respectively). At the

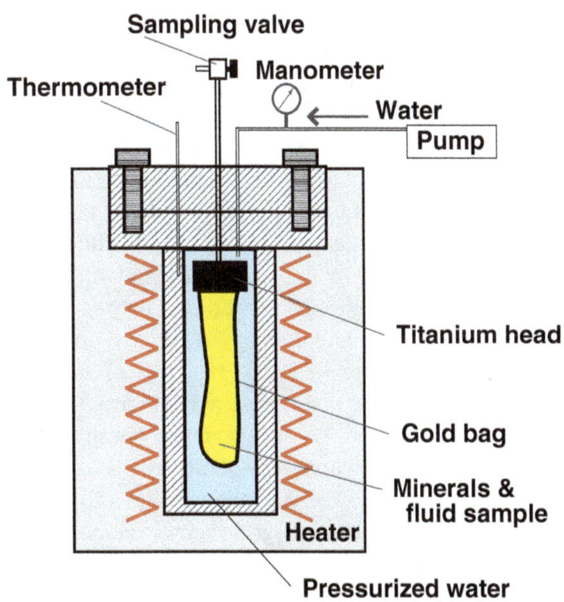

Fig. 2 Schematic illustration of the Inconel alloy autoclave used in this study (modified after Yoshizaki et al. 2009). The synthesized komatiite and hypothetical Hadean seawater were sealed in the gold bag

beginning of the experiments, the water/rock mass ratios were adjusted to approximately 5 because water/rock ratios are commonly below 5 in the high-temperature regions of natural subseafloor hydrothermal systems (Wetzel and Shock 2000). The water/rock mass ratio decreased to ~3 at the end of the experiments due to the multiple fluid samplings during the experiments.

Sampling and analytical methods

During the experiment, approximately 3–4 g of fluid samples was collected several times through a gold-lined sampling tube. For CO_2 and H_2 analyses, 0.5 mL of fluid sample was directly introduced to each Ar-purged, sealed vial at room temperature. Especially, for the CO_2 analysis, the sampled fluid was acidified (pH <2) by adding HCl to ensure the complete extraction of the dissolved bicarbonate and carbonate ions. Quantitative analysis for gas species was conducted by gas chromatography. The overall detection precision for the analyses of ΣCO_2 and H_2 concentrations in the fluid were both better than 5%. For the analysis of other dissolved species, 0.2 mL of the fluid was collected in two vials, to which either HNO_3 or NaOH was immediately added to avoid the precipitation of transition metals (Fe and Mn) and silicic acid species, respectively. The fluid samples were analyzed using inductively coupled plasma optical emission spectrometry and ion chromatography. The analytical precision (2σ) was approximately 2% for Cl and Na, and 5% for the other elements.

The pH of the fluid samples was determined using a pH meter at 25 °C under atmospheric conditions. The significant digits and precision of the pH measurement are 2 digits and ±0.1 units. This measurement was performed 1 h after sampling to allow the stabilization of the pH against CO_2 degassing. The $pH_{25\,°C}$ of the fluid cannot be measured directly because it rapidly changes due to CO_2 degassing after the sampling of fluid from the reactor. The amount of degassed CO_2 was determined from the ΣCO_2 concentrations of the fluid sample directly introduced into the sealed vial and of the fluid degassed under atmospheric conditions, then the original $pH_{25\,°C}$ was calculated based on the amount of degassed CO_2. We calculated $pH_{in\text{-}situ}$ of the high-temperature fluid in the reaction cell with the Geochemist's Workbench computer code (Bethke 2008) based on the pH at room temperature ($pH_{25\,°C}$) and concentrations of dissolved elements/species and gases. In the $pH_{in\text{-}situ}$ calculations, charge balance was constrained by $pH_{25\,°C}$, while Na was used to compensate for imbalanced charges derived from analytical errors. The required thermodynamic database was generated by the SUPCRT92 computer program (Johnson et al. 1992), using thermodynamic data of minerals, aqueous species, and complexes reported in Shock and Helgeson (1988), Shock and Koretsky (1995), Shock et al. (1989, 1997), Sverjensky et al. (1997), and McCollom and Shock (1997). The B-dot activity model was used in the calculations (Helgeson 1969; Helgeson and Kirkham 1974). The temperature-dependent activity coefficient for aqueous CO_2 was derived from the empirical relationship established by Drummond (1981), and the temperature-dependent activity of water in NaCl solution was derived from the formulation of Bethke (2008). Cleverley and Bastrakov (2005) provided useful temperature-dependent polynomial functions for both of these last-mentioned parameters.

After the experiments, alteration products were retrieved from the reaction cell and dried at 80 °C for 12 h. The solid alteration product of each experiment was analyzed with XRD and EPMA to determine the assemblage and composition of the alteration minerals. Powdery alteration products from the experiments at both temperatures were embedded in Epofix resin, and thin sections were made of these. The analytical conditions of the EPMA were 15 kV of accelerating voltage, 10 nA of specimen current, and 60–80 s of counting time. The precision of the elemental concentration was less than 10%, which was based on duplicate measurements. The detection limit for the elemental concentration was 0.01 wt.%. Moreover, the starting solid material (unaltered komatiite) was analyzed with XRD, XRF, and EPMA.

Results and Discussion

Fluid chemistry

The Cl concentration in the fluid was kept relatively constant throughout both experiments because Cl-bearing minerals are generally rare in hydrothermally altered rocks (Fig. 3 and Table 2). In contrast, the concentrations of all other components (Na, K, Mg, Ca, Si, Fe, Mn, and ΣCO_2) changed immediately after the beginning of the experiments and reached a near steady state within 2760 h at 250 °C and 1530 h at 350 °C. The Mg concentration in Exp-250 (36–40 mmol/kg) was 30–40 times higher than that in Exp-350 (<1.2 mmol/kg). The Fe concentration in Exp-250 (0.49–0.97 mmol/kg) was approximately half than that in Exp-350 (1.53–1.80 mmol/kg). No large difference between Exp-250 and Exp-350 was identified in the concentrations of other dissolved elements. Na (820–1036 mmol/kg across both experiments) had the second highest concentration next to Cl. The concentration of Ca increased and converged at 39–49 mmol/kg in Exp-250 and 37–42 mmol/kg in Exp-350. The concentrations of Si and K were 1.6–6.1 and 0.78–1.24 mmol/kg, respectively, across both experiments. The concentration of Mn was the lowest among all analyzed elements (0.15–0.73 mmol/kg across both experiments).

The ΣCO_2 concentration decreased from ca. 400 mmol/kg at the beginning of the experiments to approximately

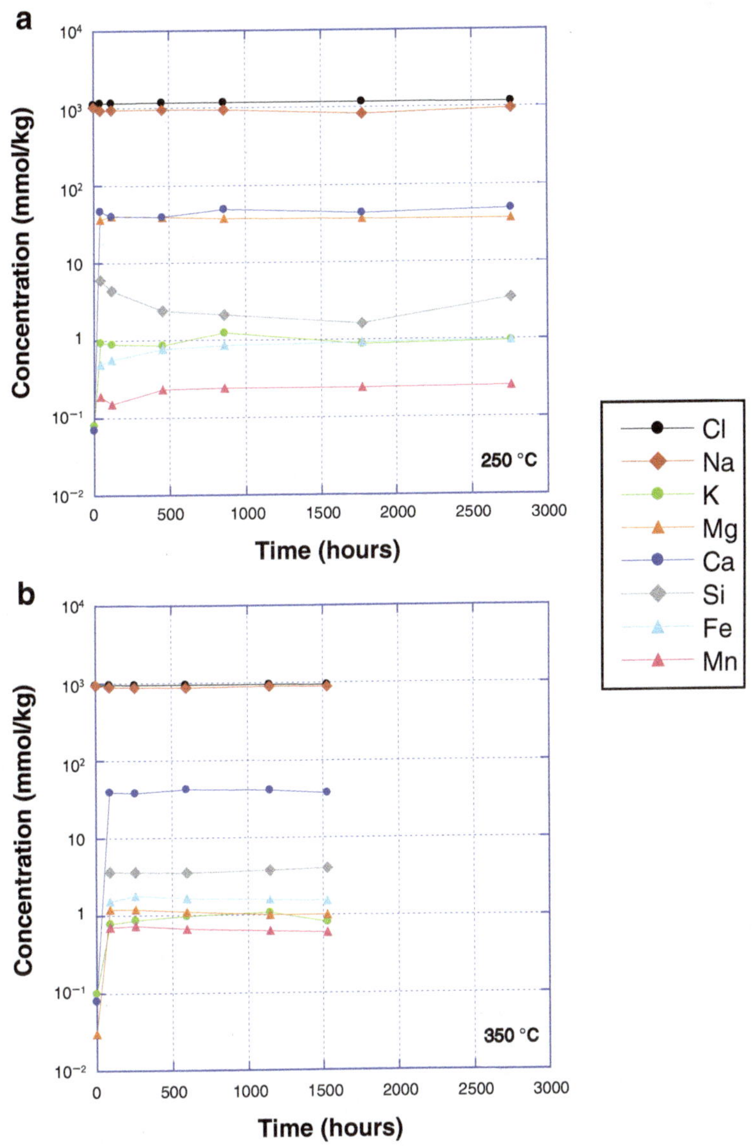

Fig. 3 Concentrations of the dissolved species in an aqueous fluid coexisting with komatiite powder and its alteration products as a function of reaction time in the experiments at **a** 250 and **b** 350 °C, 500 bars

33 mmol/kg in Exp-250 and approximately 171 mmol/kg in Exp-350 (Fig. 4a and Table 2). This trend of the steady-state ΣCO_2 concentration in the higher-temperature experiment being higher than that in the lower-temperature experiment is consistent with the results of experiments simulating high-temperature reactions between basalt and CO_2-rich seawater (Shibuya et al. 2013b). On the other hand, the $pH_{in\text{-}situ}$ values were higher than the $pH_{25\ °C}$ values at the beginning of the experiments due to the temperature dependence of the ion product of water and speciation of CO_2 species (e.g., Shibuya et al. 2010). As the reaction proceeded, the $pH_{in\text{-}situ}$ values finally decreased to 4.8 in Exp-250 and 5.7 in Exp-350, which are broadly consistent with those of high-temperature hydrothermal fluids in modern oceans ($pH_{in\text{-}situ}$ = approximately 5) (Seyfried et al. 1991).

In both experiments, the H_2 concentration in the fluid suddenly increased and temporarily reached a high value just after the start of the experiments, then slightly decreased in each experiment (Fig. 4b). Subsequently, the H_2 concentration remained almost constant in Exp-350, whereas it gradually increased with time in Exp-250. Finally, it converged to reach a steady state at up to 0.024 mmol/kg in Exp-250 and 2.9 mmol/kg in Exp-350 (Table 2). This trend indicates that the H_2 concentration was unstable due to extensive water/rock reactions just

Table 2 Composition of the sampled fluids in the experiments (mmol/kg)

Experiment	Time (h)	$pH_{25\,°C}$ [a]	ΣCO_2 [b]	$pH_{25\,°C}$ [c]	$pH_{in\text{-}situ}$ [d]	H_2	ΣCO_2	Cl	Na	K	Mg	Ca	Si	Fe	Mn
250 °C	0	n.a.	n.a.	4.9	6.0	0.003	396	1128	1036	0.08	n.d.	0.073	n.d.	n.d.	n.d.
	43	6.3	0.9	3.7	4.1	0.010	161	1174	939	0.94	36.2	46.5	6.1	0.49	0.19
	119	5.8	1.9	4.1	4.5	0.014	67	1168	951	0.89	40.4	39.7	4.4	0.56	0.15
	452	6.6	1.0	4.3	4.7	0.004	35	1186	948	0.85	38.4	39.2	2.4	0.77	0.23
	859	6.4	2.3	4.6	5.0	0.007	36	1178	937	1.24	37.2	48.7	2.1	0.85	0.24
	1771	7.1	1.0	4.4	4.8	0.021	35	1186	820	0.88	36.4	43.6	1.6	0.93	0.24
	2760	6.8	1.2	4.5	4.8	0.024	33	1175	949	0.95	36.4	48.8	3.5	0.97	0.25
350 °C	0	n.a.	n.a.	4.9	7.2	n.d.	396	966	946	0.10	0.03	0.078	n.d.	n.d.	n.d.
	90	6.1	1.9	3.8	5.6	6.0	176	971	902	0.78	1.2	39.4	3.7	1.53	0.71
	258	6.4	1.7	3.9	5.6	2.8	163	968	882	0.86	1.2	38.0	3.7	1.80	0.73
	594	6.4	1.7	3.9	5.6	2.9	179	963	877	0.98	1.1	42.2	3.6	1.65	0.67
	1146	6.5	1.8	3.9	5.7	2.7	174	969	900	1.08	1.0	41.1	3.8	1.61	0.63
	1530	6.6	1.7	3.9	5.7	2.2	171	953	884	0.81	1.0	37.3	4.1	1.55	0.60

n.a., not analyzed
n.d., not detected
[a]Measured pH after CO_2 degassing
[b]Measured ΣCO_2 concentration after CO_2 degassing
[c]Calculated $pH_{25\,°C}$ at 1 bar before CO_2 degassing at room temperature
[d]Calculated $pH_{in\text{-}situ}$ at high-temperatures, 500 bars

after the start of the experiments, and then gradually reached a steady state. The steady-state H_2 concentration in Exp-350 was approximately 100 times higher than that in Exp-250, thus indicating that differences in temperature significantly affect hydrogen generation during water-rock reactions.

Alteration products

The XRD analysis revealed that the synthetic komatiites were extensively altered throughout the experiments (Fig. 5). The results show that the alteration product in Exp-250 contains dolomite, serpentine, and a smectite/chlorite mixture (Fig. 5b), whereas calcite, serpentine, and a smectite/chlorite mixture were present as major alteration minerals in Exp-350 (Fig. 5c). The existence of serpentine and carbonate minerals in the run products indicated komatiite was altered by both serpentinization and carbonation. The precipitated carbonate mineral differed for the two temperatures.

Secondary electron imaging (SEI) and EPMA spot analysis provided more detailed information on mineral assemblage and composition (Fig. 6 and Table 3). In Exp-250, EPMA analyses of the products revealed that dolomite crystals contained a relatively high content of iron (3–9 wt.% FeO_{total}), whereas the serpentine and smectite/chlorite mixture had FeO_{total} contents of 13–15.5 wt.% (Fig. 6a–d; Table 3). Siderite (including approximately 40 wt.% FeO_{total}) and olivine were present as minor minerals in the alteration products in Exp-250 (Fig. 6a, c; Table 3). In Exp-350, the EPMA analyses of the alteration products revealed that calcite crystals had a relatively low FeO_{total} content (0.1–0.8 wt.%) and that the serpentine and smectite/chlorite mixture had FeO_{total} contents of 14–17 wt.% (Fig. 6e; Table 3). Olivine and magnetite were also identified as minor minerals; however, according to the SEI observations in many places over the thin sections, their abundances were likely greater in Exp-350 than in Exp-250. Comparing the compositions of serpentine and the smectite/chlorite mixture between Exp-250 and Exp-350, no large difference in FeO_{total} content was identified. On the other hand, on average, the FeO_{total} content in Fe-bearing dolomite in Exp-250 was ten times higher than that in calcite in Exp-350 (Table 3).

Dissolved CO_2 concentration and pH

The ΣCO_2 concentrations in fluids decreased from approximately 400 to 33 mmol/kg in Exp-250 and 171 mmol/kg in Exp-350 (Fig. 4a and Table 2). These differing steady-state ΣCO_2 concentrations indicate that carbonate minerals are destabilized with increasing temperature in the komatiite-CO_2-H_2O system. This trend is consistent with previous basalt experiments under CO_2-rich conditions at 250 and 350 °C, where the final ΣCO_2 concentrations were 1 mmol/kg at 250 °C and 108 mmol/kg at 350 °C (Shibuya et al. 2013b). Comparing results for the same temperature, the ΣCO_2 concentration in our komatiite experiments is higher than that in the basalt experiments of Shibuya et al. (2013b), thus indicating that komatiite has a lesser ability than basalt to absorb CO_2 through water-rock reactions. This would be derived from the lower CaO content of

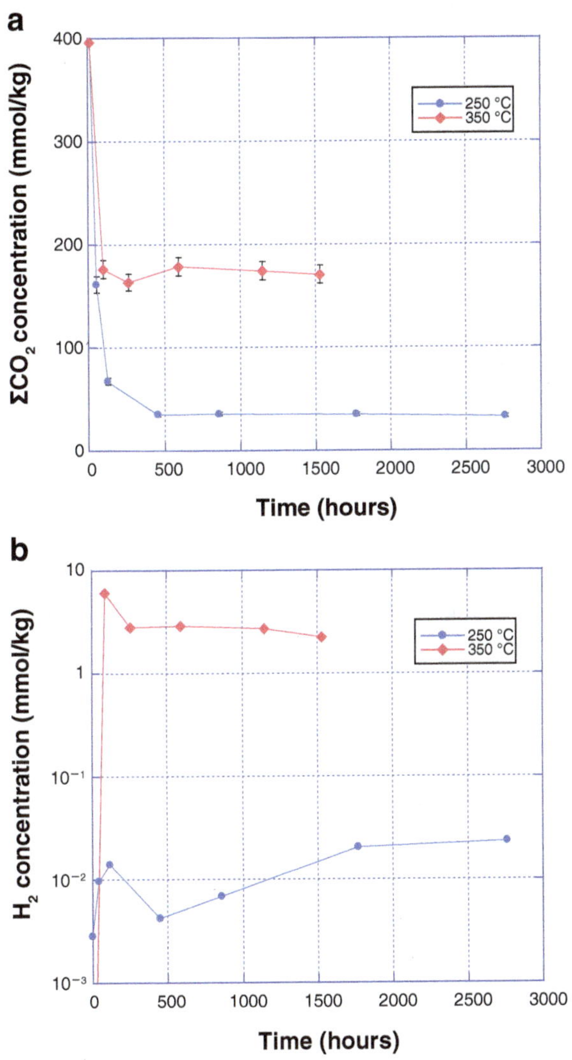

Fig. 4 Changes in **a** dissolved CO_2 and **b** H_2 concentrations in the sampled fluids as a function of reaction time

komatiite (compared with basalt) because CaO is the main component in the alteration carbonate (mainly calcite or dolomite) in the basalt and komatiite experiments (Shibuya et al. 2013b and the present study).

The final $pH_{in\text{-}situ}$ values of fluids in the present experiments ($pH_{in\text{-}situ} = 4.8$ at 250 °C and 5.7 at 350 °C) are clearly lower than those of the previous basalt-CO_2-NaCl fluid experiments ($pH_{in\text{-}situ} = 6.6$ at 250 °C and 7.2 at 350 °C) (Shibuya et al. 2013b). This discrepancy is probably derived from the lower $pH_{25\ °C}$ of the initial solution ($pH_{25\ °C} = 4.9$) in our experiments than the value of 6.5 in the basalt experiments and/or compositional differences between komatiite and basalt. In general, under CO_2-rich and near-neutral pH conditions at room temperature, substantial HCO_3^- is dissolved in the solution. With increasing temperature, the dissolved HCO_3^- is converted to $CO_{2(aq)}$ and OH^-, which elevates the $pH_{in\text{-}situ}$ value of the hydrothermal fluid (Shibuya et al. 2010). Although reactions with rocks generally cause fluid $pH_{in\text{-}situ}$ to converge near 5.5 (neutral pH at 250–350 °C), previous experiments have revealed that the alkalinization effect due to the increase in temperature under CO_2-rich conditions potentially elevates the $pH_{in\text{-}situ}$ value beyond a neutral pH value at high temperature, even if the fluid reacts with rocks (Shibuya et al. 2013b). In this study, however, the assumed $pH_{25\ °C}$ of the initial solution was 4.9, which was likely too low to form substantial HCO_3^- in the initial solution at room temperature. This implies that if seawater pH was lower than approximately 5 (CO_{2aq} was the main species), metal-rich black smoker-type hydrothermal fluids would have been generated in the Hadean komatiite hydrothermal systems in contrast to the metal-poor, alkaline hydrothermal fluids in the Archean basalt-hosted system. Although the difference in whole rock composition between komatiite and basalt should be experimentally evaluated, seawater pH was likely an important factor controlling the composition of hydrothermal fluids in the early Earth.

Serpentinization of komatiites

In this section, we compare our experimental results with conditions in natural environments and thermodynamic calculations in order to characterize the reaction occurring in the reaction cell. The EPMA analysis of alteration products revealed the presence of a small quantity of olivine in both experiments, but the amount of olivine in Exp-350 was greater than that in Exp-250. Thermodynamic calculations for serpentinization of olivine and harzburgite (80 wt.% of olivine, 15 wt.% of orthopyroxene, and 5 wt.% of clinopyroxene) indicate that olivine becomes stable and is one of the major minerals at temperatures above approximately 315–350 °C in the peridotite- or olivine-H_2O system (e.g., McCollom and Bach 2009; Klein et al. 2013). In analogy with the calculation, the experiments in this study suggest that olivine stabilizes with increasing temperature, even during komatiite serpentinization at 250–350 °C. Nevertheless, olivine was observed as a minor phase even in Exp-350. This is probably due to the composition of komatiite that contains a certain level of Al_2O_3 would stabilize serpentine and/or chlorite/smectite because smectite becomes stable as bulk Al_2O_3 content increases in ultramafic rocks such as komatiite (Shibuya et al. 2015). However, it was difficult to determine whether the igneous olivine still remains in the absence of any reaction with a fluid or if the alteration olivine was newly created through serpentinization.

Brucite commonly occurs in serpentinized rocks in modern oceanic plates (e.g., Bach et al. 2006; Klein et al.

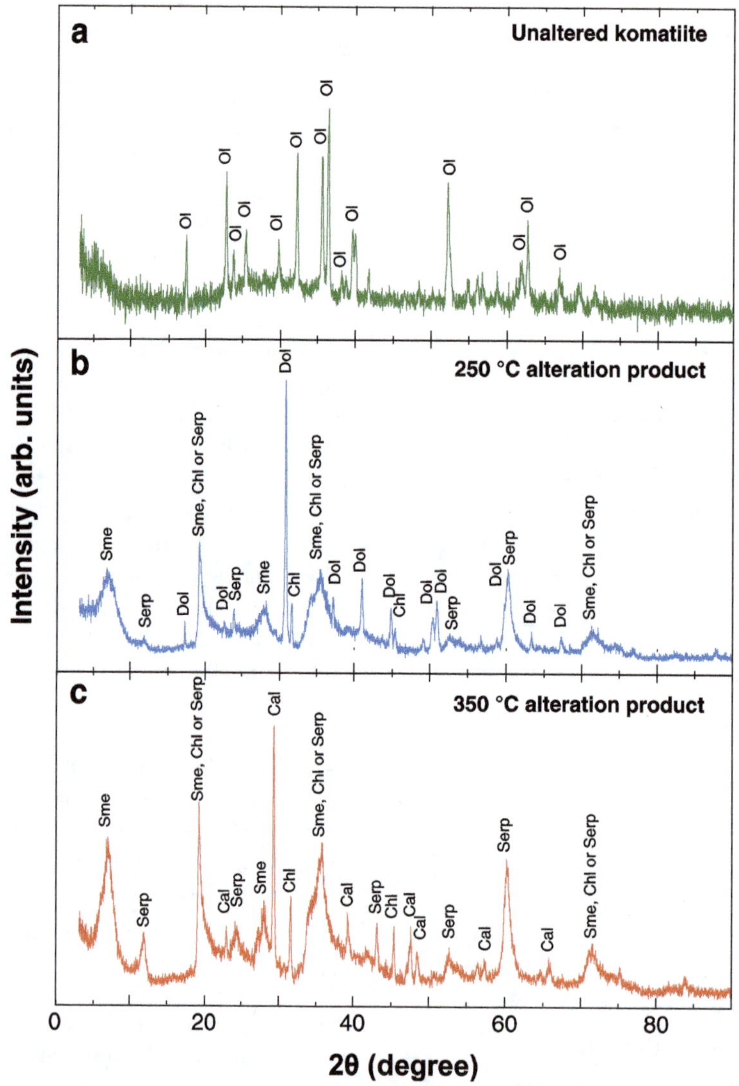

Fig. 5 Representative XRD patterns (CuKα) of **a** the unaltered synthetic komatiite and the alteration products of **b** Exp-250 and **c** Exp-350. *Dol* dolomite, *Cal* calcite, *Sme* smectite, *Chl* chlorite, *Ol* olivine, and *Serp* serpentine

2014) but was not found in the alteration products in the present experiments. The absence of brucite in Exp-350 is consistent with the thermodynamic calculations for the serpentinization of olivine and harzburgite under CO_2-poor conditions, since brucite is destabilized at temperatures above ~350 °C (McCollom and Bach 2009; Klein et al. 2013). On the other hand, the instability of brucite in Exp-250 is likely due to the CO_2-rich condition in the experiment because previous experiments revealed that Mg-bearing carbonate mineral is more stable than brucite under CO_2-rich conditions (Zhao et al. 2010; Hövelmann et al. 2012; Klein and McCollom 2013). Alternatively, high dissolved Si content in the fluid (3.7–4.1 mmol/kg) possibly prohibited brucite formation, which is supported by thermodynamic calculations (e.g., Klein and McCollom 2013).

Carbonation of komatiites

The steady-state ΣCO_2 concentrations in both experiments indicate that the amount of carbonate mineral was greater in Exp-250 than in Exp-350 (Fig. 4a and Table 2). The formation of dolomite in Exp-250 is characteristically different from the basalt-CO_2 experiments that yielded only calcite as carbonate in previous studies at 250 and 350 °C (Shibuya et al. 2013b). Comparing the experimental conditions of this study to those in the previous study, compositional differences between komatiite and basalt and/or the difference in initial pH (4.9 in this study, and 6.5 in Shibuya et al. (2013b)) likely caused the difference in precipitated carbonate species in the alteration products at 250 °C. Furthermore, it was revealed that the precipitated carbonate species strongly affected the Mg concentration in the hydrothermal fluid, i.e., Mg concentration in Exp-250

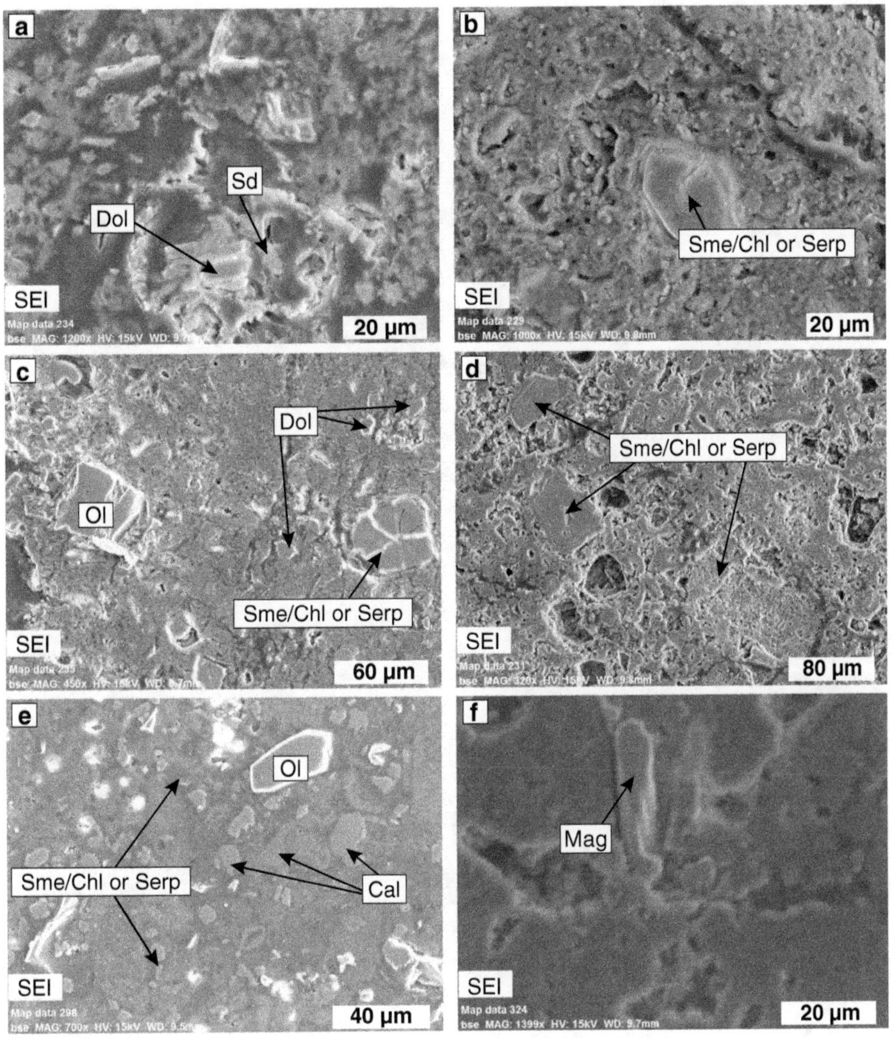

Fig. 6 Representative SEI images of alteration products (**a**–**d**) in the experiment at 250 °C and (**e**, **f**) in the experiment at 350 °C. *Sd* siderite, *Dol* dolomite, *Cal* calcite, *Sme* smectite, *Chl* chlorite, *Ol* olivine, *Serp* serpentine, and *Mag* magnetite

(carbonate as dolomite) is 30–40 times higher than that in Exp-350 (carbonate as calcite), while fluid Ca concentrations in both experiments are similar.

Previously, carbonate species were reported to correlate with the concentration ratio between Mg and Ca ($mMg^{2+}/(mMg^{2+} + mCa^{2+})$) and the temperature of the fluid (Fig. 7) (Rosenberg and Holland 1964; Rosenberg et al. 1967; Tribble et al. 1995). In Exp-350, carbonate was precipitated as calcite, whereas the $mMg^{2+}/(mMg^{2+} + mCa^{2+})$ value of the fluid was approximately 0.03; thus, it clearly falls within the stability field of calcite. On the other hand, the $mMg^{2+}/(mMg^{2+} + mCa^{2+})$ value of the fluid in Exp-250, where dolomite precipitated, was approximately 0.43, which is near the dolomite/magnesite stability boundary. Therefore, the extremely high fluid Mg concentration in Exp-250 is consistent with the precipitation of dolomite in the alteration product. However, it is difficult to discuss the anteroposterior relation between dolomite formation and enrichment of Mg in fluids because the decrease in ΣCO_2 concentration and the Mg enrichment in the fluid were already confirmed at the first sampling. It is widely accepted that seafloor hydrothermal systems play a significant role as Mg sinks in the modern ocean because basaltic oceanic crusts incorporate seawater Mg into alteration minerals such as smectite and chlorite through hydrothermal alteration at near mid-ocean ridges (e.g., Alt 1995). In addition, Charlou et al. (2002) reported that ultramafic rock-hosted high-temperature hydrothermal systems provide fluids with little Mg into seawater and generally act as Mg sinks in the ocean. On the other hand, weathering of ultramafic rocks exposed at seafloor below 150 °C and hydrothermal circulation at high water/rock ratios provide Mg into seawater (Snow and Dick 1995). Our results hypothesize that dolomite formation during the serpentinization of komatiite would be a source

Table 3 Composition of run products (wt.%)

Experiment-250 °C								Experiment-350 °C							
Sample	Serp or Sme/Chl mixture[b]			Dolomite			Siderite	Serp or Sme/Chl mixture[b]			Calcite			Olivine[c]	Magnetite
SiO_2	41.37	40.89	33.38	0.00	0.00	0.00	3.80	38.55	35.52	38.87	0.00	0.00	0.00	41.60	2.76
TiO	0.44	0.43	0.64	0.01	0.05	0.15	2.96	0.52	0.47	0.17	0.00	0.00	0.00	0.01	0.18
Al_2O_3	7.00	6.88	11.56	0.06	0.00	0.01	0.13	6.82	6.84	6.38	0.07	0.00	0.01	0.06	0.02
Cr_2O_3	0.47	0.43	0.29	0.11	0.06	0.12	1.46	0.37	0.47	0.39	0.04	0.00	0.02	0.14	0.23
FeO_{total}[a]	13.32	13.65	15.55	3.22	6.66	9.05	39.55	16.84	16.51	13.77	0.18	0.37	0.77	5.33	85.06
MnO	0.16	0.14	0.18	0.53	0.49	0.49	0.12	0.19	0.34	0.20	0.13	0.36	0.72	0.05	0.40
MgO	24.05	23.48	24.10	17.69	17.39	17.43	1.32	23.13	25.21	25.98	0.17	0.74	0.79	51.46	0.36
CaO	0.83	0.81	0.17	26.01	29.47	23.75	0.26	0.28	0.13	0.46	56.29	54.95	58.68	0.32	0.29
Na_2O	0.06	0.08	0.03	0.03	0.11	0.04	0.09	0.16	0.07	0.15	0.01	0.00	0.00	0.00	0.02
K_2O	0.09	0.11	0.02	0.03	0.02	0.02	0.05	0.04	0.01	0.03	0.01	0.01	0.01	0.02	0.03
NiO	0.17	0.20	0.12	0.00	0.00	0.08	0.10	0.10	0.00	0.16	0.01	0.00	0.05	0.45	0.10
Total	87.96	87.10	86.05	47.67	54.25	51.13	49.84	86.98	85.59	86.56	56.91	56.43	61.04	99.43	89.45
Mg# [d]	76.30	75.40	73.43	90.73	82.31	77.43	5.63	71.00	73.13	77.08	61.52	78.23	64.50	94.51	0.74

[a]Total iron as FeO
[b]Serpentine or smectite/chlorite mixture
[c]Alterlation or relict olivine
[d]Mg# = [Mg / (Mg+Fe)] × 100

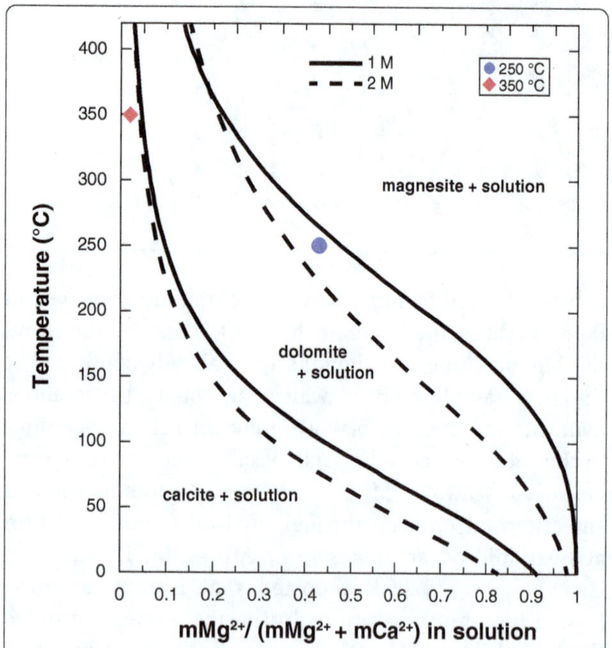

Fig. 7 The $mMg^{2+}/(mMg^{2+} + mCa^{2+})$ values of hydrothermal fluids in Exp-250 and Exp-350 plotted on the calcite-dolomite-magnesite stability fields estimated from experiments at 275–420 °C and their extrapolation to 0 °C (modified after Tribble et al. 1995). The *dotted* and *solid curves* are for the solutions with 2 and 1 M of Ca-Mg chloride, respectively. The *blue circle* and *pink diamond* indicate the values in Exp-250 and Exp-350, respectively

of Mg for the Hadean ocean, even when the temperature range of hydrothermal reaction zones was near 250 °C. Further experiments will be necessary to justify this hypothesis.

The dolomite crystals formed in this study contain up to 9 wt.% FeO_{total} in Exp-250 because Fe substitutes Mg in dolomite during its precipitation owing to the similarity in ion radius between Fe and Mg (Mg^{2+}: 0.72 Å; Fe^{2+}: 0.61 Å) (Jia 1991). A similar Fe-bearing carbonate mineral was also reported in a hydrothermal alteration experiment for komatiite conducted under CO_2-rich conditions (water 50.1 mg + CO_2 7.3 mg: about 3300 mmol/kg at the starting condition) at 300 °C and 500 bars (Hao and Li 2015). Furthermore, Klein and McCollom (2013) reported the precipitation of Fe-bearing magnesite (FeO = 6.78 wt.%) through the serpentinization of olivine under CO_2-rich conditions at 230 °C. Although differences in the composition of the starting materials between the previous studies and our experiments would have caused the difference in carbonate species, it seems likely that carbonate minerals formed during serpentinization under CO_2-rich conditions contain substantial FeO, probably at temperatures below approximately 300–350 °C.

Effect of seawater CO_2 on H_2 concentration in fluid

The maximum steady-state fluid H_2 concentrations were 0.024 mmol/kg in Exp-250 and 2.9 mmol/kg in Exp-350 (Fig. 4b), and the results are consistent with the analysis of solid alteration products. Magnetite, whose formation is related to H_2 generation (McCollom and Bach 2009), was observed in Exp-350 but not in Exp-250. Nevertheless, the H_2 concentrations in both experiments were clearly lower than the steady-state H_2 concentration (~20 mmol/kg) in fluid generated by the reactions between komatiite and a CO_2-free NaCl fluid at 300 °C and 500 bars (Shibuya et al. 2015). Although direct comparison is difficult owing to the difference in temperature conditions, two possible factors explain the differences between the previous and present experiments in terms of the steady-state H_2 concentrations from the hydrothermal reaction of komatiite.

The first factor is the temperature dependence of the stability of magnetite in ultramafic systems. At temperatures below 315–350 °C, the amount of magnetite formation increases with increasing temperature, but it drastically decreases at temperatures above 315–350 °C because olivine becomes stable (McCollom and Bach 2009; Klein et al. 2013). Moreover, Klein et al. (2013) reported that within the temperature range of 25–400 °C, the amount of hydrogen generation caused by the reaction with olivine (Fo 90) reaches a maximum at 322 °C. Thus, the temperature dependence of the stability of olivine and magnetite would be a primary cause of the difference in fluid H_2 concentrations between the previous CO_2-free komatiite experiment at 300 °C and our experiments.

The other factor is the presence of CO_2 in the experimental system. Based on thermodynamic calculations simulating the relative amount of magnetite formation at 50–400 °C under 500 bars, Klein et al. (2013) expected that the amount of hydrogen generation caused by the reaction with olivine under CO_2-free conditions is greater at 250 °C than at 350 °C. In our experiments, however, the steady-state fluid H_2 concentration was higher in Exp-350 than in Exp-250. In previous experiments that simulated the reactions between olivine and a CO_2-rich fluid at 230 °C and 35 MPa, ferrous iron-bearing talc and magnesite were generated, while magnetite formation and H_2 generation were limited (Klein and McCollom 2013). Under CO_2-free conditions, the iron is only partitioned into magnetite and ferrous iron-bearing silicate in alteration products,

$$a(FeO)_{rock} + bH_2O \rightarrow d(Fe_2O_3)_{rock} + dH_2 + (a\text{-}2d)(FeO)_{rock}.$$

In contrast, the interaction between CO_2-rich fluid and rock leads to carbonate mineral precipitation and ferrous iron incorporation therein. The reaction occurring under a CO_2-rich condition is the following:

$$a(FeO)_{rock} + bH_2O + cCO_2 \rightarrow d(Fe_2O_3)_{rock} + dH_2 + e(FeO)_{rock} + (a\text{-}2d\text{-}e)(FeCO_3)_{rock} + (c\text{-}a + 2d + e)CO_2.$$

In both equations, the value of "*d*" is the amount of H_2 generation and depends on temperature. Therefore, it was suggested that the lack of H_2 generation is due to the direct incorporation of ferrous iron into carbonate minerals without its oxidation during the hydrothermal alteration. In our experiments, although a large difference in the composition of serpentine and smectite/chlorite mixture was not identified between Exp-250 and Exp-350, the Fe content of dolomite in Exp-250 (3–9 wt.% FeO_{total}) was clearly higher than that of calcite in Exp-350 (Table 3). The amount of generated magnetite in Exp-350 was greater than that in Exp-250. These data suggest that the amount of ferrous iron incorporated into the carbonate mineral strongly affected the extent of magnetite formation and resulting hydrogen generation in the fluid. Although such a trend in the effect of CO_2 on H_2 generation has been reported in some previous experiments at temperatures below 230 °C (Jones et al. 2010; Klein and McCollom 2013; Neubeck et al. 2014), our experiments show that the suppression of H_2 generation by the presence of CO_2 in the system occurs even at higher temperatures (250 °C) and in the CO_2-H_2O-komatiite system. In addition, a small amount of ferrous iron incorporated into calcite (up to 0.8 wt.%) may have also slightly suppressed magnetite formation in Exp-350.

Implications for the Hadean H_2-rich hydrothermal environments

It has been considered that the serpentinization of komatiite would have served as the most ubiquitous geological process for hosting H_2-rich hydrothermal environments in the Hadean ocean (e.g., Takai et al. 2006). The present experiments constrained the potential for H_2 generation during the serpentinization of komatiite under CO_2-rich conditions. This work provides further important insights into our interpretation of hatcheries for the emergence and early evolution of life in the Hadean Earth. Exp-350 showed that the fluid H_2 concentration increased up to 2.9 mmol/kg during the serpentinization and carbonation of komatiite. This value is lower than the maximum H_2 concentration of the vent fluids in modern peridotite-hosted hydrothermal systems; the H_2 and ΣCO_2 concentrations in modern peridotite-hosted hydrothermal systems are 12 and 10.1 mM at Logatchev field, 16 and 16 mM at Rainbow field (Charlou et al. 2002), and <1–15 mmol/kg and <0.8 mmol/kg at Lost City field (Kelley et al. 2005), respectively. However, the H_2 concentration in Exp-350 broadly falls within the range of the fluid H_2 concentrations in the Kairei field hosted by both basalt and troctolite (H_2 = 2.5–8.5 mmol/kg and ΣCO_2 < 10.1 mmol/kg; Takai et al. 2004; Gallant and Von Damm 2006; Kumagai et al. 2008; Nakamura et al. 2009). This suggests that in the Hadean, the komatiite still had great potential to generate H_2-rich hydrothermal fluids at high-temperature, even under CO_2-rich conditions. Furthermore, considering that hydrogenotrophic methanogens—long believed to be one of the most probable candidates for the earliest living forms and primary producers on Earth (e.g., Takai et al. 2006; Martin et al. 2008)—indeed dominate in the microbial communities associated with hydrothermal fluids in the Kairei field (Takai et al. 2004), the komatiite-hosted seafloor hydrothermal systems would have fully prepared the energetic basis of hatcheries for the emergence and early evolution of hydrogenotrophic living forms in the Hadean ocean.

In contrast, the fluid H_2 concentration in Exp-250 (0.024 mmol/kg) was much lower than that in Exp-350 and is comparable to the H_2 concentration level in the typical fluids of modern basalt-hosted hydrothermal fields (Charlou et al. 1996, 2000; Gallant and Von Damm 2006). Given that the H_2 generation potential of ultramafic rocks decreases with decreasing temperature within the unstable region of olivine (McCollom and Bach 2009; Klein et al. 2013), the serpentinization of komatiite under CO_2-rich conditions might not have afforded the H_2-rich hydrothermal fluids at temperatures lower than 250 °C. More importantly, carbonate minerals become stable with decreasing temperature in H_2O-CO_2-rock systems (e.g., Shibuya et al. 2013b), which indicates that the carbonate formation and resulting limitation of H_2 generation would occur more significantly at lower temperatures (e.g., <100 °C). Therefore, the possible H_2-rich hydrothermal environments may have been created only in the deep ocean floor because high hydrostatic pressure can elevate the temperature of hydrothermal reaction zones (fluid-rock reaction zones). In other words, other potential candidates, such as the hydrothermal systems in on-land and shallow submarine environments, may not have fully prepared the energetic basis (abundant H_2 availability) of hatcheries for the emergence and early evolution of life in the Hadean Earth due to low-temperature hydrothermal reactions.

Conclusions

High-temperature and high-pressure experiments using komatiite and a CO_2-rich NaCl solution revealed that different hydrothermal reaction temperatures caused differences in the carbonate species in alteration products (iron-rich dolomite at 250 °C and calcite at 350 °C) during the serpentinization of komatiites. The hydrothermal fluid coexisting with dolomite at 250 °C showed high Mg concentrations (up to ca. 40 mmol/kg), which is markedly higher than those in typical modern high-temperature hydrothermal fluids (generally less than 1 mmol/kg). This suggests that, in contrast to modern seafloor hydrothermal systems, the dolomite-bearing komatiite-hosted hydrothermal systems may have served as a source of Mg in the Hadean ocean. The steady-state H_2 concentration in Exp-350 (up to 2.9 mmol/kg) was approximately 100 times higher than that in Exp-250 (up to 0.024 mmol/kg) because the amount of ferrous iron incorporated into carbonate was greater in Exp-250 than in Exp-350. These results perhaps suggest that only high-temperature komatiite-hosted hydrothermal systems have the potential to generate H_2-rich hydrothermal environments, even in the CO_2-abundant Hadean ocean. Further experiments based on a more precise estimation of the ancient atmospheric/oceanic CO_2 levels from geological records will clarify the geochemical nature of komatiite-hosted hydrothermal environments and other potential hatcheries for the emergence and early evolution of life in the Hadean Earth because the ΣCO_2 concentration in the fluid may significantly affect the mineral assemblage of serpentinized komatiites and hydrothermal reactions taking place under various conditions.

Abbreviations

BSE: Backscattered electron; EPMA: Electron probe microanalyzer; Exp-250: Experiment at 250 °C; Exp-350: Experiment at 350 °C; QFM: Quartz-fayalite-magnetite; SEI: Secondary electron imaging; XRD: X-ray diffraction; XRF: X-ray fluorescence

Acknowledgements

We thank two anonymous reviewers for many helpful comments that greatly improved our manuscript. We appreciate Prof. Hodaka Kawahata for editorial handling. We are grateful to Y. Ueno for discussions.

Funding

This work was supported by Grants-in-Aid for Scientific Research from the Japan Society for Promotion of Science (JSPS) (No. 15K13583) and the Ministry of Education, Culture, Sports, Science and Technology (MEXT), Japan.

Authors' contributions

HU, TS, KT, and SM proposed and designed the study. HU and TS carried out the experiments. HU, TS, and MS prepared the starting materials. HU, TS, MS, and YS analyzed the fluid samples and solid materials. All authors interpreted the data, read, and approved the final manuscript.

Competing interests

The authors declare that they have no competing interests.

Author details

[1]Department of Earth and Planetary Sciences, Tokyo Institute of Technology, 2-12-1 Ookayama, Meguro-ku, Tokyo 152-8551, Japan. [2]Laboratory of Ocean-Earth Life Evolution Research (OELE), Japan Agency for Marine-Earth Science and Technology (JAMSTEC), 2-15 Natsushima-cho, Yokosuka 237-0061, Japan. [3]Research and Development Center for Submarine Resources, Japan Agency for Marine-Earth Science and Technology (JAMSTEC), 2-15 Natsushima-cho, Yokosuka 237-0061, Japan. [4]Project Team for Development of New-generation Research Protocol for Submarine Resources, Japan Agency for Marine-Earth Science and Technology (JAMSTEC), 2-15 Natsushima-cho, Yokosuka 237-0061, Japan. [5]Department of Subsurface Geobiological Analysis and Research (SUGAR), Japan Agency for Marine-Earth Science and Technology (JAMSTEC), 2-15 Natsushima-cho, Yokosuka 237-0061, Japan. [6]Earth-Life Science Institute (ELSI), Tokyo Institute of Technology, 2-12-1 Ookayama, Meguro-ku, Tokyo 152-8551, Japan.

References

Allen DE, Seyfried WE (2003) Compositional controls on vent fluids from ultramafic-hosted hydrothermal systems at mid-ocean ridges: an experimental study at 400 °C, 500 bars. Geochim Cosmochim Acta 67:1531–1542

Alt JC (1995) Subseafloor processes in mid-oceanic ridge hydrothermal systems. In: Humphris SE, Zierenberg RA, Mullineaux LS, Thomson RE (eds) Seafloor hydrothermal systems: physical, chemical, biological, and geological interactions. Geophysical monograph. American Geophysical Union, Washington, DC, pp 85–114

Amend JP, McCollom TM (2009) Energetics of biomolecule synthesis on early Earth. In: Zaikowski L, Friedrich JM, Seidel SR (eds) Chemical evolution II: from the origins of life to modern society. American Chemical Society, Washington, DC, pp 63–94

Bach W, Paulick H, Garrido CJ, Ildefonse B, Meurer WP, Humphris SE (2006) Unraveling the sequence of serpentinization reactions: petrography, mineral chemistry, and petrophysics of serpentinites from MAR 15°N (ODP Leg 209, Site 1274). J Geophys Res 33, L13306. doi:10.1029/12006GL025681

Bethke CM (2008) Geochemical and biogeochemical reaction modeling. Cambridge Univ. Press, Cambridge, U. K

Canil D (1997) Vanadium partitioning and the oxidation state of Archaean komatiite magma. Nature 389:842_845

Charlou JL, Donval JP, Douville E, Jean-Baptiste P, Radford-Knoery J, Fouquet Y, Dapoigny A, Stievenard M (2000) Compared geochemical signatures and the evolution of Menez Gwen (37°50′N) and Lucky Strike (37°17′N) hydrothermal fluids, south of the Azores Triple Junction on the Mid-Atlantic Ridge. Chem Geol 171:49–75

Charlou JL, Donval JP, Fouquet Y, Jean-Baptiste P, Holm N (2002) Geochemistry of high H_2 and CH_4 vent fluids issuing from ultramafic rocks at the Rainbow Hydrothermal Field (36°14′N, MAR). Chem Geol 191:345–359

Charlou JL, Donval JP, Jean-Baptiste P, Dapoigny A, Rona PA (1996) Gases and helium isotopes in high temperature solutions sampled before and after ODP Leg 158 Drilling at TAG Hydrothermal Field (26° N, MAR). Geophys Res Lett 23:3491–3494

Cleverley JS, Bastrakov EN (2005) K2GWB: utility for generating thermodynamic data files for The Geochemist's Workbench® at 0-1000 °C and 1-5000 bar from UT2K and the UNITHERM database. Comput Geosci 31(6):756–767

Drummond SE (1981) Boiling and mixing of hydrothermal fluids: chemical effects on mineral precipitation. Dissertation, Pennsylvania State University

Gallant RM, Von Damm KL (2006) Geochemical controls on hydrothermal fluids from the Kairei and Edmond Vent Fields, 23°–25°S, Central Indian Ridge. Geochem Geophys Geosyst 7, Q06018. doi:10.1029/2005GC001067

Hao X, Li Y (2015) Hexagonal plate-like magnetite nanocrystals produced in komatiite-H_2O-CO_2 reaction system at 450 °C. Int J Astrobiol 14:547–553

Helgeson HC (1969) Thermodynamics of hydrothermal systems at elevated temperatures and pressures. Am J Sci 267:729_804

Helgeson HC, Kirkham DH (1974) Theoretical prediction of the thermodynamic behavior of aqueous electrolytes at high pressures and temperatures: II. Debye-Huckel parameters for activity coefficients and relative partial molal properties. Am J Sci 274:1199–1261

Hövelmann J, Putnis CV, Ruiz-Agudo E, Austrheim H (2012) Direct nanoscale observations of CO_2 sequestration during brucite [$Mg(OH)_2$] dissolution. Environ Sci Technol 46:5253–5260

Jia YQ (1991) Crystal radii and effective ionic radii of the rare earth ions. Journal of Solid State Chemistry 95:184–187

Johnson JW, Oelkers EH, Helgeson HC (1992) SUPCRT92: a software package for calculating the standard molal thermodynamic properties of minerals gases, aqueous species, and reactions from 1 to 5000 bar and 0 to 1000 °C. Comput Geosci 18:899–947

Jones LC, Rosenbauer R, Goldsmith JI, Oze C (2010) Carbonate control of H_2 and CH_4 production in serpentinization systems at elevated P-Ts. Geophys Res Lett 37, L14306

Kasting JF (1993) Earth's early atmosphere. Science 259(5097):920–926

Kelley DS, Karson JA, Fruh-Green GL, Yoerger DR, Shank TM, Butterfield DA, Hayes JM, Schrenk MO, Oison EJ, Proskurowski G, Jakuba M, Bradley A, Larson B, Ludwig K, Glickson D, Buckman K, Bradley AS, Brazelton WJ, Roe K, Elend MJ, Delacour A, Bernasconi SM, Lilley MD, Baross JA, Summons RE, Sylva SP (2005) A serpentinite-hosted ecosystem: the lost city hydrothermal field. Science 307:1428–1434

Klein F, McCollom TM (2013) From serpentinization to carbonation: new insights from a CO_2 injection experiment. Earth Planet Sci Lett 379:137–145

Klein F, Bach W, McCollom TM (2013) Compositional controls on hydrogen generation during serpentinization of ultramafic rocks. Lithos 178:55–69

Klein F, Bach W, Humphris SE, Kahl W-A, Jöns N, Moskowitz B, Berquó TS (2014) Magnetite in seafloor serpentinite—some like it hot. Geology 42:135–138

Knauth LP (2005) Temperature and salinity history of the Precambrian ocean: implications for the course of microbial evolution. Palaeogeogr Palaeoclimatol Palaeoecol 219:53–69

Komiya T, Maruyama S, Hirata T, Yurimoto H (2002) Petrology and geochemistry of MORB and OIB in the Mid-Archean North Pole region, Pilbara Craton, Western Australia, implications for the composition and temperature of the upper mantle at 3.5 Ga. Int Geol Rev 44:988–1016

Komiya T (2004) Material circulation model including chemical differentiation within the mantle and secular variation of temperature and composition of the mantle. Phys Earth Planet Int 146:333–367

Kumagai H, Nakamura K, Toki T, Morishita T, Okino K, Ishibashi JI, Tsunogai U, Kawagucci S, Gamo T, Shibuya T, Sawaguchi T, Neo N, Joshima M, Sato T, Takai K (2008) Geological background of the Kairei and Edmond hydrothermal fields along the Central Indian Ridge: implications of their vent fluids' distinct chemistry. Geofluids 8:239–251

Kump LR, Seyfried WE (2005) Hydrothermal Fe fluxes during the Precambrian: effect of low oceanic sulfate concentrations and low hydrostatic pressure on the composition of black smokers. Earth Planet Sci Lett 235:654–662

Lazar C, McCollom TM, Manning CE (2012) Abiogenic methanogenesis during experimental komatiite serpentinization: implications for the evolution of the early Precambrian atmosphere. Chem Geol 326–327:102–112

Lowe DR, Tice MM (2004) Geologic evidence for Archean atmospheric and climatic evolution: fluctuating levels of CO_2, CH_4, and O_2 with an overriding tectonic control. Geology 32:493–496

Martin W, Baross J, Kelley D, Russell MJ (2008) Hydrothermal vents and the origin of life. Nat Rev Microbiol 6:805–814

Macleod G, McKeown C, Hall AJ, Russell MJ (1994) Hydrothermal and oceanic pH conditions of possible relevance to the origin of life. Orig Life Evol Biosph 24:19–41

McCollom TM, Bach W (2009) Thermodynamic constraints on hydrogen generation during serpentinization of ultramafic rocks. Geochim Cosmochim Acta 73:856–875

McCollom TM, Shock EL (1997) Geochemical constraints on chemolithoautotrophic metabolism by microorganisms in seafloor hydrothermal systems. Geochim Cosmochim Acta 61:4375–4391

Moores EM (2002) Pre-1 Ga (pre-Rodinian) ophiolites: their tectonic and environmental implications. GSA Bulletin 114:80–95

Nakamura K, Morishita T, Bach W, Klein F, Hara K, Okino K, Takai K, Kumagai H (2009) Serpentinized troctolites exposed near the Kairei hydrothermal field Central Indian Ridge: insights into the origin of the Kairei hydrothermal fluid supporting a unique microbial ecosystem. Earth Planet Sci Lett 280:128–136

Nakamura K, Takai K (2014) Theoretical constraints of physical and chemical properties of hydrothermal fluids on variations in chemolithotrophic microbial communities in seafloor hydrothermal systems. Progr Earth Planet Sci 1(1):1–24

Neubeck A, Duc NT, Hellevang H, Oze C, Bastviken D, Bacsik Z, Holm NG (2014) Olivine alteration and H_2 production in carbonate-rich, low temperature aqueous environments. Planet Space Sci 96:51–61

Ohmoto H, Watanabe Y, Kumazawa K (2004) Evidence from massive siderite beds for a CO_2-rich atmosphere before ~1.8 billion years ago. Nature 429:395–399

Ohta H, Maruyama S, Takahashi E, Watanabe Y, Kato Y (1996) Field occurrence, geochemistry and petrogenesis of the Archean mid-oceanic ridge basalts (AMORBs) of the Cleaverville area, Pilbara Craton, Western Australia. Lithos 37:199–221

Rosenberg PE, Holland HD (1964) Calcite-dolomite-magnesite stability relations in solutions at elevated temperatures. Science 145:700–701

Rosenberg PE, Burt DM, Holland HD (1967) Calcite-Dolomite-Magnesite stability relations in solutions: the effect of ionic strength. Geochim Cosmochim Acta 31:391–396

Russell MJ, Barge LM, Bhartia R, Bocanegra D, Bracher PJ, Branscomb E, Kidd R, McGlynn S, Meier DH, Nitschke W, Shibuya T, Vance S, White L, Kanik I (2014) The drive to life on wet and icy worlds. Astrobiology 14:308–343

Russell MJ, Hall AJ (1997) The emergence of life from iron monosulphide bubbles at a submarine hydrothermal redox and pH front. J Geol Soc Lond 154:377–402

Russell MJ, Hall AJ, Martin W (2010) Serpentinization as a source of energy at the origin of life. Geobiology 8:355–371

Seyfried WE, Ding K, Berndt ME (1991) Phase equilibria constraints on the chemistry of hot spring fluids at mid-ocean ridges. Geochim Cosmochim Acta 55:3559–3580

Seyfried WE, Gordon PC, Dickson FW (1979) A new reaction cell for hydrothermal solution equipment. Am Mineral 64:646–649

Seyfried WE, Foustoukos DI, Fu Q (2007) Redox evolution and mass transfer during serpentinization: an experimental and theoretical study at 200 °C, 500 bar with implications for ultramafic-hosted hydrothermal systems at mid-ocean ridges. Geochim Cosmochim Acta 71:3872–3886

Shibuya T, Kitajima K, Komiya T, Terabayashi M, Maruyama S (2007) Middle Archean ocean ridge hydrothermal metamorphism and alteration recorded in the Cleaverville area, Pilbara Craton, Western Australia. J Metamorph Geol 25:751–767

Shibuya T, Komiya T, Nakamura K, Takai K, Maruyama S (2010) Highly alkaline, high-temperature hydrothermal fluids in the early Archean ocean. Precambrian Res 182:230–238

Shibuya T, Russell MJ, Takai K (2016) Free energy distribution and hydrothermal mineral precipitation in Hadean submarine alkaline vent systems: Importance of iron redox reactions under anoxic conditions. Geochim Cosmochim Acta 175:1–19

Shibuya T, Tahata M, Kitajima K, Ueno Y, Komiya T, Yamamoto S, Igisu M, Terabayashi M, Sawaki Y, Takai K, Yoshida N, Maruyama S (2012) Depth variation of carbon and oxygen isotopes of calcites in Archean altered upper oceanic crust: implications for the CO_2 flux from ocean to oceanic crust in the Archean. Earth Planet Sci Lett 321–322:64–73

Shibuya T, Tahata M, Ueno Y, Komiya T, Takai K, Yoshida N, Maruyama S, Russell MJ (2013a) Decrease of seawater CO_2 concentration in the Late Archean: an implication from 2.6 Ga seafloor hydrothermal alteration. Precambrian Res 236:59–64

Shibuya T, Yoshizaki M, Masaki Y, Suzuki K, Takai K, Russell MJ (2013b) Reactions between basalt and CO_2-rich seawater at 250 and 350 °C, 500 bars: implications for the CO_2 sequestration into the modern oceanic crust and the composition of hydrothermal vent fluid in the CO_2-rich early ocean. Chem Geol 359:1–9

Shibuya T, Yoshizaki M, Satoh M, Shimizu K, Nakamura K, Omori S, Suzuki K, Takai K, Hideo T, Maruyama S (2015) Hydrogen-rich hydrothermal environments in the Hadean ocean inferred from serpentinization of komatiites at 300 °C and 500 bar. Progr Earth Planet Sci 2:1–11. doi:10.1186/s40645-015-0076-z

Shock EL, Koretsky CM (1995) Metal-organic complexes in geochemical processes: estimation of standard partial molal thermodynamic properties of aqueous complexes between metal cations and monovalent organic acid ligands at high pressures and temperatures. Geochim Cosmochim Acta 59:1497–1532

Shock EL, Helgeson HC (1988) Calculation of the thermodynamic and transport properties of aqueous species at high pressures and temperatures: correlation algorithms for ionic species and equation of state predictions to 5 kb and 1000 °C. Geochim Cosmochim Acta 52:2009–2036

Shock EL, Helgeson HC, Sverjensky DA (1989) Calculation of the thermodynamic and transport properties of aqueous species at high pressures and temperatures: standard partial molal properties of inorganic neutral species. Geochim Cosmochim Acta 53:2157–2183

Shock EL, Sassani DC, Willis M, Sverjensky DA (1997) Inorganic species in geologic fluids: correlations among standard molal thermodynamic properties of aqueous ions and hydroxide complexes. Geochim Cosmochim Acta 61:907–950

Sleep NH, Zahnle K (2001) Carbon dioxide cycling and implications for climate on ancient Earth. J Geophys Res 106:1373–1399

Smith HS, Erlank AJ, Duncan AR (1980) Geochemistry of some ultramafic komatiite lava flows from the Barberton Mountain Land, South Africa. Precambrian Res 11:399–415

Snow JE, Dick HJB (1995) Pervasive magnesium loss by marine weathering of peridotite. Geochim Cosmochim Acta 59:4219–4235

Suzuki K, Shibuya T, Yoshizaki M, Hirose T (2015) Experimental hydrogen production in hydrothermal and fault systems: significance for habitability of subseafloor H_2 chemoautotroph microbial ecosystems. In: Ishibashi J, Okino K, Sunamura M (eds) Subseafloor biosphere linked to hydrothermal systems. TAIGA concept. Springer, Tokyo, pp 87–94

Sverjensky DA, Shock EL, Helgeson HC (1997) Prediction of the thermodynamic properties of aqueous metal complexes to 1000 °C and 5 kb. Geochim Cosmochim Acta 61:1359–1412

Takai K, Gamo T, Tsunogai U, Nakayama N, Hirayama H, Nealson KH, Horikoshi K (2004) Geochemical and microbiological evidence for a hydrogen-based, hyperthermophilic subsurface lithoautotrophic microbial ecosystem (HyperSLiME) beneath an active deep-sea hydrothermal field. Extremophiles 8:269–282

Takai K, Nakamura K, Suzuki K, Inagaki F, Nealson KH, Kumagai H (2006) Ultramafics-hydrothermalism-hydrogenesis-hyperslime (ultra H^3) linkage: a key insight into early microbial ecosystem in the Archean deep-sea hydrothermal systems. Paleontol Res 10:269–282

Tribble JS, Arvidson RS, Lane M, Mackenzie FT (1995) Crystal chemistry, and thermodynamic and kinetic properties of calcite, dolomite, apatite, and biogenic silica: applications to petrographic problems. Sediment Geol 95:11–51

Walker JCG (1985) Carbon dioxide on the early earth. Orig Life Evol Biosph 16:117–127

Wei K, Tronnes RG, Scarfe CM (1990) Phase relations of aluminum-undepleted and aluminum-depleted komatiites at pressures of 4–12 GPa. J Geophys Res 95:15817–15827

Wetzel LR, Shock EL (2000) Distinguishing ultramafic- from basalt-hosted submarine hydrothermal systems by comparing calculated vent fluid compositions. J Geophys Res 105:8319–8340

Yanagawa H, Kojima K (1985) Thermophilic microspheres of peptide-like polymers and silicates formed at 250 °C. J Biochem 97:1521–1524

Yoshizaki M, Shibuya T, Suzuki K, Shimizu K, Nakamura K, Takai K, Omori S, Maruyama S (2009) H_2 generation by experimental hydrothermal alteration of komatiitic glass at 300 °C and 500 bars: a preliminary result from on-going experiment. Geochem J 43:e17–e22

Zhao L, Sang L, Chen J, Ji JF, Teng HH (2010) Aqueous carbonation of natural brucite: relevance to CO_2 sequestration. Environ Sci Technol 44:406–411

14

Geochemistry of fine-grained sediments in the Yangtze River and the implications for provenance and chemical weathering in East Asia

Mengying He[1], Hongbo Zheng[1*], Peter D. Clift[2], Ryuji Tada[3], Weihua Wu[4] and Chao Luo[4]

Abstract

In order to interpret the marine clastic record preserved in the sedimentary basins of the East Asian marginal seas, it is important to understand how sediment transport and chemical weathering affect the composition of sediment enroute to its sink. Here we present a new data set for fine-grained sediment (<63 μm) from the Yangtze River and its major tributaries, which represents a baseline for interpreting sediment in the East China Sea. We demonstrate that there is no significant coherent downstream variation in the major element contents, which are generally more enriched than the average upper continental crust, except for water-soluble elements including Sr, Rb, Na, and K. Nd isotopes show that most of the sediment comes from the eastern and middle Yangtze Craton, as well as the Songpan-Garze Terrane. Chemical weathering varies significantly across the basin with upstream tributary sediments being relatively unweathered compared to those in the lower reaches. However, sediments in the main Yangtze stream show no trend in chemical weathering along its course, with some of the least weathered materials being found closest to the delta. Grain size and the abundance of hydrodynamically sorted heavy minerals affect the bulk geochemistry, especially the rare earth elements (REEs).

Keywords: The Yangtze River; Fine-grained sediments; Geochemistry; Provenance; Chemical weathering

Background

The Yangtze River is the largest river in eastern Asia and supplies sediments to the marginal seas of the Western Pacific, as well as filling sedimentary basins onshore between Tibet and the modern delta (e.g., Clark et al. 2004; Zheng et al. 2013). Understanding what controls the intensity of chemical weathering and the strength of bedrock erosion are important goals if we are to quantify how climate and tectonic activity influence the development of the Earth's surface and the composition of sediments deposited on the surrounding continental margins. If we are to use ancient sediments in the East China Sea to examine past environments in central China or the uplift history of the Tibetan Plateau, then an understanding of where the modern river derives this sediment is fundamental.

* Correspondence: zhenghb@njnu.edu.cn
[1]School of Geography Science, Nanjing Normal University, Nanjing 210046, China
Full list of author information is available at the end of the article

The Yangtze River provides a good opportunity to test how climate and tectonic compete with one another to control erosion because this river spans a wide range of landscapes that experience quite different climatic and tectonic forcing (e.g., Chen et al. 2001a, b; Clark et al. 2004). Understanding where sediment is being derived from, what its composition is and how that changes during transport due to abrasion and chemical weathering is essential if the marine sediment archive is to be exploited (e.g., He et al. 2013b).

In theory, the bulk composition of the sediment may provide a way of determining the provenance of this material if there is sufficient diversity in the bedrock compositions. However, sediment compositions may be changed from the original bedrock as a result of transport and chemical weathering. The geochemistry of bedload and suspended sediments has been used by many workers to determine the provenance, weathering, and tectonics of several river basins (e.g., Bhuiyan et al.

2011; Condie et al. 1992; Shao et al. 2012; Singh 2009, 2010; Taylor and McLennan 1985; Wu et al. 2011; Yang et al. 2004b). However, it is unclear if their interpretations are valid because of disruption to the signal caused by transport processes.

There have been several geochemical studies of Yangtze River sediments, including analysis of the rare earth element (REE) composition (e.g., Yang et al. 2002; Yang et al. 2013; Yang and Wang 2011), major element chemistry (e.g., Chen et al. 2002; Yang et al. 2004a), and the Sr-Nd isotope characteristics (e.g., Meng et al. 2008; Yang et al. 2007). However, there has been little integration of major/trace element chemistry and Sr-Nd isotope composition, especially for sediments finer than 63 μm, which are often considered to be tracers of the average REE composition of the source regions and dominate the volumetric flux to the ocean (e.g., Yang and Wang 2011). These authors argue for a strong link between the REE chemistry of the river sands and that of the dominant sources, which they infer to be in the upper reaches of the river based on the similarity of the REE composition in the Jinshajiang (upper reaches) and delta sands. Yang et al. (2002) also showed that REEs are generally enriched in clay and silt fractions, but depleted in the sand fraction, suggesting that grain size may be crucial control on sediment chemistry, something that has been documented in other SE Asian rivers (e.g., Clift et al. 2008).

In this paper, we examine the bed sediment chemistry of a series of Yangtze River sediments, including the major and trace elements, as well as the Sr-Nd isotope composition of the material finer than 63 μm, collected between the headwaters of the Yangtze on the Tibetan Plateau and the delta in easternmost China to investigate how effective the major and trace element chemistry can be in determining provenance and developing a sediment budget for the river system and thus help constrain its influence on basins in the East China Sea. We also endeavor to see how chemical weathering affects the composition of sediment in the river. It has been suggested that intensified chemical weathering, following the uplift of the Himalayas and Tibetan Plateau, has exerted an important influence in controlling Cenozoic climate since the start of India-Asia collision around 50 Ma (e.g., Berner and Berner 1997; Raymo and Ruddiman 1992). However, this theory is not universally accepted because the potential impact of chemical weathering on climate has yet to be properly quantified. In doing so, we provide a benchmark against which to compare older deposits within the East China Sea and other Chinese sedimentary basins supplied by the Yangtze.

The Yangtze River is influenced by the warm, wet climate of the East Asian monsoon and because it is recognized that warmth and moisture are important for accelerating rates of mineral alteration (e.g., West et al. 2005; White and Blum 1995). We would expect that transport and exposure to a monsoonal climate would result in a coherent increase in chemical weathering and sorting downstream and a decrease in grain size, assuming that most of the sediment load is generated in the upper reaches. No such pattern would emerge if sediment production was strong in other parts of the drainage.

In this study, we assess the effect of transport in controlling the chemical weathering proxies, via its effect on grain size as well as chemical breakdown that occurs during the transport. Degree of alteration can be calculated based on major and trace element compositions. Typically this involves tracking the relative abundance of a water-soluble element compared to an insoluble element in order to assess the degree of alteration (e.g., Nesbitt et al. 1980). We also examine whether chemical weathering and grain size exert important influences on the trace element composition of the fine-grained sediment. If these effects dominate, then it seems unlikely that rare earth element (REE) or other trace element groups can be used as effective provenance tools. Garzanti et al. (2010) have demonstrated that in South Asia the REE character of sediment is strongly controlled by the presence of a small volume of ultra-dense minerals, such as zircon and monazite and that these in turn are subject to hydrodynamic sorting and concentration so that differences between sediments from different rivers may be small compared to differences between different grain size fractions in a single point bar deposit.

Geological setting

The Yangtze River drainage is located from 33° 25′ N 91° 10′ E to 31° 23′ N 121° 58′ E and is more than 6300 km long, with a catchment area of 1.8×10^6 km^2 and an annual average discharge of ~9.6×10^{11} m^3 (e.g., Chen et al. 2001a, b; Ding et al. 2004). The river originates on the Tibetan Plateau at an elevation higher than 5000 m and finally reaches the East China Sea after merging with several large tributaries (Fig. 1). The river catchment can be divided into five broad physiographic provinces, including the northeast Tibetan Plateau, the high mountains of the Longmen Shan and associated ranges, the Sichuan Basin, mixed mountain and basin terrains (broadly referred to as the Three Gorges area), and the eastern lowlands. Generally, the Yangtze River can be divided into three parts: upper, middle, and lower reaches. However, the upstream reaches can further be divided into two segments: the Jinshajiang and the Chuanjiang (Fig. 1). The Yangtze River drainage covers several tectonic units, with the main body of the Yangtze Craton and the Qamdo Block, the Songpan-Garze terrane, the Qinling-Dabie orogenic belts, and the Cathaysia Block forming the primary potential sources (e.g., He et al. 2014). The basement

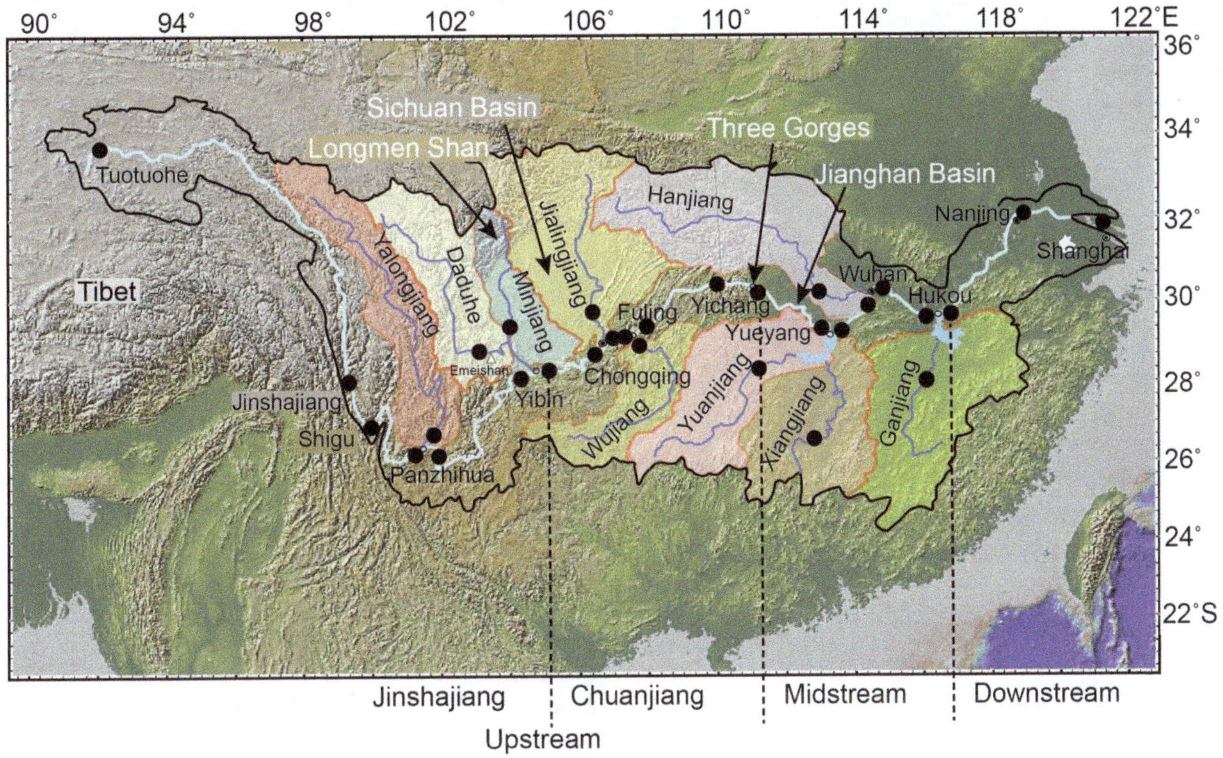

Fig. 1 Shaded topographic map showing the mainstream, major tributaries, and the sample location points within the Yangtze River drainage basin. Topography is from the Shuttle Radar Topography Mission (SRTM) plotted by GeoMapApp (http://www.geomapapp.org) (e.g., Ryan et al. 2009)

sources consists of diverse rock compositions and strata, including widely distributed carbonate rock, continental sandstone, volcanic rocks, and gneiss (e.g., He et al. 2013b).

The Yangtze River catchment is located in the humid, sub-tropical climate zone, dominated by the East Asian monsoon system, with seasonal alternation between the warm and wet summer monsoon, and a cold and dry winter monsoon. There are differences in the intensity of monsoon precipitation between the upper and lower streams. Annual precipitation tends to decrease westward from about 1000 mm in the eastern lowlands to about 700 mm in the Sichuan Basin, but rises to over 1700 mm on the eastern flanks of the central Longmen Shan. West of the Longmen Shan, precipitation decreases across the Tibetan Plateau, from about 600 mm in the middle reaches of the Yalongjiang to about 400 mm at the head of the Jinshajiang (e.g., He et al. 2014; 2013b; Yang et al. 2004b).

Methods

We collected 30 river bed sediment samples along the Yangtze River, including the mainstream and major tributaries. The sampling sites cover the whole drainage, from the head at Tuotuohe to the river mouth (Fig. 1 and Table 1). The samples from the major tributaries were taken just above their confluence with the mainstream in order to characterize what each tributary is contributing to the main stream.

All the sand samples were filtered to <63 μm after removing carbonate and organic materials, using 0.5 mol/L CH_3COOH and H_2O_2 for 24 h. Major element contents in the sediments were measured using the X-ray Fluorescence Spectrometer (ARL-9800) in the Center of Modern Analysis of Nanjing University. The samples were decomposed by pressurized HF-HNO_3 mixture in sealed Teflon beakers to determine the trace element content. Trace elements, including REEs, were determined by ICP-MS (Finnigan MAT Element-II) in the State Key Laboratory for Mineral Deposits Research, Nanjing University. Twenty-seven samples were analyzed for Sr and Nd isotopes (excluding Fuling 1, Xiangjiang, and Yueyang 1), which were dissolved in Teflon beakers with a solution of HF-HNO_3, and then separated by conventional ion-exchange techniques. The isotopic measurements were performed on a Thermo Finnigan Triton Tl Thermal Ionization Mass Spectrometer (TIMS) and Thermo Fisher Scientific Neptune MC-ICP-MS at the State Key Laboratory for Mineral Deposits Research, Nanjing University. Measured Sr and Nd isotopic ratios were normalized using a $^{86}Sr/^{88}Sr$ value of 0.1194 and a $^{146}Nd/^{144}Nd$ value of 0.7219, respectively. Analyses of

Table 1 Sample location of the Yangtze River

No.	Sample	Longitude	Latitude	Date
1	Tuotuohe	91° 5′ 9.20″	34° 17′ 43.30″	05/2009
2	Jinshajiang	99° 22′ 14″	28° 11′ 9″	09/2011
3	Shigu	99° 58′ 52.00″	26° 52′ 14.00″	05/2010
4	Panzhihua 1	101° 29′ 25.80″	26° 35′ 16.20″	03/2009
5	Yalongjiang	101° 49′ 16.20″	26° 46′ 2.40″	03/2009
6	Panzhihua 2	101° 49′ 13.80″	26° 35′ 28.80″	03/2009
7	Yibin 1	103° 45′ 18″	29° 37′ 28″	03/2009
8	Daduhe	103° 45′ 42.00″	29° 33′ 16.00″	03/2009
9	Minjiang 1	103° 45′ 18.00″	29° 37′ 28.00″	03/2009
10	Yibin 2	104° 41′ 26.40″	28° 47′ 0.00″	03/2009
11	Chongqing 1	106° 29′ 22″	29° 23′ 35″	03/2009
12	Jialingjiang	106° 29′ 1.80″	29° 33′ 22.80″	03/2009
13	Chongqing 2	106° 37′ 0.60″	29°37′11.40″	03/2009
14	Fuling 1	107° 21′ 9″	29° 44′ 1″	03/2009
15	Wujiang	107° 23′ 33.80″	29° 36′ 20.80″	03/2009
16	Fuling 2	107° 24′ 37.20″	29° 44′ 8.40″	03/2009
17	Badong	110° 20′ 46″	30° 3′ 3″	05/2009
18	Yichang	111° 18′ 43.70″	30° 39′ 49.50″	05/2009
19	Yueyang 1	112° 55′ 7.60′	29° 32′ 49.50″	05/2009
20	Yuanjiang	111° 41′ 11.20″	29° 1′ 25.90″	05/2009
21	Xiangjiang	112° 56′ 55.20″	28° 8′ 51.60″	05/2009
22	Yueyang 2	113° 11′ 27.30″	29° 29′ 28.50″	05/2009
23	Wuhan 1	114° 14′ 32″	30° 28′ 31″	05/2009
24	Hanjiang	113° 25′ 57.80″	30° 23′ 33.50″	05/2009
25	Wuhan 2	114° 25′ 32.90″	30° 40′ 19.10″	05/2009
26	Jiujiang	115° 54′ 30″	29° 43′ 8″	05/2009
27	Ganjiang	115° 51′ 21.50″	28° 38′ 57.90″	05/2009
28	Hukou	116° 18′ 26.80″	29° 46′ 3.10″	05/2009
29	Nanjing	118° 44′ 40.10″	32° 6′ 47.40″	05/2009
30	Changxing Island	121° 58′ 45.00″	31° 11′ 16.00′	05/2009

standards during the period of analysis are as follows: NBS987 for Sr isotopes ($^{87}Sr/^{86}Sr = 0.710235 \pm 4$) and JNdi-1 for Nd isotopes ($^{143}Nd/^{144}Nd = 0.512120 \pm 3$). The results of the geochemical analyses are provided in Tables 2, 3, 4, 5, and 6.

Results

We use <63 μm fraction sediment major element chemistry to assess the basic characteristics of the fine-grained sediments that we have sampled from the Yangtze River. Discrimination schemes from Herron (1988) show that the sands to silt from the Yangtze cluster between wacke and more iron-rich shales. It is clear that the sediments of the Yangtze are not very geochemically mature because they plot far from the quartz arenite field (Additional file 1: Figure S1). Thus, compared to many sediments in basins worldwide, those from the Yangtze are quite primitive and suggest relatively limited amounts of alteration from the bedrock sources.

We can also assess how the major element compositions compare with one another to assess the processes that control the sediment composition. Figure 2 shows how major elements vary compared to the silica contents, which are largely a reflection of the quartz content of the sediment. These figures show a negative correlation of CaO, Fe_2O_3, and MgO with SiO_2. There is also a rough negative correlation between SiO_2 and TiO_2. However, there is no correlation between the SiO_2 and Al_2O_3 or K_2O concentrations.

There is very little coherent variation downstream between the headwaters and the delta in terms of the major element composition, which may represent the effects of sorting in dominating the sediment composition at any one particular point, masking any general trend that might exist. Some of the samples show either very high or low values at certain points but show no clear evolution downstream. We note that there is a steady increase in MgO from the headwaters as far as Yibin, which is also broadly paralleled by an increase in K_2O from Shigu to Yibin but little else is apparent (Table2).

We examine the REE composition of sediments using a standard, multi-element diagram normalized against chondrite (Fig.3, Tables 3-5). We plot the samples taken from the different segments in the mainstream separately from those samples from the major tributaries. All the sediments show a similar type of pattern, with an enrichment in the light rare earth elements (LREE) and a relatively flat pattern in the heavy REEs (HREE; Gd to Lu). All the samples show a clear relative depletion in Eu, which is typical of the upper continental crust (e.g., Taylor and McLennan 1985). In the following discussion, we quantify the degree of LREE enrichment using the ratio La/Sm, while the relative enrichment in HREEs is represented by Tb/Yb, although the difference is small. None of the samples shows a particularly unique pattern that would show the influence of one particular tributary or source dominating the sediment in any one place.

$^{87}Sr/^{86}Sr$ ratios vary from 0.711549 ± 3 to 0.740534 ± 5, and $^{143}Nd/^{144}Nd$ ratios from 0.511860 ± 3 to 0.512216 ± 3, corresponding to εNd values ranging from −15.2 to −8.2 (Table6). High εNd values, representing to more erosion of primitive, younger crust (e.g., DePaolo and Wasserburg 1976), are found mainly in the samples from the Jinshajiang, which have low $^{87}Sr/^{86}Sr$ ratios. In the tributaries, the Jialingjiang has both low $^{87}Sr/^{86}Sr$ ratios and relatively low εNd values (i.e., old, evolved crust), whereas the Yuanjiang and Ganjiang have higher $^{87}Sr/^{86}Sr$ ratios but lower εNd values (Fig. 4).

Table 2 Major element composition of the sediments in the Yangtze River

		Al_2O_3	CaO	Fe_2O_3	K_2O	MgO	MnO	Na_2O	P_2O_5	TiO_2	SiO_2	CIA
1	Tuotuohe	12.37	12.82	5.69	2.52	2.56	0.10	1.61	0.24	0.81	60.91	60.66
2	Jinshajiang	12.45	10.46	9.08	1.95	4.29	0.11	1.51	0.30	1.88	58.07	63.69
3	Shigu	10.62	8.23	4.89	1.70	2.78	0.07	1.90	0.21	0.85	68.53	56.69
4	Panzhihua 1	12.65	9.58	7.75	2.02	4.34	0.10	1.70	0.26	1.38	59.80	61.92
5	Yalongjiang	15.61	4.44	12.65	2.15	5.19	0.21	1.96	0.34	2.81	54.88	63.95
6	Panzhihua 2	14.63	6.58	9.85	2.06	4.40	0.14	1.70	0.32	2.43	57.78	65.15
7	Yibin 1	12.34	9.64	9.75	2.51	6.80	0.14	1.32	0.30	2.03	55.09	63.59
8	Daduhe	12.96	9.09	7.36	2.55	6.06	0.10	2.20	0.30	1.25	58.24	56.45
9	Minjiang	15.44	4.54	6.57	2.75	3.03	0.08	2.05	0.31	0.86	64.35	61.37
10	Yibin 2	13.15	13.21	7.42	2.75	6.58	0.10	1.32	0.26	1.20	53.65	64.21
11	Chongqing 1	12.87	2.58	15.45	2.05	2.65	0.31	1.63	0.41	2.06	60.47	62.96
12	Jialingjiang	14.71	4.01	6.53	2.65	2.45	0.07	1.66	0.24	0.93	66.62	63.77
13	Chongqing 2	11.87	12.24	7.94	2.21	5.18	0.13	1.64	0.28	1.44	57.14	60.38
14	Fuling 1	12.97	11.46	9.98	2.49	6.21	0.13	1.31	0.32	2.01	52.96	64.97
15	Wujiang	18.06	6.83	8.65	3.24	3.18	0.12	0.53	0.27	1.03	58.14	77.50
16	Fuling 2	13.53	3.05	6.83	2.40	3.20	0.08	1.66	0.22	1.23	67.39	62.63
17	Badong	19.53	1.53	8.16	3.05	1.74	0.12	0.82	0.11	0.91	63.85	76.46
18	Yichang	13.26	20.19	6.42	2.61	4.43	0.10	1.23	0.27	0.84	50.29	65.81
19	Yueyang 1	11.61	6.27	6.96	2.08	3.96	0.09	1.60	0.25	1.59	65.15	60.70
20	Yuanjiang	13.37	0.81	4.80	2.69	1.87	0.06	0.93	0.15	0.88	74.25	69.14
21	Xiangjiang	20.55	0.26	9.27	2.90	1.48	0.11	0.59	0.28	1.08	63.63	80.14
22	Yueyang 2	12.08	3.04	6.58	2.22	3.13	0.08	1.29	0.22	1.34	70.00	64.48
23	Wuhan 1	17.17	3.36	8.36	2.90	3.64	0.10	1.32	0.28	1.29	61.32	69.60
24	Hanjiang	15.90	4.43	8.51	2.61	4.04	0.11	1.98	0.29	1.21	60.74	62.97
25	Wuhan 2	17.05	3.17	8.43	2.90	3.95	0.41	1.45	0.26	1.28	60.93	68.24
26	Jiujiang	14.52	4.31	7.79	2.53	4.00	0.11	1.54	0.24	1.45	63.71	65.01
27	Ganjiang	14.10	0.43	5.09	3.01	0.94	0.06	0.71	0.14	0.90	74.33	71.50
28	Hukou	13.22	9.99	12.45	2.27	8.76	0.24	1.68	0.38	0.99	50.73	62.31
29	Nanjing	10.28	8.15	12.13	1.67	4.32	0.16	1.50	0.45	3.20	58.50	60.35
30	Changxing	12.38	5.67	6.48	2.23	4.03	0.08	1.92	0.28	1.28	65.60	58.60

Discussion

Sedimentary sorting

Because Al_2O_3 or K_2O concentrations are largely controlled by the abundance of micas, clays, and feldspar, particularly potassium feldspar in case of K_2O, the lack of correlation between SiO_2 and Al_2O_3 or K_2O indicates that there is apparently no simple relationship between the amount of quartz in the sand and the abundance of feldspar. On the other hand, Fe_2O_3 and MgO are largely held in mafic, heavy minerals, such as pyroxenes and amphiboles, as well as iron oxides. It is perhaps not surprising that we see a steady decrease in these mafic-related elements as the total amount of quartz increases. These negative correlations therefore represent a dilution trend by quartz. Because the total amount of quartz in the sediment reflects not only the composition of the sources but also the sorting of the sediment as a result of current activity, and the effects of chemical weathering, it seems unlikely that we can use the concentrations of major elements as a provenance proxy, unless the sorting process can be accounted for (e.g., Garzanti et al. 2008). Great care would have to be taken in doing this because this could change both with time during the seasonal flooding and ebb of the river, as well as along the length of the river channel and across the channel with depth. This would be hard to do even in a modern river and seems unlikely to yield any sensible result when applied to older deposits.

The sediment composition is strongly influenced by local sorting of minerals, as a result of current activity both across the channel and in meanders where the current velocity changes rapidly and results in local concentration or

Table 3 Trace element composition of the sediments in the Yangtze River

Samples	Sc	Ti	V	Cr	Co	Ni	Rb	Sr	Y	Zr	Nb
Tuotuohe	5.97	1988.27	46.15	45.25	5.16	16.11	61.20	189.88	13.41	346.04	5.66
Jinshajiang	9.55	6352.46	135.79	115.41	14.83	39.03	42.96	176.50	29.50	1271.48	15.28
Shigu	11.81	5770.17	113.06	84.26	12.28	31.51	48.65	172.28	30.63	655.11	13.56
Panzhihua 1	5.92	8906.07	187.85	206.07	20.68	82.82	59.77	212.66	26.43	491.92	21.58
Yalongjiang	8.26	17752.43	253.91	134.77	34.54	69.61	35.13	182.18	35.18	719.90	27.02
Panzhihua 2	5.28	16150.15	314.23	199.79	29.41	77.42	18.29	198.64	31.17	669.88	24.66
Yibin 1	3.51	13501.08	220.83	170.60	22.26	50.70	85.75	180.15	31.28	967.38	26.93
Daduhe	5.64	8161.48	157.60	125.41	15.81	43.02	52.48	202.43	42.94	1066.66	30.12
Minjiang 1	8.15	5714.67	141.56	79.69	14.21	39.19	21.33	147.92	28.02	444.22	16.05
Yibin 2	5.83	7733.59	178.70	101.48	19.19	49.41	45.90	183.60	23.18	322.45	17.88
Chongqing 1	16.32	13889.48	248.35	97.02	23.88	39.51	48.51	65.61	57.23	1284.19	30.09
Jialingjiang	10.14	6452.94	137.38	124.50	12.88	37.07	28.15	125.75	35.22	1128.31	17.88
Chongqing 2	5.93	8734.99	152.28	132.52	16.62	35.47	40.56	179.38	25.66	955.23	20.40
Fuling 1	21.05	16803.91	267.87	196.23	25.32	62.18	109.89	234.58	39.60	850.64	29.67
Wujiang	10.48	5924.52	155.52	85.13	17.34	46.09	22.87	106.42	25.61	295.34	19.66
Fuling 2	10.03	8031.61	147.71	103.73	15.11	38.30	31.20	98.81	33.30	797.32	20.31
Badong	11.63	5972.17	169.45	90.46	24.38	49.21	51.25	68.88	16.36	267.32	17.52
Yichang	3.97	4981.18	126.59	79.90	15.39	44.86	43.01	264.56	12.68	313.54	13.67
Yueyang 1	17.53	10896.68	195.09	113.91	19.76	48.58	116.77	169.30	33.27	523.93	23.41
Yuanjiang	7.65	5719.15	107.50	66.35	9.80	26.13	46.03	50.75	26.45	592.06	18.02
Xiangjiang	18.38	8477.10	202.11	139.69	18.10	59.31	195.66	73.13	43.33	1149.21	26.60
Yueyang 2	10.27	9020.72	143.59	110.77	15.47	38.21	39.49	98.18	37.90	771.32	23.34
Wuhan 1	10.08	7707.60	171.55	100.17	18.21	48.04	16.97	80.51	27.16	395.24	20.37
Hanjiang	6.32	7223.73	180.99	73.31	22.99	46.94	19.97	133.61	35.09	471.24	20.84
Wuhan 2	7.68	7962.05	182.54	98.35	17.67	47.93	21.65	83.11	27.95	400.50	21.63
Jiujiang	7.96	8957.52	170.59	98.47	18.57	44.53	31.42	102.18	31.77	544.97	22.30
Ganjiang	6.09	5900.86	80.35	48.15	7.68	19.17	105.37	43.86	38.20	971.55	28.74
Hukou	2.65	6043.32	170.45	99.28	44.42	82.55	89.58	157.36	31.77	463.21	16.90
Nanjing	9.04	15785.47	207.20	238.47	14.80	38.31	65.59	207.21	74.69	3366.28	38.35
Changxing	6.29	9014.45	148.76	107.62	15.47	84.35	37.11	172.09	45.52	1147.71	24.68

dilution of heavy minerals, so that some geochemical characteristics are controlled mostly by local hydrodynamic processes rather than by chemical weathering or source composition.

We assess the effect that the presence of heavy minerals has on sediment chemistry by plotting key ratios against one another. Figure 5a shows a plot of Th/Al against Zr/Al. In this case, Al is used as a normalizing element, while Th is hosted in heavy minerals, especially in monazite (e.g., Ni et al. 1995), which is often associated with felsic intrusive rocks. Zr is enriched in zircon, which is also often associated with monazite. This plot (Fig. 5a) shows a positive correlation, suggesting that sediments that are rich in monazite are also rich in zircons, as might be expected from currents concentrating these dense minerals. Figure 5b, c shows the relationship between heavy mineral concentration, as represented by Th/Al, and the relative enrichment in LREE and HREE, as proxied by La/Sm and Tb/Yb. Yang and Wang (2011) argued that the LREEs primarily reside in clay minerals, while Garzanti et al. (2011) showed that monazite and allanite are also very rich in these elements. In contrast, HREEs are generally concentrated in certain heavy minerals, especially xenotime and zircon (e.g., Garzanti et al. 2011). The HREE-enriched heavy minerals significantly influence the REE compositions, reflecting the control of provenance (e.g., Rollinson 2013). We note that there is a poorly positive correlation with LREE and a negative correlation with HREE, which suggests that the LREE chemistry is partly controlled by the presence or absence of key heavy minerals, especially monazite or zircon, which in turn is controlled both by provenance and

Table 4 Trace element composition of the sediments in the Yangtze River

Samples	Cs	Ba	Hf	Ta	Pb	Th	U	La	Ce	Pr	Nd
Tuotuohe	33.48	344.12	9.39	0.51	35.21	9.74	2.28	25.49	50.90	5.67	21.05
Jinshajiang	3.34	431.12	29.43	1.23	26.73	12.66	5.65	39.62	81.95	9.32	34.89
Shigu	3.68	347.83	15.94	1.09	19.72	11.77	3.47	38.55	79.27	8.81	32.55
Panzhihua 1	4.85	464.53	12.16	1.51	27.73	7.19	3.31	37.22	78.54	8.86	33.84
Yalongjiang	4.38	1042.56	16.75	1.93	47.35	12.40	3.90	87.02	140.34	15.22	53.32
Panzhihua 2	4.92	427.76	15.61	1.74	24.87	8.12	3.59	41.38	88.21	10.11	39.01
Yibin 1	5.31	1178.82	21.96	1.73	86.48	8.39	4.57	48.82	100.45	11.07	41.76
Daduhe	2.34	456.18	23.50	2.47	55.73	19.67	10.19	56.79	107.09	12.74	46.46
Minjiang 1	4.65	505.81	11.29	1.29	23.61	10.85	2.83	39.77	80.14	9.20	34.42
Yibin 2	1.03	1074.33	9.07	1.36	60.83	10.32	4.35	39.23	81.41	9.02	34.34
Chongqing 1	4.68	358.44	32.02	2.38	21.65	22.49	7.15	73.90	131.94	16.42	63.03
Jialingjiang	4.96	756.74	25.71	1.36	23.50	16.10	3.43	56.35	105.34	12.82	46.24
Chongqing 2	1.07	800.32	22.42	1.29	44.14	10.31	4.23	45.95	95.25	9.71	36.57
Fuling 1	6.77	1119.80	24.71	2.00	81.77	24.38	6.36	68.01	135.52	15.11	57.24
Wujiang	6.31	566.85	6.87	1.40	36.06	10.05	3.38	35.28	79.81	8.45	31.71
Fuling 2	3.99	329.41	21.61	1.60	20.12	15.81	4.40	50.23	99.98	11.71	44.16
Badong	8.85	420.29	7.04	1.32	34.46	7.24	3.23	27.10	72.31	6.87	25.89
Yichang	1.15	455.99	7.95	0.96	47.99	8.72	4.09	34.53	76.50	7.86	29.80
Yueyang 1	7.11	569.57	14.62	1.62	50.25	17.72	4.65	51.09	107.43	11.85	45.67
Yuanjiang	5.48	512.91	14.43	1.36	28.29	8.50	3.19	31.46	64.23	7.30	27.01
Xiangjiang	24.61	831.80	33.05	2.55	189.50	37.43	10.37	65.80	136.58	14.90	56.00
Yueyang 2	5.77	396.93	22.10	2.01	44.79	19.15	5.26	56.45	113.87	12.55	47.87
Wuhan 1	5.01	383.33	10.30	1.64	46.70	10.44	2.97	41.03	84.44	9.54	35.35
Hanjiang	3.85	554.83	11.40	1.48	30.07	11.14	2.81	43.85	100.37	10.66	40.06
Wuhan 2	5.38	386.41	10.01	1.60	33.08	10.36	2.76	40.01	80.96	9.71	36.72
Jiujiang	4.85	367.64	14.18	1.60	36.52	12.61	3.61	45.72	93.33	10.89	40.93
Ganjiang	11.72	312.62	25.78	3.36	39.18	19.74	5.83	47.33	98.18	11.03	39.11
Hukou	7.17	595.00	11.86	1.26	73.31	9.72	4.38	50.30	126.50	12.16	47.06
Nanjing	3.90	373.04	82.84	3.39	56.16	31.31	12.05	103.11	202.50	23.37	87.56
Changxing	4.47	298.53	29.98	1.93	17.85	19.97	5.80	70.35	132.72	16.03	59.50

current sorting. The HREE is apparently less controlled by heavy mineral abundance, at least as tracked by Th/Al, but variations in HREE enrichments are quite small compared to LREEs. We note that there is no simple trend between the headwaters and the river mouth in these chemical ratios and that local hydrodynamic processes dominate in controlling heavy mineral abundance, consistent with the findings of Garzanti et al. (2008) in Himalayan rivers. Figure 5d demonstrates a negative correlation between the abundance of heavy minerals and enrichment in high-field strength elements (HFSE). Samples in the downstream area tend to be the richest in heavy minerals and these show relatively little HFSE enrichment.

We now investigate the relationship between grain size and bulk sediment chemistry by examining the relationship between SiO_2/Al_2O_3 and REEs and HFSEs (Fig. 6). SiO_2/Al_2O_3 is a proxy for sand/clay because these two elements are preferentially enriched in quartz and clay minerals, respectively. Yang et al. (2002) demonstrated that REEs were more abundant in finer-grained sediment. However, we found little linkages between Si/Al and total REE concentrations. In fact, we noted a slight, but poorly defined positive correlation between more sandy sediment and higher REE contents. This mismatch with the conclusion of Yang et al. (2002) may reflect variation within our more limited range of grain sizes (<63 μm) and higher abundance of monazite, allanite, and to the lesser extent xenotime in the coarser fraction that they considered. However, Fig. 6a shows that there is positive correlation between the degree of LREE enrichment and the presence of quartz within the sediment, with the exception

Table 5 Trace element composition of the sediments in the Yangtze River

Samples	Sm	Eu	Gd	Tb	Dy	Ho	Er	Tm	Yb	Lu
Tuotuohe	3.79	0.75	3.11	0.41	2.64	0.56	1.68	0.25	1.69	0.28
Jinshajiang	6.62	1.35	6.01	0.86	6.03	1.31	4.00	0.59	3.61	0.65
Shigu	6.27	1.27	5.64	0.79	5.56	1.21	3.46	0.51	3.22	0.52
Panzhihua 1	6.51	1.45	5.89	0.81	5.66	1.18	3.37	0.51	2.92	0.46
Yalongjiang	8.84	1.85	7.59	1.06	6.95	1.45	4.13	0.62	3.74	0.61
Panzhihua 2	7.40	1.64	6.53	0.89	6.14	1.27	3.67	0.53	3.26	0.53
Yibin 1	7.85	1.74	6.83	0.94	6.64	1.43	4.06	0.59	3.73	0.59
Daduhe	9.13	1.60	8.26	1.17	8.39	1.77	5.17	0.76	4.79	0.80
Minjiang 1	6.57	1.33	5.75	0.79	5.32	1.16	3.34	0.50	3.04	0.49
Yibin 2	6.69	1.57	5.62	0.78	5.01	1.08	3.20	0.46	2.82	0.45
Chongqing 1	12.19	2.87	10.76	1.52	10.72	2.40	6.99	1.03	6.62	1.08
Jialingjiang	8.34	1.43	7.18	0.97	6.75	1.44	4.27	0.64	3.94	0.63
Chongqing 2	7.06	1.48	5.94	0.85	5.53	1.23	3.47	0.51	3.08	0.53
Fuling 1	10.82	2.31	7.92	1.10	7.00	1.52	4.69	0.68	4.62	0.75
Wujiang	6.12	1.19	5.33	0.73	4.96	1.06	2.97	0.44	2.71	0.43
Fuling 2	8.57	1.65	7.12	0.96	6.51	1.39	4.13	0.60	4.11	0.68
Badong	5.34	1.07	4.60	0.66	4.58	0.99	2.87	0.41	2.59	0.41
Yichang	5.79	1.23	4.83	0.66	4.39	0.90	2.49	0.34	1.94	0.29
Yueyang 1	8.59	1.76	6.57	0.90	5.99	1.31	3.85	0.58	3.68	0.62
Yuanjiang	5.06	0.95	4.67	0.69	5.15	1.14	3.46	0.52	3.22	0.51
Xiangjiang	10.66	1.67	8.33	1.15	7.90	1.76	5.50	0.85	5.92	0.98
Yueyang 2	8.92	1.62	7.82	1.08	7.46	1.64	4.70	0.74	4.71	0.77
Wuhan 1	6.76	1.43	5.83	0.82	5.64	1.20	3.40	0.51	3.13	0.50
Hanjiang	7.81	1.56	7.00	0.98	6.85	1.43	4.12	0.58	3.63	0.59
Wuhan 2	7.11	1.45	5.99	0.81	5.62	1.19	3.53	0.54	3.21	0.51
Jiujiang	7.87	1.63	6.85	0.94	6.33	1.39	4.02	0.57	3.60	0.57
Ganjiang	7.29	0.78	6.59	1.00	7.42	1.72	5.25	0.83	5.31	0.91
Hukou	9.17	1.93	8.17	1.10	7.42	1.56	4.32	0.60	3.64	0.58
Nanjing	16.26	2.67	14.42	1.99	14.74	3.31	10.04	1.57	9.84	1.68
Changxing	11.13	1.99	9.39	1.28	8.75	1.93	5.60	0.86	5.35	0.89

of the Yalongjiang, which is strongly enriched, but relatively sand-poor. Figure 6b demonstrates that there is an approximate negative correlation between enrichment in HREEs and the abundance of sand, the opposite trend of that seen for the LREEs. Since HREEs are very concentrated in xenotime and to a lesser extent zircon (e.g., Garzanti et al. 2011), our observations would suggest that xenotime is more abundant in the finer sediment. Likewise, enrichment in HFSEs shows a negative correlation with less enrichment associated with sandier sediment (Fig. 6c). We note that there is no clear downstream development in any ratio, which we would expect if transport was dominating this relationship. If the mainstream was experiencing a simple development and most of the sediment was being supplied by a short upstream interval of the river, then it is overwhelmed by the addition of new sediment from the tributaries, or from local sources, during the journey between the headwaters and the river mouth.

REEs' composition is all influenced by sedimentary sorting, while HREEs and HFSE enrichment are less controlled by the abundance of heavy minerals compared to the LREEs, but there is no clear trend from the upstream to downstream along the drainage, indicating that the chemical ratios are more affected by hydrodynamic processes (i.e., sorting) rather than provenance.

Chemical weathering

In order to understand how chemical weathering intensity varies within the Yangtze, we employ several proxies, which have been applied in many situations and which are widely believed to reflect of the alteration of fresh

Table 6 Sr-Nd isotope composition of the sediments in the Yangtze River

No.	Samples	$^{143}Nd/^{144}Nd$	1σ	Epsilon Nd	$^{87}Sr/^{86}Sr$	1σ
1	Tuotuohe	0.512092	5	−10.65	0.73	7
2	Jinshajiang	0.512082	3	−10.85	0.71	11
3	Shigu	0.512090	3	−10.69	0.71	4
4	Panzhihua 1	0.512182	3	−8.9	0.71	4
5	Yalongjiang	0.512216	3	−8.23	0.71	4
6	Panzhihua 2	0.512185	4	−8.84	0.71	3
7	Yibin 1	0.512060	3	−11.28	0.72	8
8	Minjiang 1	0.512042	3	−11.63	0.72	6
9	Daduhe	0.512200	2	−8.54	0.71	3
10	Yibin 2	0.512089	3	−10.71	0.72	4
11	Chongqing 1	0.512038	2	−11.7	0.73	4
12	Jialingjiang	0.511860	3	−15.18	0.71	3
13	Chongqing 2	0.511956	3	−13.3	0.71	3
14	Wujiang	0.512025	3	−11.96	0.72	4
15	Fuling	0.512003	3	−12.39	0.72	5
16	Badong	0.512065	6	−11.18	0.72	2
17	Yichang	0.512038	4	−11.7	0.71	7
18	Yueyang	0.512013	4	−12.19	0.72	7
19	Yuanjiang	0.511916	3	−14.08	0.74	5
20	Wuhan 1	0.512037	4	−11.72	0.72	4
21	Hanjiang	0.512088	2	−10.73	0.72	8
22	Wuhan 2	0.512069	4	−11.1	0.72	6
23	Jiujiang	0.512063	4	−11.22	0.72	5
24	Ganjiang	0.511927	3	−13.87	0.74	4
25	Hukou	0.512087	3	−10.75	0.72	6
26	Nanjing	0.512028	5	−11.9	0.71	10
27	Changxing	0.512048	3	−11.51	0.72	7

rock particles. K_2O/Al_2O_3 is widely interpreted as a proxy for alteration because potassium is a water-soluble element that is released during the breakdown of potassium feldspar and has been applied to the study of sediments on East and Southeast Asian continental margins (e.g., Clift et al. 2008; Hu et al. 2012). In addition, we employ the classic proxy Chemical Index of Alteration (CIA), which also involves these elements, but also takes into account Na and Ca (e.g., Nesbitt and Young 1982). CIA was developed for use in soil profiles, but the proxy has been widely applied to sediments in rivers and in drill cores (e.g., Liu et al. 2007; Wan et al. 2010; Wan et al. 2009). In Fig. 7, we show how these proxies relate to grain size, as proxied by SiO_2/Al_2O_3. In Fig. 7a, we see that K_2O/Al_2O_3 shows a rough negative relationship with SiO_2/Al_2O_3 implying that sandier sediments are generally less altered than mud-rich sediments. Figure 7b shows the same relationship with CIA, although this is generally better correlated, with the exception of two of the tributaries, the Ganjiang and Yuanjiang. As with the other proxies we have examined, there is no clear correlation between the location of the sample within the Yangtze and the degree of alteration or the grain size. We do see however that the grain size is a primary control on the alteration proxy and that unless this can be accounted for the application of weathering proxies to river sediments is of limited use in understanding weathering in the system.

Figure 8 shows the calculated CIA values for the different samples inside a triangular diagram and demonstrates how our river samples compare with other major rivers. We note that the Yangtze plots between sediments from the Pearl River and from the Yellow River, with the Yellow River samples showing lower CIA values, indicating that are less altered. This is perhaps not surprising because the climate of the Yellow River basin is generally colder and drier than the Yangtze, while the Pearl River basin is warmer and wetter, which would favor more chemical alteration if transport times are similar (e.g., West et al. 2005). We note that the Yangtze and the Pearl Rivers have higher CIA and are more altered than sediment from the Ganges and Yellow Rivers. Although the Ganges River lies in a warm and wet environment, it derives its sediment from the tectonically active Himalaya and consequently has relatively unaltered sediments. The low alteration state also reflects rapid transport through the flood plains where most of the chemical weathering occurs (e.g., Lupker et al. 2012).

We further reconstruct the development in chemical weathering in the Yangtze from source to sink using both CIA and K_2O/Al_2O_3 (Fig. 9). It is interesting to note that although there are some similarities, these proxies do not give identical images of the weathering state in the river. The solid line represents the development of alteration state in the mainstream, with individual red squares showing the state of the major tributaries. CIA remains relatively stable through much of the mainstream, although there is one unusual point of very high alteration at Badong. Given its isolation, it is questionable whether this is a very representative data point, although why is not apparent. We also note a rise in the degree of alteration as tracked by CIA between Yueyang and the confluence with the Hanjiang, followed by a steady decrease in alteration from that point to the river mouth (Fig. 9a). The rise could be explained by sediment addition from the Yuanjiang and Xiangjiang, as could the initial decrease downstream from the Hanjiang but the continued fall cannot be caused by influx from other tributaries like the Ganjiang. The trend implies either that we were sampling progressively sandier material downstream (Fig. 9c) or that less altered material is being added to the mainstream from

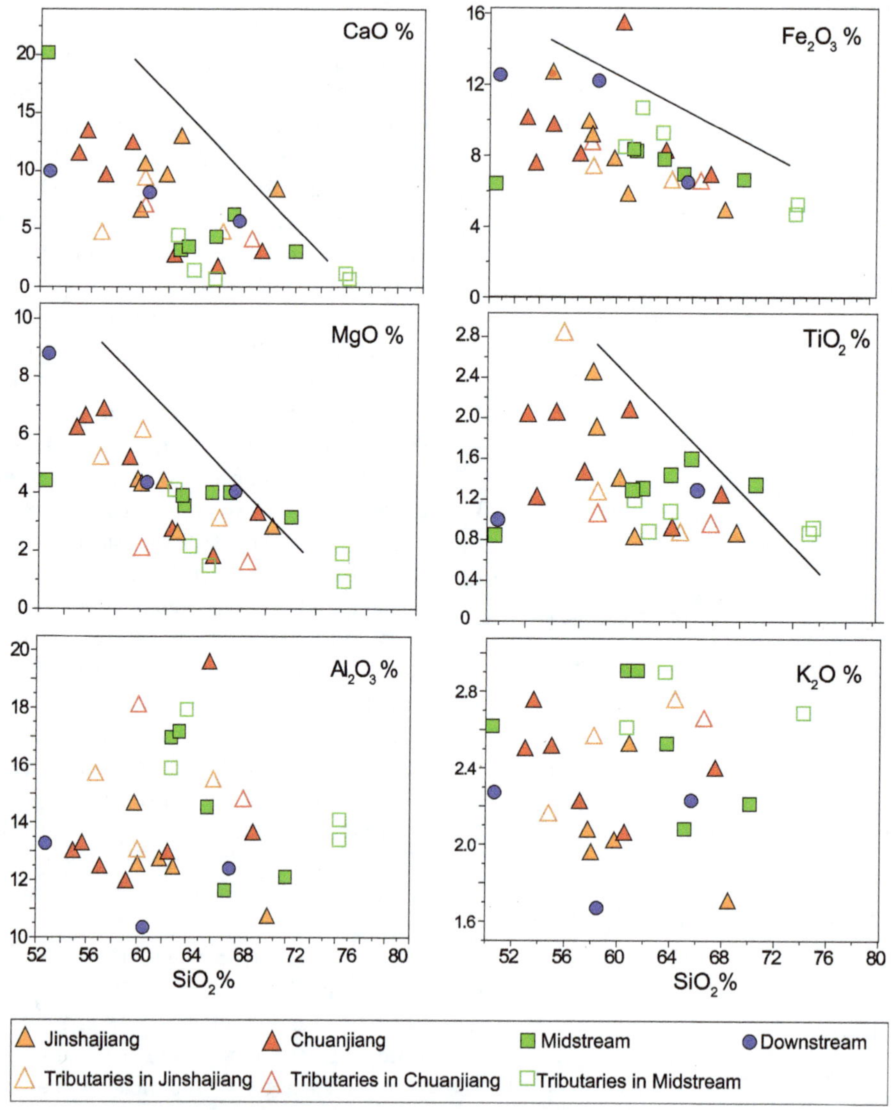

Fig. 2 Plots of major elements versus silica contents. Diagrams showing CaO, Fe_2O_3, MgO, and TiO_2 define negative correlations with SiO_2

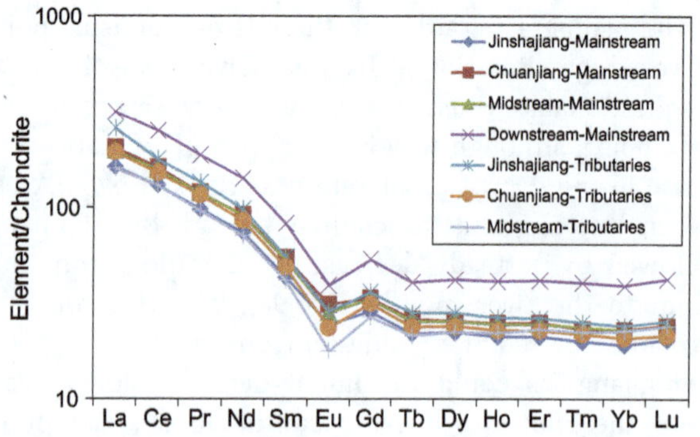

Fig. 3 Chondrite normalized REE pattern of sediments collected from different segments of the Yangtze River. The values are the average of several samples within a segment. Chondrite values from (e.g., Sun and McDonough 1989)

other sources, such as from river terraces. The trend in K_2O/Al_2O_3 (Fig. 9b) is somewhat different in being relatively stable in the lower reaches of the river, reflecting the fact that CIA is also influenced by more mobile Na as well as K. It is interesting to note that according to CIA, most of the river sediment is more altered than the sediment that we find in the river mouth, which is the opposite of what one would expect if transport resulted in more chemical weathering. It is impossible to see how sediment would be become less altered downstream with more transport and without the addition of extra sediment, unless the finer sediments were being progressively less well sampled, perhaps because they are increasingly in suspension and not deposited in the river banks where the sediments were taken. If we consider only K_2O/Al_2O_3, then we can see that the chemical weathering state at the

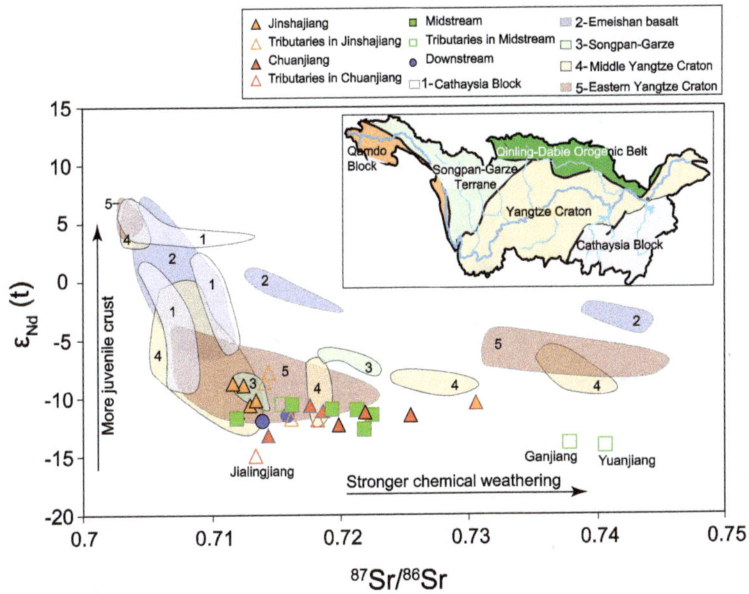

Fig. 4 $^{87}Sr/^{86}Sr$ versus εNd(t) diagram for different geological units within the Yangtze River drainage basin. Data of Eastern Yangtze Craton, Middle Yangtze Craton, Emeishan, and Cathaysia are from GEOROC (*http://georoc.mpch-mainz.gwdg.de/georoc/Start.asp*). The tectonic units map is modified from He et al. (2013a, b). Data from Songpan-Garze are from Cai et al. (2009). Sr-Nd isotopic compositions of sediment from the Yangtze River drainage shows no coherent variation downstream

river mouth is similar to that through much of the mainstream, although the sediments in the uppermost Jinshajiang do seem to be slightly more altered than most of the mainstream, despite the fact that they are close to their source regions. This a little unusual given the cold, dry conditions in the Tibetan Plateau, but higher alteration may reflect slow transport times in the upper reaches, with dilution of sediment downstream by greater supply from fresher sources. There is generally more variability of chemical weathering in the tributaries compared to the

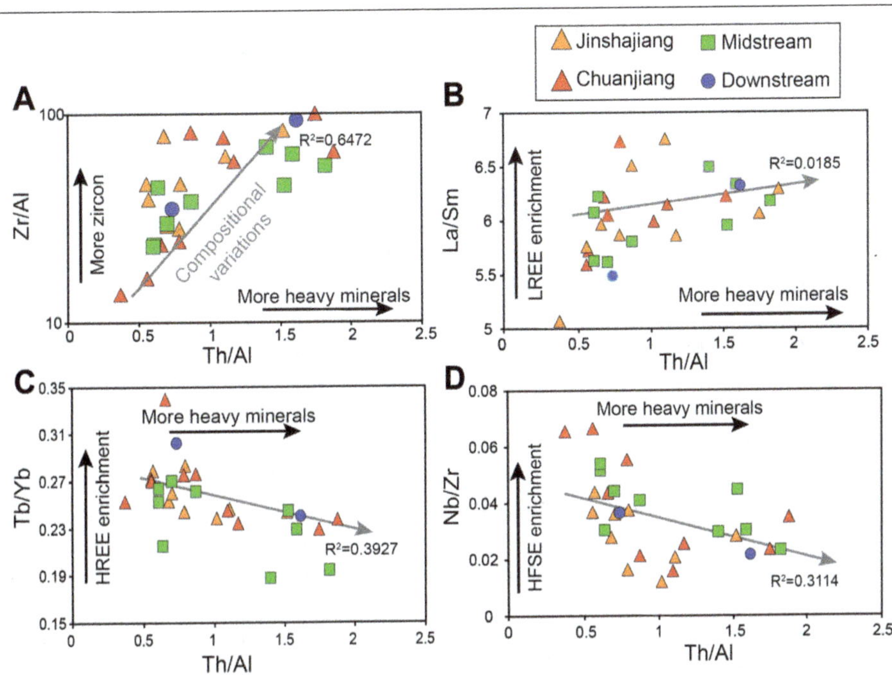

Fig. 5 Variations of trace element ratios of the Yangtze River sediments. **a** Zr/ Al_2O_3 vs. Th/Al, showing enrichment of zircon compared to heavy mineral accumulation. **b** Th/Al vs. La/Sm, indicating the relationship between heavy mineral accumulations and LREE enrichment. **c** Th/Al vs. Tb/Yb, showing heavy mineral accumulations with HREE enrichment. **d** Th/Al vs. Nb/Zr, showing heavy mineral accumulations with HFSE enrichment

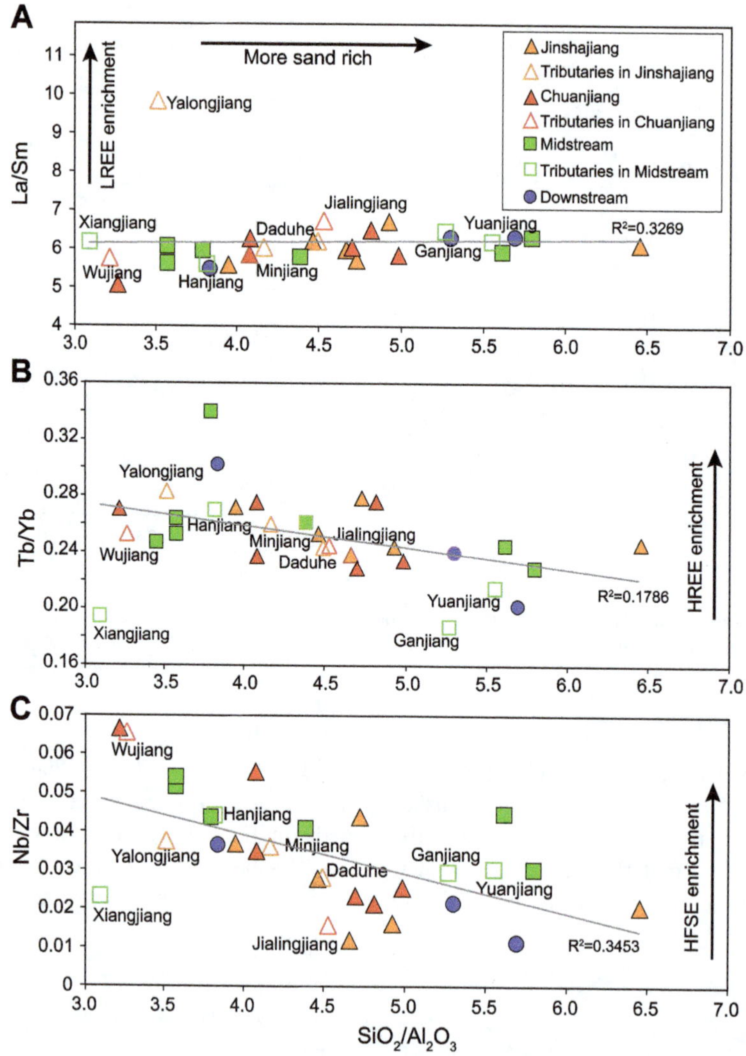

Fig. 6 Plots of SiO_2/Al_2O_3 vs. **a** LREE, **b** HREE, and **c** HFSE, showing the relationship between grain size and REEs

mainstream, especially when considering CIA, and this likely reflects the averaging of sediment within the mainstream and more local heterogeneity within the smaller tributary basins. Nonetheless, it is not clear that transport is having a significant effect on the overall state of weathering in the sands we consider.

We now discuss the effect of chemical weathering on the trace element composition of the sediments. There is no apparent relationship between the degree of chemical alteration and the extent of LREE enrichment. In contrast, Fig. 10 shows that the more altered sediments have higher degrees of enrichment in the HFSEs. Although there are a few exceptions, there is a general positive correlation between CIA and Nb/Zr. There is no link between location within the river basin and the extent of trace element enrichment. This suggests that REE chemistry might be relatively independent of chemical weathering, whereas HFSEs are affected by this process and therefore make relatively poor provenance proxies.

Provenance

We here use Sr-Nd isotopic compositions to understand the source of sediment in the river and the variations in the downstream influx of sediment both from local sources and from the larger tributary basins (Fig. 11). In the upper reaches, εNd values decrease downstream. The upper reaches, including Jinshajiang and Chuanjiang, have higher εNd values, but then decrease, especially from Panzhihua 1 to Chongqing 2, but the mainstream and tributaries in the midstream and downstream segments show lower values, which are not far from the average of εNd values (εNd = −11.4). The εNd values of the river sediments are controlled by the bedrock values of the drainage upstream of the sample point, and the monazite content of the sediments because

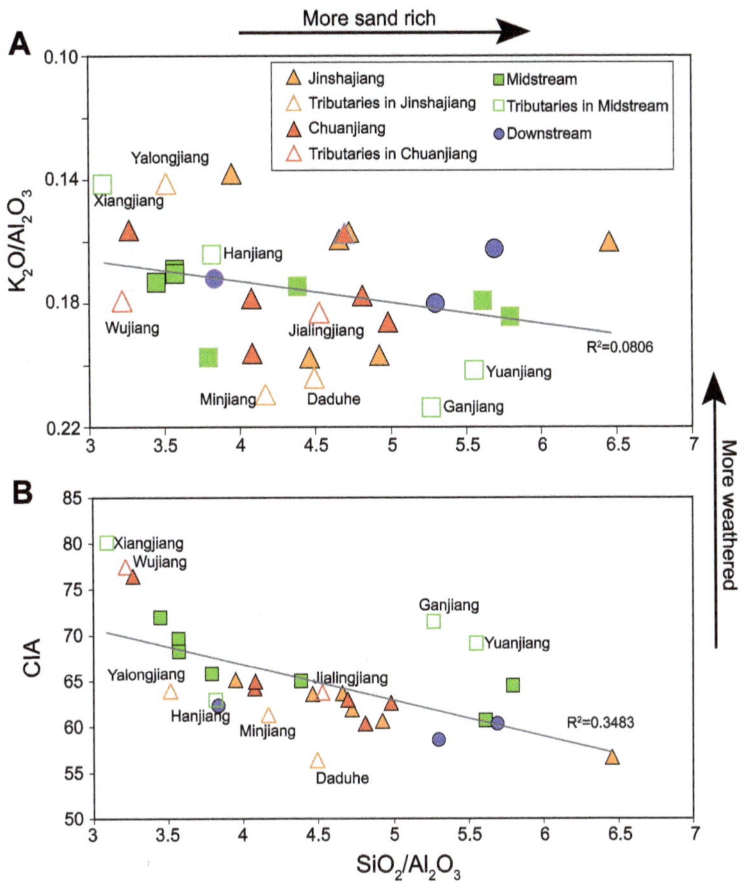

Fig. 7 Plots of SiO_2/Al_2O_3 vs. **a** K_2O/Al_2O_3, **b** CIA, showing links between chemical weathering and grain size for different sections in the Yangtze River

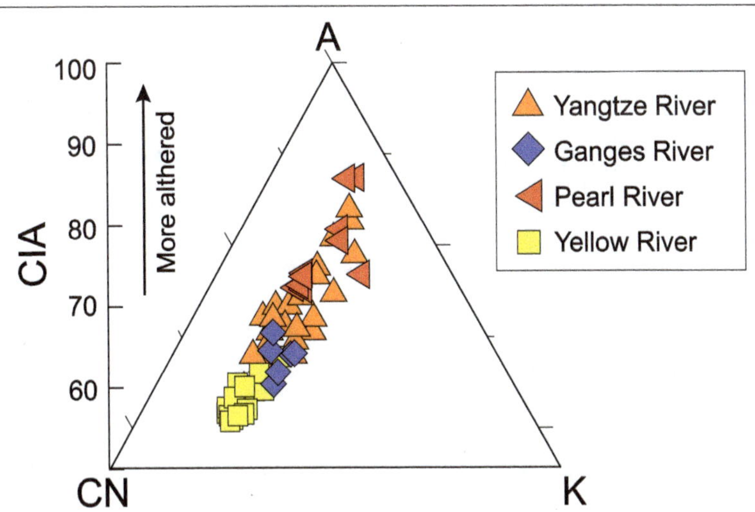

Fig. 8 A-CN-K *triangle plot* showing chemical weathering trends. Samples in the Yangtze River compared with the Ganges (e.g., Singh 2009), Pearl (e.g., Zhang and Wang 2001), and Yellow Rivers (e.g., Li 2003)

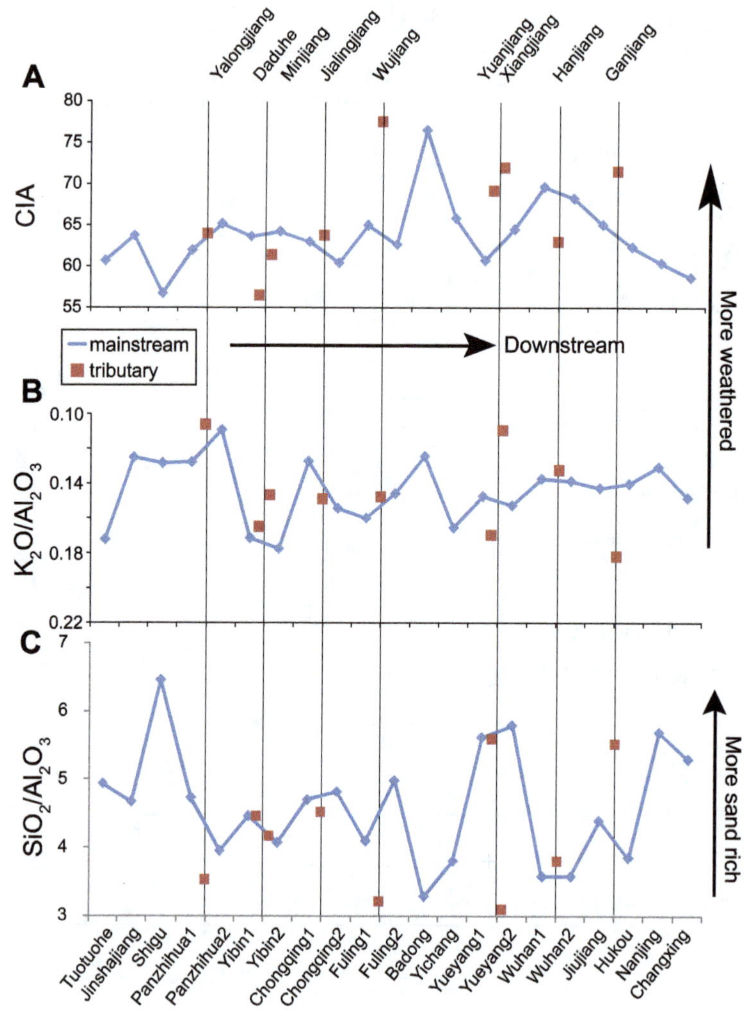

Fig. 9 Downstream variations of **a** CIA and **b** K_2O/Al_2O_3, indicating the evolving chemical weathering change from source to sink. **c** SiO_2/Al_2O_3

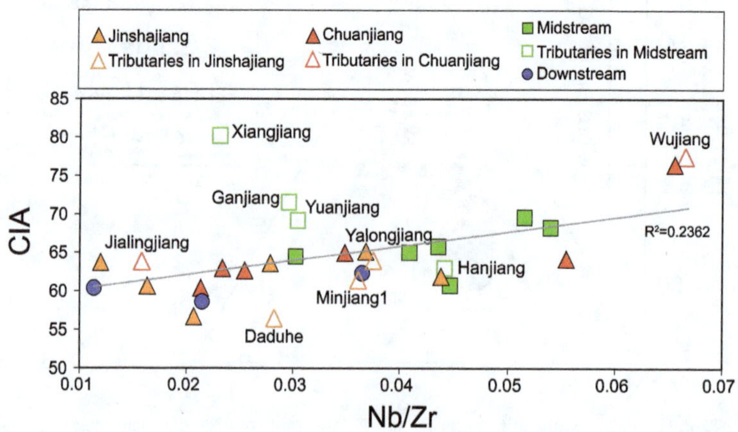

Fig. 10 Diagram showing the relationship between chemical weathering and trace element compositions, CIA vs. Nb/Zr (HFSE enrichment)

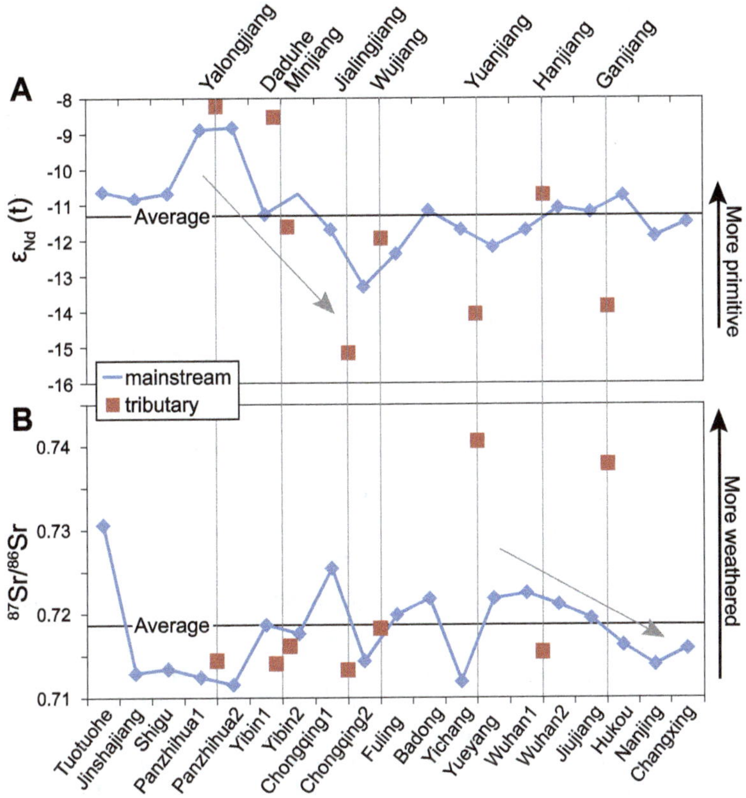

Fig. 11 Downstream variations of Sr-Nd isotopic composition of the mainstream shown as a continuous *blue solid line* and tributaries shown as *red squares*. **a** εNd(t); **b** $^{87}Sr/^{86}Sr$

Nd is especially concentrated in this mineral (e.g., Garçon et al. 2014). The Jinshajiang and Chuanjiang segments mainly incise Mesozoic and Cenozoic igneous rocks, as well as a large area of Emeishan basalts, which have high εNd values (e.g., Qi and Zhou 2008; Zhang and Wang 2003). The progressive change in εNd values from Panzhihua 1 to Chongqing 2 implies that the river is picking up a lot of new sediment along that stretch.

The mid-lower reaches mainly drain Quaternary fluvial and lacustrine sediments, together with intermediate-felsic igneous rocks and metamorphic units forming parts of the Yangtze Craton that represents much older crust originally generated at 2.5–2.8 Ga (e.g., He et al. 2013a; Iizuka et al. 2010). The lower εNd values in the downstream compared to the upstream are largely a function of this change in source bedrock and the fact that the river continues to gain sediment as it progresses downstream. Although there are basalts in the middle and lower Yangtze, they have a limited areal extent and do not affect river sediments as significantly as the Emeishan basalts (e.g., Yang et al. 2007). The moderate but steady increase in εNd values from Yueyang to Hukou implies that sediment like and including the Hanjiang must be adding sediment along that stretch and that the Yuanjiang and Ganjiang are not volumetrically significant suppliers. Sr isotope ratios vary substantially from upstream to downstream, which suggests more complex factors influencing their development compared to Nd isotopes, which is consistent with Sr being water mobile and Nd being resistant to change due to alteration. These factors could include grain size, chemical weathering, and seasonal variations, as well as the bedrock Sr isotope compositions (e.g., Goldstein and Jacobsen 1987; Ma and Liu 2001; Mao et al. 2011; Rai and Singh 2007; Tripathy et al. 2011). The $^{87}Sr/^{86}Sr$ values of the upper mainstream, excluding the very high value at Tuotuohe, are much lower than the midstream and downstream. Values from the tributaries also display the same patterns, with the Yalongjiang, Daduhe, Minjiang, Jialingjiang, and Wujiang having lower $^{87}Sr/^{86}Sr$ values, whereas the Yuanjiang and Ganjiang are much higher. This pattern reflects erosion of the low $^{87}Sr/^{86}Sr$ value Emeishan Basalts and younger igneous rocks in the upstream but higher $^{87}Sr/^{86}Sr$ values in old metamorphic rock sources and more radiogenic Sr isotope silicates in the mid- and lower reaches (Fig. 11).

As discussed above, Sr isotope ratios could vary with grain size. Here we plot $^{87}Sr/^{86}Sr$ values against the grain size, which is represented by SiO_2/Al_2O_3 (Fig. 12). It is clear that the $^{87}Sr/^{86}Sr$ values have poor positive correlation with the grain size, a little different from

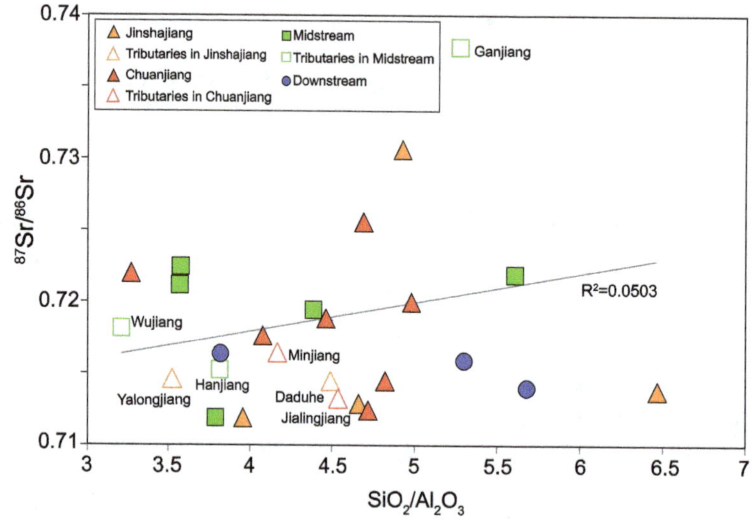

Fig. 12 Diagram showing the relationship between $^{87}Sr/^{86}Sr$ ratio and grain size

the conclusion made by some other researchers (e.g., Yang et al. 2007). This difference may be attributable to the finer grain size range we used in this study (<63 μm).

Chemical weathering is another factor in controlling the isotopic values in the modern river because unstable feldspar and epidote dominate the sediment Sr budget (e.g., Garçon et al. 2014). The $^{87}Sr/^{86}Sr$ values in the mid-lower reaches are higher than upstream, consistent with more weathered sediment in the mid-downstream of the Yangtze since alteration tends to drive $^{87}Sr/^{86}Sr$ to higher values (e.g., Derry and France-Lanord 1996). Like the CIA, the $^{87}Sr/^{86}Sr$ values imply decreasing alteration in the sediment from Yueyang to the river mouth, although some of this trend is driven by the provenance change implied by the Nd data between Yueyang and Hukou. From Fig. 11a, we can also deduce that the tributaries of the Daduhe, Minjiang, Jialingjiang, and Hanjiang contribute more sediment to the mainstream than the Yuangjiang and Ganjiang because the bulk composition of the river changes substantially downstream of each confluence towards the value of the tributary.

We summarize the Sr-Nd isotope compositions of different units in the Yangtze River drainage (Fig. 4). It is obvious that the Jinshajiang samples match bedrock compositions from the Songpan-Garze Terrane and that the samples in the Chuanjiang are consistent with strong erosion from the middle Yangtze Craton, while the samples from the mid-lower reaches fall between the two and could also reflect additional erosion from the eastern Yangtze Craton. The data are consistent with the Songpan-Garze Terrane, and the Eastern and Middle parts of the Yangtze Craton being the main sources of the Yangtze sediments. In contrast to the evidence from heavy minerals and clay mineralogy (e.g., He et al. 2013b; Yang et al. 2009), the isotope geochemistry data shows that the rocks exposed in the western Yangtze Craton, especially the Emeishan Basalts, are not the main sources of <63 μm sediment to the modern river. Not surprisingly, the river sediments have lower εNd values than the magmatic rocks from the Cathaysia Block, which mostly lies to the South within the Pearl River drainage and is not an important source. Moderate exposures of primitive magmatic rocks within the Yangtze Craton do not influence the bulk composition of the river to a significant degree. Our sediment samples have a relatively restricted range of εNd values but a wide range in $^{87}Sr/^{86}Sr$ values, which we consider to reflect the influence of chemical weathering in changing the original bedrock composition. We particularly note that the Ganjiang and Yuanjiang have very high $^{87}Sr/^{86}Sr$ values and are located in the middle-lower reaches of the river where chemical weathering might be strong as a result of slow erosion and transport rates in relatively flat topography and because of the warm and wet monsoonal climate, as indicated by their high CIA values.

The relationship between εNd values and REEs are shown in Fig. 13. The HREE (Nb/Zr) have a rough positive trend with εNd values, while LREE (La/Sm) and HFSE (Nb/Zr) have no association with εNd values. LREE and HFSE composition appears to be relatively unrelated to the average crustal age of the source rocks, whereas the degree of enrichment in HREEs increases as the average crustal age of the bedrock increases because εNd values are linked to the average age of original crustal generation and not to the last magmatic event. Yang et al. (2002) argued that the source rock is the primary control on the REE composition in river sediments but this ignores the effect of hydrodynamic sorting. We now refine this conclusion by showing that the HREEs are influenced by the source rock but that the LREEs

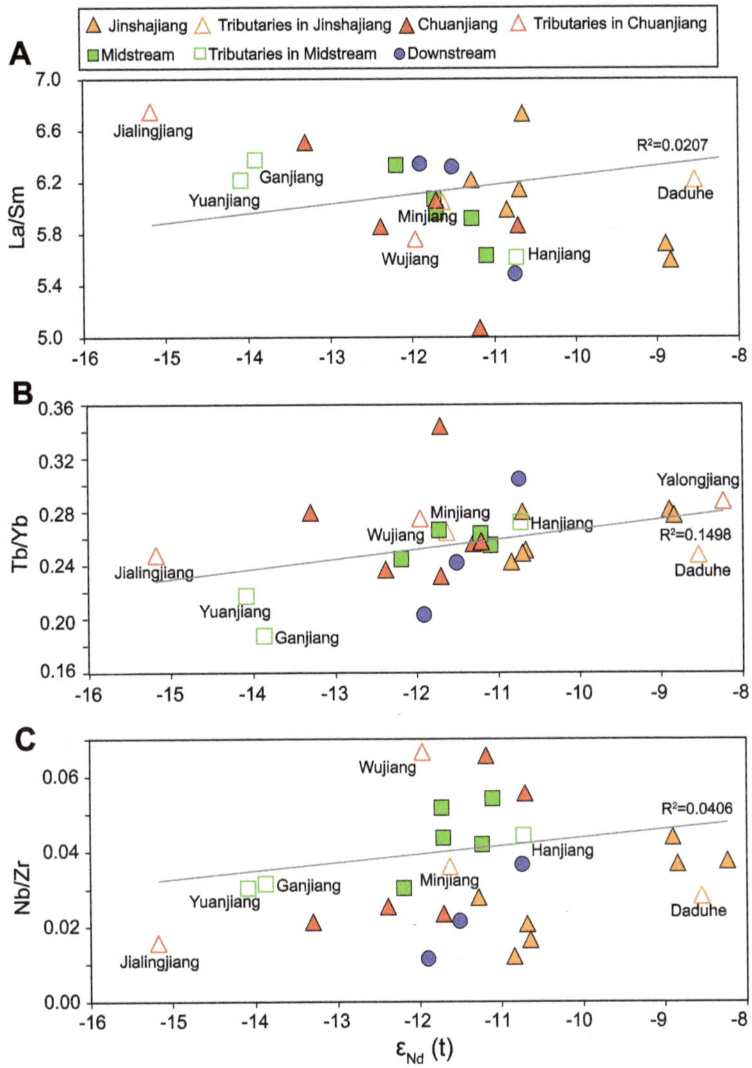

Fig. 13 εNd(t) compared with REEs and HFSEs. **a** La/Sm, **b** La/Yb, and **c** Nb/Zr

and HFSE are not affected. In addition, the fact that many of the tributaries have similar REE chemistries means that the close match between the Jinshajiang and the delta sediments cannot be used to argue that the upper reaches are the dominant source regions.

Yangtze Craton as typical continental crust

We now assess how typical the Yangtze River Basin is of continental crust at the planetary scale by comparing the trace element compositions of sediments in the river with those of average upper continental crust (UCC), as summarized by Rudnick and Fountain (1995). Figure 14a shows the trace elements normalized to the UCC from the mainstream and major tributaries. Both mainstream and tributaries define the same general patterns, with relative Rb and Sr depletion and particularly high Th, U, Zr, Hf, and Ti enrichment. We note that most of the elements lie above the average value of the UCC in both mainstream and tributaries and we suggest that this reflects the fact that the sediments we sample here are more abundant in heavy minerals and clay compared to the average of the UCC because these minerals carry many of the trace elements analyzed here (e.g., Garzanti et al. 2010). The depletions in Rb and Sr are inferred to be the product of intense chemical weathering, because these are soluble elements that have been systematically removed from the sediments during transport, which appears to exceed the provenance influence.

The major element compositions of the average samples normalized to UCC are displayed in Fig. 14b. Although the average sediment has similar SiO_2 and Al_2O_3 contents to the UCC, there is enrichment in Fe_2O_3, MnO, MgO, TiO_2, and P_2O_5, which is consistent with our interpretation that these Yangtze sediments are more concentrated

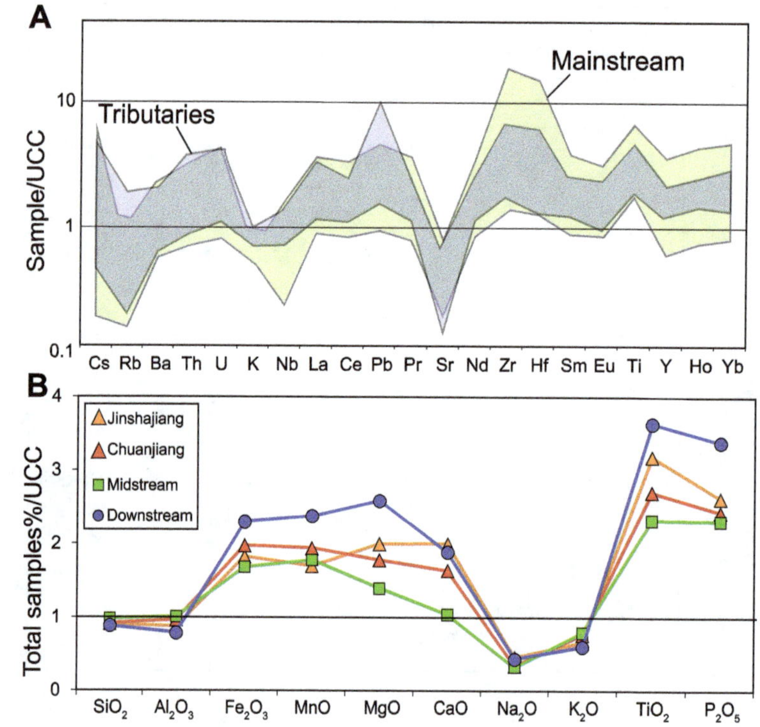

Fig. 14 Trace element normalized to the Upper Continental Crust (UCC) as defined by Rudnick and Fountain (1995). **a** Samples in the mainstream and tributaries; **b** total major element percent of sediments normalized to UCC

in heavy minerals than the average crust. We have no evidence that the Yangtze basin itself is substantially more enriched than might be expected for average crustal values. Depletions in Na_2O and K_2O concentrations are attributed to leaching by chemical weathering, similar to the Rb and Sr noted above. These elements are transported to the ocean in solution (e.g., Yuan et al. 2012). In this study, we only analyze sediment finer than 63 μm, and we inferred that the coarser material may be dominated by quartz, feldspar, and other minerals, which are relatively poor in many trace elements. Our work also did not consider the suspended sediment fraction but only the bedload, which may account for the significant differences between our average compositions and UCC (e.g., Garzanti et al. 2011). This fine-grained bedload sample can be useful for sediment provenance but is not completely representative of the crust being eroded within the basin at the present time.

Conclusions

In this study, we analyzed a series of samples from the mainstream of the Yangtze River, as well as its major tributaries, to see if there were any patterns in the major and trace element chemistry that could be used to constrain provenance and chemical weathering processes within this major river system.

No clear variations in major element chemistry were found between the source and the delta. Within the <63 μm grain size range, we see little linkage between grain size and total REE abundance, while the LREE, HREE, and HFSE show different relationship with the heavy minerals, chemical weathering, and provenance, respectively. Nd isotope data show that the major sources of sediment in the Yangtze appears to be within the lower and middle parts of the Yangtze Craton, as well as in the Songpan-Garze Terrane exposed in Eastern Tibet and Yunnan. There is no simple trend to more chemical weathering downstream in the main river according to the CIA proxy and K/Al values, but the major tributaries do show significant changes.

We conclude that the major and trace element chemistry of the sediments in the Yangtze River can provide important information about the source of the sediment and especially its chemical weathering state, which is crucial for interpreting the marine clastic record of the sedimentary basins in the East China Sea. However, caution needs to be used especially when applying REEs to provenance issues because grain size effects and the sorting of heavy minerals by hydrodynamic processes typically dominate the bulk composition of the sediment. In the East China Sea, this would be especially true because the Yangtze would not be the only source of clastic sediment. Sediment provenance can only be seriously addressed if grain size and sorting issues can be properly corrected for and by using proxies resistant to chemical weathering effects, such as Nd isotopes.

Abbreviations

CIA: Chemical Index of Alteration; HFSE: high-field strength element; HREE: heavy rare earth element; LREE: light rare earth element; REE: rare earth element; UCC: upper continental crust.

Competing interests

The authors declare that they have no competing interests.

Authors' contributions

ZHB conceived and designed the study. HMY carried out the experimental study, obtained analytical data, and wrote the manuscript with PC. RT, WWH, and LC collaborated in the sampling collection. All authors read and approved the final manuscript.

Acknowledgements

This study was supported by the "Strategic Priority Research Program" of the Chinese Academy of Sciences (Grant No. XDB03020300), the China Geological Survey: Continental Shelf Drilling Program (Project No. CSDP-2012-05), the Natural Science Foundation of China (Project No. 40930107 and 41111140016) the Natural Science Foundation of Jiangsu Province (Project No. BK20150974) and the United Nations Educational, Scientific and Cultural Organization (Project No. IGCP-581). We thank the State Key Laboratory for Mineral Deposits Research, Nanjing University and the McCord Chair in Petroleum Geology at Louisiana State University, USA, and all researchers who helped us with this article. The manuscript benefited from earlier comments by Simon Wilde and an anonymous reviewer.

Author details

[1]School of Geography Science, Nanjing Normal University, Nanjing 210046, China. [2]Department of Geology and Geophysics, Louisiana State University, Baton Rouge, LA 70803, USA. [3]Graduate School of Science, The University of Tokyo, 7-3-1 Hongo, Bunkyo-ku, Tokyo 113-0033, Japan. [4]School of Earth Sciences and Engineering, Nanjing University, Nanjing 210008, China.

References

Berner RA, Berner EK (1997) Silicate weathering and climate. Tectonic uplift and climate change. WF Ruddiman. Springer, New York, pp 353–365

Bhuiyan MAH, Rahman MJJ, Dampare SB, Suzuki S (2011) Provenance, tectonics and source weathering of modern fluvial sediments of the Brahmaputra–Jamuna River, Bangladesh: inference from geochemistry. J Geochem Explor 111:113–137

Cai H, Zhang H, Xu W (2009) U-Pb Zircon ages, geochemical and Sr-Nd-Hf isotopic compositions of granitoids in Western Songpan-Garze Fold Belt: petrogenesis and implication for tectonic evolution. J Earth Sci 20(4):681–698

Chen J, Wang F, Xia X, Zhang L (2002) Major element chemistry of the Changjiang (Yangtze River). Chem Geol 187:231–255

Chen Z, Li JF, Shen HT, Wang ZH (2001a) Yangtze River of China: historical analysis of discharge variability and sediment flux. Geomorphology 41(2-3):77–91

Chen Z, Lu LZ, Gupta A (2001b) The Yangtze River: an introduction. Geomorphology 41(2-3):73–75

Clark MK, Schoenbohm LM, Royden LH, Whipple KX, Burchfiel BC, Zhang X, Tang W, Wang E, Chen L (2004) Surface uplift, tectonics, and erosion of eastern Tibet from large-scale drainage patterns. Tectonics 23:TC1006

Clift PD, Hodges K, Heslop D, Hannigan R, Hoang LV, Calves G (2008) Greater Himalayan exhumation triggered by Early Miocene monsoon intensification. Nat Geosci 1:875–880

Condie KC, Phillip DNJ, Conway CM (1992) Geochemical and detrital mode evidence for two sources of Early Proterozoic sedimentary rocks from Tonto Basin Supergroup, central Arizona. Sediment Geol 77:51–76

DePaolo DJ, Wasserburg GJ (1976) Nd isotopic variations and petrogenetic models. Geophys Res Lett 3(5):249–252

Derry LA, France-Lanord C (1996) Neogene Himalayan weathering history and river $^{87}Sr/^{86}Sr$: impact on the marine Sr record. Earth Planet Sci Lett 142:59–74

Ding T, Wan D, Wang C, Zhang F (2004) Silicon isotope compositions of dissolved silicon and suspended matter in the Yangtze River, China. Geochim Cosmochim Acta 68(2):205–216

Garçon M, Chauvel C, France-Lanord C, Limonta M, Garzanti E (2014) Which minerals control the Nd–Hf–Sr–Pb isotopic compositions of river sediments? Chem Geol 364:42–55

Garzanti E, Andó S, France-Lanord C, Censi P, Vignola P, Galy V, Lupker M (2011) Mineralogical and chemical variability of fluvial sediments 2. Suspended-load silt (Ganga–Brahmaputra, Bangladesh). Earth Planet Sci Lett 302:107–120

Garzanti E, Andò S, France-Lanordb C, Vezzoli G, Censi P, Galy V, Najman Y (2010) Mineralogical and chemical variability of fluvial sediments 1. Bedload sand (Ganga–Brahmaputra, Bangladesh). Earth Planet Sci Lett 299:368–381

Garzanti E, Andò S, Vezzoli G (2008) Settling equivalence of detrital minerals and grain-size dependence of sediment composition. Earth Planet Sci Lett 273:138–151

Goldstein SJ, Jacobsen SB (1987) The Nd and Sr isotopic systematics of river-water dissolved material: Implications for the sources of Nd and Sr in seawater. Chem Geol 66(3-4):245–272

He M, Zheng H, Bookhagen B, Clift PD (2014) Controls on erosion intensity in the Yangtze River basin tracked by U–Pb detrital zircon dating. Earth Sci Rev 136:121–140

He MY, Zheng H, Clift PD (2013a) Zircon U-Pb Geochronology and Hf isotope data from the Yangtze River Sands: implications for major magmatic events and crustal evolution in central China. Chem Geol 360–361:186–203

He MY, Zheng HB, Huang XT, Jia JT, Li L (2013b) Yangtze River sediments from source to sink traced with clay mineralogy. J Asian Earth Sci 69:60–69

Herron MM (1988) Geochemical classification of terrigenous sands and shales from core or log data. J Sediment Petrol 58:820–829

Hu D, Böning P, Köhler CM, Hillier S, Pressling N, Wan S, Brumsack H-J, Clift PD (2012) Deep sea records of the continental weathering and erosion response to East Asian monsoon intensification since 14 ka in the South China Sea. Chem Geol 326–327:1–18

Iizuka T, Komiya T, Rino S, Maruyama S, Hirata T (2010) Detrital zircon evidence for Hf isotopic evolution of granitoid crust and continental growth. Geochim Cosmochim Acta 74(8):2450–2472

Li JY (2003) A study on the chemical weathering, mechanical denudation correlative with river water and sediment geochemistry and CO2 consumption budget and controlling factors in the major drainage basins of China. Ocean University of China

Liu Z, Colin C, Huang W, Le KP, Tong S, Chen Z, Trentesaux A (2007) Climatic and tectonic controls on weathering in south China and Indochina Peninsula: Clay mineralogical and geochemical investigations from the Pearl, Red, and Mekong drainage basins. Geochem Geophys Geosyst 8:Q05005

Lupker M, France-Lanord C, Galy V, Lave J, Gaillardet J, Gajured AP, Guilmette C, Rahman M, Singh SK, Sinha R (2012) Predominant floodplain over mountain weathering of Himalayan sediments (Ganga basin). Geochim Cosmochim Acta 84:410–432

Ma Y, Liu C (2001) Sr isotope evolution during chemical weathering of granites —— impact of relative weathering rates of minerals. Sci China Earth Sci 44(8):727–734

Mao C, Chen J, Yuan X, Yang Z, Ji J (2011) Seasonal variations in the Sr-Nd isotopic compositions of suspended particulate matter in the lower Changjiang River: Provenance and erosion constraints. Chin Sci Bull 56:2371–2378

Meng X, Liu Y, Shi X, Du D (2008) Nd and Sr isotopic compositions of sediments from the Yellow and Yangtze Rivers: implications for partitioning tectonic terranes and crust weathering of the Central and Southeast China. Front Earth Sci Chin 2(4):418–426

Nesbitt HW, Markovics G, Price RC (1980) Chemical processes affecting alkalis and alkaline earths during continental weathering. Geochim Cosmochim Acta 44:1659–1666

Nesbitt HW, Young GM (1982) Early Proterozoic climates and plate motions inferred from major element chemistry of lutites. Nature 299(5885):715–717

Ni Y, Hughes JM, Mariano AN (1995) Crystal chemistry of the monazite and xenotime structures. Am Mineral 80:21–26

Qi L, Zhou M (2008) Platinum-group elemental and Sr–Nd–Os isotopic geochemistry of Permian Emeishan flood basalts in Guizhou Province, SW China. Chem Geol 248(1-2):83–103

Rai SK, Singh SK (2007) Temporal variation in Sr and 87Sr/86Sr of the Brahmaputra: implications for annual fluxes and tracking flash floods through chemical and isotope composition. Geochem Geophys Geosyst 8:Q08008

Raymo ME, Ruddiman WF (1992) Tectonic forcing of Late Cenozoic climate. Nature 359(6391):117–122
Rollinson HR (2013) Using geochemical data: evaluation, presentation, interpretation. Routledge, London and New York
Rudnick RL, Fountain DM (1995) Nature and composition of the continental crust: a lower crustal perspective. Rev Geophys 33(3):267–309
Ryan WBF, Carbotte SM, Coplan JO, O'Hara S, Melkonian A, Arko R, Weissel RA, Ferrini V, Goodwillie A, Nitsche F, Bonczkowski J, Zemsky R (2009) Global multi-resolution topography synthesis. Geochem Geophys Geosyst 10:Q03014
Shao J, Yang S, Li C (2012) Chemical indices (CIA and WIP) as proxies for integrated chemical weathering in China: inferences from analysis of fluvial sediments. Sediment Geol 265–266:110–120
Singh P (2009) Major, trace and REE geochemistry of the Ganga River sediments: influence of provenance and sedimentary processes. Chem Geol 266:242–255
Singh P (2010) Geochemistry and provenance of stream sediments of the Ganga River and its major tributaries in the Himalayan region, India. Chem Geol 269:220–236
Sun S-s, McDonough WF (1989) Chemical and isotopic systematics of oceanic basalts: implications for mantle composition and processes. Geol Soc London Spec Publ 42:313–345
Taylor SR, McLennan SM (1985) The continental crust: its composition and evolution. Blackwell, Oxford, U. K
Tripathy GR, Singh SK, Bhushan R, Ramaswamy V (2011) Sr–Nd isotope composition of the Bay of Bengal sediments: Impact of climate on erosion in the Himalaya. Geochem J 45:175–186
Wan S, Clift PD, Li A, Li T, Yin X (2010) Geochemical records in the South China Sea: implications for East Asian summer monsoon evolution over the last 20 Ma. Monsoon Evolution and Tectonics–Climate Linkage in Asia. PD Clift, R Tada, H Zheng. London Geol Soc 342:245–263
Wan S, Kürschner WM, Clift PD, Li A, Li T (2009) Extreme weathering / erosion during the Miocene Climatic Optimum: evidence from sediment record in the South China Sea. Geophys Res Lett 36:L19706
West AJ, Galy A, Bickle MJ (2005) Tectonic and climatic controls on silicate weathering. Earth Planet Sci Lett 235:211–228
White AF, Blum AE (1995) Effects of climate on chemical weathering in watersheds. Geochim Cosmochim Acta 59(9):1729–1747
Wu W, Xu S, Lu H, Yang J, Yin H, Liu W (2011) Mineralogy, major and trace element geochemistry of riverbed sediments in the headwaters of the Yangtze, Tongtian River and Jinsha River. J Asian Earth Sci 40:611–621
Yang S, Hoi-Soo J, Li C, Dong-il L (2004a) Major element geochemistry of sediments from Chinese and Korean rivers. Geochimica 33(1):99–105
Yang S, Jiang S, Ling H, Xia X, Sun M, Wang D (2007) Sr-Nd isotopic compositions of the Changjiang sediments: Implications for tracing sediment sources. Sci China Earth Sci 50(10):1556–1565
Yang S, Jung H-S, Li C (2004b) Two unique weathering regimes in the Changjiang and Huanghe drainage basins: geochemical evidence from river sediments. Sediment Geol 164(1-2):19–34
Yang S, Jung H, Choi M, Li C (2002) The rare earth element compositions of the Changjiang (Yangtze) and Huanghe (Yellow) river sediments. Earth Planet Sci Lett 201:407–419
Yang S, Li C, Wang Z, Wang X, Shu J (2013) Heterogeneity of geochemical compositions of the Changjiang River sediments and provenance indication. Quateranry Sci 33(4):645–655
Yang S, Wang Z (2011) Rare earth element compositions of the sediments from the major tributaries and the mainstream of the Changjiang River. Bull Miner Petrol Geochem 30(1):31–39
Yang S, Wang Z, Guo Y, Li C, Cai J (2009) Heavy mineral compositions of the Changjiang (Yangtze River) sediments and their provenance-tracing implication. J Asian Earth Sci 35(1):56–65
Yuan X, Li J, Mao C, Ji J, Yang Z (2012) Geochemistry of water and suspended particulate in the lower Yangtze River: implications for geographic and anthropogenic effects. Int J Geosci 3:81–92
Zhang CS, Wang LJ (2001) Multi-element geochemistry of sediments from the Pearl River system, China. Appl Geochem 16:1251–1259
Zhang Z, Wang F (2003) Sr, Nd and Pb isotopic characteristics of Emeishan Basalt Province and discussion on their source region. Earth Sci J China Univ Geosci 28(4):431–439
Zheng H, Clift PD, Wang P, Tada R, Jia J, He M, Jourdanf F (2013) Pre-Miocene birth of the Yangtze River. Proc Natl Acad Sci 110(19):7556–7561

15

Silicate solution, cation properties, and mass transfer by aqueous fluid in the Earth's interior

Bjorn Mysen

Abstract

Aqueous fluids in the Earth's interior are multicomponent systems with silicate solubility and solution mechanisms strongly dependent on other dissolved components. Here, solution mechanisms that describe the interaction between dissolved silicate and other solutes were determined experimentally to 825 °C and above 1 GPa with in situ vibrational spectroscopy of aqueous fluid while these were at high temperature and pressure. The silicate content in Na-bearing, silicate-saturated aqueous fluid exceeds that in pure SiO_2 at high temperature and pressure. Silicate species were of Q^0 (isolated SiO_4 tetrahedra) and Q^1 (dimers, Si_2O_7) type. The temperature dependence of its equilibrium constant, $K = X_{Q1}/(X_{Q0})^2$, yields enthalpies of 22 ± 12 and 51 ± 17 kJ/mol for the SiO_2–H_2O and Na-bearing fluids. In contrast, in Ca-bearing fluids, the solubility is more than an order of magnitude lower, and only Q^0 species are present. The present data together with other published experimental information lead to the conclusion that the silicate solubility in aqueous fluids in equilibrium with mafic rocks such as amphibolite and peridotite is an order of magnitude lower than the solubility in fluids in equilibrium with felsic rocks such as andesite and rhyolite compositions (felsic gneiss) under similar temperature and pressure conditions. The silicate speciation also is more polymerized in the felsic systems. This difference is also why second critical end-points in the Earth are at lower temperature and pressure in felsic compared with mafic systems. Alkali-rich fluids formed by dehydration of felsic rocks also show enhanced high field strength element (HFSE) solubility because alkalis in such solution form oxy complexes with the HFSE cations. Fluids formed by dehydration of felsic rocks in the Earth's interior are, therefore, more efficient transport agents of silicate materials than fluids formed by dehydration of mafic and ultramafic rocks, whether for major, minor, or trace elements.

Keywords: Aqueous fluid, Solubility, Structure, Spectroscopy, Silicate

Introduction

The solubility and solution mechanism(s) of silicate components in aqueous fluids in the Earth's interior are central to our understanding of the roles of aqueous fluids in material transport within the Earth (Fockenberg et al. 2006; Newton and Manning 2009; Burchard et al. 2011). This is particularly so in and near subduction zones where water-rich fluids released from dehydrating slab materials can transport chemical components into source regions of melting and, at the same time, deplete dehydrating slab materials of fluid-soluble components (Mibe et al. 2008; Kawamoto et al. 2012). In this environment, hydrated rocks ranging from amphibolite and peridotite to felsic gneiss compositions are subjected to dehydration and release of silicate-saturated water-rich fluids.

The abundance of components known to affect silicate solubility in aqueous fluid differs significantly among those rock types in the Earth's interior. It is necessary, therefore, to determine how individual compositional variables affect the silicate solubility and structure of dissolved silicate in aqueous fluids at relevant high-temperature and high-pressure conditions.

Composition and structure of silicate-saturated aqueous fluids at high temperature and pressure have been determined experimentally for pure SiO_2–H_2O systems as well as for some more complex silicate and aluminosilicate compositions (Newton and Manning 2002; Kessel et al. 2005; Mibe et al. 2008, 2011; Mysen 2010). From those

Correspondence: bmysen@carnegiescience.edu
Geophysical Laboratory, Carnegie Institution of Washington, Washington, DC, USA

experiments, it is evident that silicate solubility in aqueous fluids depends on the presence of other components in addition to SiO_2. In the $MgO-SiO_2-H_2O$ system, for example, the solubility in the aqueous fluid decreases with increasing Mg/Si abundance ratio (Zhang and Frantz 2000). Addition of aluminosilicate components to SiO_2 also enhances the solubility in fluid (Wilke et al. 2012). However, quantitative characterization of how individual components affect silicate solubility and solution mechanisms of other components in aqueous fluids is not well developed.

Determination of fluid composition and structure at temperature and pressure conditions of the deep crust, upper mantle, and beyond is challenging because silicate-saturated fluids commonly do not retain in solution the chemical components and the fluid structure that existed at the high temperature and pressure after temperature quenching from experimental conditions to ambient temperature and pressure. Various methods have been proposed with which to overcome these obstacles. For example, Kessel et al. (2004) described a method whereby fluids were soaked into layers of pore space among diamond particles during experiments. Analysis of these diamond aggregates after termination of an experiment provided the fluid data. Others rely on accurate weighing of crystals from the experimental charges before and after an experiment at high temperature and pressure during which crystalline material equilibrated with fluid (Manning and Boettcher 1994). The weight difference is the weight of the solute.

An alternative, which does not rely on quenching to ambient conditions before analysis, is to conduct structural characterization of fluids in the equilibrium with one or more condensed phases while at the temperatures and pressures of interest. This can be accomplished with the sample contained in an externally heated hydrothermal diamond anvil cell (Bassett et al. 1993). Here, this experimental protocol was employed to assess the influence of alkalis and alkaline earth cations on SiO_2 solubility in aqueous solutions to temperatures and pressures corresponding to the deep crust and upper mantle.

Methods/Experimental

Three different starting materials were used. Those were crystalline quartz + H_2O, crystalline quartz + 0.7 m NaOH aqueous solution, and an $SiO_2 + CaO$ mixture + H_2O (SiO_2:CaO = 9:1 by mol). By using NaOH as the Na^+-bearing fluid component, it cannot be ruled out that an increased pH compared with pure H_2O could also affect the silicate solubility in the fluid.

The $CaO-SiO_2$ material was prepared by melting at ambient pressure at 1725 °C followed by crystallization at 1500 °C. At 1725 °C, this system comprises immiscible silicate liquids (Osborn and Muan 1960). In order to form the crystalline SiO_2 component desired to equilibrate with fluid, the crystallization step at 1500 °C was performed. This mixture, consisting of crystalline tridymite + glass (Ca-rich) was used together with H_2O as a starting material. During the experiments, tridymite converted to quartz and glass to wollastonite, as discussed further below.

The experiments were carried out in an externally heated, hydrothermal diamond anvil cell (HDAC) of the design originally described by Bassett et al. (1993). In the HDAC, the sample was heated with external heaters. These heaters were comprised of Mo wire wrapped around the SiC diamond seats of upper and lower diamonds that were held in place with alumina high-temperature cement. Iridium gaskets (125 μm thick) with a 500-μm sample hole were employed for sample containment. During an experiment, the gaskets shrunk to about 80 μm thickness and the hole to ~ 400 μm diameter.

MicroRaman spectroscopy and optical microscopy were used to characterize the samples while these were in the HDAC at the pressure/temperature of interest. The Raman system is a JASCO model NRS-3100 confocal microRaman spectrometer with a single monochromator and holographic gratings (1200 and 2400 grooves/mm). A long working distance (25 mm), × 50 magnification, 0.42 N.A. objective lens was used in this design, which provided a better than 5 μm spatial resolution for sample examination and spectroscopic characterization. The sampling depth was about 30 μm (Mysen 2015a). The signals were recorded with an Andor™ Model DV401-F1 1024 × 28 pixel Peltier-cooled CCD (charge-coupled detector). Energy resolution was ± 3 cm^{-1} with 1200 grooves/mm gratings, whereas with 2400 grooves/mm, spectral reproducibility was about 1 cm^{-1}.

Temperatures in the diamond cell were monitored with two K-type thermocouples that touched the diamonds at a ~ 1-mm distance from the sample chamber. A feedback mechanism was utilized for temperature control. Temperature precision was ± 1 °C, whereas the uncertainty, determined against the melting point of NaCl, was 2–3 °C. The gasket hole in the Ir gasket material deformed some during the experiments, which were not, therefore, strictly isochoric. Pressure could not, therefore, be calculated from the PVT data of pure H_2O. Furthermore, silicate components dissolved in the fluid also affect the equation of state of H_2O (Mysen 2010). Instead, the pressure was measured with the one-phonon Raman line from a chip of synthetic carbon-13 diamond placed inside the sample chamber. From this information, the pressure was determined with the calibration of Mysen and Yamashita (2010). An internal standard of Ne emission lines was employed to optimize the energy resolution. With this standard together with 2400 grooves/mm gratings, a frequency uncertainty of ± 0.1 cm^{-1} was obtained. This uncertainty translates to a pressure precision of ± 40 MPa. However, the overall pressure uncertainty is approximately ± 110 MPa when taking into considerations the uncertainties in the pressure calibration by Mysen and Yamashita (2010).

Several series of measurements at different temperatures and pressures were carried out with the three different starting materials such that for each of the three starting compositions, the sample was first taken to the highest planned temperature and pressure and left there for an hour before spectroscopic measurements. The sample was then cooled at 1 °C/s in 50–75 °C intervals to the next temperature and pressure condition where the spectra were recorded after about 60-minute dwell time. With this sample configuration, the 60-minute dwell time is sufficient to reach equilibrium (Mysen 2015b).

Results

Quartz and aqueous fluid were present in all experiments (Table 1). Additionally, in the system SiO_2–CaO–H_2O, wollastonite coexists with quartz and fluid (Fig. 1). The quartz and wollastonite, identified from their Raman spectra, grew from the starting materials during the initial hour at the highest temperature and pressure. The high-temperature spectra typically are broader and less well resolved compared with the spectra recorded at ambient temperature and pressure (Fig. 2).

The spectra of silicates and fluid at high temperature and pressure show Raman signals in a high-frequency region centered between 3500 and 3600 cm^{-1} and a low-frequency region between 300 and 1200 cm^{-1}. The high-frequency region comprises that of the fundamental O–H vibrations in OH groups, whether in molecular H_2O or as OH groups that form bonding with metal cations (Ratcliffe and Irish 1982; Walrafen et al. 1986). The low-frequency region, between about 300 and 1200 cm^{-1}, includes the Raman bands assigned to vibrations in Si–O, Al–O, and metal oxide components (Brawer and White 1975; McMillan et al. 1992).

The high-frequency region comprises a single slightly asymmetric band. The asymmetry is less pronounced in the spectra from fluid in SiO_2–CaO–H_2O than for the other two compositions (Fig. 3). The low-frequency region characteristically has a strong and sharp band between 770 and 800 cm^{-1} (Fig. 4). In the SiO_2–CaO–H_2O fluid spectra, this is the only Raman band in this frequency region (Fig. 4c), whereas for SiO_2–H_2O and SiO_2–NaOH–H_2O fluids, there is a Raman intensity on both the low- and high-frequency side of 770–800 cm^{-1} peak (Fig. 4a, b). These latter intensities are strongest in the spectra of SiO_2–NaOH–H_2O fluids (Fig. 4b). The intensity of all bands grows with the increasing temperature and pressure.

The 770–800 cm^{-1} band is assigned to Si–O^- stretch vibrations in isolated SiO_4 tetrahedra (Q^0 species) with one or more of the nonbridging oxygens forming bonds with H^+ (Zotov and Keppler 2002; Mibe et al. 2008). This is the same vibration as that gives rise to a ~ 850 cm^{-1} Raman band in anhydrous silicate melts and glasses (see review of Raman band assignments in such systems by Mysen and Richet 2005). The lower frequency of the Si–O^- stretch vibration in hydrous melts is because of the much smaller radius of H^+ bonded to nonbridging oxygens in the structure compared with alkali metals and alkaline earths. Raman intensity at a higher frequency, as observed in the spectra of SiO_2–H_2O and SiO_2–NaOH–H_2O fluid, is assigned to Si–O^- vibrations in silicate species more polymerized than isolated SiO_4 groups. Most likely, these are species with one bridging oxygen (Q^1 species) (Mysen et al. 2013). Bending vibrations of such a bridge give rise to the intensity centered near 600 cm^{-1}.

It follows, therefore, that for fluids in equilibrium with quartz in the system SiO_2–NaOH–H_2O, the silicate polymerization with Q^0 and Q^1 species is similar

Table 1 Run data

SiO_2–H_2O			SiO_2–NaOH–H_2O			SiO_2–CaO–H_2O		
Temp (°C)[a]	Press (MPa)[b]	Phases	Temp (°C)	Press (MPa)	Phases	Temp (°C)	Press (MPa)	Phases
25	0.1	Qtz + fluid	25	0.1	Qtz + fluid	25	0.1	Qtz + Wo + fluid
550	391	Qtz + fluid	550	583	Qtz + fluid	575	396	Qtz + Wo + fluid
625	542	Qtz + fluid	625	638	Qtz + fluid	625	468	Qtz + Wo + fluid
700	696	Qtz + fluid	700	851	Qtz + fluid	675	498	Qtz + Wo + fluid
775	956	Qtz + fluid	775	1012	Qtz + fluid	725	709	Qtz + Wo + fluid
800	725	Qtz + fluid	800	1134	Qtz + fluid	775	952	Qtz + Wo + fluid
						825	933	Qtz + Wo + fluid

[a]Temperature uncertainty, 2–3 °C
[b]Pressure uncertainty, ± 110 MPa

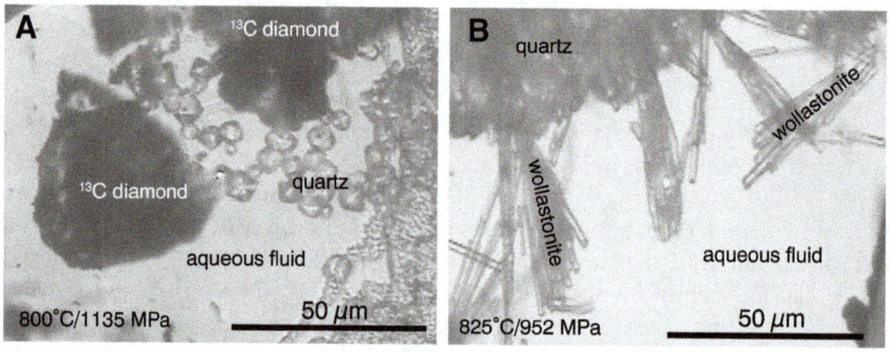

Fig. 1 Examples of microphotographs of samples in the hydrothermal diamond anvil cell at high temperature and pressure. **a** View of the sample from SiO_2–NaOH–H_2O at 800 °C and 1135 MPa. **b** View of the sample from SiO_2–CaO–H_2O at 825 °C and 925 MPa

to that in SiO_2–H_2O fluid spectra at higher pressures (Mysen et al. 2013). It also resembles the aqueous fluids in the equilibrium with alkali aluminosilicate melt under similar temperature and pressure conditions (Mysen 2010).

Discussion

The Raman spectra indicate that the silicate concentration and speciation in the fluid at given temperature and pressure is dependent on the nature of the crystalline materials in the equilibrium with the fluid.

Solubility

Silicate solubility in the aqueous fluids might be estimated by combining silica solubility in the fluid in the system SiO_2–H_2O (Manning 2004) with the intensity of relevant Raman bands in order to generate a calibration curve from the intensity of Raman bands in spectra of

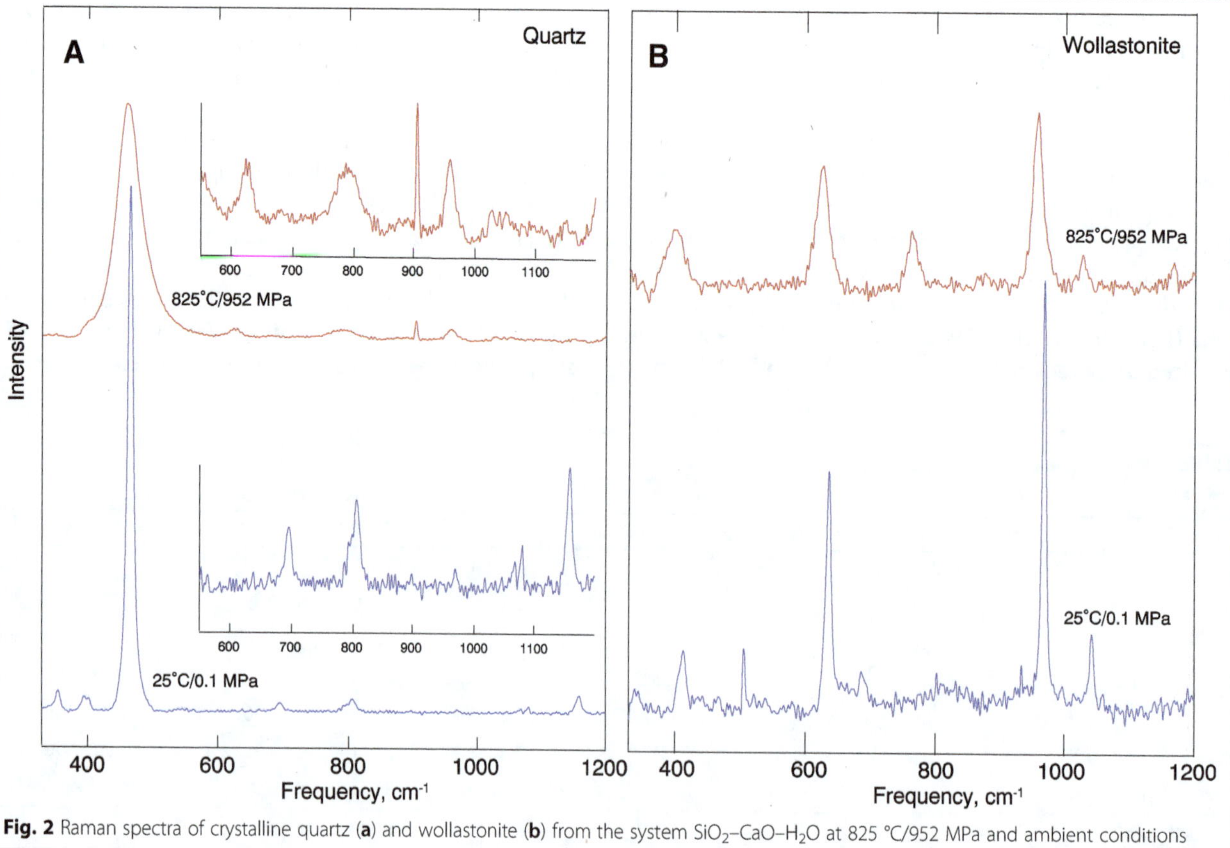

Fig. 2 Raman spectra of crystalline quartz (**a**) and wollastonite (**b**) from the system SiO_2–CaO–H_2O at 825 °C/952 MPa and ambient conditions

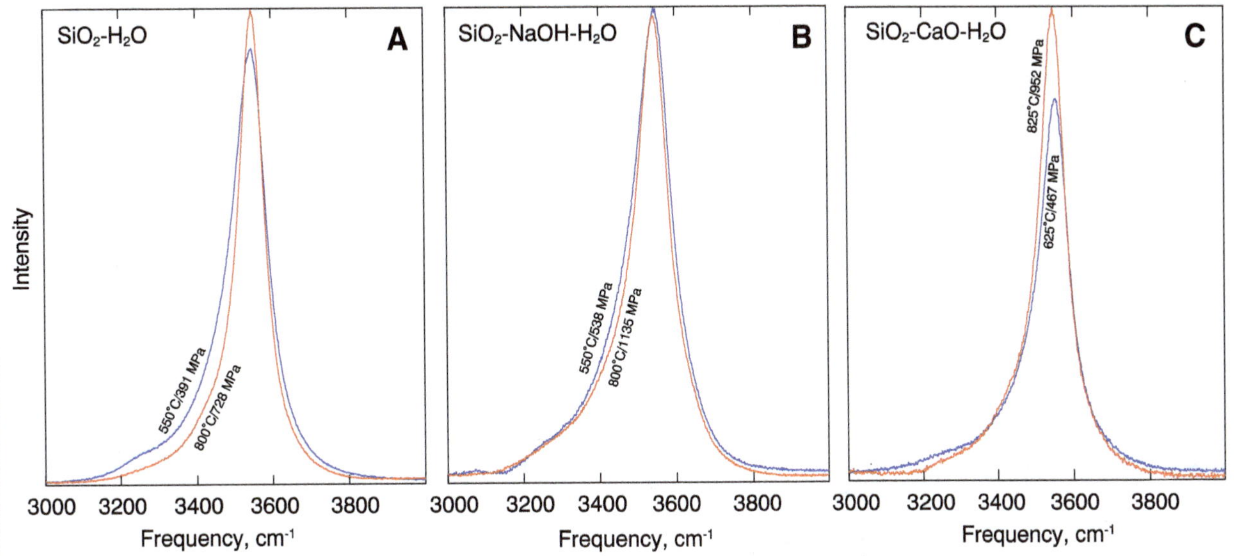

Fig. 3 Examples of high-frequency portion of Raman spectra in the systems $SiO_2–H_2O$ (**a**), $SiO_2–NaOH–H_2O$ (**b**), and $SiO_2–CaO–H_2O$ (**c**) at pressure and temperature conditions as indicated on the individual spectra

fluid. This calibration curve then may be used to compute an approximate solubility of SiO_2 in the fluids from the other two series of experiments ($SiO_2–NaOH–H_2O$ and $SiO_2–CaO–H_2O$).

The silicate contents of fluid in equilibrium with quartz in the $SiO_2–H_2O$ system were calculated with the algorithm of Manning (1994) in which the temperature and the density of pure H_2O are variables. For the density of the fluid, the calculations by Withers, using the formulism of Pitzer and Sterner (1994), were employed [URL: https://www.esci.umn.edu/people/researchers/withe012/fugacity.htm]. Those SiO_2 solubilities (Table 2) were then used to create a calibration curve based on the 770 + 850 cm^{-1} integrated Raman intensities from the fluid spectra (see Fig. 4). The data points defining the solubility relations were fitted to the function (Fig. 5):

$$\log m_{SiO2}(\text{mol/kg}) = 1.6–2.0{\bullet}\text{Int}^{-0.32}(\text{cts/s}), \qquad (1)$$

where "Int" is the integrated intensity of the 770 + 850 cm^{-1} Raman bands in counts per second, and m_{SiO2} is silica molality.

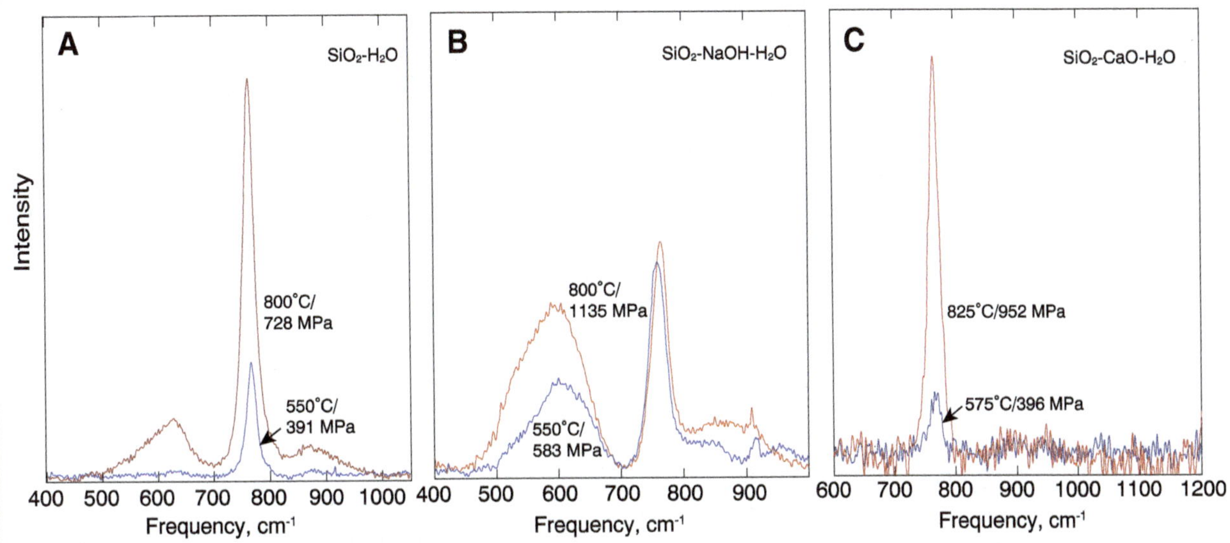

Fig. 4 Examples of low-frequency portion of Raman spectra in the systems $SiO_2–H_2O$ (**a**), $SiO_2–NaOH–H_2O$ (**b**), and $SiO_2–CaO–H_2O$ (**c**) at pressure and temperature conditions as indicated on individual spectra

Table 2 Solubility data

Temp (°C)	Press (MPa)	Density (g/cm^3)	f_{H2O} (MPa)	SiO_2 solubility, m_{SiO2} (kg/mol)[a]
SiO_2–H_2O				
800	728	0.794	852	0.0304
775	956	0.876	1440	0.0273
700	696	0.831	668	0.0082
625	542	0.813	364	0.0022
550	391	0.789	178	0.0004
SiO_2–NaOH–H_2O				
Temp (°C)	Press (MPa)	Density (g/cm^3)	f_{H2O} (MPa)	SiO_2 solubility, m_{SiO2} (kg/mol)[b]
800	1135	0.911	1978	0.0238
775	1012	0.891	1642	0.0448
700	851	0.88	999	0.0288
550	583	0.871	328	0.0092
SiO_2–CaO–H_2O				
Temp (°C)	Press (MPa)	Density (g/cm^3)	f_{H2O} (MPa)	SiO_2 solubility, m_{SiO2} (kg/mol)[b]
575	396	0.775	198	4.86525E–08
625	468	0.779	291	1.43231E–07
675	498	0.764	362	1.38469E–05
725	709	0.824	723	9.61047E–05
775	953	0.875	1430	0.000369701
825	933	0.849	1444	0.000506355

[a]Calculations using the formalism from Manning (1994)
[b]See text for calculations

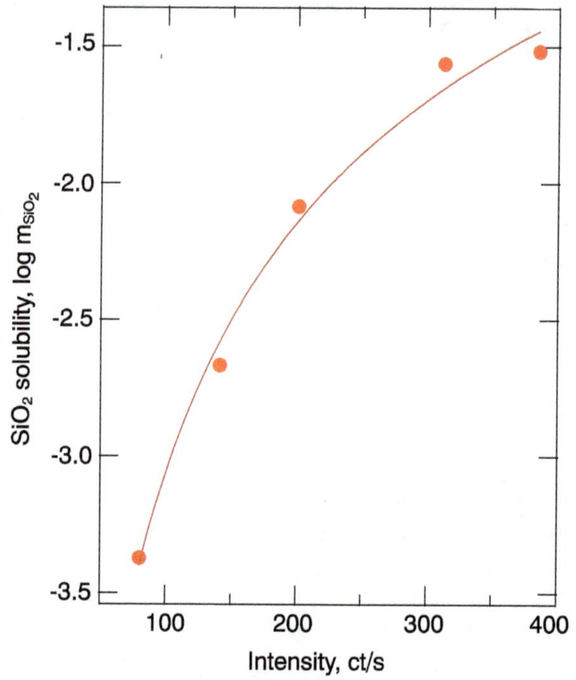

Fig. 5 Calibration curve for SiO_2 solubility as a function of 770 + 850 cm^{-1} integrated band intensity from the system –H_2O with silica solubility calculated with the algorithm of Manning (1994). The fitted curve is log m_{SiO2} (kg/mol) = 1.6–2.0 × Int$^{-0.32}$, where "Int" denotes the integrated Raman intensity and m_{SiO2} is silica molality of aqueous fluid

The SiO_2 solubility in the SiO_2–NaOH–H_2O and SiO_2–CaO–H_2O fluids was then derived from the calibration curve in Fig. 6 and the integrated Raman intensities at 770 + 850 cm^{-1} from the SiO_2–NaOH–H_2O and CaO–SiO_2–H_2O fluid spectra. In doing this, it was assumed that the different cations, H^+, Na^+, and Ca^{2+}, in the aqueous solutions had no effect on the Raman intensities and that possible temperature and pressure effects in the pressure range under study on Raman intensity can be ignored.

With the above caveats in mind, the SiO_2 solubility in SiO_2–CaO–H_2O fluids from the calibration curve in Fig. 6 is an order of magnitude, or more, lower than in the SiO_2–NaOH–H_2O and SiO_2–H_2O fluids in the f_{H2O} (H_2O fugacity) range defined by the temperature and pressure conditions of these experiments (Fig. 6; Table 2). The solubility difference does, however, decrease with increasing fugacity of H_2O.

The significant solubility depression resulting from 10 mol% CaO to SiO_2 appears similar to that from MgO in the system MgO–SiO_2–H_2O (Zhang and Frantz 2000). This effect contrasts with adding NaOH where, despite the dilution of SiO_2 concentration, the SiO_2 solubility in aqueous fluids is similar to that in the

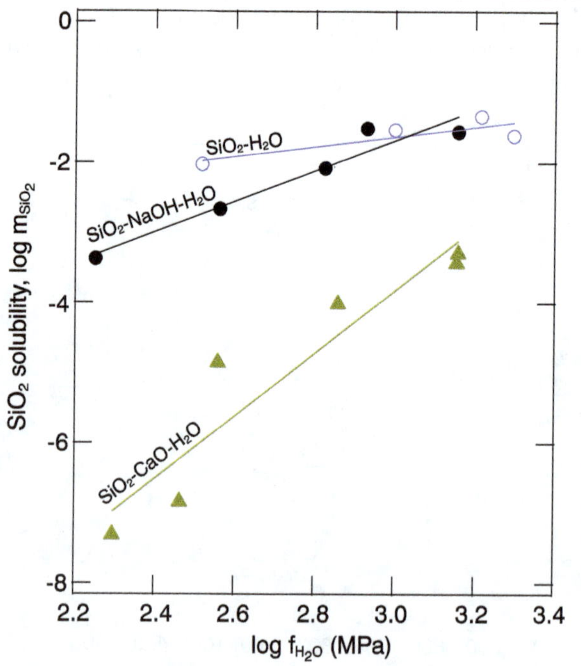

Fig. 6 Silicate solubility in the SiO_2–H_2O, SiO_2–NaOH–H_2O, and SiO_2–CaO–H_2O systems as a function of fugacity of pure H_2O (see text for details of calculations and source of f_{H2O} calculation)

simpler SiO_2–H_2O system. The solubility is, however, more sensitive to f_{H2O} in the SiO_2–NaOH–H_2O system so that at high H_2O fugacity, the SiO_2 solubility in the SiO_2–NaOH–H_2O system exceeds that in SiO_2–H_2O fluids (Fig. 6). The temperature/pressure coordinates of the solubility cross-over (750–800 °C at ~900 MPa), from linear extrapolation of the curves in Fig. 6, occur at the conditions just below the second critical point in the SiO_2–H_2O system at 1080 °C and 970 MPa (Kennedy et al. 1962). In contrast, linear extrapolation of the SiO_2–CaO–H_2O line leads to a cross-over at an f_{H2O} corresponding to ~1300 MPa pressure at 800 °C. These pressure and temperature conditions are inside the supercritical region of the SiO_2–H_2O system, which means that conditions of greater silica solubility in SiO_2–CaO–H_2O fluids than in SiO_2–H_2O and SiO_2–NaOH–H_2O cannot be realized.

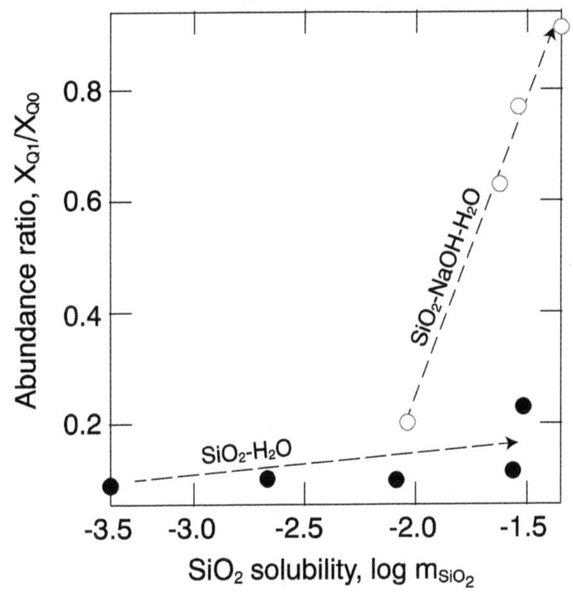

Fig. 7 Silicate species abundance ratio, X_{Q1}/X_{Q0} (X = mol fraction) as a function of SiO_2 solubility (kg/mol)

Silicate speciation in fluid

The approximate concentration of the silicate species in fluid (Q^n species) can be obtained from the integrated intensities of the 770 and 850 cm^{-1} bands of the Raman spectra of fluid. The concentrations thus obtained depend, however, on the assumption that the relative intensities of these two Raman bands are equal to the mol fraction of the species. Given the similar Si–O^- force constants for Si–O^- stretch vibrations in Q^0 and Q^1 in hydrous melts (Cody et al. 2005), the equivalence of relative intensities of the 770 and 850 cm^{-1} bands and the mol fraction of Q^0 and Q^1 species seems a reasonable assumption.

In the SiO_2–H_2O and SiO_2–NaOH–H_2O systems, silica-saturated fluids contain both Q^1 and Q^0 species, whereas in the SiO_2–CaO–H_2O system, only Q^0 species were detected in the temperature and pressure range. For the SiO_2–H_2O and SiO_2–NaOH–H_2O fluids, the Q^0 abundance decreases with increasing temperature and pressure although this decrease is more pronounced in the SiO_2–NaOH–H_2O system. This decrease is coupled with increasing Q^1 abundance. The rate of change of the abundance ratio, X_{Q1}/X_{Q0}, with temperature and pressure is considerably faster in the SiO_2–NaOH–H_2O system and, in fact, can be expressed as linear functions of SiO_2 molality, m_{SiO2} (Fig. 7). Qualitatively, this behavior resembles that for SiO_2–H_2O and SiO_2–MgO–H_2O fluids at pressures above 2 GPa (Mysen et al. 2013). This behavior also is qualitatively similar to that of increased silicate polymerization with increasing total SiO_2 concentration in silicate melts (Buckermann et al. 1992). The difference in silicate species abundance between SiO_2–H_2O and SiO_2–NaOH–H_2O fluids probably occurs because steric hindrance associated with nonbridging oxygen (NBO) bonding to H^+ is greater than Na^+–NBO bonding given the much smaller ionic radius of H^+. Protons will, therefore, favor nonbridging oxygen in Q^0 species, whereas Na^+ will form bonding with nonbridging oxygen in Q^1 species. Hence, the more rapid increase in X_{Q1}/X_{Q0} in the SiO_2–NaOH–H_2O systems than in SiO_2–H_2O.

Thermodynamic considerations

When Q^0 and Q^1 species are present in the fluid, the equilibrium is:

$$2Q^0 \Leftrightarrow Q^1, \tag{2}$$

with the equilibrium constant:

$$K_{(2)} = X_{Q1}/(X_{Q0})^2, \tag{3}$$

where X_{Q1} and X_{Q0} are mol fractions of Q^1 and Q^0, respectively. This equilibrium applies to the SiO_2–H_2O and SiO_2–NaOH–H_2O fluids in the temperature and pressure range discussed here. We note that it differs from that of SiO_2–H_2O at higher pressures (1.8–5.2 GPa) where the silica concentration is sufficiently high to stabilize even more polymerized Q^2 species as well (Mysen et al. 2013).

The silicate speciation in the SiO_2–CaO–H_2O fluids is simpler than in the other two systems as the only species throughout the temperature and pressure is Q^0 only (Fig. 4). Under such circumstances, the equilibrium constant equals concentration in the fluid, m_{SiO2}:

$$K = m_{SiO2}. \tag{4}$$

From the temperature dependence of the equilibrium constants, the enthalpy change of the reactions follows from:

$$\ln K = -\Delta H/RT + \Delta S/R, \tag{5}$$

where ΔH and ΔS are the enthalpy and entropy changes, T is absolute temperature, and R is the gas constant. It is assumed in using this equation that there is no pressure effect on the equilibria in the pressure range of these experimental results (Table 1). This conclusion is substantiated by the SiO_2–H_2O and SiO_2–MgO–H_2O fluid data of Mysen et al. (2013) who did not detect a pressure dependence in the same pressure range as the present experiments.

A linear fit to ln K versus $1/T$ yields significantly different ΔH values for the equilibria among the Q^n species in the fluid (Fig. 8). For the SiO_2–CaO–H_2O system (Fig. 8a), equilibrium (4) describes the situation. It is more sensitive to temperature ($\Delta H = 322 \pm 4$ kJ/mol) than equilibrium (2), which describes the speciation variations in the SiO_2–H_2O (22 ± 12 kJ/mol) and SiO_2–NaOH–H_2O systems (51 ± 17 kJ/mol; see Fig. 8).

Conclusions

Silicate speciation in silica-saturated aqueous fluids suggested from thermodynamic modeling indicates that species more polymerized than Si $(OH)_4$ type exist in aqueous fluids under pressure/temperature conditions relevant to upper mantle processes (Newton and Manning 2003). This suggestion is consistent with the structural observations reported here and elsewhere (Zotov and Keppler 2002; Mysen et al. 2013). We also note, however, that when NaOH instead of CaO is the third component added to SiO_2–H_2O, the silicate solubility is higher, and both Q^0 and Q^1 species occur in the SiO_2–NaOH–H_2O fluids at lower pressures than in the alkaline earth systems. We note, though, that in these experiments Na^+ was added as an NaOH + H_2O solution. We cannot rule out a contribution to the silicate solubility from the alkaline nature of the fluid such as, for example:

$$2Q^n + 2Na^+ \ + 2OH^- = 2Q^{n-1} + H_2O, \tag{6}$$

where, therefore, pH would be a contributing factor to the silicate solubility.

Furthermore, as the silicate concentration in aqueous fluids increases and silicate species becomes more polymerized with pressure and temperature, the types and abundance Q^n species more closely resemble those in hydrous silicate melts (Zotov and Keppler 1998; Cody et al. 2005; Mysen 2010). This structural evolution may lead to complete miscibility between silicate and aqueous fluid. It is likely that complete miscibility between fluid and melt occurs at lower pressure in alkali silicate systems than in those of alkaline earth silicate systems because of the greater silicate solubility in alkali silicate fluids. This conclusion accords with experimental observations that show complete miscibility near 1–1.5 GPa in alkali silicate and aluminosilicate systems, whereas pressure exceeds 3 GPa before reaching miscibility in alkaline earth systems (Bureau and Keppler 1999; Stalder et al. 2001; Mibe et al. 2007).

Experimental data indicate that solubility of geochemically important minor and trace elements in aqueous fluids can be significantly affected by other silicate components (Manning 2007; Mysen 2012a, b). This happens because the stability and concentration of minor and trace element complexes (oxy complexes) in fluids depend on the electronic properties of cations available for this purpose in the fluid (Mysen 2012b, 2015b). This means that when considering partitioning of such elements between

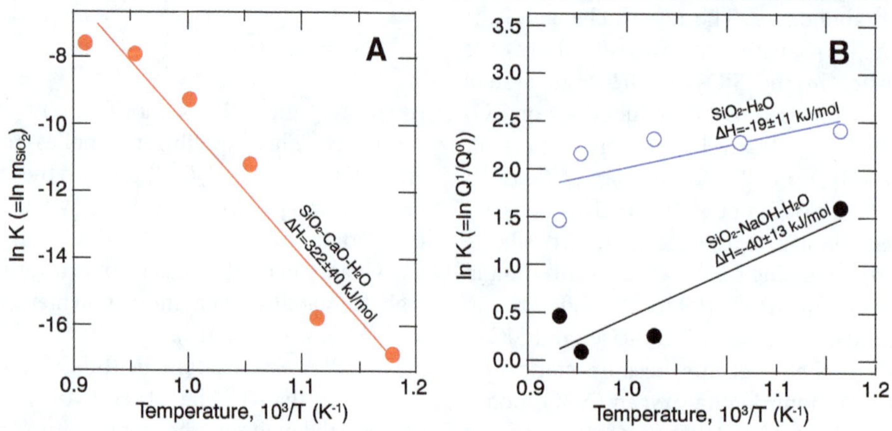

Fig. 8 Equilibrium constants for reactions (4) (**a**) and (2) (**b**) as a function of temperature. The ΔH values are from least squares fit to Eq. (3)

minerals and aqueous fluids in the upper mantle and deep crust, pressure-dependent partition coefficients likely reflect changing silicate-cation interaction in the fluid as a function of silicate solubility and speciation.

Qualitatively, this behavior also means that for fluids in equilibrium with mafic and ultramafic mineral assemblages where alkaline earths are dominating cations, the silicate solubility in fluid is less and silicate polymerization is less extensive than in fluids in the equilibrium in felsic systems such as typical gneiss compositions, for example, where alkali metals are more important. Moreover, because the fluid composition in equilibrium with such different crystalline rocks differs, minor and trace element complexing in those different fluids also differs, which leads to different fluid-melt-crystal partition coefficients as a function of rock type (mineral assemblage) with which aqueous fluids equilibrated.

Abbreviations
ΔH: Enthalpy change; ΔS: Entropy change; CCD: Charge-coupled detector; f_{H2O}: H_2O fugacity; HDAC: Hydrothermal diamond anvil cell; HFSE: High field strength element; m_{SiO2}: SiO_2 molality; R: Gas constant; X_{Q0}: Mol fraction of Q^0 species; X_{Q1}: Mol fraction of Q^1 species

Acknowledgements
Instrument, electronics, and library support was provided by Geophysical Laboratory technical staff. Their support is gratefully acknowledged.

Funding
This research was supported by grant EAR1212854 from the National Science Foundation.

Authors' contributions
All research reported in this report was carried out by the author. The author read and approved the final manuscript.

Competing interests
The author declares that he has no competing interests.

References
Bassett WA, Shen AH, Bucknum M, Chou IM (1993) A new diamond cell for hydrothermal studies to 2.5 GPa and from -190°C to 1200°C. Rev Sci Instrum 64:2340–2345. https://doi.org/10.1063/1.1143931

Brawer SA, White WB (1975) Raman spectroscopic investigation of the structure of silicate glasses. I. The binary silicate glasses. J Chem Phys 63:2421–2432. https://doi.org/10.1063/1.431671

Buckermann WA, Muller-Warmuth W, Frischat GH (1992) A further 29Si MAS NMR study on binary alkali silicate glasses. Glasstech Ber 65:18–21

Burchard M, Maresch WV, Fockenberg T, Doltsinis NL, Adeagbo WA (2011) Modelling high-pressure aqueous fluids in the system CaO-SiO_2-H_2O: a comprehensive semi-empirical thermodynamic formalism. Eur J Mineral 23: 409–424. https://doi.org/10.1127/0935-1221/2011/0023-2106

Bureau H, Keppler H (1999) Complete miscibility between silicate melts and hydrous fluids in the upper mantle; experimental evidence and geochemical implications. Earth Planet Sci Lett 165:187–196. https://doi.org/10.1016/S0012-821X(98)00266-0

Cody GD, Mysen BO, Lee SK (2005) Structure vs. composition: a solid-state H-1 and Si-29 NMR study of quenched glasses along the Na_2O-SiO_2-H_2O join. Geochim Cosmochim Acta 69:2373–2384. https://doi.org/10.1016/j.gca.2004.11.012

Fockenberg T, Burchard M, Maresch VW (2006) Experimental determination of the solubility of natural wollastonite in pure water up to pressures of 5 GPa and at temperatures of 400-800 degrees C. Geochim Cosmochim Acta 70: 1796–1806. https://doi.org/10.1016/j.gca.2005.12.017

Kawamoto T, Kanzaki M, Mibe K, Matsukage KN, Ono S (2012) Separation of supercritical slab-fluids to form aqueous fluid and melt components in subduction zone magmatism. Proc Natl Acad Sci USA 109:18695–18700. https://doi.org/10.1073/pnas.1207687109

Kennedy GC, Wasserburg GJ, Heard HC, Newton RC (1962) The upper three-phase region in the system SiO_2-H_2O. Am J Sci 260:501–521. https://doi.org/10.2475/ajs.260.7.501

Kessel R, Schmidt MW, Ulmer P, Pettke T (2005) Trace element signature of subduction-zone fluids, melts and supercritical liquids at 120-180 km depth. Nature 437:724–727. https://doi.org/10.1038/nature03971

Kessel R, Ulmer P, Pettke T, Schmidt MW, Thompson AB (2004) A novel approach to determine high-pressure high-temperature fluid and melt compositions using diamond-trap experiments. Am Mineral 89:1078–1086

Manning CE (1994) The solubility of quartz in H_2O in the lower crust and upper mantle. Geochim Cosmochim Acta 58:4831–4840. https://doi.org/10.1016/0016-7037(94)90214-3

Manning CE (2004) The chemistry of subduction-zone fluids. Earth Planet Sci Lett 223:1–16. https://doi.org/10.1016/j.epsl.2004.04.030

Manning CE (2007) Solubility of corundum+kyanite in H_2O at 700°C and 10 kbar: evidence for Al-Si complexing at high pressure and temperature. Geofluids 7: 258–268. https://doi.org/10.1111/j.1468-8123.2007.00179.x

Manning CE, Boettcher S (1994) Rapid quench hydrothermal experiments at mantle pressures and temperatures. Am Mineral 79:1153–1158

McMillan PF, Wolf GH, Poe BT (1992) Vibrational spectroscopy of silicate liquids and glasses. Chem Geol 96:351–366. https://doi.org/10.1016/0009-2541(92)90064-C

Mibe K, Chou IM, Bassett WA (2008) In situ Raman spectroscopic investigation of the structure of subduction-zone fluids. J Geophys Res 113. https://doi.org/10.1029/2007jb005179

Mibe K, Kanzaki M, Kawamoto T, Matsukage KN, Fei Y, Ono S (2007) Second critical endpoint in the peridotite-H_2O system. J Geophys Res 112. https://doi.org/10.1029/2005JB004125

Mibe, K., Kawamoto, T., Matsukage, K.N., Fei, Y., and Ono, S. (2011) Slab melting versus slab dehydration in subduction-zone magmatism. Proc. Nat'l. Acad. Sci. https://doi.org/10.1073/pnas.1010968108

Mysen BO (2010) Structure of H_2O-saturated peralkaline aluminosilicate melt and coexisting aluminosilicate-saturated aqueous fluid determined in-situ to 800 C and 800 MPa. Geochim Cosmochim Acta 74:4123–4139. https://doi.org/10.1016/j.gca.2010.04.024

Mysen BO (2012a) Silicate-COH melt and fluid structure, their physicochemical properties, and partitioning of nominally refractory oxides between melts and fluids. Lithos 148:228–246. https://doi.org/10.1016/j.lithos.2012.06.005

Mysen BO (2012b) High-pressure/-temperature titanium solution mechanisms in silicate-saturated aqueous fluids and hydrous silicate melts. Am Mineral 97: 1241–1251. https://doi.org/10.2138/am.2012.4084

Mysen BO (2015a) Carbon speciation in silicate-C-O-H as a function of redox conditions: an experimental study, in-situ to 1.7 GPa and 900°C. Am Mineral 100:872–882. https://doi.org/10.2138/am-2015-4976

Mysen BO (2015b) An in-situ experimental study of Zr^{4+} transport capacity of water-rich fluids in the temperature and pressure range of the deep crust and upper mantle. Prog Earth Planet Sci 2(1):38. https://doi.org/10.1186/s40645-015-0070-5

Mysen BO, Mibe K, Chou IM, Bassett WA (2013) Structure and equilibria among silicate species in aqueous fluids in the upper mantle: experimental SiO_2-H_2O and MgO-SiO_2-H_2O data recorded in situ to 900 degrees C and 5.4GPa. J Geophys Res 118:6076–6085. https://doi.org/10.1002/2013jb010537

Mysen BO, Richet P (2005) Silicate glasses and melts—properties and structure. Elsevier, New York, p 548

Mysen BO, Yamashita S (2010) Speciation of reduced C-O-H volatiles in coexisting fluids and silicate melts determined in-situ to ~1.4 GPa and 800°C. Geochim Cosmochim Acta 74:4577–4588

Newton RC, Manning CE (2002) Solubility of enstatite+forsterite in H_2O in deep crust/upper mantle conditions: 4 to 15 kbar and 700 to 900°C. Geochim Cosmochim Acta 66:4165–4176. https://doi.org/10.1016/S0016-7037(02)00998-5

Newton RC, Manning CE (2003) Activity coefficient and polymerization of aqueous silica at 800°C, 12 kbar, from solubility measurements and SiO_2-buffering mineral assemblages. Contrib Mineral Petrol 146:135–146. https://doi.org/10.1007/s00410-003-0483-9

Newton RC, Manning CE (2009) Hydration state and activity of aqueous silica in H_2O-CO_2 fluids at high pressure and temperature. Am Mineral 94:1287–1290. https://doi.org/10.2138/am.2009.3287

Osborn EF, Muan A (1960) Plate 2. The system CaO-MgO-SiO_2: phase equilibrium diagrams of oxide systems. American Ceramic Society

Pitzer KS, Sterner SM (1994) Equations of state valid continuously from zero to extreme pressures for H_2O and CO_2. J Chem Phys 101:3111–3116. https://doi.org/10.1063/1.467624

Ratcliffe CI, Irish DE (1982) Vibrational studies of solutions at elevated temperatures and pressures. 5. Raman studies of liquid water up to 300 °C. J Phys Chem 86:4897–4905. https://doi.org/10.1021/j100222a013

Stalder R, Ulmer P, Thompson AB, Gunther D (2001) High pressure fluids in the system MgO-SiO_2-H_2O under upper mantle conditions. Contrib Mineral Petrol 140:607–618. https://doi.org/10.1007/s004100000212

Walrafen GE, Fisher MR, Hokmabadi MS, Yang WH (1986) Temperature dependence of the low- and high-frequency Raman scattering from liquid water. J Chem Phys 85:6970–6982. https://doi.org/10.1063/1.451384

Wilke M, Schmidt C, Dubrail J, Appel K, Borchert M, Kvashnina K, Manning CE (2012) Zircon solubility and zircon complexation in H_2O+Na_2O+SiO_2±Al_2O_3 fluids at high pressure and temperature. Earth Planet Sci Lett 349-350:15–25. https://doi.org/10.1016/j.epsl.2012.06.054

Zhang Y-G, Frantz JD (2000) Enstatite-forsterite-water equilibria at elevated temperatures and pressures. Am Mineral 85:918–925. https://doi.org/10.2138/am-2000-0705

Zotov N, Keppler H (1998) The influence of water on the structure of hydrous sodium tetrasilicate glasses. Am Mineral 83:823–834

Zotov N, Keppler H (2002) Silica speciation in aqueous fluids at high pressures and high temperatures. Chem Geol 184:71–82. https://doi.org/10.1016/S0009-2541(01)00353-9

16

An in situ experimental study of Zr^{4+} transport capacity of water-rich fluids in the temperature and pressure range of the deep crust and upper mantle

Bjorn Mysen

Abstract

Throughout the Earth's history, mass transport involved fluids. In order to address the circumstances under which Zr^{4+} may have been transported in this manner, its solubility behavior in aqueous fluid with and without NaOH and SiO_2 in equilibrium with crystalline ZrO_2 was determined from 550 to 950 °C and 60 to 1200 MPa. The measurements were carried out in situ while the samples were at the temperatures and pressures of interest. In ZrO_2-H_2O and ZrO_2-SiO_2-H_2O fluids, the Zr^{4+} concentration ranges from ≤10 to ~70 ppm with increasing temperature and pressure. Addition of SiO_2 to the ZrO_2-H_2O system does not affect these values appreciably. In these two environments, Zr^{4+} forms simple oxide complexes in the H_2O fluid with $\Delta H \sim 40$ kJ/mol for the solution equilibrium, ZrO_2(solid) = ZrO_2(fluid). The Zr^{4+} concentration in aqueous fluid increases about an order of magnitude upon addition of 1 M NaOH, which reflects the formation of zirconate complexes. The principal solution mechanism is $ZrO_2 + 4NaOH = Na_4ZrO_4 + 2H_2O$ with $\Delta H \sim 200$ kJ/mol. Addition of both SiO_2 and NaOH to ZrO_2-H_2O enhances the Zr^{4+} by an additional factor of about 5 with the formation of partially protonated alkali zircon silicate complexes in the fluid. The principal solution mechanism is $2ZrO_2 + 2NaOH + 2SiO_2 = Na_2Zr_2Si_2O_9 + H_2O$ with $\Delta H \sim 40$ kJ/mol. These results, in combination with other published experimental data, imply that fluid released during high-temperature/high-pressure dehydration of hydrous mineral assemblages in the Earth's interior under some circumstances may carry significant concentrations of Zr and probably other high field strength elements (HFSEs). This suggestion is consistent with the occurrence of Zr-rich veins in high-grade metamorphic eclogite and granulite terranes. Moreover, aqueous fluids transported from dehydrating oceanic crust into overlying mantle source rocks of partial melting also may carry high-abundance HFSE of fluids released from dehydrating slabs and transported to the source rock of partial melting in the overlying mantle wedge. These processes may have been operational in the Earth's history within which subduction resembling that observed today was operational.

Keywords: Mass transfer, Aqueous fluid, High pressure, Structure, Solubility, Spectroscopy

Background

Mass transfer in the deep crustal and upper mantle involves fluids, supercritical fluids, and melts. In the young Earth's mantle and crust, stable isotope evidence is consistent with liquid water interacting with zircon-bearing rocks (e.g., Compston and Pidgeon 1986, Mojzsis et al. 2001, Valley et al. 2005). Early mass transfer processes were not, however, necessarily similar to more recent activities (Kröner and Layer 1992, Hamilton 2007). For example, subduction zone processes, an area of major mass transfer in more recent terrestrial environments, may not have begun until some 1–2 billion years later (Hamilton, 2007). From this time in the Earth's history until the present day, fluid was transferred from dehydrating hydrous minerals in subduction zone materials to the overlying mantle wedge where partial melting may occur (e.g., Bebout 2007). These magmatic liquids would carry element signatures of the source of those

Correspondence: bmysen@carnegiescience.edu
Submission to SPEPS: High-Pressure Earth and Planetary Science in the last and next Decade
Geophysical Laboratory, Carnegie Instn. Washington, Washington, DC, USA

fluids (Gill 1981, Wyllie and Sekine 1982, Tatsumi and Eggins 1995, Hermann et al. 2006).

It is often found, however, that the solubility of high field strength elements (HFSEs) such as Ti, Hf, Zr, Nb, and Ta in pure H_2O fluids is in the ppm range (Brenan et al. 1994, Audetat and Keppler 2005, Antignano and Manning 2008, Newton et al. 2010, Bernini et al. 2013), which would suggest that H_2O alone is not an efficient HFSE transport medium at least in the temperature/pressure range of the crust and upper mantle. There is, however, evidence from veins in high-grade metamorphic terranes that Zr and other HFSEs can be significantly mobile (Giere 1986, Philippot and Selverstone 1991, Möller et al. 2002). The latter observations have led to suggestions that complexing of HFSEs with major-element solutes in aqueous fluids would greatly enhance the HFSE solubility. Scattered solubility data of some HFSEs in water-rich multi-component fluids at high temperature and pressure are consistent with such ideas (Ayers and Watson 1993, Antignano and Manning 2008, Mysen 2012, Ayers et al. 2012, Louvel et al. 2013). Effects of major element composition on HFSE solubility in silicate melts have also been reported (Watson 1976, 1979, Ellison and Hess 1986, Linnen and Keppler 2002).

The majority of experimental studies have focused on zircon ($ZrSiO_4$) solubility in fluids and typically on its solubility in pure H_2O or in aqueous solution as a function of additional solutes (SiO_2, NaOH, Na-silicates, etc.). That approach limits the interpretation of the solution mechanism of Zr^{4+} in fluids because the presence of SiO_2 in those experiments restricts identification of effects of individual components on Zr^{4+} solution mechanisms. Moreover, most, although not all, characterizations of experimental charges were carried out on fluids after returning the sample to ambient temperature and pressure. This process often leads to compositional and structural changes of the fluid because solubility in aqueous fluid typically is temperature- and pressure-dependent (Pascal and Anderson 1989, Manning 1994) and, thus, changes along the pressure-temperature path of the quenching path.

In the present study, experiments were carried out in order to address some of the unanswered questions regarding the solubility and solution mechanism(s) of Zr^{4+} in aqueous solutions. Experimental complexities resulting from compositional and structural changes during quenching of the aqueous fluid to ambient conditions after an experiment were avoided by examination of the Zr-saturated fluids in situ while these were at the temperatures and pressures of interest. The effects of specific solutes on Zr^{4+} solubility and solution mechanisms in aqueous solution were addressed.

Methods

Aqueous fluids (double-distilled, deionized H_2O and 1 M NaOH aqueous solutions) together with crystalline spectroscopically pure ZrO_2 (Alfa Aesar™) and SiO_2 (Alfa Aesar™) were used as starting materials. The grain size of the oxides was less than 5 μm. Crystalline Na_2ZrO_3 was synthesized from spectroscopically pure $Na_2CO_3 + SiO_2$ mixtures at ambient pressure by using the method of multiple heat cycles (to ~1000 °C at ambient pressure) described by Ampian (1968). Results from X-ray diffractometry indicated the complete reaction of $Na_2CO_3 + SiO_2$ to crystalline Na_2ZrO_3.

In situ, high-temperature/high-pressure experiments were carried out in externally heated, hydrothermal diamond anvil cells (HDAC) (Bassett et al. 1996). Iridium was used as gasket material in order to avoid chemical reaction between the gasket and aqueous solutions during the experiments. Gasket thickness was 125 μm with a 500 μm diameter sample hole. Temperatures were measured with two K-type thermocouples touching the upper and lower diamonds at a distance of about 1 mm from the sample. The temperature variations within the cell, as indicated by NaCl melting at ambient pressure, are 2–3 °C.

In the hydrothermal diamond cell experiments, pressure was generated by increasing the temperature of the aqueous solution + solid sample contained in the sample hole of the Ir gasket. Pressure is, therefore, a temperature-dependent variable. This temperature-dependent pressure is governed by the PVT properties of the fluid (e.g., Bassett et al. 1996). The hydrothermal diamond cell experimental protocol does not, therefore, allow for independent control of pressure. Pressure was determined with the one-phonon shift in the Raman spectra by using a 10–20 μm chip of synthetic carbon-13 diamond located within the sample region itself (Schiferl et al. 1997, Mysen and Yamashita 2010). Synthetic carbon-13 diamond was used for this purpose to avoid interference of the Raman signal from the natural diamond anvils in the diamond anvil cell. The frequency difference between the one-phonon Raman shifts from carbon-12 and carbon-13 is about 80 cm^{-1} (Hanfland et al. 1985).

A confocal JASCO™ Model NRS-3100 microRaman spectrometer with the 490 nm line of a Coherent MX-Series solid-state laser operating at ~40 mW at the sample for Raman excitation was used for all measurements. The Raman spectrometer is equipped with a single monochromator, holographic notch filter, and holographic gratings. For pressure calibration with a ^{13}C diamond, gratings with 2400 grooves/mm density were used in order to optimize frequency accuracy of the spectrometer. Because of the relatively small pressure-induced shift of the Raman signal from diamond (~1.8 cm^{-1}/GPa at ambient temperature; Hanfland et al. 1985), additional improvement in spectrometer accuracy was accomplished by using Ne emission lines as an internal standard. A ±0.1 cm^{-1} uncertainty thus attained translates to a ±40 MPa precision of pressure measurements. The overall pressure uncertainty, which also includes the uncertainty in the calibration of the one-phonon Raman shift of the carbon-13 diamond from Mysen and Yamashita (2010), is ±110 MPa. This uncertainty (and precision) was the same in all experiments.

Raman spectra of the samples (fluids and crystalline materials) were recorded with 1200 grooves/mm grating density, which in the frequency range of interest in the present study yields a spectral range of ~1900 cm^{-1} per CCD window (decreasing slightly with distance from the excitation line) and a frequency uncertainty of ~±3 cm^{-1}. This uncertainty is not an issue given the large width of the Raman bands of fluids and crystals at the high temperatures of the experiments (full width at half height (FWHH) ≥ 30 cm^{-1}).

Experiments were conducted in several temperature/pressure series. The proportion of fluid and solid phases in the various experimental series differs. The pressure-temperature trajectories, therefore, also differ between individual experimental series.

In each experimental series, the sample was initially brought to the highest planned temperature and pressure, left there for about an hour followed by recording the spectra of fluid and coexisting solid materials. The spectra of carbon-13 diamond for pressure determination also were recorded at each temperature to determine the pressure. The temperature was then lowered at 1 °C/s to the next temperature/pressure condition where the same spectroscopic acquisition protocol was used. A 60-minute dwell time at each temperature and pressure prior to analysis is considered sufficient to reach equilibrium given that similar experimental duration has been shown sufficient to establish isotopic equilibrium between fluid and condensed phases at similar temperatures (Horita 1988, Mysen 2015). Moreover, Wilke et al. (2012), using a similar experimental protocol, found a constant Zr concentration in fluid coexisting with crystalline ZrO_2 after a 15–30-minute experimental duration.

Results

In the HDAC experiments, there is a general increase in pressure with increasing temperature (Table 1; see also Fig. 1). The "average pressure" shown in the subsequent figures is that corresponding to the intersection of the temperature with a single straight line fitted to all the data points (Fig. 1; see Tables 2 and 3 for regression parameters). The exact pressure for each data point can be extracted from the data in Table 1.

Table 1 Experimental conditions

Temperature °C	Pressure MPa	Comment
ZrO_2-H_2O		
25[a]	0.1[b]	ZrO_2 + fluid
550	299	ZrO_2 + fluid
625	431	ZrO_2 + fluid
700	637	ZrO_2 + fluid
775	842	ZrO_2 + fluid
850	1195	ZrO_2 + fluid
900	1187	ZrO_2 + fluid
925	697	ZrO_2 + fluid
ZrO_2-SiO_2-H_2O		
25	0.1	ZeO_2 + quartz + fluid
575	544	ZrO_2 + quartz + fluid
650	701	ZrO_2 + quartz + fluid
725	783	ZrO_2 + quartz + fluid
800	1013	ZrO_2 + quartz + fluid
ZrO_2-1 M NaOH-H_2O		
25	0.1	ZrO_2 + fluid
600	59	ZrO_2 + fluid
650	183	ZrO_2 + fluid
700	304	ZrO_2 + fluid
750	457	ZrO_2 + fluid
800	680	ZrO_2 + fluid
850	419	ZrO_2 + fluid
900	559	ZrO_2 + fluid
950	622	ZrO_2 + fluid
ZrO_2-SiO_2-1 M NaOH-H_2O		
25	0.1	ZrO_2 + quartz + fluid
575	138	ZrO_2 + quartz + fluid
650	183	ZrO_2 + quartz + fluid
725	266	ZrO_2 + quartz + fluid
775	435	ZrO_2 + quartz + fluid
850	677	ZrO_2 + quartz + fluid
SiO_2-H_2O		
25	0.1	Quartz + fluid
575	211	Quartz + fluid
650	257	Quartz + fluid
725	487	Quartz + fluid
800	1087	Quartz + fluid

[a]Temperature uncertainty: 2–3 °C (see text)
[b]Pressure uncertainty: ±110 MPa and precision ±40 MPa (see text) in all experiments

Raman spectroscopy

Raman spectra were collected in the ~300–2000 cm^{-1} and ~3000–4100 cm^{-1} frequency ranges. The higher frequency range comprises signals from fundamental OH vibrations. In the low-frequency range, Raman signals from vibrations in Zr-O and Si-O bonding environments can be detected.

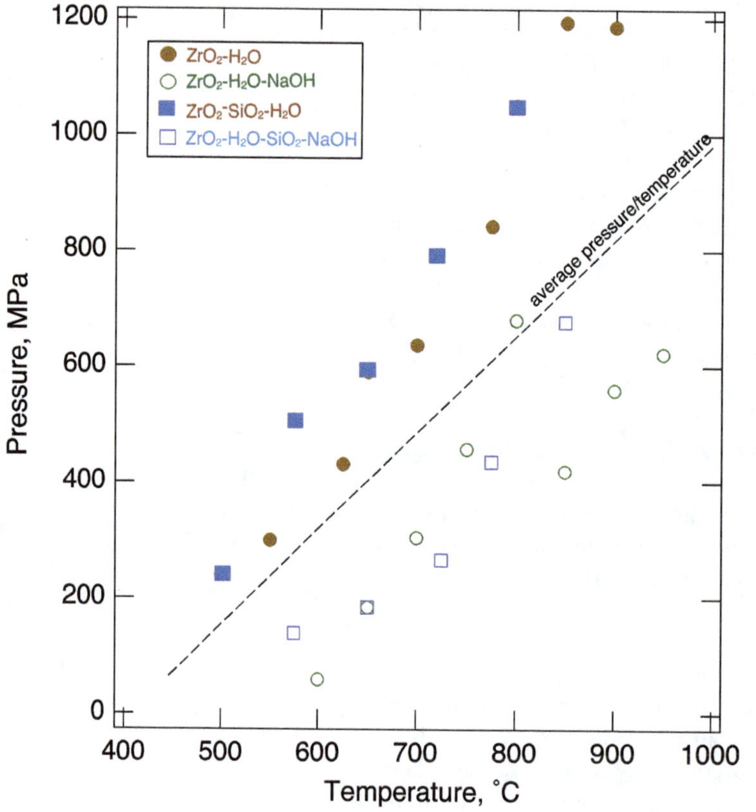

Fig. 1 Summary of temperature/pressure conditions of all experiments. The *dashed line* represents a simple linear fit to all the data points (see Tables 2 and 3 for regression coefficients). See also text for additional discussion of this fit

The spectra in the high-frequency range characteristically show a sharp maximum near 3600 cm^{-1} with an asymmetry toward lower frequency, in particular at the lower temperatures (Fig. 2). For all compositions, the FWHH of this peak decreases with increasing temperature and pressure (Fig. 3). These features are qualitatively similar to the changes in the same frequency region of the Raman spectra of other aqueous fluids systems (Walrafen et al. 1986, 1999, Frantz et al. 1993).

The FWHH vs. temperature relations in the spectra of fluids in the ZrO_2-H_2O and ZrO_2-SiO_2-H_2O systems are identical within the error of the measurements (Fig. 3). The FWHH is more sensitive to temperature with NaOH added to the H_2O fluid (Tables 2 and 3). Addition of SiO_2 (ZrO_2-SiO_2-NaOH-H_2O) results in FWHH that may be slightly narrower than without SiO_2 and up to >40 cm^{-1} narrower than in fluid spectra from the SiO_2-free ZrO_2-NaOH-H_2O system (Fig. 3).

Table 2 Regression coefficients for figures, $y = a + bx$

Comment	a	b	r^2
Figure 1: pressure (MPa) = $a + b \cdot$ temperature (°C)	−677 ± 308	1.7 ± 0.4	0.42
Figure 3: FWHH(cm^{-1}) = $a + b \cdot$ temperature (°C)			
ZrO_2-H_2O	175 ± 4	0.116 ± 0.006	0.995
ZrO_2-NaOH-H_2O	255 ± 14	−0.19 ± 0.02	0.95
ZrO_2-SiO_2-H_2O	176 ± 6	−0.120 ± 0.008	0.993
ZrO_2-SiO_2-NaOH-H_2O	170 ± 15	−0.12 ± 0.02	0.96

Note that data from the system ZrO_2-SiO_2-H_2O are not shown because only two compositions at nearby temperatures were determined, and for those, there was no discernible temperature effect (see also Fig. 3)

See also text for detailed description and discussion of these equilibria

Table 3 Regression coefficients for figures, $y = a + bx$

Comment	a	b	r^2	ΔH (kJ/mol)
Figure 6: ln $K = a + b/T(K^{-1})$				
ZrO_2-H_2O (ZH)	−2.4 ± 1.9	−5121 ± 1940	0.78	43 ± 16
ZrO_2-NaOH-H_2O (low-dens) (ZHN-lo)	32 ± 3	−23,436 ± 2910	0.98	195 ± 24
ZrO_2-NaOH-H_2O (high-dens) (ZHN-hi)	39 ± 4	−27,866 ± 3540	0.95	232 ± 29
ZrO_2-NaOH-H_2O (ZHNS)	19 ± 1	−11,467 ± 1350	0.98	94 ± 11
ZrO_2-NaOH-H_2O (ZHN-Wilke et al. 2012)	44 ± 1	−33,845 ± 175	0.999	282 ± 15
ZrO_2-SiO_2-NaOH-H_2O (ZHNS)	5.7 ± 0.5	−3655 ± 473	0.95	30 ± 4

Abbreviations: *ZH*: ZrO_2-H_2O; *ZHN (-lo, -hi, -Wilke et al. 2012, and ZHNS)*: ZrO_2-NaOH-H_2O with low- and high-density (pressure) trajectories, interpolation from Wilke et al. (2012), and the ZHN equilibrium in the more complete system ZrO_2SiO_2-NaOH-H_2O; *ZHNS*: ZrO_2SiO_2-NaOH-H_2O

Note that data from the system ZrO_2-SiO_2-H_2O are not shown because only two compositions at nearby temperatures were determined, and for those, there was no discernible temperature effect (see also Fig. 3)

See also text for detailed description and discussion of these equilibria

The vibrations in this high-frequency range (Fig. 2) are assigned to fundamental O-H stretching possibly in several locally different environments (Scholze 1960, Mysen and Virgo 1986, Walrafen et al. 1986, Le Losq et al. 2011). Their frequencies reflect which cation or cations form bonds with OH groups. Here, these cations could be H^+, Na^+, Si^{4+}, and Zr^{4+}. In addition, hydrogen bonding may play a role, in particular at temperatures below about 600 °C (Walrafen et al. 1999, Foustoukos and Mysen 2012). Linkage of OH groups to Na^+ lowers the OH fundamental frequency compared with the spectrum of pure H_2O (Mysen and Virgo 1986). This could appear as an asymmetry toward lower frequency as observed in the spectra of the ZrO_2-NaOH-H_2O and

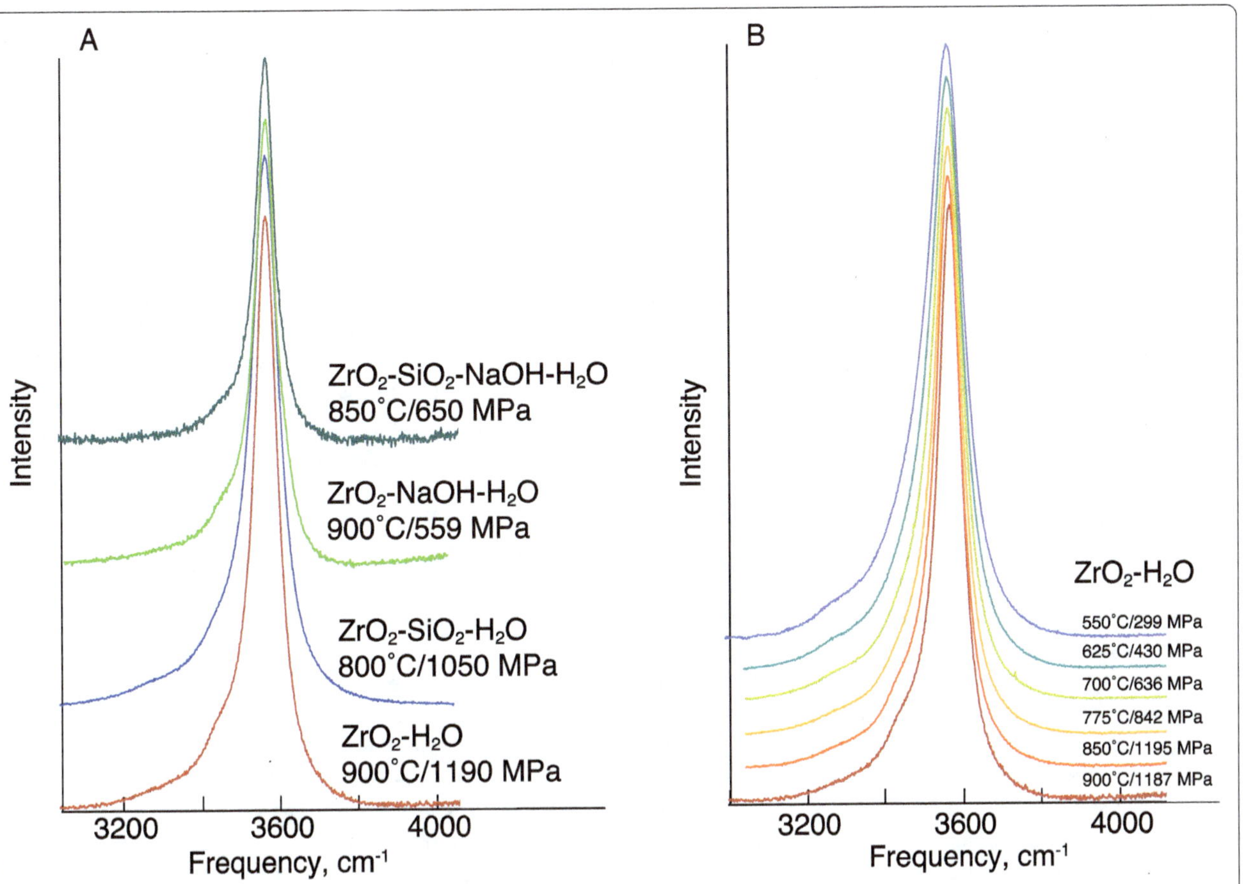

Fig. 2 Examples of Raman spectra of high-frequency intensity envelope. **a** Spectra at approximately the same temperature conditions for each of the compositions studied. **b** Temperature/pressure evolution example from the ZrO_2-H_2O system. All spectra were normalized to the total integrated area of the 3600 cm^{-1} envelope. Note that individual spectra are offset for clarity

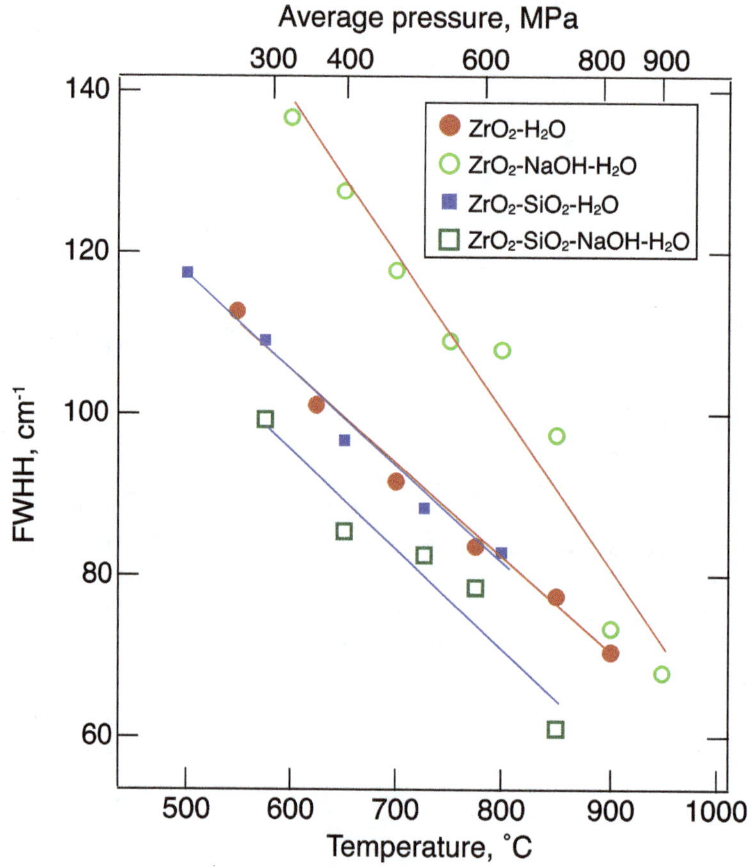

Fig. 3 Temperature/pressure evolution of full width at half height (FWHH) of the high-frequency envelope of the spectra of fluids from the experiments. Average pressure is that calculated from the regression line in Fig. 2. For actual pressure, see Table 1

ZrO_2-SiO_2-NaOH-H_2O fluid (Fig. 2a). The difference in FWHH in the spectra of NaOH+ SiO_2-bearing and SiO_2-absent but NaOH-bearing compositions likely reflects the formation of Si-OH linkages in systems with SiO_2. This is because Si-OH linkage tends to shift the Raman bands assigned to the fundamental OH vibrations to a higher frequency than that of OH in pure H_2O.

The Raman spectra of fluids and crystals in the low-frequency regime characteristically comprise one or more intensity envelopes between 300 and 800 cm^{-1} (Fig. 4). From the similar Raman spectra of fluid in the ZrO_2-H_2O and ZrO_2-SiO_2-H_2O systems with two bands between 550 and 650 cm^{-1}, it appears that the structure of dissolved Zr-containing complexes in those fluids resembles one another. These two bands also are at nearly the same frequency as the dominant bands of crystalline ZrO_2 at the same temperature and pressure. These latter spectral features resemble those of tetragonal ZrO_2 in which the oxygen coordination number is 7 (Keramidas and White 1974, Smith and Newkirk 1965). However, Wilke et al. (2012) interpreted their Zr X-ray near-edge spectra (XANES) to be consistent with eightfold coordinated Zr^{4+} in ZrO_2-H_2O solutions. The present Raman spectra of ZrO_2-H_2O and ZrO_2-SiO_2-H_2O fluids cannot be used to distinguish between Zr^{4+} in seven- or eightfold coordination with oxygen with more precision than to state that the short-range structure of Zr^{4+} dissolved in H_2O resembles that of coexisting crystalline ZrO_2. Fluids in the ZrO_2-SiO_2-H_2O system also contain dissolved silicate. This is in the form of Q^o, as evidenced by the strong band near 770 cm^{-1} (Fig. 4; see also Zotov and Keppler 2002, Mibe et al. 2008).

The spectra of NaOH-bearing fluids characteristically have an additional strong band centered near 350 cm^{-1} and one or two bands near 500 cm^{-1}. These were not observed in the spectra of fluids without NaOH. It is also noted that the spectra of ZrO_2-SiO_2-NaOH-H_2O fluids differ from those of ZrO_2-NaOH-H_2O in that the bands in the 400–600 cm^{-1} region of the ZrO_2-SiO_2-NaOH-H_2O spectra are narrower and better defined than those of the ZrO_2-NaOH-H_2O fluid and that the 650 cm^{-1} band in the ZrO_2-NaOH-H_2O fluid spectra is absent in the spectra of ZrO_2-SiO_2-NaOH-H_2O (Fig. 4). The Raman spectra of ZrO_2-SiO_2-NaOH-H_2O fluids are notable in their absence of Raman signals that could be

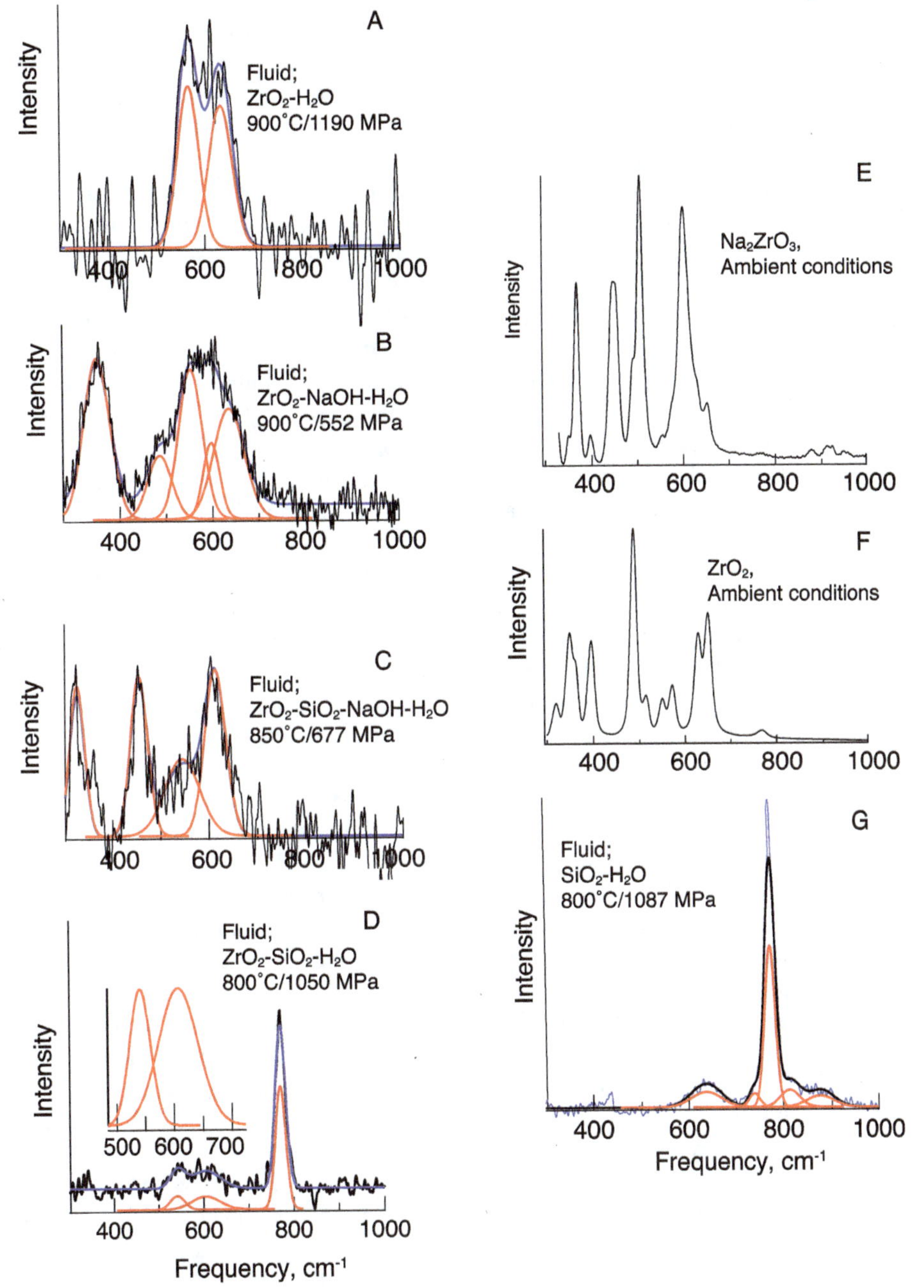

Fig. 4 Examples of the Raman spectra of fluids and crystals at conditions as indicated on the individual spectra. Panels a-d) show spectra of fluids from systems and under conditions indicated on the individual spectra. Panels (e) and (f) show examples of crystalline, Zfr-bearing materials, and panel (g) shows an example of the spectrum of fluid from the SiO2-H2O system at temperature and pressure indcated. Note the insert in (**d**), which shows an expanded version of the 500–700 cm^{-1} frequency range so as to better show the similarity between this region and the same region in the spectra of ZrO_2-H_2O fluids in (**a**)

assigned to vibrations in silicate tetrahedra in the structure (e.g., the absence of the Q^o species with its diagnostic 770 cm^{-1} band seen in the spectra of SiO_2-H_2O and ZrO_2-SiO_2-H_2O fluids; see Fig. 4). In comparison with the Raman spectra of crystalline Zr-compounds, the spectrum of Na-zirconate Na_2ZrO_3 characteristically has two strong bands near 500 cm^{-1} and a single strong band near 350 cm^{-1}. The spectrum resembles that of $BaZrO_3$ with a sixfold coordinated Zr^{4+} (Dixit et al. 2004). Data, mostly X-ray-based, from Zr-bearing silicate glasses have been interpreted to indicate that Zr^{4+} is in sixfold coordination with oxygen (Farges et al. 1991, Galoisy et al. 1999; Farges and Rossano 2000). From this limited information, we conclude that Zr^{4+} in ZrO_2-

$NaOH-H_2O$ fluids coexisting with crystalline ZrO_2 likely forms zirconate complexes possibly with local structural similarity to crystalline zirconate. This structural interpretation also is consistent with published XANES and EXAFS (Extended X-ray Absorption Fine Structure) spectra of Zr-bearing Na-silicate aqueous fluids, where structures resembling zirconosilicates were proposed (Wilke et al. 2012, Louvel et al. 2013). Alkali zirconosilicate complexing also has been suggested for hydrous, peralkaline aluminosilicate melts (Watson 1979).

The composition control of how Zr^{4+} can form complexes with other solutes in an aqueous solution in aqueous fluids shows some similarities with the behavior of Ti^{4+} in aqueous fluids at similar temperatures and pressures. In the TiO_2-H_2O system, for example, the solubility is in the ppm range under similar temperature and pressure conditions (e.g., Audetat and Keppler 2005), and the local structural environment resembles that of TiO_2 with Ti^{4+} in sixfold coordination with oxygen (Mysen 2012). In Na-silicate-bearing solutions, on the other hand, the Ti^{4+} solubility is several orders of magnitude greater (Antignano and Manning 2008), and Ti^{4+} is dissolved in aqueous fluid as TiO_4- or TiO_3-type complexes with Ti^{4+} attaining a coordination number near 4 (Mysen 2012).

Solubility

The zirconium concentration in a fluid in equilibrium with crystalline ZrO_2 may be estimated by comparing the intensities of Raman bands from the fluid and from the crystalline ZrO_2 recorded at the same temperature and pressure. In such a calculation, it is assumed that the integrated Raman intensity of spectra and fluid and melt are related to one another via the same correction factor because the local structure of the Zr complexes in such fluids resembles that of crystalline ZrO_2 at the same temperature and pressure. This approach to solubility determination is similar to that for Ti solubility in aqueous fluid in equilibrium with crystalline TiO_2 (Mysen 2012) where Ti solubility determined in this manner compared well with values obtained by other methods (Audetat and Keppler 2005, Antignano and Manning 2008).

The Zr^{4+} concentration in ZrO_2-H_2O fluid ranges from ~10 ppm at 600 °C/60 MPa to about 40 ppm Zr at 900 °C and 560 MPa (Fig. 5). Two data points from the $ZrO_2-SiO_2-H_2O$ system at 725 °C/783 MPa and 800 °C/~1050 MPa indicate solubilities near 60 ppm or about 25 ppm higher than that in the absence of SiO_2. This slightly higher solubility likely is because of the higher pressures in the experiments in the $ZrO_2-SiO_2-H_2O$ system because Zr^{4+} solubility in aqueous fluids is positively correlated with pressure (Ayers et al. 2012, Wilke et al. 2012).

The Zr^{4+} concentration in aqueous fluid is sensitive to alkali and alkali + silica content. For example, addition of 1 M NaOH to H_2O raises the Zr^{4+} content of the aqueous fluid by approximately a factor of 10. Addition of SiO_2 to NaOH-bearing systems increases the solubility by an additional factor of 5 or more (Fig. 5). Addition of Na-silicate component also results in major zircon ($ZrSiO_4$) solubility increases (Ayers et al. 2012, Wilke et al. 2012). This feature may be analogous to the solution behavior of Zr^{4+} in hydrous and peralkaline aluminosilicate melt where the solubility is a positively correlated peralkalinity of the melt (Watson 1979). This concept would also be consistent with the XANES data on silicate glasses by Farges and Rossano (2000). We note, however, that addition of SiO_2 to ZrO_2-H_2O without NaOH does not have significant influence on the Zr^{4+} solubility. The solubility increase with NaOH qualitatively is similar to that reported elsewhere for zircon ($ZrSiO_4$) and for ZrO_2 solubility in aqueous fluid (e.g., Wilke et al. 2012). However, in order to quantitatively compare Zr^{4+} concentration in fluid equilibrated with ZrO_2 from the experiments by Wilke et al. (2012), their data were extrapolated linearly to the temperature/pressure conditions of the present experiments. The resulting solubility compares well with the results reported here (Fig. 5).

Discussion

It has been suggested that pH is an important variable governing Zr solubility in aqueous fluid based on the positive correlation of Zr^{4+} solubility with NaOH concentration in the fluid. The formation of hydroxylated Zr complexes such as $Zr(OH)^{3+}$, $Zr(OH)_2{}^{2+}$, $Zr(OH)_3{}^{3+}$, $Zr(OH)_4$, and $Zr(OH)_5{}^-$ could then be an important mechanism by which Zr^{4+} can dissolve in aqueous fluid (e.g., Ayers et al. 2012). The interpretation of the Raman signals assigned to fundamental OH vibrations in the 3000–4100 cm^{-1} frequency range from fluids in the ZrO_2-H_2O and $ZrO_2-SiO_2-H_2O$ systems does not require the existence of Zr-OH complexes in the fluids as there is no clear evidence for Zr-OH bonding in the spectra. However, crystalline $Zr(OH)_4$ exists, so one cannot rule out undetected OH-bearing Zr complexes. That caveat notwithstanding, for these simple ZrO_2-H_2O and $ZrO_2-SiO_2-H_2O$ systems, a simple solution mechanism is proposed:

$$ZrO_2(\text{xtal}) = ZrO_2(\text{fluid}), \tag{1}$$

with the associated equilibrium constant:

$$K_{(1)} = mZrO_2(\text{fluid}). \tag{2}$$

Here, $mZrO_2$(fluid) is the ZrO_2 molality of the fluid. The $mZrO_2$(fluid) was calculated from the weight fraction of ZrO_2 according to the expression:

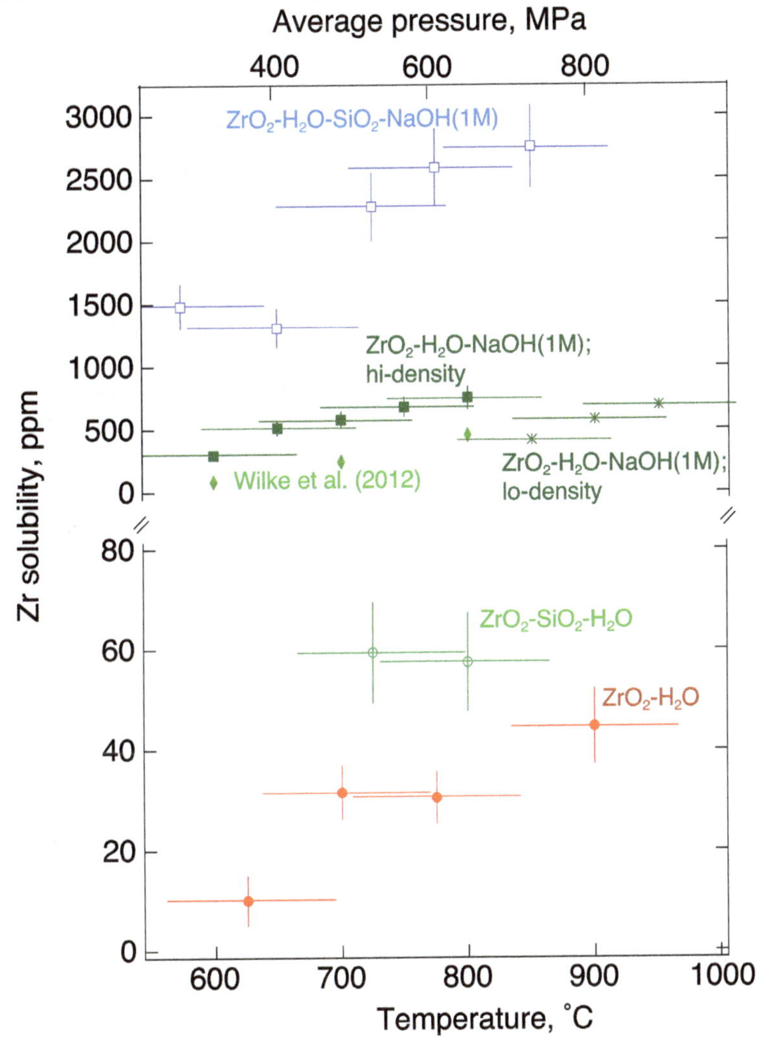

Fig. 5 Zr concentration in fluids as a function of temperature and pressure for the systems indicated. Average pressure is that calculated from the regression line in Fig. 2. For actual pressure, see Table 1. Only two data points (at 725 and 800 °C) are shown for fluids in the ZrO_2-SiO_2-H_2O system because at a lower temperature (and pressure) Raman signals that could be assigned to vibrations in Zr-bearing complexes in the fluid were not detected above background noise. *Horizontal error bars* represent calculated pressure uncertainty (±110 MPa; see text)

$$m_{ZrO_2} = \left(\frac{1 - \left(\frac{X^{wt}_{ZrO_2} \cdot M_{H_2O}}{M_{ZrO_2} - (X^{wt}_{ZrO_2} \cdot M_{ZrO_2} - X^{wt}_{ZrO_2} \cdot M_{H_2O})} \right)}{\left(\frac{X^{wt}_{ZrO_2} \cdot M_{H_2O}}{M_{ZrO_2} - (X^{wt}_{ZrO_2} \cdot M_{ZrO_2} - X^{wt}_{ZrO_2} \cdot M_{H_2O})} \right)} \right) \cdot 55.56, \tag{3}$$

where $M_{subscript}$ denotes molecular weight of the subscripted entity, and $X^{wt}{}_{subscript}$ is the corresponding weight fraction (in g ZrO_2/kg H_2O) that was derived from the Raman spectra as discussed above.

The temperature-dependent equilibrium constant, $K_{(1)}$, (Fig. 6), yields $\Delta H = 43 \pm 16$ kJ/mol (Tables 2 and 3). In these (and succeeding) calculations, it is assumed that the pressure change associated with changing temperature is not a factor. In other words, the ΔV values of the equilibrium reactions are considered negligible.

The Zr^{4+} solubility behavior in the ZrO_2-SiO_2-H_2O system appears similar to that in the ZrO_2-H_2O (Fig. 5). However, only two data points (at 725 and 800 °C) were obtained for the ZrO_2-SiO_2-H_2O system, and these were within 75 °C of each other because a Raman signal above background noise could be detected at lower temperatures. No clear temperature-dependence could be discerned. Given the data uncertainty, ΔH for reaction (1) in the ZrO_2-SiO_2-H_2O system could not be obtained, but with the similar solubility in fluid in the two systems, the Zr^{4+} solution mechanism and its energetics likely also are similar.

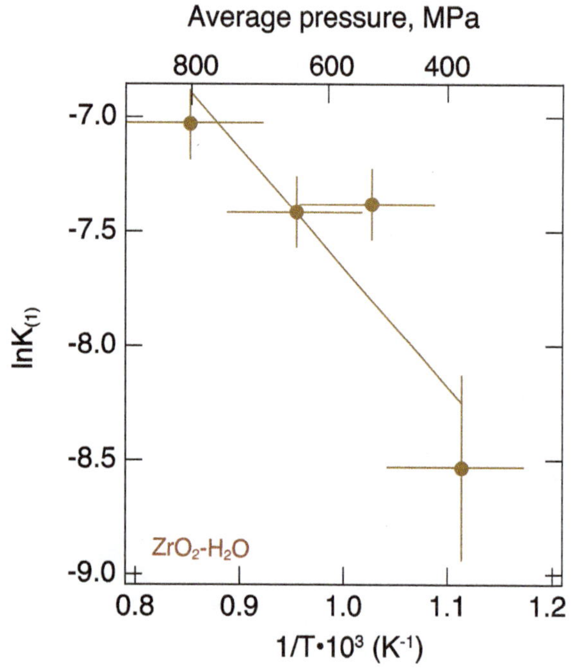

Fig. 6 Temperature-dependence of the equilibrium constant, $K_{(1)}$, for equilibrium (1). Regression coefficients are shown in Tables 2 and 3. Average pressure is that calculated from the regression line in Fig. 2. For actual pressure, see Table 1. *Horizontal error bars* represent calculated pressure uncertainty (±110 MPa; see text)

Addition of 1 M NaOH to the aqueous solution results in a Zr^{4+} solubility increase of approximately an order of magnitude compared with the NaOH-free solution (Fig. 5). This positive correlation between solubility and NaOH concentration of the fluid was why Ayers et al. (2012) suggested that Zr^{4+} dissolved as hydroxylated Zr complexes. However, enhanced solubility could just as easily be ascribed to Na-zirconate complexing. Na-zirconate type of complexing is consistent with the available XANES data (Louvel et al. 2013). It is also consistent with the Raman spectra of solutions and crystalline Na-zirconate discussed above (Fig. 4).

Detailed stoichiometry of the zirconate complex(es) cannot be inferred from existing data. The simplest possible zirconate group is ZrO_4^{4-} so that a complex resembling Na_4ZrO_4 could be considered. The solution mechanism then can be described with the expression:

$$ZrO_2 + 4NaOH = Na_4ZrO_4 + 2H_2O, \tag{4}$$

with the equilibrium constant:

$$K_{(4a)} = \frac{m_{Na_4ZrO_4} \cdot (f_{H_2O})^2}{(m_{NaOH})^2}. \tag{5}$$

In equilibrium with pure ZrO_2, the ZrO_2 activity equals unity. The molality of NaOH also equals 1 so that Eq. (5) reduces to:

$$K_{(4)} = m_{Na_4ZrO_4} \cdot (f_{H_2O})^2. \tag{6}$$

The $m_{Na_4ZrO_4}$ is calculated from the solubility weight fraction of Na_4ZrO_4 by substituting $M_{Na_4ZrO_4}$ and $X^{wt}_{Na_4ZrO_4}$ for M_{ZrO_2} and $X^{wt}_{ZrO_2}$ in Eq. (3).

The regression of ln $K_{(4)}$ vs. 1/T for the ZrO_2-NaOH-H_2O fluids is divided into high density and low density (Fig. 6) because the three highest temperature data points for this composition were recorded at a lower pressure as the result of extensive deformation of the sample hole of the Ir gasket (see also Table 1 and Fig. 1). The linear slope of these two curves does agree within error with the ΔH value derived from this temperature-dependence between 195 ± 24 kJ/mol and 232 ± 29 kJ/mol (Fig. 7; Tables 2 and 3). There is, however, an offset of the solubility axis of that figure, which likely reflects the positive correlation of solubility with pressure at any temperature. The $K_{(4)}$ values calculated from the experimental data for ZrO_2-NaOH-H_2O from Wilke et al. (2012) in the same temperature and pressure range fall nearly within the range of the present data (Fig. 7) with a slightly higher ΔH value (282 kJ/mol; see Tables 2 and 3). It is somewhat difficult to establish the error for those latter data because of the way they were extracted from Wilke et al. (2012) by interpolation and

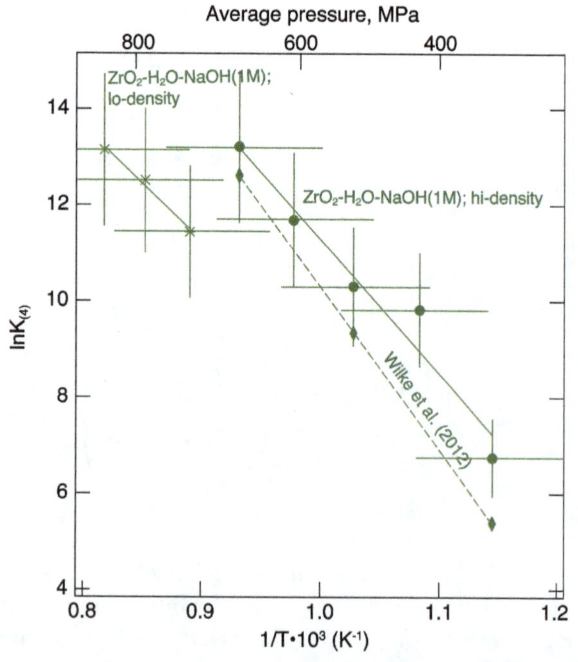

Fig. 7 Temperature-dependence of the equilibrium constant, $K_{(4)}$, for equilibrium (4) for the compositions indicated. Regression coefficients are shown in Tables 2 and 3. Average pressure is that calculated from the regression line in Fig. 2. For actual pressure, see Table 1. *Horizontal error bars* represent calculated pressure uncertainty (±110 MPa; see text)

extrapolation of data obtained at different pressure than those of the present experiments.

The differences in the spectral topology of fluids below 800 cm^{-1} in (Fig. 4) suggest a different structural behavior of Zr^{4+} in the ZrO_2-NaOH-H_2O and ZrO_2-SiO_2-NaOH-H_2O fluids. The 3000–4100 cm^{-1} intensity envelope also indicates structural differences where at any temperature the FWHH is considerably smaller in spectra of fluids in the ZrO_2-SiO_2-NaOH-H_2O system compared with those in the ZrO_2-NaOH-H_2O system (Fig. 3). Such narrowing of the Raman band would be consistent with fundamental OH vibrations from OH groups forming Si-OH bonds perhaps linked to Q^o species (SiO_4 groups). However, the lack of a detectable 770 cm^{-1} band, which is assignable to protonated orthosilicate (SiO_4) groups (e.g., Mibe et al., 2008), makes that possibility less likely. An Si-OH linkage could, nevertheless, exist as a part a zirconosilicate complex in the fluid. This latter suggestion would be consistent with the EXAFS- and XANES-based structural interpretation of fluids in alkali silicate zirconate environments (Wilke et al. 2012, Louvel et al. 2013).

From existing X-ray data and the Raman data reported here, a number of reactions involving zirconosilicate or separate silicate and zirconate complexes could be considered. These possibilities include, but are not unique, an expression such as:

$$2ZrO_2 + 2NaOH + 2SiO_2 = Na_2Zr_2Si_2O_9 + H_2O. \quad (7)$$

Given the interpretation of the high-frequency envelope of the fluids in ZrO_2-SiO_2-NaOH-H_2O suggesting some Si-OH bonding, it is possible that a portion of Na^+- in $Na_2Zr_2Si_2O_9$ might be replaced with H^+, which leads to a possible modification of Eq. (7);

$$2ZrO_2 + NaOH + 2SiO_2 = NaZr_2Si_2O_8(OH) + H_2O, \quad (7b)$$

with the equilibrium constant:

$$K_{(7b)} = \frac{m_{NaZr_2Si_2O_8(OH)}}{(a_{ZrO_2})^2 \bullet (a_{SiO_2})^2 \bullet m_{NaOH}}. \quad (8)$$

In Eq. (8), the activity of SiO_2 and ZrO_2 equals 1 as these are pure crystalline phases. The m_{NaOH} also equals 1. Therefore, $K_{(7b)}$ can be simplified to:

$$K_{(7b)} = m_{NaZr_2Si_2O_8(OH)}. \quad (8b)$$

From the temperature-dependent $K_{(7b)}$, ΔH is 30 ± 4 kJ/mol (Fig. 8; Tables 2 and 3). In other words, the Zr solubility is less temperature-dependent than fluids in SiO_2-free environments.

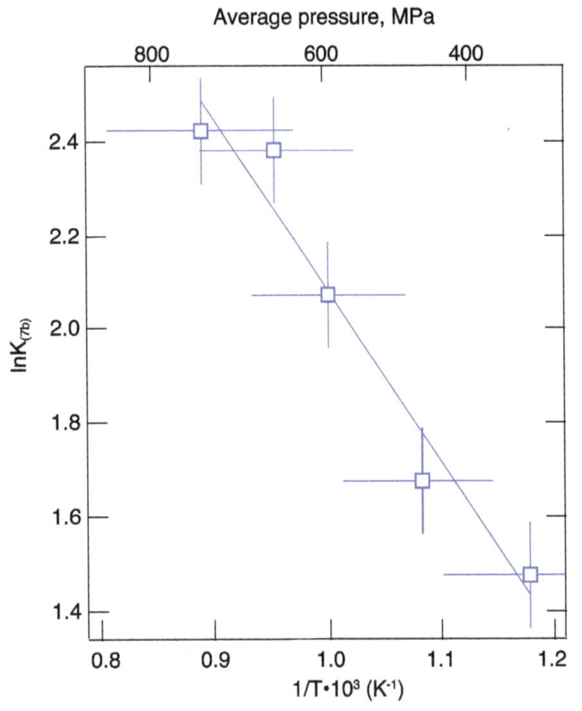

Fig. 8 Temperature-dependence of the equilibrium constant, $K_{(7b)}$, for equilibrium (7b). Regression coefficients are shown in Tables 2 and 3. Average pressure is that calculated from the regression line in Fig. 2. For actual pressure, see Table 1. *Horizontal error bars* represent calculated pressure uncertainty (±110 MPa; see text)

Conclusions

The experimental Zr^{4+} solubility data for water-rich fluids as a function of other solutes illustrate how an essentially insoluble high field strength cation such as Zr^{4+} can be made quite soluble in aqueous fluid under deep crustal and upper mantle temperature and pressure conditions through the stabilization of zirconate complexes that incorporate other solutes. This conclusion implies, for example, that fluid released during high-temperature dehydration of hydrous mineral assemblages such as those of oceanic crust descending into the upper mantle in subduction zones may carry significant (perhaps thousands of ppm) Zr and likely other HFSEs. Such fluids will alter the environment through which they migrate. This is consistent with natural evidence for Zr^{4+} transport such as that observed in veins in high-grade metamorphic eclogite and granulite (e.g., Philippot and Selverstone 1991, Möller et al. 2002). Moreover, aqueous fluids transported from dehydrating oceanic crust can also carry this signature to the source rock of partial melting. These processes may have been operational through for at least the last 2 billion years of the Earth's history where subduction resembling that observed today was operational. Fluid-mediated HFSE transport likely also occurred earlier in the Earth's history, but those transport mechanisms are less certain.

Abbreviations

ΔH: change of enthalpy; ΔV: change of volume; CCD: charge-coupled detector; FWHH: full width at half height; HDAC: hydrothermal diamond anvil cell; HFSE: high field strength element; NSF: National Science Foundation; XANES: X-ray near-edge spectroscopy.

Competing interests

The author declares that he/she has no competing interests.

Acknowledgements

Support from NSF grant EAR-1212754 made this research possible. The W. M. Keck Foundation supported acquisition of analytical instruments used. The carbon-13 diamonds were produced by Mr. J. Lai. His assistance is gratefully acknowledged. All data used in this manuscript are available from the author upon request.

References

Ampian SG. X-ray and optical crystallographic data for Na_2ZrO_3. J Am Ceram Soc. 1968;51:607–8.

Antignano A, Manning CE. Rutile solubility in H_2O, H_2O –SiO_2, and H_2O-NaAlSi $_3O_8$ fluids at 0.7-2.0 GPa and 700–1000 °C; implications for mobility of nominally insoluble elements. Chem Geol. 2008;255:283–93.

Audetat A, Keppler H. Solubility of rutile in subduction zone fluids, as determined by experiments in the hydrothermal diamond anvil cell. Earth Planet Sci Lett. 2005;232:393–402.

Ayers JC, Watson EB. Rutile solubility and mobility in supercritical aqueous fluids. Contrib Mineral Petrol. 1993;112:321–30.

Ayers JC, Zhang L, Luo Y, Peters TJ. Zircon solubility in alkaline aqueous fluids at upper crustal conditions. Geochim Cosmochim Acta. 2012;96:18–28.

Bassett WA, Wu T-C, Chou I-M, Haselton T, Frantz JD, Mysen BO. The hydrothermal diamond anvil cell (DAC) and its applications. In: Dyar MD, McCammon CA, Schaefer M, editors. Mineral spectroscopy: a tribute to Roger G. Burns. Houston: The Geochemical Society; 1996. p. 261–72.

Bebout GE. Metamorphic chemical dynamics of subduction zones. Earth Planet Sci Lett. 2007;260:905–20.

Bernini D, Audetat A, Dolejs D, Keppler H. Zircon solubility in aqueous fluids at high temperatures and pressures. Geochim Cosmochim Acta. 2013;119:178–87.

Brenan JM, Shaw HF, Phinney DL, Ryerson RJ. Rutile-aqueous fluid partitioning of Nb, Ta, Hf, Zr, U and Th: implications for high field strength element depletions in island-arc basalts. Earth Planet Sci Lett. 1994;128:327–39.

Compston W, Pidgeon RT. Jack Hills, evidence of more very old detrital zircons in Western Australia. Nature. 1986;321:766_9.

Dixit A, Majumder SB, Dobal PS, Katiyar RS, Bhalla AS. Phase transition studies of sol–gel deposited barium zirconate titanate thin films. Thin Solid Films. 2004;447–448:284–8.

Ellison AJ, Hess PC. Solution behavior of +4 cations in high silica melts; petrologic and geochemical implications. Contrib Mineral Petrol. 1986;94:343–51.

Farges F, Rossano S. Water in Zr-bearing synthetic and natural glasses. Eur J Mineral. 2000;12:1093–107.

Farges F, Ponader CW, Brown CW, Jr GE. Structural environments of incompatible elements in silicate glass/ melt systems: I. Zirconium at trace levels. Geochim Cosmochim Acta. 1991;55:1563–74.

Foustoukos DI, Mysen BO. D/H isotopic fractionation in the H_2-H_2O system at supercritical water conditions: composition and hydrogen bonding effects. Geochim Cosmochim Acta. 2012;86:88–102.

Frantz JD, Dubessy J, Mysen BO. An optical cell for Raman spectroscopic studies of supercritical fluids and its applications to the study of water to 500°C and 2000 bar. Chem Geol. 1993;106:9_26.

Galoisy L, Pelegrin E, Arrio ME, Ildefonse P, Calas G. Evidence for 6-coordinated zirconium in inactive nuclear waste glasses. J Am Ceram Soc. 1999;82:2219–24.

Giere R. Zirconolite, allanite, and hoegbomite in a marble skarn from the Bergell contact aureole: implications for mobility of Ti, Zr, and REE. Contrib Mineral Petrol. 1986;93:459_70.

Gill JB. Orogenic andesites and plate tectonics. Berlin: Springer; 1981. p. 391.

Hamilton W. Earth's first two billion years - the era of internally mobile crust. Geol Soc Am Mem. 2007;200:233–96. doi:10.1130/2007.1200(13).

Hanfland M, Syassen K, Fahy S, Louie SG, Cohen ML. Pressure-dependence of the first-order Raman mode in diamond. Phys Rev B. 1985;31:6896–9.

Hermann J, Spandler C, Hack A, Korsakov AV. Aqueous fluids and hydrous melts in high-pressure and ultra-high pressure rocks: implications for element transfer in subduction zones. Lithos. 2006;92:399–417.

Horita J. Hydrogen isotope analyses of natural waters using an H_2-water equilibration method: a special implication to brines. Chem Geol. 1988;72:89–94.

Keramidas VG, White WB. Raman scattering study of the crystallization and phase transformations of ZrO_2. J Am Ceram Soc. 1974;57:22–4.

Kröner A, Layer PW. Crust formation and plate motion in the early archean. Science. 1992;256:1405–11.

Le Losq CR, Moretti R, Neuville DR. Water speciation in silicate melts investigated by Raman spectroscopy; implication for volcanic processes. Mineral Mag. 2011;75:1296.

Linnen RL, Keppler H. Melt composition control on Zr/Hf fractionation in magmatic processes. Geochim Cosmochim Acta. 2002;66:3293–302.

Louvel M, Sanchez-Valle C, Malfait WJ, Testemale D, Hazemann J-L. Zr complexation in high pressure fluids and silicate melts and implications for the mobilization of HFSE in subduction zones. Geochim Cosmochim Acta. 2013;104:281–99.

Manning CE. The solubility of quartz in H_2O in the lower crust and upper mantle. Geochim Cosmochim Acta. 1994;58:4831–40.

Mibe K, Chou I-M, Bassett W A. In situ Raman spectroscopic investigation of the structure of subduction-zone fluids. J Geophys Res. 2008; doi:10.1029/2007JB005179.

Mojzsis SJ, Harrison TM, Pidgeon RT. Mass transfer in the Earth's mantle and crust is documented on rocks dating to the first few hundred million years of the Earth's history. Nature. 2001;409:178–81.

Möller A, O'Brian PJ, Kennedy A, Kröner A. Polyphse zircon in ultrahigh-temperature granulites (Rogaland, SW. Norway): constraints from Pb diffusion in zircon. J Metamorph Geol. 2002;20:727–40.

Mysen BO. High-pressure/-temperature titanium solution mechanisms in silicate-saturated aqueous fluids and hydrous silicate melts. Am Mineral. 2012;97:1241–51.

Mysen BO. Carbon speciation in silicate-C-O-H melt and fluid as a function of redox conditions: an experimental study, in-situ to 1.7 GPa and 900 °C. Am Mineral. 2015;100:872–82.

Mysen BO, Virgo D. Volatiles in silicate melts at high pressure and temperature. 1. Interaction between OH groups and Si^{4+}, Al^{3+}, Ca^{2+}, Na^+ and H^+. Chem Geol. 1986;57:303–31.

Mysen BO, Yamashita S. Speciation of reduced C-O-H volatiles in coexisting fluids and silicate melts determined in-situ to ~1.4 GPa and 800 °C. Geochim Cosmochim Acta. 2010;74:4577–88.

Newton RC, Manning CE, Hanchar JM, Colasanti CV. Free energy of formation of zircon based on solubility measurements at high temperature and pressure. Am Mineral. 2010;95:52–8.

Pascal ML, Anderson GM. Speciation of Al, Si, and K in supercritical solutions: experimental study and interpretation. Geochim Cosmochim Acta. 1989;53:1843–56.

Philippot P, Selverstone J. Trace-element-rich brines ion eclogitic veins: implications for fluid composition and transport during subduction. Contrib Mineral Petrol. 1991;106:417–30.

Schiferl D, Nicol M, Zaug JM, Sharma SK, Cooney TF, Wang S-Y, et al. The diamond $^{13}C/^{12}C$ isotope Raman pressure sensor system for high temperature/pressure diamond-anvil cells with reactive samples. J Appl Phys. 1997;82:3256–65.

Scholze H. Zur frage der unterschiedung zwischen H_2O-molekulen und OH-gruppen in glasern und mineralen. Naturwissenschaften. 1960;47:226–7.

Smith DK, Newkirk HW. Crystal structure of baddeleyite (monoclinic ZrO_2) and its relation to polymorphism of ZrO_2. Acta Crystallogr A. 1965;18:983–91.

Tatsumi Y, Eggins S. Subduction zone magmatism. Cambridge: Blackwell Scientific Publications; 1995.

Valley JW, Lackey JS, Cavoisie AJ, Clenchenko CC, Spicuzza MJ, Basel MAS, et al. 4.4 billion years of crustal maturation: oxygen isotope ratios of magmatic zircon. Contrib Mineral Petrol. 2005;150:561–80.

Walrafen GE, Fisher MR, Hokmabadi MS, Yang W-H. Temperature dependence of the low- and high-frequency Raman scattering from liquid water. J Chem Phys. 1986;85:6970–82.

Walrafen GE, Yang W-H, Chu YC. Raman spectra from saturated water vapor to the supercritical fluid. J Chem Phys B. 1999;103:1332–8.

Watson EB. Two-liquid partition coefficients; experimental data and geochemical implications. Contrib Mineral Petrol. 1976;56:119–34.

Watson EB. Zircon saturation in felsic liquids; experimental results and applications to trace element geochemistry. Contrib Mineral Petrol. 1979;70:407–19.

Wilke M, Schmidt C, Dubrail J, Appel K, Borchert M, Kvashnina K, et al. Zircon solubility and zircon complexation in H_2O+Na_2O+SiO_2±Al_2O_3 fluids at high pressure and temperature. Earth Planet Sci Lett. 2012;349–350:15–25.
Wyllie PJ, Sekine T. The formation of mantle phlogopite in subduction zone hybridization. Contrib Mineral Petrol. 1982;79:375–80.
Zotov N, Keppler H. Silica speciation in aqueous fluids at high pressures and high temperatures. Chem Geol. 2002;184:71–82.

Permissions

All chapters in this book were first published in PEPS, by Springer; hereby published with permission under the Creative Commons Attribution License or equivalent. Every chapter published in this book has been scrutinized by our experts. Their significance has been extensively debated. The topics covered herein carry significant findings which will fuel the growth of the discipline. They may even be implemented as practical applications or may be referred to as a beginning point for another development.

The contributors of this book come from diverse backgrounds, making this book a truly international effort. This book will bring forth new frontiers with its revolutionizing research information and detailed analysis of the nascent developments around the world.

We would like to thank all the contributing authors for lending their expertise to make the book truly unique. They have played a crucial role in the development of this book. Without their invaluable contributions this book wouldn't have been possible. They have made vital efforts to compile up to date information on the varied aspects of this subject to make this book a valuable addition to the collection of many professionals and students.

This book was conceptualized with the vision of imparting up-to-date information and advanced data in this field. To ensure the same, a matchless editorial board was set up. Every individual on the board went through rigorous rounds of assessment to prove their worth. After which they invested a large part of their time researching and compiling the most relevant data for our readers.

The editorial board has been involved in producing this book since its inception. They have spent rigorous hours researching and exploring the diverse topics which have resulted in the successful publishing of this book. They have passed on their knowledge of decades through this book. To expedite this challenging task, the publisher supported the team at every step. A small team of assistant editors was also appointed to further simplify the editing procedure and attain best results for the readers.

Apart from the editorial board, the designing team has also invested a significant amount of their time in understanding the subject and creating the most relevant covers. They scrutinized every image to scout for the most suitable representation of the subject and create an appropriate cover for the book.

The publishing team has been an ardent support to the editorial, designing and production team. Their endless efforts to recruit the best for this project, has resulted in the accomplishment of this book. They are a veteran in the field of academics and their pool of knowledge is as vast as their experience in printing. Their expertise and guidance has proved useful at every step. Their uncompromising quality standards have made this book an exceptional effort. Their encouragement from time to time has been an inspiration for everyone.

The publisher and the editorial board hope that this book will prove to be a valuable piece of knowledge for researchers, students, practitioners and scholars across the globe.

List of Contributors

Fumiya Noritake and Katsuyuki Kawamura
Graduate School of Environmental and Life Science, Okayama University, 3-1-1, Tsushimanaka, Kita-Ku, Okayama 700-8530, Japan

Junichi Miyazaki and Ken Takai
Department of Subsurface Geobiological Analysis and Research (D-SUGAR), Japan Agency for Marine-Earth Science and Technology (JAMSTEC), 2-15 Natsushima-cho, Yokosuka 237-0061, Japan

Shinsuke Kawagucci
Department of Subsurface Geobiological Analysis and Research (D-SUGAR), Japan Agency for Marine-Earth Science and Technology (JAMSTEC), 2-15 Natsushima-cho, Yokosuka 237-0061, Japan
Institute of Geochemistry and Petrology, ETH Zürich, Zürich, Switzerland

Yuki Morono
Kochi Institute for Core Sample Research, Japan Agency for Marine-Earth Science and Technology (JAMSTEC), Nankoku, Kochi 783-8502, Japan

Jeff S. Seewald
Department Marine Chemistry and Geochemistry (MC&G), Woods Hole Oceanographic Institution (WHOI), Woods Hole, MA 02543, USA

C. Geoff Wheat
Global Undersea Research Unit, University of Alaska Fairbanks, P. O. Box 475, Moss Landing, CA 95039, USA

Catherine McCammon, Vasily Potapkin and Leonid Dubrovinsky
Bayerisches Geoinstitut, Universität Bayreuth, Bayreuth 95440, Germany

Anastasia Kantor
Bayerisches Geoinstitut, Universität Bayreuth, Bayreuth 95440, Germany European Synchrotron Radiation Facility, Grenoble Cedex 9 38043, France

Ryosuke Sinmyo
Bayerisches Geoinstitut, Universität Bayreuth, Bayreuth 95440, Germany
ELSI, Tokyo Institute of Technology, Tokyo 152-8550, Japan

Clemens Prescher
Bayerisches Geoinstitut, Universität Bayreuth, Bayreuth 95440, Germany
Institut für Geologie und Mineralogie, Universität zu Köln, Köln 50969, Germany

Ilya Kupenko
Bayerisches Geoinstitut, Universität Bayreuth, Bayreuth 95440, Germany
European Synchrotron Radiation Facility, Grenoble Cedex 9 38043, France
Institut für Mineralogie, Universität Münster, Münster 48149, Germany

Razvan Caracas
Laboratoire de Géologie de Lyon UMR CNRS 5276, Ecole Normale Supérieure de Lyon, Lyon Cedex 07 69364, France

Konstantin Glazyrin
Deutsches Elektronen-Synchrotron, Hamburg 22607, Germany

Aleksandr Chumakov
European Synchrotron Radiation Facility, Grenoble Cedex 9 38043, France
National Research Center "Kurchatov Institute", Moscow 123182, Russia

Takazo Shibuya and Ken Takai
Department of Subsurface Geobiological Analysis and Research (D-SUGAR), Japan Agency for Marine-Earth Science and Technology (JAMSTEC), 2-15 Natsushima-cho, Yokosuka 237-0061, Japan
Research and Development Center for Submarine Resources, Japan Agency for Marine-Earth Science and Technology (JAMSTEC), 2-15 Natsushima-cho, Yokosuka 237-0061, Japan
Project Team for Development of New-Generation Research Protocol for Submarine Resources, Japan Agency for Marine-Earth Science and Technology (JAMSTEC), 2-15 Natsushima-cho, Yokosuka 237-0061, Japan

Tsuyoshi Komiya
Department of Earth Science and Astronomy, The University of Tokyo, 3-8-1 Komaba, Meguro-ku, Tokyo 153-8902, Japan

Shigenori Maruyama
Earth-Life Science Institute, Tokyo Institute of Technology, 2-12-1 Ookayama, Meguro-ku, Tokyo 152-8551, Japan
Novosibirsk State University, Pirogova 1 630090, Russia

Michael J. Russell
Jet Propulsion Laboratory, California Institute of Technology, 4800 Oak Grove Drive, Pasadena, CA 91108, USA

Bjorn Mysen
Geophysical Laboratory, Carnegie Instn. Washington, Washington DC, USA

Jun Kawano
Creative Research Institution (CRIS), Hokkaido University, N21 N10, Kita-ku, Sapporo, Hokkaido 001-0021, Japan
Faculty of Science, Hokkaido University, N10 W8, Kita-ku, Sapporo, Hokkaido 060-0810, Japan

Takaya Nagai
Faculty of Science, Hokkaido University, N10 W8, Kita-ku, Sapporo, Hokkaido 060-0810, Japan

Hiroshi Sakuma
Environment and Energy Materials Division, National Institute for Materials Science, 1-1 Namiki, Tsukuba, Ibaraki 305-0044, Japan

Tetsuji Onoue, Katsuhito Soda and Mitsutaka Miura
Department of Earth and Environmental Sciences, Kumamoto University, Kumamoto 860-8555, Japan

Naoto Takahata and Yuji Sano
Atmosphere and Ocean Research Institute, The University of Tokyo, Chiba 277-8564, Japan

Honami Sato
Ocean Resources Research Center for Next Generation, Chiba Institute of Technology, Chiba 275-0016, Japan

Akira Ishikawa
Department of Earth and Planetary Sciences, Tokyo Institute of Technology, Tokyo 152-8550, Japan

Yukio Isozaki
Department of Earth Science and Astronomy, The University of Tokyo, Tokyo 153-8902, Japan

Ryuichi Nomura and Youmo Zhou
Geodynamics Research Center, Ehime University, 2-5 Bunkyo, Matsuyama, Ehime 790-8577, Japan

Tetsuo Irifune
Geodynamics Research Center, Ehime University, 2-5 Bunkyo, Matsuyama, Ehime 790-8577, Japan
Earth-Life Science Institute, Tokyo Institute of Technology, 2-12-1 Ookayama, Meguro, Tokyo 152-8550, Japan

Kiyoshi Fujimoto
Faculty of Policy Studies, Nanzan University, 18 Yamazatocho, Showa-Ku, Nagoya 466-8673, Japan

Masafumi Miura
Graduate School of Frontier Sciences, The University of Tokyo, Kashiwa, Japan
Kasukabe City Office, Kasukabe, Japan

Shigeo Kobayashi
Graduate School of Asian and African Area Studies, Kyoto University, Kyoto, Japan

Herwint Simbolon
Research Center for Biology, The Indonesian Institute of Sciences, Cibinong, Indonesia

Stefano Poli
Dipartimento di Scienze della Terra, Università degli Studi di Milano, Via Botticelli 23, Milan 20133, Italy

Célia Dalou
Department of Geological Sciences, Jackson School of Geosciences, The University of Texas at Austin, 1 University Station C1100, Austin, TX 78712, USA
Université Blaise Pascal, Laboratoire Magmas et Volcans, Clermont Université, Clermont-Ferrand, BP 10448, F-63000, France
CNRS, UMR 6524, LMV, Clermont-Ferrand F-63038, France
IRD, R 163, LMV, Clermont-Ferrand F-63038, France

Kenneth T Koga
Université Blaise Pascal, Laboratoire Magmas et Volcans, Clermont Université, Clermont-Ferrand, BP 10448, F-63000, France
CNRS, UMR 6524, LMV, Clermont-Ferrand F-63038, France
IRD, R 163, LMV, Clermont-Ferrand F-63038, France

Marion Le Voyer
Department of Terrestrial Magnetism, Carnegie Institution of Washington, Washington, DC 20015, USA
Department of Mineral Sciences, National Museum of Natural History, Smithsonian Institution, Washington, DC 20013, USA

Nobumichi Shimizu
Woods Hole Oceanographic Institution, Woods Hole, MA 02453, USA

Hiroshi Fujioka, Hideko Takayanagi and Yasufumi Iryu
Department of Earth Science, Graduate School of Science, Tohoku University, Sendai 980-8578, Japan

Koshi Yamamoto
Department of Earth and Environmental Sciences, Nagoya University, Nagoya 464-8601, Japan

Yusuke Sawaki
Department of Earth and Planetary Sciences, Tokyo Institute of Technology, 2-12-1 Ookayama, Meguro-ku, Tokyo 152-8551, Japan

Hisahiro Ueda
Department of Earth and Planetary Sciences, Tokyo Institute of Technology, 2-12-1 Ookayama, Meguro-ku, Tokyo 152-8551, Japan
Laboratory of Ocean-Earth Life Evolution Research (OELE), Japan Agency for Marine-Earth Science and Technology (JAMSTEC), 2-15 Natsushima-cho, Yokosuka 237-0061, Japan

Takazo Shibuya and Masafumi Saitoh
Laboratory of Ocean-Earth Life Evolution Research (OELE), Japan Agency for Marine-Earth Science and Technology (JAMSTEC), 2-15 Natsushima-cho, Yokosuka 237-0061, Japan
Research and Development Center for Submarine Resources, Japan Agency for Marine-Earth Science and Technology (JAMSTEC), 2-15 Natsushima-cho, Yokosuka 237-0061, Japan
Project Team for Development of New-generation Research Protocol for Submarine Resources, Japan Agency for Marine-Earth Science and Technology (JAMSTEC), 2-15 Natsushima-cho, Yokosuka 237-0061, Japan

Ken Takai
Laboratory of Ocean-Earth Life Evolution Research (OELE), Japan Agency for Marine-Earth Science and Technology (JAMSTEC), 2-15 Natsushima-cho, Yokosuka 237-0061, Japan
Research and Development Center for Submarine Resources, Japan Agency for Marine-Earth Science and Technology (JAMSTEC), 2-15 Natsushima-cho, Yokosuka 237-0061, Japan
Project Team for Development of New-generation Research Protocol for Submarine Resources, Japan Agency for Marine-Earth Science and Technology (JAMSTEC), 2-15 Natsushima-cho, Yokosuka 237-0061, Japan
Department of Subsurface Geobiological Analysis and Research (SUGAR), Japan Agency for Marine-Earth Science and Technology (JAMSTEC), 2-15 Natsushima-cho, Yokosuka 237-0061, Japan

Shigenori Maruyama
Earth-Life Science Institute (ELSI), Tokyo Institute of Technology, 2-12-1 Ookayama, Meguro-ku, Tokyo 152-8551, Japan

Mengying He and Hongbo Zheng
School of Geography Science, Nanjing Normal University, Nanjing 210046, China

Peter D. Clift
Department of Geology and Geophysics, Louisiana State University, Baton Rouge, LA 70803, USA

Ryuji Tada
Graduate School of Science, The University of Tokyo, 7-3-1 Hongo, Bunkyo-ku, Tokyo 113-0033, Japan

Weihua Wu and Chao Luo
School of Earth Sciences and Engineering, Nanjing University, Nanjing 210008, China

Index

O

P

R

S

T

U

V

www.ingramcontent.com/pod-product-compliance
Lightning Source LLC
LaVergne TN
LVHW061510220926
848485LV00017B/611